T0238557

Lecture Notes in Artificial Intelligence 8176

Subseries of Lecture Notes in Computer Science

Patrice Perny   Marc Pirlot
Alexis Tsoukiàs (Eds.)

Algorithmic Decision Theory

Third International Conference, ADT 2013
Bruxelles, Belgium, November 13-15, 2013
Proceedings

 Springer

Volume Editors

Patrice Perny
UPMC, LIP6, 75005 Paris, France
E-mail: patrice.perny@lip6.fr

Marc Pirlot
UMONS Faculty of Engineering, Mathematics
and Operations Research
7000 Mons, Belgium
E-mail: marc.pirlot@umons.ac.be

Alexis Tsoukiàs
CNRS, LAMSADE, Université Paris Dauphine
75016 Paris, France
E-mail: tsoukias@lamsade.dauphine.fr

ISSN 0302-9743   e-ISSN 1611-3349
ISBN 978-3-642-41574-6   e-ISBN 978-3-642-41575-3
DOI 10.1007/978-3-642-41575-3
Springer Heidelberg New York Dordrecht London

Library of Congress Control Number: 2013950618

CR Subject Classification (1998): I.2, H.3, F.1, H.4, G.1.6, F.4.1-2, C.2

LNCS Sublibrary: SL 7 – Artificial Intelligence

Typesetting: Camera-ready by author, data conversion by Scientific Publishing Services, Chennai, India

Printed on acid-free paper

Springer is part of Springer Science+Business Media (www.springer.com)

Preface

This volume contains the proceedings of ADT 2013, the Third International Conference on Algorithmic Decision Theory held at ULB (Université Libre de Bruxelles), Belgium, November 12–14, 2013.

ADT seeks to bring together researchers and practitioners coming from diverse areas such as articial intelligence, database systems, operations research, decision theory, discrete mathematics, game theory, multiagent systems, computational social choice, and theoretical computer science in order to improve the theory and practice of modern decision support and automation systems.

ADT provides a multi-disciplinary forum for sharing knowledge in these areas with a special focus on algorithmic issues in decision theory. The two first International Conference on Algorithmic Decision Theory (ADT 2009, 2011) brought together researchers and practitioners from diverse areas of computer science, economics, and operations research from around the globe, with proceedings published in LNAI 5783 and LNAI 6992. ADT 2013 sought to continue this tradition and presented 33 technical research papers concerning preferences in reasoning and decision making, uncertainty and robustness in decision making, multi-criteria decision analysis and optimization, collective decision making, learning and knowledge extraction for decision support.

There were more than 70 submissions of abstracts, and finally 60 full papers. Each submission was reviewed by at least two Program Committee members. The committee decided to accept 33 papers for the proceedings. We also accepted six oral presentations not submitted to the proceedings. In addition to the contributed papers, the conference proposed various invited talks including talks by Matthias Ehrgott (Lancaster University) on "Multiobjective Optimisation," Itzhak Gilboa (Tel Aviv University and HEC Paris) on "Decision Making Under Uncertainty," and Arkadii Slinko (University of Auckland) on "Social Choice."

We wish to thank all authors who submitted papers to this conference, as well as the Program Committee members and external reviewers for their involvement in the reviewing process. ADT 2013 was made possible thanks to the support of CNRS (the French national research center) through the International Research Group Algodec, FNRS (Research foundation of the Federation Wallonia-Brussels, Belgium), EURO (Association of European Operational Research Societies), LIP6, LAMSADE, ULB, and UMONS (Université de Mons).

We would also like to acknowledge the support of Easychair in the management of submitted papers and in the preparation of the proceedings.

July 17, 2013

Patrice Perny
Marc Pirlot
Alexis Touskiàs

Organization

Program Committee

Leila Amgoud	CNRS, Université Toulouse III, France
Craig Boutilier	University of Toronto, Canada
Ronen Brafman	Ben-Gurion University, Israel
Paolo Ciaccia	University of Bologna, Italy
Matthias Ehrgott	The University of Auckland, New Zealand
Helene Fargier	CNRS, University Toulouse III, France
Judy Goldsmith	University of Kentucky, USA
Michel Grabisch	University of Paris I, France
Frank Hsu	Fordham University, New York, USA
Eyke Hüllermeier	University of Marburg, Germany
Ulrich Junker	-
Werner Kiessling	University of Augsburg, Germany
Christian Klamler	University of Graz, Austria
Jérôme Lang	CNRS, Université Paris Dauphine, France
Thierry Marchant	Universiteit Gent, Belgium
Nicolas Maudet	University of Paris 6, France
Thomas Dyhre Nielsen	Aalborg University, Denmark
Wlodzimierz Ogryczak	Warsaw University of Technology, Poland
Sasa Pekec	Duke University, USA
Patrice Perny	University of Paris 6, France
Marc Pirlot	Université de Mons, Belgium
David Rios	Universidad Rey Juan Carlos, Spain
Fred Roberts	DIMACS, USA
Francesca Rossi	University of Padova, Italy
Scott Sanner	NICTA, Australia
Arkadii Slinko	The University of Auckland, New Zealand
Roman Slowinski	Poznan University of Technology, Poland
Olivier Spanjaard	University of Paris 6, France
Alexis Tsoukias	CNRS, Université Paris Dauphine, France
Kristen Brent Venable	Tulane University and IHMC, USA
Paolo Viappiani	CNRS, University of Paris 6, France
Toby Walsh	NICTA and UNSW, Australia
Nic Wilson	4C, UCC, Cork, Ireland
Michael Wooldridge	University of Oxford, UK

Additional Reviewers

Aziz, Haris
Darmann, Andreas
Mattei, Nicholas

Roocks, Patrick
Wuillemin, Pierre-Henri

Table of Contents

Two Agents Competing for a Shared Machine

Alessandro Agnetis[1], Gaia Nicosia[2], Andrea Pacifici[3], and Ulrich Pferschy[4]

[1] Dipartimento di Ingegneria dell'Informazione e Scienze Matematiche,
Università degli Studi di Siena, Italy
agnetis@dii.unisi.it

[2] Dipartimento di Ingegneria, Università degli studi "Roma Tre", Italy
nicosia@dia.uniroma3.it

[3] Dipartimento di Ingegneria Civile e Ingegneria Informatica,
Università degli Studi di Roma "Tor Vergata", Italy
pacifici@disp.uniroma2.it

[4] Department of Statistics and Operations Research, University of Graz, Austria
pferschy@uni-graz.at

Abstract. In this paper we address a deterministic scheduling problem in which two agents compete for the usage of a single machine. The agents submit their tasks in successive steps to an external coordinator, who sequences them according to a known priority rule. We introduce the problem for three different shop configurations, namely when the agents' parts are transferred to the machine through two distinct linear conveyor belts, when they are transferred through circular conveyor belts, and when parts can be freely picked from the two agents' buffer. We consider the problem from different perspectives. First, we look at the problem from a centralized point of view as a bicriteria optimization problem and characterize the set of Pareto optimal solutions from the computational complexity perspective. Then, we address the problem from one agent's point of view. In particular, we propose algorithms suggesting to an agent how to sequence its own tasks in order to optimize its own objective function, regardless of the other agents objectives.

Keywords: scheduling, multi-agent optimization, bicriteria optimization.

1 Introduction

Classical scheduling problems deal with situations in which a set of *tasks* has to be processed on some processing *resource*. In addition, in *two-agent scheduling problems* there are two agents, each task belongs to only one agent, and each agent is interested in optimizing his/her own performance index. Although these problems can be viewed as a special case of bicriteria scheduling problems [7], their specific properties and applications have spurred a considerable amount of research since the seminal work by Agnetis et al. [1] and Baker and Smith [2].

In this paper we consider a new problem: There are two agents, A and B, each owning a set of nonpreemptive tasks that require a single machine to be

P. Perny, M. Pirlot, and A. Tsoukiàs (Eds.): ADT 2013, LNAI 8176, pp. 1–14, 2013.

processed. There is also one coordinator (e.g., the machine owner), who is interested in maximizing the machine throughput, which can be defined as the number of processed tasks per time unit. The coordinator gives precedence to shorter tasks, among those that are promoted for execution. More precisely, to regulate access to the machine, the coordinator defines a *selection mechanism*, consisting of the iterative application of the following steps.

1. Each agent submits one of its tasks.
2. The coordinator selects the *shortest* between submitted tasks for processing[1].

Each repetition of the above steps is a *round*, and the selected task is referred to as the *winner* of the round. The two steps are repeated until all tasks of one agent have been processed (the remaining tasks of the other agent are appended thereafter).

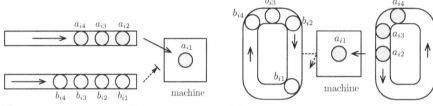

(a) Linear conveyors. Task b_{i1} is the next B task in the schedule.   (b) Circular conveyors. Task b_{i1} is moved to the end of the queue.

Fig. 1. Layout of different conveyor types. Agents submit tasks $a_{i1} < b_{i1}$: a_{i1} is selected.

For the task submitted in each round, we consider three distinct situations, which may be interpreted as corresponding to different shop configurations [6]:

Linear conveyor: Two linear conveyor belts, one for each agent, transport parts to the machine. In this configuration, each agent sequences the parts on the conveyor, implying that, at each round, one of the two candidate tasks is the loser of the preceding round. In other words, each task is submitted for possible processing, in the given order, until it wins.

Circular conveyor: Two circular conveyor belts, one for each agent, transport parts to the machine. In this configuration, each agent sequences the parts on the conveyor, however differently from the previous scenario, since the two belts move simultaneously, at each round the unselected task is moved on and therefore placed at the end of the (current) sequence.

Flexible processing: In this scenario, there are no queues at the machine and any part from the two agents' buffers can be picked up and submitted for possible processing. Hence, in this case, the agents are free to choose any available task for submission at each round, independently from the outcome of the previous round.

---

[1] Ties are broken by giving preference to one of the agents, e.g. agent A.

We denote by $shop(f^A, f^B)$ the above problem, where $shop \in \{line, circ, flexi\}$ refers to one of the three above described shop configurations, and f^A and f^B are the two agents' objective functions.

We consider the problem from different perspectives.

Centralized perspective. This analysis aims at characterizing the set of Pareto optimal feasible schedules in terms of size and computational complexity.

Agent perspective. Considering an agent (say, B), its *strategy* consists in deciding which task to submit at each round, taking into account its own objective function. As in [3–5] we consider two different settings concerning the information available to agent B.

Offline. Agent B knows in advance the length of each task of the opponent, as well as its submission sequence. In this case, B wants to determine how to sequence its tasks under such an advantageous asymmetry of information.

Online. Agent B knows the length of each task of the opponent (agent A), but B has no information at all on A's strategy. In this case, B may want to select a strategy that minimizes its solution cost in the worst possible case, i.e., for any strategy of A. This corresponds to what is usually called *minimax strategy* in game theory.

Also, in this context one is interested in assessing the performance of some classical single-agent sequencing rules in the two-agent setting.

In this paper we consider that the agents pursue the minimization of the most commonly used objective functions in scheduling problems, i.e., makespan, total flow time and total weighted flow time.

2   Problem Formulation and Notation

Let A and B denote the two agents. Each agent owns a set of n nonpreemptive tasks to be performed on a single machine. Tasks have nonnegative deterministic processing times $a_1 \leq a_2 \leq \ldots \leq a_n$ for agent A (A-tasks) and $b_1 \leq b_2 \leq \ldots \leq b_n$ for agent B (B-tasks). All processing times are known to both agents.

Each agent chooses a *strategy*, i.e., a submission sequence of its tasks. The application of the coordinator selection mechanism to the submitted sequences results in a schedule σ. Each agent wants to optimize its own objective function, which only depends on the completion times of its tasks: $f^X(\sigma) = f^X(C_1^X(\sigma), \ldots, C_n^X(\sigma))$, where $C_i^X(\sigma)$ is the completion time of task i of agent X ($i = 1, \ldots, n$, $X = A, B$) in σ, i.e. the point in time when task i is completely processed. In this paper, we will consider objective function pairs (f^A, f^B) consisting of makespan (i.e. the latest completion time over all tasks of A, resp. B), total flow time (i.e. the sum over all completion times), and total weighted flow time, where the completion time of each task i is weighted by a relevance factor w_i.

The machine can process only one task at a time and each agent wants to optimize its own cost function. The decision process is divided into $2n$ *rounds* in which each agent submits one task — if available — for possible processing. The *shortest* between the two submitted tasks is selected and appended at the end of the current schedule, which is initially empty.

At the beginning, each agent decides which task to submit for the first round. Then, the winning agent submits a new task for the second round, while the losing task may be subject to specific constraints depending on the shop configuration (see Section 1). The process goes on until all tasks are scheduled. For a given shop configuration we call a schedule *feasible*, if it can be obtained through the application of such mechanism in that particular setting.

2.1   Our Contribution

The first shop configuration we consider is the linear conveyor (Section 3). In this scenario, we provide results on the complexity of the problem from a centralized perspective. In particular, when the agents' objective is the makespan minimization, we show that there is only one Pareto optimal feasible schedule, while when minimizing the total flow time the number of Pareto optimal solutions becomes exponential. In the latter case, we also prove that it is $\mathcal{NP}$-complete to decide whether a feasible solution achieving given objectives for the two agents exists (Section 3.1). Then, in Section 3.2 we deal with the design of algorithms suggesting to an agent which task to submit in each round in order to optimize its objective function. In particular, standard greedy algorithms are optimal when minimizing makespan or total flow time, while a more complex local search algorithm is needed when the objective is the weighted total flow time minimization. In Section 4 some negative results concerning the performances of simple natural strategies are briefly reported for the circular conveyor case. Moreover, we provide a lower bound on the competitive ratio of any single agent online strategy aiming at makespan minimization. Finally, in Section 5 we summarize results on the flexible processing scenario. From the centralized point of view, these results are similar to those of the linear conveyor case, whereas from a single agent perspective the problem of designing optimal strategies becomes harder. In particular, we limit ourselves to the analysis of the performance of the well-known SPT submission list.

3   Linear Conveyors

3.1   Centralized Perspective

In this section, we address the problem of characterizing the set $\mathcal{P}$ of Pareto optimal solutions in the case of linear conveyors for various objective functions.

If both agents want to minimize their makespan, it is easy to observe that, since the largest task can never win against any opponent's task, there is only *one* Pareto optimal solution, precisely the one in which the largest task is played in the first round. As a consequence, the other agent wins the first n rounds (regardless of its submission sequence).

Consider now problem $line(\sum C_j^A, \sum C_j^B)$: It is possible to show that (i) there are exponentially many Pareto optimal solutions and (ii) it is $\mathcal{NP}$-complete to recognize a Pareto optimal solution.

Regarding the characterization of Pareto optimal solutions, it is easy to note that, when both agents submit their tasks in SPT order, i.e. in increasing order of processing times, the outcome is a Pareto optimal solution. Note that because of the selection mechanism, this results in an overall SPT schedule, which is well-known to be the schedule minimizing $\sum_j C_j^A + \sum_j C_j^B$. Consider the following instance POEXP where M is a very large number and each agent owns $n = 2m$ tasks. Agent A's tasks are $a_{2i-1} = 1 + M^i$, $a_{2i} = 4 + M^i$, while B's tasks are $b_{2i-1} = 2 + M^i$ and $b_{2i} = 3 + M^i$, for $i = 1, \ldots, m$. It is possible to show that, for instance POEXP, there exist at least 2^m nondominated schedules. Hence, the following result holds.

Theorem 1. *There can be exponentially many Pareto optimal solutions for problem $line(\sum C_j^A, \sum C_j^B)$.*

Consider now the problem of deciding whether a certain given objective for each agent can be achieved. Formally, given two positive values P_A and P_B, we want to know whether there exists a feasible schedule σ^* such that $\sum C_j^A(\sigma^*) \leq P_A$ and $\sum C_j^B(\sigma^*) \leq P_B$. We refer to the latter problem as RECOGNITION.

It is possible to reduce EVEN-ODD PARTITION to our problem, hence the following result (which trivially extends to $line(\sum w_j^A C_j^A, \sum w_j^B C_j^B)$) holds.

Theorem 2. RECOGNITION *for $line(\sum C_j^A, \sum C_j^B)$ is $\mathcal{NP}$-complete.*

3.2 Single Agent Perspective

We now focus our attention to the design of algorithms suggesting to an agent, say agent B, which task to submit at each round, in order to optimize its objective function.

Minimization of Makespan or Total Flow Time
When B's objective is the minimization of the makespan or sum of completion times, finding an optimal strategy for B is trivial. Since in every round the shorter between the two candidate tasks wins, there is no advantage in deviating from the SPT sequence. More precisely, independently from A's submission sequence, by standard pairwise interchange arguments, it is easy to show that an optimal solution for agent B is attained by submitting tasks in SPT order. Hence, SPT is both a minimax strategy and an optimal offline algorithm for agent B for problems $line(f^A, C_{\max}^B)$ and $line(f^A, \sum C_j^B)$ for any objective f^A.

This argument does not hold anymore for the minimization of the *weighted* sum of completion times, for which different aspects will be considered in the subsequent section.

Minimization of Weighted Total Flow Time: Offline Case
In the following, we assume that the sequence of tasks played by agent A is fixed and known by agent B, i.e. we are in the offline scenario. Hereafter, we

propose an algorithm that builds a sequence minimizing B's objective, for a given sequence of submissions of A.

Scheduling all the tasks of B in SPT order has the advantage that the tasks are scheduled as early as possible which was beneficial for the previous objectives. However weights of tasks are not taken into account. It is well known that in a single agent problem sorting tasks by the WSPT rule, i.e. in nonincreasing order of ratios w_j/b_j, minimizes $\sum_j w_j C_j$. However, sequencing the tasks of B in WSPT order may result in a schedule that is not optimal for B due to the presence of the A tasks, which may delay the heavier (and possibly long) tasks.

First observe that, given the submission sequences of A and B, the resulting schedule can be represented by a sequence of blocks: $\langle A_1, B_1, A_2, B_2, \ldots A_q, B_q \rangle$

An A-block A_i (resp., B-block B_j) is a maximal subsequence of consecutive A-tasks (resp., B-tasks). Each A-block starts with some task a' which wins against some task b', i.e. $a' \leq b'$. Since b' remains submitted until it wins, the A-block may contain a number of additional tasks also winning against b'. The block ends when A submits an item losing against b', which now starts a new B-block. Therefore, the first task of each block is longer than all the tasks in the preceding block. The first and the last blocks A_1 and B_q are possibly empty.

Also note that by an exchange argument a solution minimizing B's total weighted completion time is such that, in each block of B, all the tasks of the block are in nonincreasing order of the ratio w_i/b_i, i.e. in WSPT order.

Denote as BLOCKWSPT the schedule built as follows: B submits its tasks in SPT order (while those of A are submitted in the given order). Let the resulting schedule be arranged in the blocks $\langle A_1, B_1, A_2, B_2, \ldots A_q, B_q \rangle$. Sort the tasks within each block of B in WSPT order. Note that this schedule is feasible. In fact, by construction, all tasks of a block B_i have shorter processing times than the first task of block A_{i+1} and longer than all tasks of block A_i.

Let us define the *density* ρ_Δ of a sequence Δ of consecutive tasks as the ratio between the total weight of B's tasks in Δ and the total processing time of all tasks in Δ (including A's tasks), that is

$$\rho_\Delta = \frac{w_\Delta}{t_\Delta} = \frac{\sum_{j \in \Delta \cap B} w_j}{\sum_{j \in \Delta \cap A} a_j + \sum_{j \in \Delta \cap B} b_j}.$$

Clearly, if Δ is a single B-task x, then $\rho_x = w_x/b_x$.

A key property of an optimal solution is that if two B-tasks are not in relative WSPT order, then the task with higher density cannot come earlier because of its large processing time. This can be proved through the following lemma.

Lemma 3. *In a schedule σ_1, let x and y be two tasks such that $\rho_x > \rho_y$, x belongs to block B_j, y to B_i with $j > i$, and b_x is smaller than the first A-task in block A_{i+1}, i.e., x might feasibly be scheduled in block B_i. Let σ_2 be the schedule in which, with respect to σ_1, x is moved before y and let σ_3 be the schedule in which y is moved after x. Then either σ_2 or σ_3 is better than σ_1 (cf. Figure 2).*

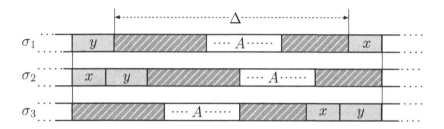

Fig. 2. Pictorial representation of schedules σ_1, σ_2, and σ_3 of Lemma 3

Proof. Let F_i indicate the objective function value of schedule σ_i, $i = 1, 2, 3$. It is quite easy to see that the relation between the different objective function values is as follows.

$$F_2 = F_1 + (w_y + w_\Delta)b_x - w_x(b_y + t_\Delta) \tag{1}$$
$$F_3 = F_1 + w_y(b_x + t_\Delta) - (w_x + w_\Delta)b_y \tag{2}$$

Assume first that $F_3 > F_1$, we must show that $F_2 < F_1$. By (2), $w_y(b_x + t_\Delta) - (w_x + w_\Delta)b_y > 0$ and hence, since $w_y b_x - w_x b_y < 0$, it must be $w_y t_\Delta > w_\Delta b_y$. Therefore, $\rho_\Delta < \rho_y < \rho_x$ and hence,

$$F_2 - F_1 = \underbrace{w_y b_x - w_x b_y}_{<0} + \underbrace{w_\Delta b_x - w_x t_\Delta}_{<0} < 0.$$

On the contrary, if $F_2 > F_1$, then we show that $F_3 < F_1$. In fact, by (1), $(w_y + w_\Delta)b_x - w_x(b_y + t_\Delta) > 0$ and since $w_y b_x - w_x b_y < 0$, then $\rho_\Delta > \rho_x > \rho_y$. As a consequence,

$$F_3 - F_1 = \underbrace{w_y b_x - w_x b_y}_{<0} + \underbrace{w_y t_\Delta - w_\Delta b_y}_{<0} < 0.$$

The thesis follows. □

The local search algorithm (MOVE) that we present hereafter starts from the BLOCKWSPT schedule and iteratively "moves" one task from its position to one later in the schedule if this operation results in an improvement of B's objective. For instance, suppose we move a task y of agent B in position i to a later position j in the schedule. Denote by Δ the set of consecutive tasks between positions i and j. Then the change of the objective function of agent B is given by the quantity $w_y t_\Delta - w_\Delta b_y$, which is smaller than 0 if $\rho_\Delta > \rho_y$. If task y can be beneficially relocated to more than one position, e.g. positions P and Q in Figure 3(a), then by doing some easy calculations, we observe that position Q is better than P if the following holds:

$$\frac{w_{\Delta_Q} - w_{\Delta_P}}{t_{\Delta_Q} - t_{\Delta_P}} > \frac{w_y}{b_y} = \rho_y. \tag{3}$$

Algorithm 1. Algorithm of agent B for problem $line(f^A, \sum w_j C_j^B)$ offline case

---

MOVE

1: Compute the BLOCKWSPT schedule.
2: For all tasks x in increasing order of ρ_x (reverse WSPT order)
   Try to move x from its current block B_h to block B_ℓ , $\ell > h$, where B_ℓ is the block maximizing the improvement of B's objective, if an improvement is possible.
3: Return the resulting schedule σ.

This relation can be used to establish the best position for moving y. Note that when a task y is moved from its current position to a different block B_ℓ, then y's best position within B_ℓ corresponds to a WSPT sequencing of the tasks in the block.

Observe that σ is built in $O(n)$ moving steps, where each step costs $O(n)$ by checking all possible insertion positions each in constant time. The next proposition is useful for reducing the computational costs of MOVE.

Proposition 4. *In* MOVE *algorithm, if it is not beneficial to move the last task of a B-block to a later position, then this applies also for the tasks preceding it.*

Proof. Consider three feasible schedules where the B-blocks are WSPT ordered and that are identical but for the positions of two B-tasks, x and y, with

$$\rho_x \geq \rho_y, \tag{4}$$

and a piece Δ consisting of B and A tasks. The first schedule σ_1 sequences (in order) x, y, and Δ consecutively. The second schedule σ_2 sequences y, Δ, and x consecutively. The third schedule σ_3 sequences Δ first and then x and y, consecutively. The remainder of the schedule is identical for the three schedules. In the following, w_Δ and t_Δ indicate the total weight of B-tasks in the sub-schedule Δ and its length, respectively.

The thesis is equivalent to show that if schedule σ_2 is better than σ_1, then σ_3 is better than σ_2. In other words, if moving the second last task x of a B block is beneficial, then it is even better to postpone the last two tasks of a block. If F_i is the total weighted completion time of B tasks in schedule σ_i, $i = 1, 2, 3$, then it is easy to see that:

$$F_2 = F_1 - (w_\Delta + w_y)b_x + w_x(t_\Delta + b_y) \tag{5}$$
$$F_3 = F_2 - (w_\Delta + w_x)b_y + w_y(t_\Delta + b_x). \tag{6}$$

Suppose then that F_2 improves upon F_1 but, by contradiction, that F_3 is strictly worse than F_2:

$$w_\Delta b_x + w_y b_x \geq w_x t_\Delta + w_x b_y \tag{7}$$
$$w_\Delta b_y + w_x b_y < w_y t_\Delta + w_y b_x. \tag{8}$$

Since, from (4), $w_x b_y \geq w_y b_x$, from (7) we obtain $w_\Delta b_x \geq w_x t_\Delta$ and from (8) we obtain $w_\Delta b_y < w_y t_\Delta$ which imply $\frac{w_y}{b_y} > \frac{w_\Delta}{t_\Delta} \geq \frac{w_x}{b_x}$. This contradicts (4), and the thesis follows. □

Proposition 4 together with Lemma 3 have a strong impact on the computational costs of MOVE. In fact, (i) if during the execution of the algorithm no beneficial move is possible for a certain task y, then it is not beneficial to move any other task preceding y in the same block; (ii) whenever a certain task y is moved to a block B_ℓ, then all tasks in B_ℓ preceding y shall not be considered in the successive steps of the algorithm because they would certainly not benefit from a move in the schedule. Hence, the number of tasks to be considered for a beneficial move can be substantially reduced although we cannot get below the $O(n^2)$ running time complexity.

Theorem 5. *Given the sequence of submissions of agent A, algorithm* MOVE *finds an optimal solution of problem* $line(f^A, \sum w_j C_j^B)$ *for agent B.*

Proof. In the following we show that if MOVE moves a task y to its best position at a certain iteration, then it is never beneficial to move it again in a successive step. This implies that each task is moved at most once by MOVE and that it is moved to the optimal position. Hence, MOVE finds an optimal solution for B.

First, we observe that by construction each task belonging to a block B_i in the BLOCKWSPT schedule, in an optimal schedule will either be in the same block or in a block B_j succeeding block B_i, i.e. with $j > i$. Let us refer to the blocks of the optimal schedule as *optimal blocks*.

In what follows we show that whenever a task is moved in MOVE, then it is moved to its optimal block. We will do so by observing that whenever a task y is moved to its best position at a certain iteration of MOVE, then even if several changes in the structure of the schedule occur after y has been moved, then it is never beneficial to move y again in a successive step.

To prove the thesis, we assume that after y is relocated some other task x, with $\rho_x > \rho_y$, is moved and we analyze how the relocation of task x affects the contribution of y to the objective function. In particular, we distinguish two cases depending on the position of x with respect to y in the BLOCKWSPT schedule:

Case 1. y is moved to a certain position and then some other task x located *after* y in the BLOCKWSPT schedule is moved.

Assume first that, when moving y, the best improvement in the objective function is attained in a position Q located after x in the BLOCKWSPT schedule (see Figure 3(a)). Due to Lemma 3 if y is moved to position Q, then it is never beneficial to have x in any position R after Q in the schedule. On the other hand, if x is moved to any position before Q in the schedule, the density of Δ_Q does not change. So, in both cases the block containing Q is the optimal block for y.

Assume now that MOVE moves task y to position P corresponding to the best improvement in the objective function (see Figure 3(a)). Assume also that, in a later step of the algorithm, x is moved to position R. It can be shown that in this

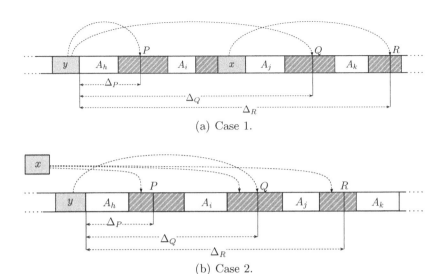

(a) Case 1.

(b) Case 2.

Fig. 3. Illustration of cases 1 and 2 in the proof of Theorem 5

case it not beneficial to move again task y. In fact, position Q is not beneficial for y due to Lemma 3, while any position after R is worse for y than the current position P. In fact, the densities of intervals Δ_P and Δ_R do not change when moving x to R, so recalling (3) P remains the best choice for y.

Case 2. y is moved to a certain position and then some other task x located *before* y in the BlockWspt schedule is moved. Let Q be the position in which y is moved by Move (that is the position with the best improvement in the objective function). Similarly to Case 1, consider different possibilities for x. If x is moved in any position before the position of task y in the BlockWspt schedule, then this change in the schedule does not affect y's contribution to the objective function.

If x is moved in any position before P (see Figure 3(b)), then one can show that this does not affect the y contribution to the objective function. Due to (3), if Q is better than P for y, then $\frac{w_{\Delta_Q} - w_{\Delta_P}}{t_{\Delta_Q} - t_{\Delta_P}} > \frac{w_y}{b_y}$. So, if x is moved in a position before P, both t_{Δ_Q} and t_{Δ_P} and their total weights increase by the same quantity and hence the best position for task y remains Q.

If x is moved to any position between P and Q, it is easy to observe that Q remains the best position for y. In fact, due to Lemma 3, y could not be in position P nor in its original block. On the other hand, R remains worse than Q since equation (3) holds when applied to R and Q. Finally, note that due to Lemma 3, it is never beneficial to move x after position Q.

In conclusion, whenever a task y is moved it is placed in its optimal block and this proves the theorem. $\qquad\square$

We have observed before that SPT is a minimax strategy of agent B for problems $line(f^A, C_{\max}^B)$ and $line(f^A, \sum C_j^B)$. For the case of $\sum_j w_j C_j^B$ we can characterize the worst-case situation for B as follows.

Lemma 6. *For any strategy σ^B played by B, the maximum value of B's objective $\sum w_j C_j^B$ is attained if A plays its tasks in SPT order.*

Now Lemma 6 implies that a minimax strategy of B for problem $line(f^A, \sum w_j C_j^B)$ can be attained by assuming that A submits its tasks in SPT order and applying the offline algorithm MOVE for B.

4   Circular Conveyors

In the circular conveyor setting solving our problem becomes much harder. In fact, since losing tasks are moved to the end of the sequence, the outcome of the process in the first n rounds "freezes" the sequence in the last n rounds. This consideration is particularly interesting in the on-line setting when devising strategies for agent B. The cyclic conveyor scenario can be seen as a mixture of settings: the first n rounds are indeed online, i.e, one has to see what agent A does, but then all the remaining n rounds are fixed. In the following, we provide a few negative results but the most important questions pertaining the circular conveyor setting remain indeed open.

We start with the minimization of the makespan, i.e., problem $circ(C_{\max}^A, C_{\max}^B)$. In this case, even determining the best *offline* strategy for agent B against any given strategy of agent A is a non trivial task and remains an open problem. In particular, it can be proven that it could be beneficial for B to voluntarily lose a round (even if B could win) to obtain a better matching in a successive stage. Hence, no greedy-type algorithm can be optimal. This applies even when A submits its tasks in SPT order. Similar negative results hold for the online strategies. Hereafter, we provide a lower bound on the competitive ratio of any online strategy.

Theorem 7. *No on-line strategy of agent B against an arbitrary strategy of agent A can have a competitive ratio smaller than $\frac{3}{2}$ for problem $circ(C_{\max}^A, C_{\max}^B)$.*

Proof. To prove the theorem we consider an instance with $n = 4$ jobs for each agent and
$$a_1 < a_2 < b_1 < b_2 < \varepsilon \ll M < a_3 < b_3 < b_4 < a_4.$$
where ε and M are suitably given parameters.

We will consider two strategies of agent A: In both cases A starts by submitting $\langle a_1, a_2 \rangle$, thus winning, in the first two rounds. In rounds 3 and 4, strategy S_1 submits $\langle a_3, a_4 \rangle$ while strategy S_2 submits $\langle a_4, a_3 \rangle$. Given the first two items a_1 and a_2, agent B is forced to lose and may decide to do so with, say b_1 and b_3 while keeping b_2 and b_4 to win against a_3 and a_4.

In Case 1, B submits $\langle b_3, b_1 \rangle$ losing the first two rounds. Then A plays strategy S_1 and B can win the subsequent two rounds with $\langle b_2, b_4 \rangle$. After that the sequences for both agents are fixed and B completes the processing of its jobs with a makespan $C_{\max}^B = P_B + a_1 + a_2 + a_3$, as depicted in the following table (boxed items refer to winning jobs).

Rounds	1	2	3	4	5	6	7	8
A	a_1	a_2	a_3	a_4	a_3	a_4	a_4	a_4
B	b_3	b_1	b_2	b_4	b_3	b_1	b_3	

However, an optimal response for B against strategy S_1 would play $\langle b_1, b_3 \rangle$ in the first two rounds and therefore obtain a makespan $C_{\max}^B = P_B + a_1 + a_2$.

Rounds	1	2	3	4	5	6	7	8
B_{OPT}	b_1	b_3	b_2	b_4	b_1	b_3		

In Case 2, B submits $\langle b_1, b_3 \rangle$ in the first two rounds but then A plays strategy S_2, and we have exactly the same values for the actual and optimal makespan of agent B.

In conclusion we get a worst-case competitive ratio of $\frac{P_B+a_1+a_2+a_3}{P_B+a_1+a_2} = \frac{4\varepsilon+3M}{4\varepsilon+2M}$ which also gives a lower bound of $\frac{3}{2}$.   □

Note that in the proof above, agent B may or may not know the submission of A in each round in advance. So, the theorem can be extended to prove that even in the case of partial information, B is not able to devise an optimal *response* strategy. These negative results easily extend to the minimization of total completion times.

5   Flexible Processing

We now consider a different shop configuration that allows the agents to freely choose any available task for submission at each round. In the linear conveyor case a strategy of an agent is completely described by a submission list. In the flexible processing case this is not true anymore. In fact, when a task loses a round it is not mandatory to resubmit it again in the next round: this increased level of flexibility makes it possible for an agent to adopt what we call an *adaptive* strategy. In particular any agent strategy can be described by an algorithm (not by a submission list).

In this scenario, with increased degree of freedom, finding an algorithm for a single agent against its opponent becomes a harder task. In fact, our results here are limited to the performance analysis of a natural submission strategy. However, from a centralized point of view, things remain roughly the same.

5.1   Centralized Perspective

Similarly to the linear conveyor case, in the flexible processing setting we have:

- $flexi(C_{\max}^A, C_{\max}^B)$: there is a single PO solution. The owner of the longest job cannot avoid finishing this job last.

- $flexi(\sum C_j^A, \sum C_j^B)$: the recognition version of this problem is $\mathcal{NP}$-complete. In fact, consider an instance of $flexi(\sum C_j^A \leq W_A, \sum C_j^B \leq W_B)$ where A and B have $n+1$ tasks each and the processing times of the first n A-tasks are equal to the n items of PARTITION while those of the first n B-tasks are (almost) equal to their A counterparts (e.g., $a_i + \varepsilon = b_i$ for small ε.) a_{n+1} and b_{n+1} are very large. Since the subsequences a_i, b_i or b_i, a_i are both feasible—the $n+1$-th task can be used to lose against any of the first n tasks of the opponent—we may use a proof analogous to [1, Theorem 9.2].
- The latter result implies also the $\mathcal{NP}$-completeness of the recognition version of $flexi(\sum w_j^A C_j^A, \sum w_j^B C_j^B)$.

5.2 Single Agent Perspective

Differently from the centralized perspective, most of the results for the linear conveyor architecture do not extend to the flexible processing scenario. It can be shown that when B submits its tasks in SPT order, in general, this is not optimal even when B wants to minimize its makespan due to the possible erratic behavior of A. Hereafter, we report a result on the worst-case performance of an SPT strategy for B in which losing tasks are submitted until they win as in the linear conveyor case.

Theorem 8. *The SPT strategy for B against an arbitrary adaptive strategy of A has performance bounds $r(SPT)$, with*

- $r(SPT) = n$ *for problem* $flexi(f^A, C_{\max}^B)$;
- $r(SPT) = n$ *for problem* $flexi(f^A, \sum C_j^B)$;
- $n \leq r(SPT) \leq 2n$ *for problem* $flexi(f^A, \sum w_j^B C_j^B)$.

6 Future Research

We have studied a two agent scheduling problem in different shop configurations. Most of the results concern the linear conveyor case. However, from a mathematical point of view the circular conveyor scenario poses an intriguing challenge which we only started to consider. Also for the flexible processing case it would be interesting to construct and analyze more advanced algorithms. A first interesting problem to be addressed is the design and analysis of an algorithm for the minimization of weighted total flow time, which —differently from the SPT heuristic here considered— explicitly takes weights into account.

Acknowledgments. Ulrich Pferschy was supported by the Austrian Science Fund (FWF) [P 23829-N13].

References

1. Agnetis, A., Mirchandani, P.B., Pacciarelli, D., Pacifici, A.: Scheduling Problems with Two Competing Agents. Operations Research 52(2), 229–242 (2004)
2. Baker, K., Smith, J.C.: A multiple criterion model for machine scheduling. Journal of Scheduling 6(1), 7–16 (2003)

3. Marini, C., Nicosia, G., Pacifici, A., Pferschy, U.: Strategies in Competing Subset Selection. Annals of Operations Research 207(1), 181–200 (2013)
4. Nicosia, G., Pacifici, A., Pferschy, U.: Subset Weight Maximization with Two Competing Agents. In: Rossi, F., Tsoukias, A. (eds.) ADT 2009. LNCS, vol. 5783, pp. 74–85. Springer, Heidelberg (2009)
5. Nicosia, G., Pacifici, A., Pferschy, U.: Competitive subset selection with two agents. Discrete Applied Mathematics 159(16), 1865–1877 (2011)
6. Pu, P., Hughues, J.: Integrating AGV schedules in a scheduling system for a flexible manufacturing environment. In: IEEE International Conference on Robotics and Automation, vol. 4, pp. 3149–3154 (1994)
7. T'Kindt, V., Billaut, J.-C.: Multicriteria scheduling. Theory, models and algorithms. Springer (2006)

Identification of a 2-Additive Bi-Capacity by Using Mathematical Programming

Julien Ah-Pine[1], Brice Mayag[2], and Antoine Rolland[1]

[1] ERIC Lab, University of Lyon, Bron, France
{julien.ah-pine,antoine.rolland}@univ-lyon2.fr
[2] LAMSADE, University Paris-Dauphine, Paris, France
brice.mayag@dauphine.fr

Abstract. In some multi-criteria decision making problems, it is more convenient to express the decision maker preferences in bipolar scales. In such cases, the bipolar Choquet integral with respect to bi-capacities was introduced. In this paper, we address the problem of eliciting a bipolar Choquet integral with respect to a 2-additive bi-capacity. We assume that we are given a set of examples with (i) their scores distribution in regard to several criteria and (ii) their overall scores. We propose two types of optimization problems that allow identifying the parameters of a 2-additive bi-capacity such that the inferred bipolar Choquet integral is consistent with the given examples as much as possible. Furthermore, since the elicitation process we study has many relationships with problems in statistical machine learning, we also present the links between our models and concepts developed in the latter field.

Keywords: 2-additive bi-capacity identification, Bipolar Choquet integral, Preference elicitation.

1 Introduction

Multi-criteria decision making (MCDM) aims at representing the preferences of a decision maker (DM) over a set of options (or alternatives) and in regard to several criteria. It then seeks to formalize the DM's decision process through mathematical tools in order to help him make decisions over the set of alternatives. The DM's decision process is assumed to be guided by the importance and the relationships he wants to take into account regarding the criteria. Concerning the preferences representations, one possible model is the Multi-Attribute Utility Theory (MAUT) which assumes that each attribute (or criterion) provides a utility value (or score) over the set of alternatives. Then, an aggregation function is used to combine, for each option, its scores distribution (or profile) in an overall score. The latter global utility values are then employed to make decisions. There are many types of aggregation functions to model a decision process. The Choquet integral has been proved to be a versatile tool to construct overall scores (see for example [1–4]). This aggregation function is intimately based on the concept of a capacity (or fuzzy measure). In particular, it assumes that partial utilities belong to non-negative or unipolar scales.

P. Perny, M. Pirlot, and A. Tsoukiàs (Eds.): ADT 2013, LNAI 8176, pp. 15–29, 2013.

Unipolar scales are not always appropriate to represent the DM's preferences (see the motivating example in [5]). In some problems, bipolar scales are more convenient. This type of scales is typically composed of a negative, a positive and a neutral part which respectively allow representing a negative, a positive and a neutral affect towards an option. To apply the Choquet integral in the case of bipolar scales, the bipolar Choquet integral (BCI) was introduced in [6] and [7]. In this paper, we particularly focus on BCI which use the concept of a bi-capacity (BC) introduced in [6] and which was further studied in [8, 9] and in [10, 11]. The BCI typically requires the DM to set $3^n - 1$ values where n is the number of attributes. When n exceeds some units, it is impossible for the DM to set all parameters of his decision model. In order to better cope with this combinatorial burden, the BCI with respect to (w.r.t.) a 2-additive bi-capacity (2A-BC) was introduced in the following papers [10, 12]. The 2-additivity property implies that only the interactions between at most two criteria are taken into account in a BC, and it enables reducing the number of parameters from $3^n - 1$ to $2n^2 + 1$. This has facilitated the use of this aggregation function in practice.

Even though there have been many papers studying 2A-BC, most of them have focused on theoretical aspects. In this contribution, we study the practical problem of identifying the parameters of a 2A-BC on the basis of information provided by the DM. This problem is also known as preference elicitation. There are different contexts in which we can proceed to the elicitation of the preference model of a DM. In our case, we assume that the DM provides the bipolar scores for a subset of (real or fictitious) options w.r.t. all criteria of the decision problem. In addition, he provides the overall bipolar scores of the same set of alternatives. These evaluated examples constitute the only data we have at our disposal. Then, the elicitation model consists in inferring the parameters of a 2A-BC such that the associated BCI is consistent with the preferences given by the DM on these examples. We propose optimization models that address this kind of preference elicitation problems.

Eliciting preference models is a research topic that has been studied by many researchers (see for example [13, 14]). However, the BCI w.r.t. 2A-BC has not been studied very much so far. To our knowledge, the only paper that addressed this exact problem is [15]. Yet, in the latter paper, the authors assumed an elicitation process in which the DM was asked to provide cardinal information on trinary actions. This setting is different from the one considered in this paper. Our approach is also in line with the work detailed in [14][1] about the preference elicitation using unipolar Choquet integrals with mathematical programming. In this latter work, the authors also assume that the only information provided are examples evaluated by the DM.

The elicitation process we deal with has many relationships with the problems addressed in statistical machine learning. The interconnections between preference elicitation on the one hand and machine learning on the other hand were highlighted in [16]. There has been a growing interest for the last years about cross-fertilizing these two domains by studying how the concepts developed in

---

[1] And with the papers cited therein.

one field can be applied in the other one. In line with this research topic, we also discuss the links between the concepts of these two domains that our models involve.

The rest of the paper is organized as follows. We recall in section 2 some basic definitions about BC and the properties of 2A-BC by using the bipolar Möbius transform defined by [10]. We then propose in section 3 two identification methods of a 2A-BC using linear programming and quadratic programming. In order to illustrate our proposals, we apply the different methods to a numerical example. Next, in section 4, we underline the relationships of our approaches with concepts developed in the field of machine learning. We finally conclude this paper and sketch some future works in section 5.

2   Bi-capacities and Bipolar Choquet Integrals

Let us denote by $N = \{1, \ldots, n\}$ a finite set of n criteria and $X = X_1 \times \cdots \times X_n$ the set of possible alternatives, where $X_1, \ldots, X_n$ represent the attributes. For all $i \in N$, the function $u_i : X_i \to \mathbb{R}$ is called a utility function. Given an element $x = (x_1, \ldots, x_n)$, we denote by $U(x) = (u_1(x_1), \ldots, u_n(x_n))$, the element's profile or its scores distribution. We will often write ij, ijk instead of $\{i, j\}$ and $\{i, j, k\}$ respectively.

2.1   2-Additive Bi-capacities

Let us denote by $2^N := \{S \subseteq N\}$ the set of subsets of N and $3^N := \{(A, B) \in 2^N \times 2^N : A \cap B = \emptyset\}$ the set of couples of subsets of N with an empty intersection. We define on 3^N the following relation $\sqsubseteq$, $\forall (A_1, A_2), (B_1, B_2) \in 3^N$:

$$(A_1, A_2) \sqsubseteq (B_1, B_2) \Leftrightarrow [A_1 \subseteq B_1 \text{ and } B_2 \subseteq A_2]$$

Definition 1 (Bi-capacity (BC) [9], [5]). *A function* $\nu : 3^N \to \mathbb{R}$ *is a BC on* 3^N *if it satisfies the following two conditions :*

$$\nu(\emptyset, \emptyset) = 0 \tag{1}$$

$$\forall (A_1, A_2), (B_1, B_2) \in 3^N : [(A_1, A_2) \sqsubseteq (B_1, B_2) \Rightarrow \nu(A_1, A_2) \leq \nu(B_1, B_2)] \tag{2}$$

Note that (2) is called the monotonicity condition.

In addition, a BC is said to be normalized if it satisfies :

$$\nu(N, \emptyset) = 1 \text{ and } \nu(\emptyset, N) = -1 \tag{3}$$

A BC is also said to be additive if the following relation holds :

$$\forall (A_1, A_2) \in 3^N : \nu(A_1, A_2) = \sum_{i \in A_1} \nu(i, \emptyset) + \sum_{j \in A_2} \nu(\emptyset, j) \tag{4}$$

An additive BC assumes that the attributes are independent from each other and this kind of BC boils down to linear decision models.

In order to better formalize some of the properties of BC, the following definition of a (bipolar) Möbius transform[2] of a BC was proposed.

Definition 2 (Bipolar Möbius Transform of a Bi-capacity [10, 17]). *Let ν be a BC on 3^N. The bipolar Möbius transform of ν is a set function $b : 3^N \to \mathbb{R}$ defined for any $(A_1, A_2) \in 3^N$ by :*

$$b(A_1, A_2) := \sum_{\substack{B_1 \subseteq A_1 \\ B_2 \subseteq A_2}} (-1)^{|A_1 \setminus B_1| + |A_2 \setminus B_2|} \nu(B_1, B_2) \tag{5}$$

$$= \sum_{(\emptyset, A_2) \sqsubseteq (B_1, B_2) \sqsubseteq (A_1, \emptyset)} (-1)^{|A_1 \setminus B_1| + |A_2 \setminus B_2|} \nu(B_1, B_2)$$

Conversely, for any $(A_1, A_2) \in 3^N$, it holds that :

$$\nu(A_1, A_2) := \sum_{\substack{B_1 \subseteq A_1 \\ B_2 \subseteq A_2}} b(B_1, B_2). \tag{6}$$

Note that using b, (1) is equivalent to :

$$b(\emptyset, \emptyset) = 0 \tag{7}$$

BC on 3^N generally require $3^n - 1$ parameters. In order to reduce this number, [8, 9] and [5] proposed the concept of k-additivity of a BC. This concept translates as follows in terms of the bipolar Möbius transform.

Proposition 1 ([17]). *Given a positive integer $k < n$, a BC ν is k-additive if and only if the two following conditions are satisfied :*

$$\forall (A_1, A_2) \in 3^N : |A_1 \cup A_2| > k \Rightarrow b(A_1, A_2) = 0 \tag{8}$$

$$\exists (A_1, A_2) \in 3^N : |A_1 \cup A_2| = k \wedge b(A_1, A_2) \neq 0 \tag{9}$$

To avoid a heavy notation, we use the following shorthands for all $i, j \in N$, $i \neq j$:

- $\nu_{i|} := \nu(i, \emptyset)$, $\nu_{|j} := \nu(\emptyset, j)$, $\nu_{i|j} := \nu(i, j)$, $\nu_{ij|} := \nu(ij, \emptyset)$, $\nu_{|ij} := \nu(\emptyset, ij)$,
- $b_{i|} := b(i, \emptyset)$, $b_{|j} := b(\emptyset, j)$, $b_{i|j} := b(i, j)$, $b_{ij|} := b(ij, \emptyset)$, $b_{|ij} := b(\emptyset, ij)$.

Whenever we use i and j together, it always means that they are different.

Using the above definitions, we propose the following properties of a 2A-BC ν and its bipolar Möbius transform b :

---

[2] Note that [12] was the first paper to define the Möbius transform of a BC. Their definition is different from the one given in [10]. However, there is a one-to-one correspondence between the two Möbius transform definitions. This equivalence was established in [11].

Proposition 2.

1. *Let ν be a 2A-BC and b its bipolar Möbius transform. For any $(A_1, A_2) \in 3^N$ we have :*

$$\nu(A_1, A_2) = \sum_{i \in A_1} b_{i|} + \sum_{j \in A_2} b_{|j} + \sum_{\substack{i \in A_1 \\ j \in A_2}} b_{i|j} + \sum_{\{i,j\} \subseteq A_1} b_{ij|} + \sum_{\{i,j\} \subseteq A_2} b_{|ij} \quad (10)$$

2. *If the coefficients $b_{i|}$, $b_{|j}$, $b_{i|j}$, $b_{ij|}$, $b_{|ij}$ are given for all $i, j \in N$, then the necessary and sufficient conditions to get a 2A-BC generated by (10) are :*

$$\forall (A, B) \in 3^N, \forall k \in A : b_{k|} + \sum_{j \in B} b_{k|j} + \sum_{i \in A \setminus k} b_{ik|} \geq 0 \quad (11)$$

$$\forall (A, B) \in 3^N, \forall k \in A : b_{|k} + \sum_{j \in B} b_{j|k} + \sum_{i \in A \setminus k} b_{|ik} \leq 0 \quad (12)$$

3. *The inequalities (11) and (12) can be respectively reformulated in terms of the BC ν as follows :*

$$\forall (A, B) \in 3^N, \forall k \in A : \sum_{j \in B} \nu_{k|j} + \sum_{i \in A \setminus k} \nu_{ik|} \geq (|B| + |A| - 2)\nu_{k|} + \sum_{j \in B} \nu_{|j} + \sum_{i \in A \setminus k} \nu_{i|}$$

$$\forall (A, B) \in 3^N, \forall k \in A : \sum_{j \in B} \nu_{j|k} + \sum_{i \in A \setminus k} \nu_{|ik} \leq (|B| + |A| - 2)\nu_{|k} + \sum_{j \in B} \nu_{j|} + \sum_{i \in A \setminus k} \nu_{|i}$$

Proof. (Sketch of)

1. Because ν is 2-additive, the proof of (10) is given by using the relation (6) between ν and b.
2. The proof of the second point is based on the expression of $\nu(A_1, A_2)$ given in (10) and on these equivalent monotonicity properties (which are easy to check) : $\forall (A, B) \in 3^N$ and $\forall A \subseteq A'$,
   (a) $\nu(A, B) \leq \nu(A', B) \Leftrightarrow \{\forall k \in A : \nu(A \setminus k, B) \leq \nu(A, B)\}$;
   (b) $\nu(B, A') \leq \nu(B, A) \Leftrightarrow \{\forall k \in A : \nu(B, A) \leq \nu(B, A \setminus k)\}$.
3. These inequalities are obtained departing from (11) and (12) and by using the relation (6) between ν and b.
   □

Hence, according to proposition 2 and (10), the computation of a 2A-BC ν only requires the values of b on the elements $(i, \emptyset)$, $(\emptyset, i)$, (i, j), $(ij, \emptyset)$, $(\emptyset, ij)$, $\forall i, j \in N$. However, in order to satisfy the monotonicity condition given in (2) a 2A-BC should also satisfy the inequalities (11) and (12). Moreover, we have the following conditions in order to obtain a normalized 2A-BC :

$$\nu_{N|} = \sum_{i \in N} b_{i|} + \sum_{\{i,j\} \subseteq N} b_{ij|} = 1 \text{ and } \nu_{|N} = \sum_{i \in N} b_{|i} + \sum_{\{i,j\} \subseteq N} b_{|ij} = -1 \quad (13)$$

2.2   Bipolar Choquet Integral w.r.t. a 2-Additive Bi-capacity

Definition 3 (Bipolar Choquet integral (BCI) (w.r.t. a BC) [9]). *Let ν be a BC on 3^N and $x = (x_1, \ldots, x_n) \in \mathbb{R}^n$. The expression of the BCI of x w.r.t. ν is given by*

$$\mathcal{C}_\nu(x) := \sum_{i=1}^{n} |x_{\sigma(i)}| \left[\nu(N_{\sigma(i)} \cap N^+, N_{\sigma(i)} \cap N^-) - \nu(N_{\sigma(i+1)} \cap N^+, N_{\sigma(i+1)} \cap N^-) \right]$$

(14)

where $N^+ = \{i \in N | x_i \geq 0\}$, $N^- = N \setminus N^+$, $N_{\sigma(i)} := \{\sigma(i), \ldots, \sigma(n)\}$ and σ is a permutation on N such that $|x_{\sigma(i)}| \leq |x_{\sigma(i+1)}| \leq \cdots \leq |x_{\sigma(n)}|$.

We also have the following equivalent expression of the BCI w.r.t. b, given by [11] :

$$\mathcal{C}_b(x) = \sum_{(A_1, A_2) \in 3^N} b(A_1, A_2) \left(\bigwedge_{i \in A_1} x_i^+ \wedge \bigwedge_{j \in A_2} x_j^- \right)$$

(15)

where $\begin{cases} x_i^+ = x_i & \text{if } x_i > 0 \\ x_i^+ = 0 & \text{if } x_i \leq 0 \end{cases}$ and $\begin{cases} x_i^- = -x_i & \text{if } x_i < 0 \\ x_i^- = 0 & \text{if } x_i \geq 0 \end{cases}$.

Note that $\mathcal{C}_\nu(x) = \mathcal{C}_b(x)$ and the subscript is meant to clarify whether it is ν or b which is used in the calculation. Besides, the BCI of x w.r.t. a 2A-BC represented by b reduces to :

$$\mathcal{C}_b(x) = \sum_{i=1}^{n} b_{i|} x_i^+ + \sum_{i=1}^{n} b_{|i} x_i^- + \sum_{i,j=1}^{n} b_{i|j} (x_i^+ \wedge x_j^-)$$

$$+ \sum_{\{i,j\} \subseteq N} b_{ij|} (x_i^+ \wedge x_j^+) + \sum_{\{i,j\} \subseteq N} b_{|ij} (x_i^- \wedge x_j^-)$$

(16)

We have introduced the basic tools related to 2A-BC. In the next section, we focus on the problem of identifying a 2A-BC.

3   Identifying a 2-Additive Bi-capacity

We establish mathematical programming problems that enable the identification of a 2A-BC. First, we detail the type of elicitation process we are concerned with. Next, we state all the constraints that allow the representation of a 2A-BC. Then, we define objective functions that reflect the quality of the identified 2A-BC in regard to the information provided by the DM. We end this section by illustrating the results obtained with the proposed models on a numerical example.

3.1   Elicitation Process

In MCDM, there are two types of paradigms for elicitation processes : direct and indirect methods. In the former case, the DM is able to provide the parameters of his decision model directly. However, when using a BCI w.r.t. a BC, the direct method is infeasible if the number of attributes n exceeds some units (typically 4), since a BC requires $3^n - 1$ values to be set. We argued in the introduction that in order to reduce this complexity, 2A-BC were introduced. Nevertheless, this latter case cannot be applied in practice neither since, even if the number of parameters reduces to $2n^2 + 1$, the complexity remains very high. Moreover, a BC is a too complex aggregation operator to ensure that a DM will understand the influence of each parameter on the final result. Even with more simple aggregation rules, it has been shown that there is no clear link between the parameters values provided by the DM and the way these values are used in the decision model [18].

Therefore, we follow the indirect paradigm. In that case, the DM does not give information about his decision model, instead, he provides information on the outputs of his decision model. In our setting, we suppose that the DM gives for some examples $x \in X' \subseteq X$, their partial utilities for all criteria $(U(x))$ and also their overall scores $(S(x))$. We then assume, that there is no further interaction with the DM. Given the judged examples, we have to infer a decision model based on the BCI w.r.t. a 2A-BC. The estimated BCI should predict overall scores, $\mathcal{C}_b(x)$, that are consistent with the preference relations provided by the DM. In other words, if $S(x) \geq S(x')$, which means that x is preferred or equivalent to x', then the inferred decision model should also satisfy $\mathcal{C}_b(x) \geq \mathcal{C}_b(x')$.

However, it might happen that this condition is not fulfilled for some pairs (x, x'). There are two main reasons for such situations : either the judgements provided by the DM himself are not consistent or the restriction of the decision model to 2A-BC does not allow fitting the DM preferences correctly. In MCDM, inconsistencies are usually treated in an interaction loop with the DM. It is assumed that the DM preferences can change in order to fix these incoherences when they are encountered. In our setting, the interaction loop is not permitted. Consequently, in order to cope with incoherences, we propose two versions of our models : the first one does not deal with inconsistencies and thus will return that the problem is infeasible if any incoherence is encountered whereas the second one allows inconsistencies and attempts to infer a model that minimizes errors due to such situations as much as possible.

3.2   Mathematical Programming Problems

We propose two types of optimization problems to identify a 2A-BC in the context we have described in the previous paragraph. We base our work on some of the elicitation methods detailed in [14] in the case of unipolar Choquet integral. Before introducing the objective functions of our optimization problems, we start by enumerating the different sets of constraints that need to be satisfied.

We represent the unknown 2A-BC ν, *via* its associated bipolar Möbius transform b. There are two reasons for this. Firstly, equations (5) and (6) state that it is equivalent to work with either ν or b. Secondly, since we restrict the BC to be 2-additive and since this property, given in (8) and (9), is defined in terms of b, it is thus necessary to use the latter representation in our optimization problems.

However, we do not take (9) into account in our set of constraints. This equation ensures that ν is exactly 2-additive and by discarding it, we explicitly allow b to be either 2-additive or simply additive. To summarize this first set of constraints, we need to integrate the following relations in our optimization problems in order to have a normalized 2A-BC in terms of b : (7), (8) with $k = 2$, (11), (12) and (13).

Next, we have to take into account the preference relations provided by the DM on the subset of examples X'. If $S(x) \geq S(x')$ then the BCI should be in concordance with this inequality. Accordingly, we have the following second set of constraints :

$$\forall x, x' \in X', x \neq x' : S(x) - S(x') \geq 0 \Rightarrow \mathcal{C}_b(x) - \mathcal{C}_b(x') \geq \delta_c \qquad (17)$$

where δ_c is a non-negative indifference threshold which is a parameter of the model. Note that this set of constraints does not allow incoherences. Indeed, the inferred 2A-BC b could not be flexible enough to satisfy $\mathcal{C}_b(x) - \mathcal{C}_b(x') \geq \delta_c$ for some pairs (x, x'). In that case the optimization problem is infeasible.

As discussed previously, in order to overcome this drawback, we transform the previous constraints as follows :

$$\forall x, x' \in X', x \neq x' : S(x) - S(x') \geq 0 \Rightarrow \mathcal{C}_b(x) - \mathcal{C}_b(x') \geq \delta_c - \xi_{xx'} \qquad (18)$$

where $\xi_{xx'}$ are non-negative slack variables which allow inconsistencies. However, we want $\xi_{xx'}$ to be has low as possible and thus there should be a term in the objective function seeking to minimize $\sum_{x,x':S(x) \geq S(x')} \xi_{xx'}$. Note that when the latter term is null, it means that the inferred model does not produce any incoherence. On the contrary, if for some pairs (x, x'), $\xi_{xx'} > \delta_c$ then the optimal solution has not been able to satisfy the preference relations on these pairs.

The third set of constraints is related to the computation of the BCI. Indeed, in (17) or (18), we need to calculate $\mathcal{C}_b(x)$ for each $x \in X'$. As a consequence, we need to add the constraints provided by (16) in our models. Note that despite the fact that the latter equations involve the minimum function, we can pre-compute the terms $(x_i^{\pm} \wedge x_j^{\pm})$ since they are parameters of the models. Consequently, the constraints (16) are linear equations.

The fourth set of constraints is optional. It simply consists in adding upper and lower bounds for the BCI values :

$$\forall x \in X' : lb \leq \mathcal{C}_b(x) \leq ub \qquad (19)$$

where lb and ub are two real parameters.

After having introduced the constraints, we now focus on the different objective functions and the resulting optimization problems.

We propose two kinds of optimization models. In the first approach, we extend the maximum split method introduced in [19]. This model assumes the following constraints in place of (17) :

$$\forall x, x' \in X', x \neq x' : S(x) - S(x') \geq 0 \Rightarrow \mathcal{C}_b(x) - \mathcal{C}_b(x') \geq \delta_c + \varepsilon \qquad (20)$$

where ε is a variable of the problem unlike δ_c which is a parameter. The objective function consists in maximizing ε. In other words, we want to maximize the difference (split) $\mathcal{C}_\nu(x) - \mathcal{C}_\nu(x')$ for any $x \neq x' \in X'$ such that $S(x) \geq S(x')$. We refer to the following optimization problem as the *split* method : $\max \varepsilon$ subject to (7), (8) with $k = 2$, (11), (12), (13), (16), (19) and (20).

However, the *split* model does not address incoherences. Hence, as explained previously, we propose the *split flex* approach which uses the following third set of constraints instead of (20) :

$$\forall x, x' \in X', x \neq x' : S(x) - S(x') \geq 0 \Rightarrow \mathcal{C}_b(x) - \mathcal{C}_b(x') \geq \delta_c + \varepsilon - \xi_{xx'} \quad (21)$$

More formally, the *split flex* model is defined by : $\max \varepsilon - \sum_{x,x':S(x)\geq S(x')} \xi_{xx'}$ subject to (7), (8) with $k = 2$, (11), (12), (13), (16), (19) and (21). Note that the *split* and *split flex* optimization problems have linear objective functions and linear constraints. Therefore, there are linear programs.

We now present the second type of model for 2A-BC identification. This approach is a regression-like method and yields to quadratic programs. We propose to minimize the sum of square errors between S and $\mathcal{C}_b$ which results in the following objective function : $\min \sum_{x,x' \in X'} (S(x) - \mathcal{C}_b(x))^2$. Accordingly, we named *rss* (for Residual Sum of Square) the following problem : $\min \sum_{x,x' \in X'} (S(x) - \mathcal{C}_b(x))^2$ subject to (7), (8) with $k = 2$, (11), (12), (13), (16) and (17). Similarly to *split*, the *rss* method does not permit incoherences.

As a consequence, we introduce a flexible version of *rss* that we call *rss flex* and which is given by : $\min \sum_{x,x' \in X'} (S(x) - \mathcal{C}_b(x))^2 + \sum_{x,x':S(x)\geq S(x')} \xi_{xx'}$ subject to (7), (8) with $k = 2$, (11), (12), (13), (16) and (18).

3.3   An Illustrative Example

We applied the four different mathematical programming problems defined previously on a numerical example taken from [14]. It concerns the grades (utilities) obtained by 7 students (alternatives) for $n = 5$ subjects (attributes) : statistics (S), probability (P), economics (E), management (M), and English (En). The grades globally belong to $[0, 20]$ but in this example, the scores only vary in $[11, 18]$. In our perspective, we transformed them in order to have a bipolar scale by simply applying a translation of -14 to the original scores. Therefore, in this bipolar scale, the scores belong to $[-3, 4]$. Suppose that a student is delivered his diploma with honors providing that his overall grade is greater or equal to 14. Hence, the translated scores in the bipolar scale allow us to deal with the decision problem of delivering honors as follows : the student is attributed the honors if and only if his overall grade in the bipolar scale is non-negative.

The performance table is given in Table 1 (a). In Table 1 (b), the first column S corresponds to the (translated) overall grades as given in [14]. Then, in the subsequent columns of Table 1 (b), we show the different estimated scores. Note that for all models we set $\delta_c = 0.5$, $lb = -3$ and $ub = 4$. For *split* and *split flex*, even if the inferred overall scores are not the same, the two solutions are actually equivalent since they give the same objective function value (the problem is not strictly concave). However, in terms of decisions, the sign of the overall grade for c is not the same for the two models. This example exhibits some limits

Table 1. (a) Performance table; (b) Results obtained with the original (translated) overall score S; (c) Results obtained with the modified overall score S' that presents inconsistencies

Student	S P E M En	S	split	split flex	rss	rss flex	S'	split	split flex	rss	rss flex
a	4 -3 -3 -3 4	1	1.68	1.02	1	1	1	.	0.22	.	1.12
b	4 -3 4 -3 -3	0.5	1.04	0.38	0.5	0.5	0.5	.	-0.28	.	0.62
c	-3 -3 4 -3 4	0	0.41	-0.25	0	0	0	.	-0.78	.	0.12
d	4 4 -3 -3 -3	-0.5	-0.23	-0.89	-0.5	-0.5	-0.5	.	-1.28	.	-0.5
e	-3 -3 4 4 -3	-1	-0.86	-1.53	-1	-1	-1	.	-1.78	.	-1
f	-3 -3 4 -3 -3	-1.5	-1.5	-2.16	-1.5	-1.5	-1.5	.	-2.28	.	-1.5
g	-3 -3 -3 -3 4	-2	-2.14	-2.8	-2	-2	**0.5**	.	-0.78	.	0.12
(a)		(b)					(c)				

of this type of model. Besides, we precise that in *split flex* outputs, all slack variables $\xi_{xx'}$ are null which means that there is no inconsistency. Regarding *rss* and *rss flex*, we obtain a null objective function value and in the latter case, the slack variables are also all null as expected.

To illustrate the case with incoherences, we modified the overall score S into S'. We simply change the global grade of g from -2 to 0.5. This new score is an example of inconsistent preferences provided by the DM since his decision model is not monotonic in that case. Indeed, if we compare the profiles of c and g in regard to their overall score $S'(c)$ and $S'(g)$, we can observe that g is preferred to c while the scores distribution of the former student is Pareto dominated by the latter one. This situation is not consistent with a rational decision. Table 1 (c) presents the estimated BCI. As expected, *split* and *rss* returned an infeasible problem. On the contrary, *split flex* and *rss flex* provide interesting results. In both cases, we precise that $\xi_{gc} = 0.5$ while the other slack variables are null.

We present in Table 2 the elicited bipolar Möbius transform b for each optimization problem when S is the targeted overall grades vector. In Table 3, it is the estimated b for the *split flex* and *rss flex* methods when S' is put in place of S, which are shown. Notice that an empty cell in these two tables means that the solver returned a null value for the corresponding elements.

We show the bipolar Möbius transform b and not the 2A-BC ν because the latter set function requires $3^5 - 1 = 242$ non-null values which represents a too large table. In contrast, because ν is 2-additive, we have at most $2 \times 5^2 + 1 = 51$ values for its related bipolar Möbius transform b.

The utility of presenting Tables 2 and 3, is that they allow one to check that the constraints (7), (8) with $k = 2$, (11), (12), (13) are indeed satisfied. We thus obtain a 2A-BC. However, it is difficult to interpret the bipolar Möbius transform b with regard to the underlying elicited preference model. Moreover, from this illustrative example, it is not straightforward to understand the impact of taking into account inconsistencies either when we compare the regular and the *flex* versions of *split* and *rss* models in Table 2, or when we look at the obtained b for S in Table 2 and the one estimated for S' in Table 3. Accordingly, further

Table 2. Values of the elicited bipolar Möbius transform b when S is the overall score

A_1	A_2	split $b_{A_1|A_2}$	$b_{A_2|A_1}$	split flex $b_{A_1|A_2}$	$b_{A_2|A_1}$	rss $b_{A_1|A_2}$	$b_{A_2|A_1}$	rss flex $b_{A_1|A_2}$	$b_{A_2|A_1}$
∅	∅								
∅	S		0.26	-0.145		-0.039	0.077		0.056
∅	P					-0.077	0.038	-0.056	
∅	E				0.145	-0.038	0.077		0.056
∅	M					-0.08	0.038		
∅	En			-0.066		-0.039	0.008	-0.056	
∅	SP			-0.56		-0.33		-0.24	
∅	SE	-0.076			0.11	-0.004	0.11		0.19
∅	SM	-0.21					-0.008		
∅	SEn	-0.29	0.48		0.43	0.039	0.47		0.53
∅	PE					-0.29		-0.43	
∅	PM						-0.03		
∅	PEn						-0.008		
∅	EM	-0.42		-0.23		0.004			
∅	EEn		0.26		0.32	-0.14	0.23	-0.056	0.17
∅	MEn							-0.17	
S	P								
S	E					-0.07	-0.039		-0.056
S	M				0.145	0.004		-0.02	
S	En					0.039	0.039	0.056	
P	E								
P	M						0.039		
P	En						0.038		0.056
E	M					0.004			
E	En			-0.145		-0.039	0.034		
M	En			0.066			0.073		

experiments should be undertaken in order to have a better understanding of the behaviors of the elicited ν and b for each proposed optimization problem but such an experimental study is out of the scope of this paper.

4 Relationships with Machine Learning

The preference elicitation problem have many common points with supervised learning problems (SL) in statistical machine learning (ML). In the latter field, we are given a training set which consists of items described in a feature space and each element also comes with a value in regard to a target variable. In ML, the goal of SL is to infer from the training set a mapping from the feature space to the target variable. This description if similar to the preference elicitation setting we have described previously : the set of examples with their profiles $U(x)$ and their overall scores $S(x)$ are the equivalent of the training set in ML and identifying the parameters of the decision model in order to reproduce the

Table 3. Values of the elicited bipolar Möbius transform b when S' is the overall score with incoherences

A_1	A_2	split flex		rss flex	
		$b_{A_1\mid A_2}$	$b_{A_2\mid A_1}$	$b_{A_1\mid A_2}$	$b_{A_2\mid A_1}$
∅	S			-0.077	0.47
∅	P	-0.19		-0.53	0.077
∅	E			-0.077	0.023
∅	M		0.031	-0.14	0.077
∅	En	-0.031			0.19
∅	SP	-0.26		0.077	0.09
∅	SE		0.19		0.263
∅	SM		0.73		
∅	SEn	-0.16	0.055		-0.018
∅	PE				
∅	PM			0.14	
∅	PEn	0.031			-0.077
∅	EM				
∅	EEn	-0.24		-0.39	-0.023
∅	MEn	-0.16			-0.077
S	P			0.06	
S	E				
S	M		-0.031	-0.33	
S	En				
P	E				
P	M				0.14
P	En		0.19	-0.077	0.31
E	M				
E	En				0.077
M	En			-0.077	

preference relations given by the DM is the same as inferring a mapping from a feature space (X) and a target variable (S). Despite these straightforward similarities, the roots of these two domains are distinct, and it is only recently, that there has been a growing interest in applying concepts or/and techniques developed in MCDM to ML problems and *vice-versa* [16].

To contribute in that direction, we discuss the relationships between our work and some concepts defined in ML.

Firstly, in MCDM, incoherences are not generally allowed from the DM viewpoint who has to fix them during the course of the elicitation procedure. In ML, on the contrary, such situations are typically observed in real-world applications. Thus, the models *split* and *rss* are typical of MCDM whereas *split flex* and *rss flex* are closer to SL problems.

Secondly, it is noteworthy that the restriction to 2A-BC can be interpreted as an explicit regularization of the decision model based on the BCI. Indeed, in SL, regularization is a concept that aims at dealing with the bias-variance trade-off of predictive models. Typically, in order to avoid an over-fitting effect, we allow the estimated predictive model to make more errors on the training set (bias) but in return we want it to be less variable (variance). The goal of regularization is to enhance the ability of the inferred model to predict correctly the overall score of observed examples (the training set) but also and in particular the global score of unseen examples (the test set). In our case, 2A-BC are less flexible than unconstrained BC and this could lead to more inconsistencies as explained beforehand. However, the 2-additivity property makes the BC less complex and we intuitively expect 2A-BC to be less variable than unconstrained BC.

Thirdly, we stated that the constraint (9) ensuring that ν is exactly 2-additive was not part of our set of constraints. In that case ν could thus be either 2-additive or simply additive. We can make the correspondence between this approach and the Occam's Razor principle often used in ML. This concept states that one should prefer simpler models than more complex ones because they allow a better understanding of the phenomenon under study. In our case, we can transpose this statement as follows : if an additive BC fits better the DM judgements than a 2A-BC then we should go for the former one. By discarding (9) from our constraints we make the latter statement possible. However, if it happens that a 2A-BC and an additive BC yield to the same optimal objective function value then it is not guaranteed that the optimization solver will provide the simplest solution.

Finally, in order to make the models *split* and *rss* more flexible regarding inconsistencies, we have proposed to integrate slack variables in the constraints and in the objective function as well. Our approach is inspired from the Support Vector Machine (SVM) method developed in SL in order to deal with non linearly separable cases in binary classification [20]. However, it is noteworthy that the UTA (UTilités Additives) framework is a MCDM methodology that also addresses inconsistencies by integrating overestimation and underestimation error variables in the elicitation model (see for example [21]). This approach is similar to adding slack variables.

5   Conclusion

We have proposed optimization problems that allow the identification of the parameters of a 2A-BC. The decision model is inferred from examples that the DM evaluated both regarding their partial utilities and their global scores. We have considered the traditional preference elicitation setting where no inconsistency is allowed. But, we have also extended the models to the more flexible case where we have to cope with such incoherences. In this context, we have emphasized the relationships between our approaches and concepts in ML.

In our future work, we intend to integrate other kinds of information provided by the DM such as the importance of criteria and the interactions between them. Moreover, as pointed out previously, additional experiments should be conducted in order to better characterize the elicited bipolar Möbius transform b and its associated 2A-BC ν, we obtain for each type of mathematical program.

As regard to our ongoing work, we are investigating optimization problems for preferences learning that integrate other concepts developed in the ML literature.

We are currently working on objective functions that involve a penalty term that favors sparse BC. Such an approach adds a trade-off between the accuracy of the model and its simplicity (Occam's Razor principle). The expected advantage is to elicit preference models that are easier to interpret and such a feature is of great importance in MCDM.

References

1. Choquet, G.: Theory of capacities. Annales de l'institut Fourier 5, 131–295 (1954)
2. Grabisch, M.: k-order additive discrete fuzzy measures and their representation. Fuzzy Sets and Systems 92(2), 167–189 (1997)
3. Grabisch, M., Labreuche, C.: Fuzzy measures and integrals in MCDA. In: Multiple Criteria Decision Analysis: State of the Art Surveys. Int. Series in Op. Res. & Manag. Sci., vol. 78, pp. 563–604. Springer, New York (2005)
4. Meyer, P., Pirlot, M.: On the expressiveness of the additive value function and the choquet integral models. In: DA2PL 2012 From Multiple Criteria Decision Aid to Preference Learning (2012)
5. Grabisch, M., Labreuche, C.: A decade of application of the Choquet and Sugeno integrals in multi-criteria decision aid. 4OR 6(1), 1–44 (2008)
6. Grabisch, M., Labreuche, C.: Bi-capacities for decision making on bipolar scales. In: Proc. of the EUROFUSE 2002 Workshop on Information Systems, pp. 185–190 (2002)
7. Greco, S., Matarazzo, B., Slowinski, R.: Bipolar Sugeno and Choquet integrals. In: Proceedings of the EUROFUSE 2002 Workshop on Information Systems (2002)
8. Grabisch, M., Labreuche, C.: Bi-capacities-I: definition, Möbius transform and interaction. Fuzzy Sets and Systems 151(2), 211–236 (2005)
9. Grabisch, M., Labreuche, C.: Bi-capacities-II: the Choquet integral. Fuzzy Sets and Systems 151(2), 237–259 (2005)
10. Fujimoto, K.: New characterizations of k-additivity and k-monotonicity of bi-capacities. In: Joint 2nd Int. Conf. on Soft Computing and Intelligent Systems and 5th International Symposium on Advanced Intelligent Systems (2004)
11. Fujimoto, K., Murofushi, T.: Some characterizations of k-monotonicity through the bipolar möbius transform in bi-capacities. JACIII 9(5), 484–495 (2005)
12. Grabisch, M., Labreuche, C.: The choquet integral for 2-additive bi-capacities. In: EUSFLAT Conf., pp. 300–303 (2003)
13. Jacquet-Lagreze, E., Siskos, J.: Assessing a set of additive utility functions for multicriteria decision-making, the UTA method. European Journal of Operational Research 10(2), 151–164 (1982)
14. Grabisch, M., Kojadinovic, I., Meyer, P.: A review of methods for capacity identification in Choquet integral based multi-attribute utility theory: Applications of the kappalab R package. EJOR 186(2), 766–785 (2008)
15. Mayag, B., Rolland, A., Ah-Pine, J.: Elicitation of a 2-additive bi-capacity through cardinal information on trinary actions. In: Greco, S., Bouchon-Meunier, B., Coletti, G., Fedrizzi, M., Matarazzo, B., Yager, R.R. (eds.) IPMU 2012, Part IV. CCIS, vol. 300, pp. 238–247. Springer, Heidelberg (2012)
16. Waegeman, W., Baets, B.D., Boullart, L.: Kernel-based learning methods for preference aggregation. 4OR 7(2), 169–189 (2009)
17. Fujimoto, K., Murofushi, T., Sugeno, M.: k-additivity and c-decomposability of bi-capacities and its integral. Fuzzy Sets Syst. 158(15), 1698–1712 (2007)

18. Bouyssou, D., Marchant, T., Pirlot, M., Tsoukias, A., Vincke, P.: Evaluation and Decision models with multiple criteria: stepping stones for the analyst. Int. Series in Op. Res. & Manag. Sci., vol. 86. Springer (2006)
19. Marichal, J.L., Roubens, M.: Determination of weights of interacting criteria from a reference set. EJOR 124(3), 641–650 (2000)
20. Cortes, C., Vapnik, V.: Support-vector networks. Mach. Learn. 20(3), 273–297 (1995)
21. Siskos, Y., Grigoroudis, E., Matsatsinis, N.: UTA methods. In: Multiple Criteria Decision Analysis: State of the Art Surveys. International Series in Operations Research & Management Science, vol. 78, pp. 297–334. Springer, New York (2005)

How to Put through Your Agenda
in Collective Binary Decisions[*]

Noga Alon[1], Robert Bredereck[2], Jiehua Chen[2], Stefan Kratsch[2],
Rolf Niedermeier[2], and Gerhard J. Woeginger[3]

[1] School of Mathematical Sciences, Tel Aviv University, Israel
[2] Institut für Softwaretechnik und Theoretische Informatik, TU Berlin, Germany
[3] Department of Mathematics and Computer Science, TU Eindhoven,
The Netherlands

Abstract. We consider the following decision scenario: a society of voters has to find an agreement on a set of proposals, and every single proposal is to be accepted or rejected. Each voter supports a certain subset of the proposals–the *favorite ballot* of this voter–and opposes the remaining ones. He accepts a ballot if he supports more than half of the proposals in this ballot. The task is to decide whether there exists a ballot approving a set of selected proposals (agenda) such that all voters (or a strict majority of them) accept this ballot.

On the negative side both problems are NP-complete, and on the positive side they are fixed-parameter tractable with respect to the total number of proposals or with respect to the total number of voters. We look into further natural parameters and study their influence on the computational complexity of both problems, thereby providing both tractability and intractability results. Furthermore, we provide tight combinatorial bounds on the worst-case size of an accepted ballot in terms of the number of voters.

1 Introduction

Consider the following decision scenario which may occur in contexts like coalition formation, the design of party platforms, the change of statutes of an association, or the agreement on contract issues: A leader has an agenda, that is, a set of proposals she wants to get realized. However, a set of proposals has to be approved or disapproved as a whole by a set of voters. Each voter has his favorite proposals he wants to support. A set of proposals is acceptable to a voter if he supports more than half of these proposals. Now, the leader is searching for a set of proposals containing her personal agenda such that a majority of

---

[*] NA is supported in part by an ERC advanced grant, by a USA-Israeli BSF grant, by an ISF grant and by the Israeli I-Core program. RB is supported by the DFG, research project PAWS, NI 369/10. JC is supported by the Studienstiftung des Deutschen Volkes. SK is supported by the DFG, research project PREMOD, KR 4286/1. GW is supported by DIAMANT (a mathematics cluster of the Netherlands Organization for Scientific Research NWO) and, while staying at TU Berlin (October 2012 - June 2013), by the Alexander von Humboldt Foundation, Bonn, Germany.

P. Perny, M. Pirlot, and A. Tsoukiàs (Eds.): ADT 2013, LNAI 8176, pp. 30–44, 2013.

voters accepts this set. Can the leader efficiently find such a successful set of proposals realizing her agenda? What about when this set of proposals has to be acceptable to *all* voters and not just to a majority?

Mathematical Model. Let $\mathcal{V}$ be a society of n voters and $\mathcal{P}$ be a set of m proposals. Each voter may support any number of proposals in $\mathcal{P}$ and rejects all the others. Subsets of $\mathcal{P}$ are called *ballots*. The favorite ballot $B_i \subseteq \mathcal{P}$ of a voter i ($1 \leq i \leq n$) consists of all proposals he supports.

The voters evaluate a ballot $Q \subseteq \mathcal{P}$ according to the size of the intersection of Q and their favorite ballots. More precisely, voter i *accepts* Q if and only if a strict majority of proposals from Q is also contained in his favorite ballot, that is,

$$|B_i \cap Q| > |Q|/2.$$

We say that in this case voter i is *happy* with Q.

The central question is whether there exists a ballot Q that (a) contains a *given* agenda and that (b) is acceptable to the society. The agenda in (a) is a set $Q_+ \subseteq \mathcal{P}$ of proposals that have to be contained in Q, that is, $Q_+ \subseteq Q$. The society's acceptance in (b) might be a *unanimous* acceptance or a *majority* acceptance. This leads to the following two problems which only differ in the respective questions asked.

> UNANIMOUSLY ACCEPTED BALLOT (UNAAB)
> **Input:** A set $\mathcal{P}$ of m proposals; a society $\mathcal{V}$ of n voters with favorite ballots $B_1, \ldots, B_n \subseteq \mathcal{P}$; an agenda $Q_+ \subseteq \mathcal{P}$.
> **Question:** Is there a ballot $Q_+ \subseteq Q \subseteq \mathcal{P}$ which *every* single voter i accepts (that is, $|B_i \cap Q| > |Q|/2$)?

> MAJORITYWISE ACCEPTED BALLOT (MAJAB)
> **Input:** A set $\mathcal{P}$ of m proposals; a society $\mathcal{V}$ of n voters with favorite ballots $B_1, \ldots, B_n \subseteq \mathcal{P}$; an agenda $Q_+ \subseteq \mathcal{P}$.
> **Question:** Is there a ballot $Q_+ \subseteq Q \subseteq \mathcal{P}$ which a *strict majority* of the voters accepts (that is, $|B_i \cap Q| > |Q|/2$)?

One important special case of UNAAB or MAJAB is when the agenda is empty, that is, $Q_+ = \emptyset$. In that case, the only question is whether there is a ballot acceptable to the society.

Interestingly, the following example demonstrates that the solutions sizes to our problems are *not* monotone, that is, a solution ballot of size h does not imply a solution of a size smaller or larger than h. This is in notable contrast to many natural decision problems, such as all problems we reduce from in this paper.

Example 1. Consider the society $\mathcal{V} = \{1, 2, 3, 4\}$ of voters and the set $\mathcal{P} = \{p_1, p_2, p_3, p_4, p_5\}$ of proposals. The favorite ballots are given as $B_1 = \{p_2, p_3, p_4\}$, $B_2 = \{p_1, p_3, p_5\}$, $B_3 = \{p_1, p_2, p_4\}$, and $B_4 = \{p_1, p_2, p_3\}$.

Suppose that the hidden agenda is empty. Then the only unanimously accepted ballots are $\{p_1, p_2, p_3\}$ of size three and $\{p_1, p_2, p_3, p_4, p_5\}$ of size five. This shows that the set of the sizes of all solution ballots may contain gaps.

With regard to majority acceptance, if the hidden agenda is $\{p_5\}$, then ballots $\{p_1, p_2, p_5\}$, $\{p_2, p_3, p_5\}$, and $\{p_1, p_2, p_3, p_4, p_5\}$ are the only ballots that are acceptable to a strict majority of voters. Again, the set of the sizes of all solution ballots contains gaps. If the hidden agenda is empty, then ballot $\{p_1, p_2, p_3, p_4\}$ is also acceptable to a strict majority of voters.

Related Work. While the two problems we introduce and study seem to be new, the investigation of situations where a society has to decide upon binary (that is, yes-or-no) issues is common within the theory and practice of decision making. For instance, Laffond and Lainé [26] recently investigated the conditions under which issue-wise majority voting allows for reaching several types of compromise. An alternative to issue-wise evaluation is to compare issue sets (which correspond to ballots in our setting) using the *symmetric difference from a voter's favorite issue set* [8,25,26]. A small symmetric difference is good, and a large symmetric difference is bad. This way of comparing issue sets is very close to the way we study in our paper: A voter accepts a ballot Q if and only if the symmetric difference from his favorite ballot B to Q is smaller than the symmetric difference from B to the empty ballot. Typically, the studies in this context focus on proving desirable properties or on showing how to deal with certain paradoxes. Computational complexity studies are established for related binary decision making problems like judgment aggregation [4,14], lobbying [7,9,15,16], or control of multiple referenda [10]. In the context of judgment aggregation, Alon et al. [2] investigated the computational complexity of control by bundling issues which is also related to "vote on bundled proposals" as considered in this paper.

The scenario considered in our work is also (weakly) related to the concepts of collective domination [13] and proportional representation [28]—in both cases one has to select certain alternatives (proposals in our context) that provide a "good representation" of the voter's will. Herein, extending the Condorcet winner principle to Condorcet winner sets plays a central role. In our work, we also deal with "collectively winning ballots", namely more than half of the proposals in such a ballot are supported by a voter.

Finally, we mention in passing that central computational complexity results of our work are cast within the framework of parameterized complexity analysis, which due to its refined view on algorithmic (in)tractability fits particularly well with voting and related problems [5].

Our Contributions. We analyze the combinatorial and algorithmic behavior of UNANIMOUSLY ACCEPTED BALLOT and MAJORITYWISE ACCEPTED BALLOT. In particular, we investigate the role of the following natural parameters:

- the number m of proposals,
- the number n of voters,
- the size h of the solution ballot Q, that is, $h = |Q|$,
- the maximum size $b_{\max}$ of favorite ballots, that is, $b_{\max} = \max_{i \in \mathcal{V}} |B_i|$, and
- the difference b_{gap} between $\lceil (m + 1)/2 \rceil$ and the minimum size of favorite ballots, that is, $b_{\mathrm{gap}} = \lceil (m + 1)/2 \rceil - \min_{i \in \mathcal{V}} |B_i|$.

Table 1. Parameterized complexity results on two central problems. An entry "ILP-FPT" means fixed-parameter tractability based on a formulation as an integer linear program. Note that all our "intractability" results also hold for the case of $Q_+ = \emptyset$.

Parameters	UnaAB	MajAB
Number m of proposals	FPT, no polynomial kernel (Thm. 2)	
Number n of voters	ILP-FPT, no polynomial kernel (Thm. 3)	
Parameter h	W[2]-complete (Thm. 4)	W[2]-hard (Thm. 4)
Parameter $b_{\max}$	FPT, no polynomial kernel (Thm. 5)	in W[1] (Thm. 5)
Parameter b_{gap}	NP-complete already for $b_{\mathrm{gap}} = 1$ (Thm. 6)	

The parameter b_{gap} measures how far a given instance is from being trivial in terms of the number of proposals: If each voter's favorite ballot contains at least $\lceil (m+1)/2 \rceil$ proposals, then choosing $Q = \mathcal{P}$ makes every voter happy, so the instance is a trivial yes-instance. While the parameters n and m are naturally related to the "dimensions" of the input, the parameters h, $b_{\max}$, and b_{gap} measure certain degrees of contradiction or inhomogeneity in an instance.

Section 2 is devoted to computational complexity results. The main picture is summarized in Table 1. Not too much of a surprise, UNANIMOUSLY ACCEPTED BALLOT and MAJORITYWISE ACCEPTED BALLOT turn out to be NP-complete. More surprisingly, this remains so even when the input ballots are almost trivial, that is, $b_{\mathrm{gap}} = 1$. Namely, if $|B_i| \geq \lceil (m+1)/2 \rceil$ for all voters i, then all voters accept the ballot $\mathcal{P}$. But if every voter i only satisfies the slightly weaker condition $|B_i| \geq \lfloor m/2 \rfloor$, then both problems already become NP-complete. Next, formulating the problems as integer linear programs (ILPs) where the number of variables only depends (exponentially) on n implies fixed-parameter tractability with respect to the parameter n. Using simple brute-force search, one easily obtains that both problems are fixed-parameter tractable with respect to the parameter m. As to efficient and effective preprocessing by polynomial-time data reduction, however, we show that neither for parameter n nor for parameter m polynomial-size problem kernels exist unless an unlikely collapse in complexity theory occurs. As to the parameter h, we prove parameterized intractability— more precisely, W[2]-completeness for UNANIMOUSLY ACCEPTED BALLOT and W[2]-hardness for MAJORITYWISE ACCEPTED BALLOT. While the two problems behave in almost the same way with respect to the parameters n, m, and h, the situation may change for the parameter $b_{\max}$: While UNANIMOUSLY ACCEPTED BALLOT is shown fixed-parameter tractable, for MAJORITYWISE ACCEPTED BALLOT we only could show containment in W[1] and leave hardness as an open question.

In Section 3, we provide an in-depth combinatorial analysis concerning the dependence of the size of a solution ballot on the parameter n. In particular, we show the upper bound $(n+1)^{(n+1)/2}$ and the lower bound $n^{n/2-o(n)}$ for UNANIMOUSLY ACCEPTED BALLOT with $Q_+ = \emptyset$, thus achieving asymptotically almost matching bounds. Analogous results hold for MAJORITYWISE ACCEPTED BALLOT. In Section 4, we conclude with some open questions for future research.

Due to the lack of space, we only sketch the ideas of the proofs for some of our results.

Parameterized Complexity Preliminaries. The concept of parameterized complexity was pioneered by Downey and Fellows [12] (see also [18,27] for more recent textbooks). A parameterized problem is a language $L \subseteq \Sigma^* \times \Sigma^*$, where Σ is an alphabet. The second component is called the *parameter* of the problem. Typically, the parameter or the "combined" ones are non-negative integers. A parameterized problem L is *fixed-parameter tractable* if there is an algorithm that decides in $f(k) \cdot |x|^{O(1)}$ time whether $(x, k) \in L$, where f is an arbitrary computable function depending only on k. Correspondingly, FPT denotes the class of all fixed-parameter tractable parameterized problems. A core tool in the development of fixed-parameter algorithms is polynomial-time preprocessing by *data reduction* [6,22]. Here, the goal is to transform a given problem instance (x, k) in polynomial time into an equivalent instance (x', k') with parameter $k' \leq k$ such that the size of (x', k') is upper-bounded by some function g only depending on k. If this is the case, we call instance (x', k') a (problem) *kernel* of size $g(k)$. If g is a polynomial, then we say that this problem has a *polynomial-size problem kernel*, in short, *polynomial kernel*.

Fixed-parameter intractability under some plausible complexity-theoretic assumptions can be shown by means of *parameterized reductions*. A parameterized reduction from a parameterized problem P to another parameterized problem P' is a function that, given an instance (x, k), computes in $f(k) \cdot |x|^{O(1)}$ time an instance (x', k') (with k' only depending on k) such that (x, k) is a yes-instance for P if and only if (x', k') is a yes-instance for P'. The two basic complexity classes for fixed-parameter intractability are W[1] and W[2]. A parameterized problem L is W[1]- or W[2]-hard if there is a parameterized reduction from a W[1]- or W[2]-hard problem to L. For instance, both INDEPENDENT SET and HITTING SET are known to be NP-complete [20]. However, when parameterized by the solution size, INDEPENDENT SET is W[1]-complete while HITTING SET is W[2]-complete [12].

2   Computational Complexity

The following observation is used many times in our proofs.

Observation 1. *Let i and j be two voters that are both happy with some $Q \subseteq \mathcal{P}$.*

(i) Then $B_i \cap B_j \neq \emptyset$.
(ii) If $B_i \cap B_j = \{p\}$, then $p \in Q$ and furthermore $|B_i \cap Q| = |B_j \cap Q|$.

The next observation basically says that UNAAB can be many-one reduced in polynomial time to MAJAB with the same agenda. This implies that the "majority problem" is computationally at least as hard as the "unanimous problem".

Observation 2. *Let I_{una} be a* UNAAB *instance with n voters, and let I_{maj} be a* MAJAB *instance with $2n - 1$ voters such that*

- *I_{una} and I_{maj} both have the same proposal set $\mathcal{P}$ and the same agenda Q_+,*
- *the voters from I_{una} and the first n voters from I_{maj} have the same favorite ballots $B_1, \ldots, B_n$, and*
- *the remaining $n - 1$ voters from I_{maj} support no proposals.*

Then, $Q \subseteq \mathcal{P}$ is a solution for I_{una} if and only if Q is a solution for I_{maj}.

We will use the NP-complete HITTING SET (HS) problem [20] to show many of our intractability results. Given a finite set U, subsets $S_1, \ldots, S_r$ of U, and a nonnegative integer k, HS asks whether there is a *hitting set* of size k, that is, whether there is a size-k set $U' \subseteq U$ such that $S_i \cap U' \neq \emptyset$, $i \in \{1, \ldots, r\}$. The following reduction from HS to UNAAB is used several times in our intractability proofs. Note that, due to Observation 2, it implies a reduction to MAJAB.

Reduction 1. Let $(U, S_1, \ldots, S_r, k)$ be an instance of HS. Construct an instance of UNAAB as follows. The proposal set $\mathcal{P}$ consists of all the elements of U, of k new dummy proposals, and of a special proposal α. There are $r + 2$ voters. For $1 \leq i \leq r$, the favorite ballot B_i consists of the elements from S_i together with all dummy proposals. Furthermore, $B_{r+1} = U \cup \{\alpha\}$ and B_{r+2} consists of α together with all dummy proposals. Finally, set $Q_+ = \emptyset$.

Lemma 1. *Reduction 1 is a parameterized reduction where the parameters h, n, and m are linearly bounded in the parameters k, r, and $|U|$, respectively. More precisely, $h = 2k + 1$, $n = r + 2$, and $m = |U| + k + 1 \leq 2|U| + 1$.*

2.1 NP-Completeness

We show that UNAAB and MAJAB are NP-complete even if $Q_+ = \emptyset$. This implies that there is no hope for fixed-parameter tractability parameterized by $|Q_+|$.

Theorem 1. *Both* UNANIMOUSLY ACCEPTED BALLOT *and* MAJORITYWISE ACCEPTED BALLOT *are* NP-complete even if $Q_+ = \emptyset$.

Proof (Sketch). Containment in NP is easy to see; the hardness result is achieved due to Observations 1 and 2 and Lemma 1.   □

2.2 Few Proposals or Few Voters

Complementing our intractability result from Theorem 1, we show that instances with few proposals or few voters are tractable. More precisely, we show that the considered problems are polynomial-time solvable for a fixed number of proposals or a fixed number of voters and the degree of the polynomial is a constant. However, we also show that under plausible complexity-theoretic assumptions these problems do not admit polynomial-time preprocessing algorithms that reduce the size of an instance to be polynomially bounded by the the the number m of proposals or the number n of voters. In other words, UNAAB and MAJAB are unlikely to allow for polynomial kernels with respect to the parameters n or m, respectively.

Theorem 2. *Parameterized by the number m of proposals,* UNANIMOUSLY ACCEPTED BALLOT *and* MAJORITYWISE ACCEPTED BALLOT *are fixed-parameter tractable. Unless* $\mathsf{NP} \subseteq \mathsf{coNP/poly}$, *both problems do not admit a polynomial kernel even if* $Q_+ = \emptyset$.

Proof (Sketch). For the fixed-parameter tractability result, one guesses a ballot Q with $Q_+ \subseteq Q \subseteq \mathcal{P}$ and checks whether this is a solution for UNAAB (resp. MAJAB). This takes $O(2^m \cdot n^c)$ time with c being a constant. As for the non-existence of a polynomial kernel for UNAAB, this is due to the non-existence of a polynomial kernel of HS parameterized by $|U|+k+1$ [11] and due to Lemma 1. Together with Observation 2, the non-existence of polynomial kernels transfers to MAJAB even if $Q_+ = \emptyset$.   □

Theorem 3. *Parameterized by the number n of voters,* UNANIMOUSLY ACCEPTED BALLOT *and* MAJORITYWISE ACCEPTED BALLOT *are fixed-parameter tractable. Unless* $\mathsf{NP} \subseteq \mathsf{coNP/poly}$, *both problems do not admit a polynomial kernel even if* $Q_+ = \emptyset$.

Proof. We first describe how to formulate MAJAB as an integer linear program (ILP) and show how to modify the ILP to also work for UNAAB. Let N_V be the number of proposals that are accepted by the voter set V, that is, $N_V := |\{j \mid (\forall i \in V : j \in B_i) \wedge (\forall i' \notin V : j \notin B_{i'})\}|$. As the proposals counted by N_V only depend on V, we refer to V as a *proposal type*. Let x_V be the number of proposals of type V in the ballot Q. Further, let N_V^+ be the number of proposals in Q_+ that are accepted by the voter set $V \subseteq \mathcal{V}$, that is, $N_V^+ := |Q_+ \cap \{j \mid (\forall i \in V : j \in B_i) \wedge (\forall i' \notin V : j \notin B_{i'})\}|$. For each voter i we introduce a binary variable z_i that may only have value 1 if voter i is happy with Q (and may have value 0 in any case). Then Q must satisfy the following constraints (1)–(3).

$$\sum_{i=1}^{n} z_i \geq \frac{n+1}{2} \tag{1}$$

$$\sum_{\substack{V \subseteq \mathcal{V}: \\ i \in V}} x_V - \sum_{\substack{V \subseteq \mathcal{V}: \\ i \notin V}} x_V \geq m(z_i - 1) + 1 \qquad \forall i \in \{1, \dots, n\} \tag{2}$$

$$N_V \geq x_V \geq N_V^+ \qquad \forall V \subseteq \mathcal{V} \tag{3}$$

Constraint (1) requires that a strict majority of voters is happy with Q. Constraint set (2) ensures that voter i is happy if variable z_i is set to 1. Constraint set (3) requires ballot Q to contain all proposals in Q_+ and restricts the number of proposals of each type in Q to those actually present.

Our ILP contains at most 2^n variables x_V and n variables z_i. The total number of constraints is at most $2^n + n + 1$. Since an ILP with ρ variables and L input bits can be solved in $O(\rho^{2.5\rho+o(\rho)}L)$ time [24,19], MAJAB is fixed-parameter tractable with respect to the number n of voters.

If we delete constraint (1) and the variables z_i, and replace the right-hand side of constraint (2) with 1, then we gain an ILP for UNAAB with at most 2^n variables and $2^n + n$ constraints. Thus, UNAAB is also fixed-parameter tractable with respect to parameter n.

Unless NP $\subseteq$ coNP/poly, even if $Q_+ = \emptyset$, both problems do not have a polynomial kernel with respect to the parameter n: Reduction 1 is a polynomial-time reduction from the NP-complete problem HITTING SET; the number n of voters in the reduced instance is linearly bounded by the number r of sets in the instance one reduces from; and $Q_+ = \emptyset$. A polynomial kernel of UNAAB with $Q_+ = \emptyset$ parameterized by n would yield a polynomial kernel for HITTING SET parameterized by r. However, this is not possible unless NP $\subseteq$ coNP/poly (e.g. [23, Lemma 14]). Thus, even if $Q_+ = \emptyset$, UNAAB does not admit a polynomial kernel. Neither does MAJAB admit a polynomial kernel even if $Q_+ = \emptyset$ due to Observation 2.   □

2.3 Small Ballots

We perform a parameterized complexity analysis concerning parameters based on the ballot sizes. We start with the size h of the solution ballot. For technical reasons, we need to assume that h is given as part of the input when dealing with the parameterized problems.

Theorem 4. *Parameterized by the size h of the solution ballot,* UNANIMOUSLY ACCEPTED BALLOT *is W[2]-complete and* MAJORITYWISE ACCEPTED BALLOT *is W[2]-hard. Both results hold even if $Q_+ = \emptyset$.*

Proof (Sketch). Reduction 1 is a parameterized reduction from the W[2]-hard HITTING SET parameterized by the size k of the hitting set to UNAAB parameterized by the size h of the solution ballot with $Q_+ = \emptyset$ (see Lemma 1). Because of Observation 2, this implies W[2]-hardness for MAJAB parameterized by h even if $Q_+ = \emptyset$. To show that UNAAB is in W[2], we reduce from UNAAB parameterized by h to the W[2]-complete INDEPENDENT DOMINATING SET parameterized by the solution size k [12].   □

The membership of MAJAB parameterized by the size h of the solution ballot for the class W[2] remains open. Note that the W[2]-hardness reduction in the proof of Theorem 4 does not rely on (an upper bound for) h being given as part of the input. That is, the problem is computationally hard also for the cases where the size of ballot Q is not explicitly required to be bounded by h.

Except for the parameter h where we only know that MAJAB is W[2]-hard while UNAAB is even W[2]-complete, all results shown so far are the same for unanimous acceptance and majority acceptance. The following theorem shows that this may change when considering the parameter $b_{\max}$ where UNAAB remains tractable but for MAJAB we only know W[1]-membership.

Theorem 5. *Parameterized by the maximum size b_{max} of the favorite ballots,* UNANIMOUSLY ACCEPTED BALLOT *can be solved in $O(b_{max}^{2b_{max}} \cdot nm)$ time implying fixed-parameter tractability; however, it admits no polynomial kernel unless* NP $\subseteq$ coNP/poly *even if $Q_+ = \emptyset$.* MAJORITYWISE ACCEPTED BALLOT *parameterized by b_{max} is in* W[1].

Proof (Sketch). To show that UNAAB is solvable in $O(b_{max}^{2b_{max}} \cdot nm)$ time, we first observe that any solution Q must satisfy $|Q| \leq 2b_{max}$. Based on this, we can design a depth-bounded search tree algorithm solving UNAAB where the number of branching possibilities in each step is at most b_{max} and the depth of the algorithm is at most $2b_{max}$.

The non-existence of a polynomial kernel for UNAAB with respect to parameter m shown in Theorem 2 also holds for parameter b_{max}, as $b_{max} \leq m$.

Finally, to show the W[1] containment, we use a theorem from [18, Theorem 6.22.] which states that a parameterized problem L with parameter k is in W[1] if and only if there is a *tail-nondeterministic k-restricted nondeterministic random access machine (NRAM)* program deciding L. The description of a tail-nondeterministic b_{max}-restricted NRAM program $\mathbb{P}$ for MAJAB is omitted due to lack of space. □

Next, we discuss the relation between the parameters "maximum size b_{max} of the favorite ballots" and "the size h_{max} of the maximum symmetric difference between any two favorite ballots". As the following proposition shows, for the cases with $Q_+ = \emptyset$, the two parameters h_{max} and b_{max} are "equivalent" in terms of parameterized complexity theory: The fact that a parameter x is linearly bounded by a parameter y implies that the parameterization by x and the parameterization by y are in the same level of the W-hierarchy and yield the same parameterized hardness results.

Proposition 1. *For any instance of* UNANIMOUSLY ACCEPTED BALLOT *or* MAJORITYWISE ACCEPTED BALLOT *it holds that $h_{max} \leq 2b_{max}$, where h_{max} denotes the size of the maximum symmetric difference between two favorite ballots and b_{max} denotes the maximum size of the given favorite ballots. Instances of* UNANIMOUSLY ACCEPTED BALLOT *or* MAJORITYWISE ACCEPTED BALLOT *are yes-instances if $h_{max} < b_{max}/2$ and $Q_+ = \emptyset$.*

We conclude this section with the following theorem which uses the fact that an instance of UNAAB or MAJAB is a trivial yes-instance if the minimum size of the favorite ballots is at least $\lceil (m+1)/2 \rceil$ where m denotes the total number of proposals in $\mathcal{P}$. However, both problems become NP-complete when this minimum size is one less than the guarantee $\lceil (m+1)/2 \rceil$, even if $Q_+ = \emptyset$. This implies that there is no hope for fixed-parameter tractability with respect to the "below guarantee parameter" b_{gap} which is the difference between $\lceil (m+1)/2 \rceil$ and the minimum size of the favorite ballots.

Theorem 6. *An instance of* UNANIMOUSLY ACCEPTED BALLOT *(resp.* MAJOR-ITYWISE ACCEPTED BALLOT*) is a yes-instance if each voter i satisfies $|B_i| > m/2$.* UNANIMOUSLY ACCEPTED BALLOT *(resp.* MAJORITYWISE ACCEPTED BALLOT*) is* NP-complete even if $Q_+ = \emptyset$ and each voter i satisfies $|B_i| > m/2 - 1$.

Proof. As for the first statement, choosing $Q = \mathcal{P}$ makes every voter happy. To show the second statement, we many-one reduce from the NP-complete VERTEX COVER (VC) problem. Given an undirected graph $G = (U, E)$ and an integer $k \leq |U|$, VC asks whether there is a *vertex cover* of at most k vertices, that is, whether there is a set $U' \subseteq U$ with $|U'| \leq k$ and $e \cap U' \neq \emptyset, \forall e \in E$.

Let $I = ((U, E), k)$ with vertex set $U = \{u_1, \ldots, u_r\}$ and edge set $E = \{e_1, \ldots, e_s\}$ be a VC instance. We first reduce from it to an instance I' for UNAAB and then extend this reduced instance I' to an instance I'' for MAJAB.

Both instances I' and I'' have the same proposal set $\mathcal{P}$. It consists of one special proposal α, of all vertices in U, of k dummy proposals β_j $(1 \leq j \leq k)$, and of $r - k$ additional dummy proposals $\gamma_{j'}$ $(1 \leq j' \leq r - k)$. Thus, $|\mathcal{P}| = 2r + 1$.

Instance I' contains four types of voters: one voter v_0, one voter $\overline{v}_0$, s *edge voters*, and $r - k$ *vertex haters*. Voter v_0 favors proposal α and all the r dummy proposals. Voter $\overline{v}_0$ also favors proposal α, and all the vertices in U. For $1 \leq i \leq s$, the ith edge voter's favorite ballot A_i consists of the two vertices in e_i, of all the k dummy proposals β_j, and of $r - k - 2$ arbitrarily chosen dummy proposals from $\{\gamma_1, \ldots, \gamma_{r-k}\}$. For $1 \leq i' \leq r - k$, the favorite ballot $B_{i'}$ of vertex hater i' consists of α and of all dummy proposals but $\gamma_{i'}$. In total, the number of voters in I' is $s + r - k + 2$, with each voter supporting at least $r = \lfloor |\mathcal{P}|/2 \rfloor$ proposals. Set $Q_+ = \emptyset$. Obviously, this reduction runs in polynomial time.

To show the reduction's correctness, we have to show that I has a vertex cover of size at most k if and only if there is a ballot $Q \subseteq \mathcal{P}$ that all the voters in I are happy with.

For the "only if" part, suppose that $U' \subseteq U$ with $|U'| \leq k$ is a vertex cover. We show that every voter is happy with $Q = \{\alpha\} \cup \{\beta_j \mid 1 \leq j \leq |U'|\} \cup U'$. First, the size of Q is $2|U'| + 1$. To make a voter happy, at least $|U'| + 1$ of his favorite proposals must be also in Q. Obviously, voters v_0, $\overline{v}_0$ and all vertex haters are happy with Q. For each $i \in \{1, \ldots, s\}$, $Q \cap A_i$ contains all dummy proposals β_j with $1 \leq j \leq |U'|$ and at least one vertex proposal $v_{j'}$ with $v_{j'} \in e_i \cap U'$ since U' is a vertex cover. This sums up to at least $|U'| + 1$ proposals. Hence, every edge voter is also happy with Q.

For the "if" part, by applying Observation 1(ii) to the ballots of voters v_0 and $\overline{v}_0$, ballot Q must contain α, and furthermore, Q contains an equal number x of vertex proposals and dummy proposals. For each $i' \in \{1, \ldots, r - k\}$, ballot Q cannot contain dummy proposal $\gamma_{i'}$ since otherwise $|B_{i'} \cap Q| = x < \lfloor |Q|/2 \rfloor + 1$. Thus, vertex hater i' would not be happy. Therefore, the x dummy proposals must come from $\{\beta_1, \ldots, \beta_k\}$ and $x \leq k$. To make the ith edge voter happy, ballot Q must satisfy the condition $|Q \cap A_i| \geq x + 1$. But since no edge voter favors proposal α, ballot Q must contain at least one proposal $u_j \in A_i$. By definition of A_i, the corresponding vertex u_j is incident to edge e_i. This implies that the x vertices in Q form a vertex cover for (U, E).

Next, we extend instance I' to instance I'' for MAJAB by adding $r - k$ *vertex lovers* who have the same favorite ballot U, and s *edge-inverse voters* such that for $1 \leq i \leq s$, edge-inverse voter i's favorite ballot $C_i = (U \cup \{\gamma_1, \ldots, \gamma_{r-k}\}) \setminus A_i$. Thus, C_i and A_i are disjoint. In total, I'' has $2(s + r - k) + 2$ voters. Since each of the newly added voters favors exactly r proposals, the constraint that each voter's proposal set has at least $r = \lfloor |\mathcal{P}|/2 \rfloor$ holds. This extension also runs in polynomial time.

Now we show the correctness of the extended reduction, that is, I has a vertex cover of size at most k if and only if there is a ballot $Q \subseteq \mathcal{P}$ which more than half of the voters in I'' are happy with.

For the "only if" part, the ballot Q as constructed in the "only if" part above makes all voters in I' happy. This sums up to $s + r - k + 2$. Since I'' contains all the voters from I' and has $2(s + r - k) + 2$ voters, this also means that more than half of the voters in I'' is happy with Q.

For the "if" part, for $1 \leq i \leq s$, the ith edge voter and the ith edge-inverse voter do not share a common favorite proposal. Furthermore, no vertex hater's favorite ballot intersects any vertex lover's favorite ballot. Hence, by applying Observation 1(i), any ballot can make at most s voters from the edge voters and the edge-inverse voters happy, and can make at most $r - k$ voters from the vertex haters and the newly constructed vertex lovers happy. But I'' has $2(s + r - k) + 2$ voters. This means that in order to be a solution ballot for I'', Q must make both v_0 and $\overline{v}_0$ happy. By applying Observation 1(ii), Q must then contain α, and, furthermore, Q contains the same number x of vertex proposals and dummy proposals. The ballot Q cannot make any vertex lover happy since his favorite ballot and Q have an intersection of size x which is smaller than $\lfloor |Q|/2 \rfloor + 1$. Thus, Q needs to make all vertex haters happy. Then, Q cannot contain any dummy proposal $\gamma_{i'}$ since otherwise the vertex hater i' is not happy due to $|B_{i'} \cap Q| = x < \lfloor |Q|/2 \rfloor + 1$. Hence, Q contains x dummy proposals from $\{\beta_1, \ldots, \beta_k\}$ with $x \leq k$. Then, no edge-inverse voter is happy with Q since at most x proposals from his favorite ballot are in Q. This means that all edge voters must be happy with Q. To make the ith edge voter happy, Q must intersect with A_i in at least one vertex $u_j \in A_i$. By definition of A_i, the corresponding vertex u_j is incident to edge e_i. Thus, the x vertices in Q form a vertex cover for (U, E). $\square$

3   Combinatorial Bounds on Minimal Accepted Ballots

We say that a unanimously (resp. majoritywise) accepted ballot is *minimal* if no proper subset of it is also unanimously (resp. majoritywise) accepted. In this section, we investigate the largest possible size of a minimal unanimously accepted ballot for the situation with n voters and $Q_+ = \emptyset$. We derive (almost tight) upper and lower bounds on this quantity. From this bound, a similar result can be derived for majoritywise accepted ballots.

It is not hard to see that both upper and lower bounds come down to studying the case where the set $\mathcal{P}$ of all proposals already is a minimal accepted ballot:

Such instances cannot have smaller solutions (giving a lower bound), and upper bounds directly carry over to $Q \subseteq \mathcal{P}$ by considering a restricted instance with $\mathcal{P}' := Q$. To make the question more amenable to combinatorial tools we translate it into a problem on a sequence of vectors with $\{-1, 1\}$-entries: Given n voters and m proposals we create m vectors $x_1, \ldots, x_m \in \{-1, 1\}^n$; the ith entry in vector x_j is 1 if the jth proposal is contained in the favorite ballot of voter i, else it is -1. In this formulation, a unanimously accepted ballot Q corresponds to a subset of the vectors whose vector sum is positive in each coordinate: Considering some voter i, for each proposal in $B_i \cap Q$ we incur 1, for each proposal in $Q \setminus B_i$ we incur -1. If $|B_i \cap Q| > |Q|/2$ then this gives a positive sum in coordinate i; the converse is true as well.

Let us normalize the question a little more. First of all, no minimal ballot can be of even size: Otherwise all coordinate sums would be even and hence each sum is at least 2; then however we may discard an arbitrary vector and still retain sums of at least 1 each. Secondly, it is clear that replacing $+1$ entries by -1 entries does not introduce additional subsequences with positive coordinate sums. Thus, we may restrict ourselves to the case where the coordinate sums over the minimal sequence of m vectors are all equal to 1 (all sums are odd and such a replacement lowers a sum by exactly 2).

Now, a collection of vectors is called a *minimal majority sequence of dimension* n (an n-mms for short) if all its coordinate-wise sums are 1 and no proper subsequence of the vectors has a positive sum in each coordinate. Note that an n-mms cannot contain a nonempty subsequence S whose sum is at most 0 in each coordinate, since otherwise the sum of the vectors that are in this n-mms but not in S must be positive in each coordinate—a contradiction to the minimality of an n-mms. Thus, the definition of an n-mms is equivalent to that all its coordinate-wise sums are 1 and no nonempty subsequence has sum of at most 0 in each coordinate. The *length* of the sequence is the number m of its elements. Let $f(n)$ denote the maximum possible length of an n-mms. In this section, we show that $f(n) \approx n^{n/2+o(n)}$.

Theorem 7. *The maximum possible length $f(n)$ of a minimal majority sequence of dimension n satisfies*

$$n^{n/2-o(n)} \leq f(n) \leq (n+1)^{(n+1)/2}.$$

Proof (Sketch). One way to obtain an upper bound on $f(n)$ is to apply a known result of Sevastyanov [29]. It asserts that any sequence of vectors whose sum is the zero vector, where the vectors lie in an arbitrary n-dimensional normed space R and each of them has norm at most 1, can be permuted so that all initial sums of the permuted sequence are of norm at most n. Given an n-mms $v_1, \ldots, v_m \in \{-1, 1\}^n$, append to it the vector $-\mathbf{1}$ where $\mathbf{1}$ is the all-1-vector of length n to get a zero-sum sequence of $m + 1$ vectors in R^n, where the ℓ_∞ norm of each vector is 1. By the above mentioned result there is a permutation $u_1, u_2, \ldots, u_{m+1}$ of these vectors so that the ℓ_∞-norm of each initial sum $\sum_{i=1}^{j} u_i$ is at most n. If $m + 1 > (2n + 1)^n$ then, by the pigeonhole principle, some two distinct initial sums are equal, and their difference gives a proper subsequence of the original mms with sum either the zero vector (if this difference does not include the vector -1), or $\mathbf{1}$ (if it does). In both cases, this contradicts the assumption that the original sequence is an mms. This shows that $f(n) \leq (2n + 1)^n$. See [1] for a similar argument.

The proof of the stronger upper bound stated in Theorem 7 is similar to that of a result of Huckeman, Jurkat, and Shapley (cf. [21]) and is based on some simple facts from convex geometry. The details and the proof for the lower bound are omitted.   □

The proof combines the main result of Alon and Vu [3] with arguments from Linear Algebra, Geometry, and Discrepancy Theory. Instead of turning to the proof, let us give a corollary for the effect on our two central problems.

Corollary 1. *Consider a* UNANIMOUSLY ACCEPTED BALLOT *instance with n voters. If there exists a unanimously accepted ballot, then there also exists one of size at most $(n + 1)^{(n+1)/2}$. This bound is essentially tight, as there exist choices of accepted ballots such that any unanimously accepted ballot has size at least $n^{n/2-o(n)}$. For* MAJORITYWISE ACCEPTED BALLOT, *the corresponding upper and lower bounds are respectively $(t + 1)^{(t+1)/2}$ and $t^{t/2-o(t)}$, where $t = \lceil (n + 1)/2 \rceil$ denotes the majority threshold.*

Proof. As the correspondence between favorite ballots and vector sequences has been thoroughly discussed above for the unanimous case, we now concentrate on the majority case.

To see the lower bound for the majority case, we start from a lower bound example for the unanimous case with t old voters and a minimum accepted ballot size of $t^{t/2-o(t)}$, and we add $n - t < n/2$ new voters with empty favorite ballots to it. Note that the resulting instance has a total of n voters and that its majority threshold indeed is t. Then any majoritywise accepted ballot must be unanimously accepted by the t old voters, so that the minimum majoritywise accepted ballot has size at least $t^{t/2-o(t)}$.

For the upper bound, consider any majoritywise accepted ballot Q for n voters and consider any minimal majority of t voters that (amongst themselves) unanimously accept this ballot. Then any other unanimously accepted ballot for these voters is also majoritywise accepted by all n voters, so that we get the desired upper bound of $(t + 1)^{(t+1)/2}$ on the size of Q.   □

4   Open Questions and Conclusion

We have introduced new and naturally motivated problems in computational social choice, and we studied their computational complexity and started an analysis of their combinatorial properties. We conclude this paper with a few challenges for future research.

First, recall that in Proposition 1 we stated upper bounds on $h_{\max}$ (the size of the maximum symmetric difference between two favorite ballots) in terms of linear functions in $b_{\max}$ (the maximum ballot size of voters). Hence, parameterized hardness results with respect to $b_{\max}$ transfer to the parameterization by h_{max}. In the case of empty agenda, that is, $Q_+ = \emptyset$, however, we have no good lower bounds for h_{max} in terms of b_{max}. Thus, it remains to classify the parameterized computational complexity of both UNANIMOUSLY ACCEPTED BALLOT and MAJORITYWISE ACCEPTED BALLOT using parameter h_{max}. Notably, in the cases of $Q_+ = \emptyset$ the parameters h_{max} and b_{max} are linearly related so that the same parameterized complexity results will hold for both parameterizations.

Second, with respect to parameter h (the size of the solution ballot Q), we established W[2]-hardness for MAJORITYWISE ACCEPTED BALLOT even if $Q_+ = \emptyset$, but we left open the precise location of this problem in the parameterized complexity hierarchy. It might be W[2]-complete, but all we currently know is that it is contained in W[2] (Maj), a class presumably larger than W[2] [17].

Third, the combinatorial bounds from Section 3 do not hold for instances with nonempty agenda, since such bounds cannot be independent of $|Q_+|$. For cases with nonempty agenda there are similar bounds with an extra factor of $|Q_+|$. A detailed analysis could be part of investigations of weighted variants of our problems. In this regard, weights on the voters, weights on the proposals, or weights on the acceptance threshold of the voters seem to be well-motivated.

Fourth, can we avoid Integer Linear Programs for showing fixed-parameter tractability with respect to the parameter number n of votes and provide direct combinatorial algorithms beating the ILP-based running times? In this context, the exponential lower bound on the number of proposals in ballots accepted by society from Section 3 might be relevant.

Finally, it remains a puzzling open question whether MAJORITYWISE ACCEPTED BALLOT parameterized by b_{max} is fixed-parameter tractable—we could only show containment in W[1].

References

1. Alon, N., Berman, K.A.: Regular hypergraphs, Gordon's lemma, Steinitz' lemma and invariant theory. J. Combin. Theory Ser. A 43(1), 91–97 (1986)
2. Alon, N., Falik, D., Meir, R., Tennenholtz, M.: Bundling attacks in judgment aggregation. In: Proc. 27th AAAI. AAAI Press (2013)
3. Alon, N., Vu, V.H.: Anti-Hadamard matrices, coin weighing, threshold gates, and indecomposable hypergraphs. J. Combin. Theory Ser. A 79(1), 133–160 (1997)
4. Baumeister, D., Erdélyi, G., Rothe, J.: How hard is it to bribe the judges? A study of the complexity of bribery in judgment aggregation. In: Brafman, R. (ed.) ADT 2011. LNCS, vol. 6992, pp. 1–15. Springer, Heidelberg (2011)
5. Betzler, N., Bredereck, R., Chen, J., Niedermeier, R.: Studies in computational aspects of voting—a parameterized complexity perspective. In: Bodlaender, H.L., Downey, R., Fomin, F.V., Marx, D. (eds.) Fellows Festschrift 2012. LNCS, vol. 7370, pp. 318–363. Springer, Heidelberg (2012)
6. Bodlaender, H.L.: Kernelization: New upper and lower bound techniques. In: Chen, J., Fomin, F.V. (eds.) IWPEC 2009. LNCS, vol. 5917, pp. 17–37. Springer, Heidelberg (2009)
7. Bredereck, R., Chen, J., Hartung, S., Kratsch, S., Niedermeier, R., Suchý, O.: A multivariate complexity analysis of lobbying in multiple referenda. In: Proc. 26th AAAI, pp. 1292–1298. AAAI Press (2012)

8. Çuhadaroğlu, T., Lainé, J.: Pareto efficiency in multiple referendum. Theory Dec. 72(4), 525–536 (2012)
9. Christian, R., Fellows, M., Rosamond, F., Slinko, A.: On complexity of lobbying in multiple referenda. Rev. Econ. Design 11(3), 217–224 (2007)
10. Conitzer, V., Lang, J., Xia, L.: How hard is it to control sequential elections via the agenda? In: Proc. 21st IJCAI, pp. 103–108. AAAI Press (2009)
11. Dom, M., Lokshtanov, D., Saurabh, S.: Incompressibility through colors and IDs. In: Albers, S., Marchetti-Spaccamela, A., Matias, Y., Nikoletseas, S., Thomas, W. (eds.) ICALP 2009, Part I. LNCS, vol. 5555, pp. 378–389. Springer, Heidelberg (2009)
12. Downey, R.G., Fellows, M.R.: Parameterized Complexity. Springer (1999)
13. Elkind, E., Lang, J., Saffidine, A.: Choosing collectively optimal sets of alternatives based on the Condorcet criterion. In: Proc. 22nd IJCAI, pp. 186–191. AAAI Press (2011)
14. Endriss, U., Grandi, U., Porello, D.: Complexity of judgment aggregation. J. Artif. Intell. Res. 45, 481–514 (2012)
15. Erdélyi, G., Fernau, H., Goldsmith, J., Mattei, N., Raible, D., Rothe, J.: The complexity of probabilistic lobbying. In: Rossi, F., Tsoukias, A. (eds.) ADT 2009. LNCS, vol. 5783, pp. 86–97. Springer, Heidelberg (2009)
16. Erdélyi, G., Hemaspaandra, L.A., Rothe, J., Spakowski, H.: On approximating optimal weighted lobbying, and frequency of correctness versus average-case polynomial time. In: Csuhaj-Varjú, E., Ésik, Z. (eds.) FCT 2007. LNCS, vol. 4639, pp. 300–311. Springer, Heidelberg (2007)
17. Fellows, M.R., Flum, J., Hermelin, D., Müller, M., Rosamond, F.A.: W-hierarchies defined by symmetric gates. Theory Comput. Syst. 46(2), 311–339 (2010)
18. Flum, J., Grohe, M.: Parameterized Complexity Theory. Springer (2006)
19. Frank, A., Tardos, É.: An application of simultaneous diophantine approximation in combinatorial optimization. Combinatorica 7(1), 49–65 (1987)
20. Garey, M.R., Johnson, D.S.: Computers and Intractability—A Guide to the Theory of NP-Completeness. W. H. Freeman and Company (1979)
21. Graver, J.E.: A survey of the maximum depth problem for indecomposable exact covers. In: Infinite and Finite Sets, Colloquia Mathematica Societatis János Bolyai, vol. 10, pp. 731–743. North-Holland (1973)
22. Guo, J., Niedermeier, R.: Invitation to data reduction and problem kernelization. ACM SIGACT News 38(1), 31–45 (2007)
23. Hermelin, D., Kratsch, S., Soltys, K., Wahlström, M., Wu, X.: Hierarchies of inefficient kernelizability. In: CoRR, abs/1110.0976 (2011)
24. Kannan, R.: Minkowski's convex body theorem and integer programming. Math. Oper. Res. 12(3), 415–440 (1987)
25. Laffond, G., Lainé, J.: Condorcet choice and the Ostrogorski paradox. Soc. Choice Welf. 32(2), 317–333 (2009)
26. Laffond, G., Lainé, J.: Searching for a compromise in multiple referendum. Group Decis. and Negot. 21(4), 551–569 (2012)
27. Niedermeier, R.: Invitation to Fixed-Parameter Algorithms. Oxford University Press (2006)
28. Procaccia, A.D., Rosenschein, J.S., Zohar, A.: On the complexity of achieving proportional representation. Soc. Choice Welf. 30(3), 353–362 (2008)
29. Sevastyanov, S.V.: On approximate solutions of scheduling problems. Metody Discretnogo Analiza 32, 66–75 (1978) (in Russian)

Exact Approaches for Parameter Elicitation in Lexicographic Ordering*

Noureddine Aribi[1,2] and Yahia Lebbah[1,2]

[1] Laboratoire LITIO, Université d'Oran, BP 1524, El-M'Naouer, 31000 Oran, Algérie
[2] Laboratoire I3S/CNRS, Université de Nice - Sophia Antipolis, Nice, France
{aribi.noureddine,ylebbah}@gmail.com

Abstract. In this paper, we study the use of exact approaches based on constraint programming and mixed integer programming, to tackle the parameter elicitation problem in the lexicographic ordering approach (LO). Like all multicriteria optimization methods, the LO method has the criteria order parameter that should be fixed carefully. Indeed, the criteria usually conflict with each other, and thus, finding an appropriate order between the criteria is challenging. This is why we propose elicitation methods in order to assist the Decision Maker (DM) in fixing this parameter. These methods require some prior knowledge, that the DM can give easily. We present some numerical experiments, showing the effectiveness of our approaches.

1 Introduction

Many methods for solving multicriteria optimization problems exist, and it is not so simple to choose a method well adapted to a given problem. Moreover, even after a multicriteria method has been selected, different parameters (e.g., some weights, some utility functions, ...) need to be determined, either to find the optimal solution (best tradeoff) or to rank the set of feasible solutions (alternatives). This may be even more difficult, and elicitation methods are sometimes used to assist the decision maker in this task.

In this paper, we focus on the lexicographic ordering method [9], which is appropriate when a strict dominance relation between the criteria can be established. The parameter of this method is a total ordering of the criteria. Additionally, we make the assumption that we have prior information about the preferences of the decision maker, given across an outcome vector (cf. [3]), and we focus on how to use these information, rather than how to get them. Thus, we assume that we have some alternatives along with their outcome values. For instance, consider a situation in which a seller wishes to explain/justify the price of his different products, say digital cameras, to his customers. A digital camera is characterized by several criteria: its number of mega-pixels, its weight, the maximal zoom range,... The need of the seller is to have a justification for his customers, which is both simple and consistent with its products. The customers of this seller do not understand subtle tradeoffs between criteria. Indeed, they can only

---

* This work is supported by TASSILI research program 11MDU839 (France, Algeria). This work is also supported by PNR research project 70/TIC/2011 (Algeria).

P. Perny, M. Pirlot, and A. Tsoukiàs (Eds.): ADT 2013, LNAI 8176, pp. 45–56, 2013.

understand that one criterion (e.g., zoom range) is more important than another one (e.g., weight). So, the challenge here consists to find a permutation between the criteria that best justifies the product's prices.

In the present work, we propose two exact approaches based on constraint programming (CP), and mixed integer programming (MIP) techniques. These methods are then used to get automatically the appropriate order parameter.

The rest of this paper is organized as follows. Section 2 provides some necessary preliminaries. Section 3 gives a formulation of the multicriteria elicitation problem. Section 4 illustrates our exact approaches on a motivating example. Sections 5 and 6 describe with details our exact approaches. Empirical results are given and discussed in Section 7. We present some related works in Section 8. Section 9 concludes the paper.

2 Background

Before discussing our approaches, we provide some necessary background.

Definition 1 (Lexicographic order). [14] *The criteria are ordered by the decision maker according to their perceived importance. Hence, the solution with the best value for the most important criterion is selected. Tied solutions are evaluated using their values in the second most important criterion, and so on. Formally, let $x, y \in \mathbb{R}^n$,*

$$x <_{lex} y \equiv (x_1 < y_1) \vee ((x_1 = y_1) \wedge \langle x_2, ..., x_n \rangle <_{lex} \langle y_2, ..., y_n \rangle) \qquad (1)$$

In the context of this paper, we need to measure the discrepancy between an ordering given by the lexicographic method, and an ideal ordering given by the outcomes. Thus, we suggest to use one of the most widely used distance defined as follows.

Definition 2 (Kendall tau distance). *The Kendall tau distance [13] is a metric that counts the number of pairwise disagreements between two lists (τ_1, τ_2).*

$$K(\tau_1, \tau_2) = |\{(i, j) : i \in \{1, ..., n - 1\}, j \in \{i + 1, ..., n\},$$
$$(\tau_1(i) < \tau_1(j) \wedge \tau_2(i) > \tau_2(j)) \vee (\tau_1(i) > \tau_1(j) \wedge \tau_2(i) < \tau_2(j))\}| \qquad (2)$$

When the two lists are identical, then $K(\tau_1, \tau_2) = 0$. If τ_1 is strictly increasing, and τ_2 contains the same elements as τ_1 in reverse order, then $K(\tau_1, \tau_2) = n(n - 1)/2$, where $n = |\tau|$. The time complexity of Kendall tau method is of $\mathcal{O}(n^2)$ [24].

Example 1. The Kendall tau distance between $[0, 3, 1, 6, 2, 5, 4]$ and $[1, 0, 3, 6, 4, 2, 5]$ is equal to six, i.e., $|\{(0, 1), (1, 2), (1, 4), (2, 5), (4, 5), (5, 6)\}|$, and all other pairs are in the same order [21].

2.1 Constraint Programming

Constraint Programming (CP) [19,18] is a powerful paradigm to model and solve combinatorial problems. It has been successfully applied in many different areas (e.g., resource allocation, scheduling and planning problems). This paradigm is based on the notion of *constraint network* [15], which is formally defined as a triple $\langle X, D, C \rangle$ where:

- $X = \{X_1, ..., X_n\}$ is a set of n *decision* variables;
- $D = \{D_{X_1}, ..., D_{X_n}\}$ is a set of associated domains, where D_{X_i} is a finite set of potential values for X_i;
- C is a set of constraints where, each $c \in C$, refers to a set of permitted tuples $R(c)$ over a set of variables $X(c) \subseteq X$ called its scope. The size of the scope is the arity of the constraint.

An instantiation v of a set of variables S is a function that maps each variable $x \in S$ to a value $v(x)$ from its domain D_x. A solution is a complete instantiation that satisfies all the constraints. The problem of finding a solution to a constraint network is called *Constraint Satisfaction Problem* (CSP), and is, in general, an NP-complete problem[1]. The set of solutions of a CSP is denoted by $sol(X, D, C)$. A CSP can be extended to involve optimization problems. Thus, Given a constraint network $\langle X, D, C \rangle$, and an *objective* variable $O \in X$, find a value $m \in D_O$ where $m = max\{v(O)|v \in sol(X, D, C)\}$.

Example 2. Let be the following CSP:

$$X = \{x, y, z\}$$
$$D = \{D_x, D_y, D_z\} \text{ where, } D_x = D_y = D_z = 1..10.$$
$$C = \{C_1(x, y), C_2(x, z), C_3(y, z)\} \text{ where,}$$
$$C_1(x, y) : x + y = 10,$$
$$C_2(x, z) : x + z = 8,$$
$$C_3(y, z) : |y - z| = 2.$$

One possible solution for this CSP is: $(x, y, z) = (7, 3, 1)$. If we consider the objective function "$max(x + y + z)$", then the optimal solution is $(x, y, z) = (1, 9, 7)$, where the objective value takes its max value 17.

2.2 Mixed Integer Linear Programming

Mixed-Integer Linear Programming (MILP,MIP) [23] is one of the most widely used methods for handling optimization problems, due to its rigorousness, flexibility and extensive modeling capability. A MIP program is a linear program with the added restriction that some, but not necessarily all, of the variables must be integer. Typically, a MIP model involves: (i) a set of decision variables, (ii) a set of linear constraints, where each constraint requires that a linear function of the decision variables is either equal to, less than, or more than, a scalar value, and (iii) an objective function that assesses the quality of the solution. Solving a MIP problem consists to find the best solution for the objective function in the set of solutions that satisfy the constraints. Formally, a MIP problem takes the form:

```
Maximize or minimize c^T x
        subject to Ax (≤, =, or ≥) b
                    x_i ∈ Z, i = 1..p
                    x_i ∈ R, i = p + 1..n
```

---

[1] Which means that the solving time scales exponentially as the problem size increases in the worst case.

where x represents the vector of decision variables, p is some positive integer value, $c_j, \forall j = 1, ..., n$ are referred to as objective coefficients, A is an $m \times n$ matrix of coefficients, and b is an $m \times 1$ vector of the right-hand-side values of the constraints.

3   Problem Formulation

The issue of parameter elicitation regarding the lexicographic ordering method can be described by:

- A set of criteria: $X = \{X_1, ..., X_n\}$, where $n \geq 2$;
- A set of alternatives $A = \{A_1, ..., A_m\}$, where A_{ij} is the value of the jth criterion for the ith alternative. These alternatives are ordered according to their finite outcome values Y, i.e., $Y(A_i) \leq Y(A_j), i = 1, ..., m - 1, j = i + 1, ..., m$, where $Y(A_i)$ denotes the outcome value of the ith alternative. These outcome values can be gathered either experimentally, or from responses of a questionnaire survey of preference (cf. [3]).

The problem is to find a total ordering of criteria θ, for which the lexicographic ordering induced on the set of alternatives A_θ, becomes as close as possible to the "ideal" ordering given in A. Before we state our optimality criterion, we introduce a measure of discrepancy between A_θ and A.

Definition 3   (Discrepancy measure). *Let* $\theta = [X_{\theta(1)}, ..., X_{\theta(n)}]$ *be a permutation between criteria* $[X_1, ..., X_n]$. *According to* θ, *we get a lexicographic ordering* $A_\theta = (A_{\sigma(1)}, ..., A_{\sigma(m)})$, *where* $[\sigma(1), ..., \sigma(m)]$ *is a permutation of* $[1, ..., m]$. *The discrepancy measure for* A_θ *is given by:*

$$D(A_\theta) = K([\sigma(1), ..., \sigma(m)], [1, ..., m]) \tag{3}$$

Definition 4   (Optimal permutation). *A permutation* θ^* *is said to be optimal, if and only if for any permutation* θ,

$$D(A_\theta) \geq D(A_{\theta^*}) \tag{4}$$

Note that for a multicriteria problem with n criteria, we have $n!$ possible permutations. So, the space of possible permutations has a size being exponential in the number of criteria. Therefore, the challenge consists to find the best order without going into all possibilities.

4   Motivating Example

Let us reconsider the introductory example, in which we have an outcome vector Y (the *prices*), that the seller wishes to explain/justify to his customers. A sampling data of 18 apparatus are shown in Table (1), where X_1 indicates the *weight*, X_2 the *resolution* of image sensor, and X_3 the *zoom* of the apparatus.

Table 1. Sample data for the Seller Problem

#	1	2	3	4	5	6	7	8	9	10	11	12	13	14	15	16	17	18
X_1	440	510	470	400	340	390	250	250	360	220	320	280	310	300	280	200	270	210
X_2	2	2	4	4	6	8	8	12	14	14	18	20	20	22	24	26	28	30
X_3	5	2	2	4	12	1	5	6	8	7	7	9	10	9	3	12	14	13
Y	389	416	421	425	434	449	461	465	468	473	478	484	485	488	527	529	532	566

In this example, we aim to find a permutation between criteria that minimizes the optimality criterion given by Equation (3). To evaluate efficiently the degree of the discrepancy between the ideal ordering given by the outcome vector (Y), and the lexicographic ordering of the alternatives, we have used a normalized[2] version of the Kendall tau distance given in Definition (2). In this sense, values that are close to 1 indicate a strong discrepancy; whereas values that are close to 0 indicate a weak discrepancy.

The proposed exact approaches (i.e., CP and MIP) seek to find the optimal permutation between criteria by construction, and for our case study, both have yielded the same (optimal) permutation, i.e., $\theta_{cp} = \theta_{mip} = [X_2, X_3, X_1]$, where X_2 is the most important criterion, and so forth ending with X_1. Besides, the discrepancy measure associated to this permutation is equal to[3] 0.97, which captures the goodness of the computed parameter.

5   Constraint Programming Model

Constraint Programming frameworks (e.g., CPO, Gecode, Choco,...) aim at making easy problems formulation and solving. An important ingredient of a CP model are global constraints [18] which provide an efficient reasoning mechanisms within a problem model. Two global constraints have been exploited in our CP optimization model, namely, All-different [17], and Element constraints [10]. This model is described in Algorithm (1), where:

- $\theta_i^*, i = 1..n$, is the index of the ith criterion,
- $D_{\theta_i^*}, i = 1..n$, (*inst.* 1) specifies the domain of the variable θ_i^*.
- Variables in θ^* must be all different (*inst.* 2), since any two criteria cannot be placed at the same index.
- The preferences are extracted from the outcome vector, and are handled using *Reified Constraints*[4] (*loop.* 3 − 5). So, for each preference relation $Y(A_i) < Y(A_j)$ we post the reified constraint $(A_i <_{lex} A_j) \iff b$, where $b \in \{0, 1\}$.
- When $Y(A_i) = Y(A_j)$ or A_i and A_j are indifferent, nothing is done, since each alternative (A_i and A_j) outranks the other, and hence this case does not affect the discrepancy degree. To keep the algorithm simple, this last case is not detailed in

---

[2] By dividing the Kendall tau distance by the number of all possible pairs.

[3] The underline indicates that the result is less than the significance level (α, most commonly fixed to 5%, cf. [6]). Note that a significance level indicates how likely a result is due to chance.

[4] Reified constraints reflect the validity of a constraint C into a $0/1$ finite domain.

Algorithm 1. LEXMAXCSP finds the optimal permutation between criteria

---

Input: n criteria; m alternatives: $\{A_l, ..., A_m\}$
Output: Optimal permutation θ^*

1   $D_{\theta_i^*} \leftarrow \{1, ..., n\}, i = 1..n$
2   ALL-DIFFERENT(θ^*)
3   **for** $i \leftarrow 1$ **to** $m - 1$ **do**
4   **for** $j \leftarrow i + 1$ **to** m **do**
5       $pref_k \leftarrow$ LEXLE(A_i, A_j, θ^*)
6   MAXIMIZE($\sum_{k=1}^{|pref|} pref_k$)
7   **return** θ^*

Algorithm 2. LEXLE finds a permutation θ, so that $(x <_{lex} y)$

Input: x, y : two vectors of n integer values
Output: $(x <_{lex} y)$

1   $lex_{n-1} \leftarrow (x_{\theta_{n-1}} < y_{\theta_{n-1}}) \vee ((x_{\theta_{n-1}} = y_{\theta_{n-1}}) \wedge (x_{\theta_n} < y_{\theta_n}))$
2   **for** $k \leftarrow n - 2$ **downto** 1 **do**
3     $lex_k \leftarrow (x_{\theta_k} < y_{\theta_k}) \vee ((x_{\theta_k} = y_{\theta_k}) \wedge (lex_{k-1}))$
4   **return** lex_1

Algorithm (1). It can be easily integrated by avoiding generating the preference constraint when $Y(A_j) = Y(A_i)$.

– The objective is modelled using a *cost function* (*inst.* 6), with which the number of satisfied preferences is maximized.

Besides, Algorithm (2) is proposed as an iterative version of the recursive Formula (1). It looks for a permutation between criteria, that allows to a vector X to be lexicographically less than another vector Y. In addition, Algorithm (2) contains constraints of the form x_z *op* y_z, where x, y are two vectors of integers, z a decision variable whose domain is $D_z = \{1, ..., n\}$, and $op \in \{<, =\}$. This kind of constraints is handled efficiently via the use of Element global constraint. That is,

$$(x_z \ op \ y_z) \equiv \begin{cases} \text{Element}(z, [x_1, ..., x_n], t_1) \\ \text{Element}(z, [y_1, ..., y_n], t_2) \\ t_1 \ op \ t_2 \\ \text{integer}(t_i) \text{ for } i = \{1, 2\} \end{cases} \quad (5)$$

The constraint Element($z, [x_1, ..., x_n], t_1$) is satisfied if $t_1 = x_z$. And also, the constraint Element($z, [y_1, ..., y_n], t_2$) is satisfied if $t_2 = y_z$. With these two Element constraints, we can reformulate the constrait $(x_z \ op \ y_z)$ as $t_1 \ op \ t_2$. For purpose of writing ease, we decided to preserve the notation $x_z \ op \ y_z$.

The number of generated constraints of this model is of $\mathcal{O}(n \times m^2)$ in the worst case (i.e., $m(m-1)/2$ constraints to model the preference relations, and $2n-1$ constraints for each LEXLE constraint). By construction, our CP approach computes the optimal order between criteria. In fact, the CP model finds the best permutation between criteria to satisfy at most $m(m-1)/2$ possible preference relations. Similarly, the Kendall tau criterion focuses on $m(m-1)/2$ possible relations.

6 MIP Model

In the sequel we introduce a mixed integer programming (MIP) formulation of the elicitation problem. Due to high performance of MIP solvers in handling combinatorial optimization problems, we thought that it is convenient to propose and compare the effectiveness of a MIP model against the proposed CP model.

Our idea consists to reformulate our CP model toward usage by a MIP solver. In particular, we have to linearize All-Different and Element global constraints, used in Algorithm (1).

Linearizing All-Different *constraint.* The linear formulation of this constraint is given in Algorithm (3); in which we associate to each decision variable a vector of binary variables, so that at most one variable is fixed to 1. Constraints in *loop* $(2-3)$, and *loop* $(4-5)$ ensure that variables in θ should have distinct values.

Linearizing Element *constraint.* Due to the decision variables (θ) among the indices on the x and y vectors in Algorithm (2), the later have to be reformulated toward usage by a MIP solver. To tackle constraints for variable indexing (i.e., Element constraint), we make use of a compact reformulation given in [16,11]. So, consider the constraint $x_z \ op \ y_z$, where x, y are two alternatives of constant values, z an integer variable whose domain $D_z = \{1, ..., n\}$, and $op \in \{<, >, =, \neq, \leq, \geq\}$. Here is how to reformulate this constraint into linear terms.

$$(x_z \ op \ y_z) \equiv \begin{cases} z = \sum_{i=1}^{n} i \cdot \gamma_i \\ \sum_{i=1}^{n} x_i \cdot \gamma_i \ op \ \sum_{i=1}^{n} y_i \cdot \gamma_i \\ \sum_{i=1}^{n} \gamma_i = 1 \\ \gamma_i \in \{0, 1\}, \text{for } i \in \{1, .., n\} \end{cases} \quad (6)$$

Thus, the MIP elicitation model is obtained by applying this reformulation to Algorithms (1 and 2). The number of elementary constraints of this model is of $\mathcal{O}(n \times m^2)$ in the worst case (i.e., $m(m-1)/2$ constraints to model the preference relations, and $4n-1$ constraints for each linearized LEXLE constraint). Straightforwardly, this linear reformulation preserves the optimality of the computed solutions.

Algorithm 3. ALLDIFFMIP - MIP reformulation of ALL-DIFFERENT constraint

---
Input: b: a matrix of $n \times n$ binary variables;
    a: a vector of n integers;
    θ: a vector of n integer variables.
Output: ALLDIFF(θ)

1 $D_\theta \leftarrow \{1, .., n\}$
2 **for** $i \leftarrow 1$ *to* n **do**
3    $\lfloor \sum_{j=1}^{n} b_{ij} \leq 1$
4 **for** $j \leftarrow 1$ *to* n **do**
5    $\lfloor \sum_{i=1}^{n} b_{ij} \leq 1$
6 **for** $i \leftarrow 1$ *to* n **do**
7    $\lfloor \theta_i = \sum_{j=1}^{n} b_{ij} \cdot a_j$
8 **return** θ

---

7 Experimental Evaluation

We evaluated experimentally the proposed elicitation approaches on a set of realistic instances [22,5]. We have implemented and solved our CP and MIP models using IBM ILOG Concert C++ interface, and CP Optimizer(CPO)/CPLEX solvers[5] (version 12.5). Our experiments have been performed on x86_64 machine with Intel Xeon (x5460) 4xCores @3.16 Ghz and 8 Gb of free memory. We have also used the discrepancy measure (see Equation 3) to evaluate the similarity between the ideal ordering, and the ordering given by the lexicographic method, once the parameter has been identified.

Table (2) summarizes our empirical results of both exact approaches, conducted on a data set of 11 instances with 100 alternatives for each. In this table, columns #, Cr and D_{opt} respectively indicate, the benchmark number, the number of criteria and the optimal discrepancy degree computed from the result of the CP approach.

For the CP approach, columns Vars, Cons, Mem and T_{CP} respectively designate, the number of variables and constraints for each CP model, along with the memory usage (in Mb) and the solving time of CPO solver (in seconds), with a timeout of $3,600\ sec$ in all experiments. Besides, for each MIP model, we have reported the number of constraints in rows column box; the number of integer and binary variables in both columns cols and bin; the memory usage (in Mb) in Mem column box; while the solving time (in seconds) of CPLEX solver is depicted in T_{MIP} column box, with a timeout of 1 hour for all instances.

It is worth observing from Table (2, columns T_{CP} and T_{MIP}), that the CP approach outperforms the MIP approach in all benchmarks. Here, we point out that the TO value, especially in T_{MIP} column box, indicates that CPLEX solver failed to prove the optimality of the solution within the time allotted. Thus, a feasible solution is given instead. In this particular case, the D_{opt} value can be used to measure how far is the discrepancy degree corresponding to the computed feasible solution, from the optimal discrepancy

---

[5] http://www-142.ibm.com/software/products/fr/fr/
  ibmilogcplecpopti/

Table 2. Empirical results of the CP and MIP approaches

#	Cr	D_{opt}	Vars	Cons	Mem	T_{CP}	rows	cols	bin	Mem	T_{MIP}
			CP approach				MIP approach				
1	3	0.40	3	44,551	59	0.83	2,414	904	192	0.01	4.16
2	5	0.44	5	84,151	130	2.46	5,219	6,429	3,694	0.42	78.37
3	6	0.48	6	103,951	158	71.87	4,609	9,242	4,649	3.98	439.18
4	8	0.42	8	143,551	212	2.94	4,625	9,306	4,713	166.9	896.53
5	8	0.44	8	143,551	213	3.74	4,683	9,422	4,771	494.85	1,587.10
6	9	0.47	9	163,351	276	257.71	4,861	9,865	5,048	2,273.15	1,796.53
7	11	0.44	11	202,951	342	6.65	—	—	—	—	TO
8	11	0.42	11	202,951	342	5.84	—	—	—	—	TO
9	18	0.37	18	341,551	566	30.27	—	—	—	OM	—
10	19	0.16	19	361,351	687	23.19	—	—	—	OM	—
11	44	0.41	44	856,351	800	537.30	—	—	—	OM	—

TO: timeout (3,600 sec)
OM: out of memory

degree computed using the CP optimal solution. Furthermore, from benchmark #9 to #11 (see Table 2), CPLEX solver has reported an "out of memory" (OM) status, due to a huge memory used to solve the MIP models, that exceeded, in those cases, the amount of available memory.

Tuning Parameters for Computational Performance

The results of both CPO and CPLEX solvers were obtained according to a custom parameter settings. Using these parameters, we succeed to boost the solving performances.

– **CPLEX parameters**
   It is known that CPLEX has a number of sophisticated features that drastically improve solving performance. To solve our MIP models, we have investigated the following parameters:
   1. **Priority orders** Assign higher priority to the integer variables that should be decided earlier.
   2. **Cuts** Adding cuts are one of the principal reasons of recent increases in MIP performance. In general (and particularly in our benchmarking), disabling this parameter will boost CPLEX solver.
   3. **Probing** This looks at the logical implications of fixing binary variables, which happens after presolve but before branch and bound. Here, we have applied more intensive probing, which improved the results.
– **CP Optimizer**
   We argue that there are several parameters that may enhance the performance of CP Optimizer. We examine, without being exhaustive, most of prominent parameters that we have investigated in our experiments.

1. **Search phases with selectors** This parameter explicitly indicates to the search engine the key variables, and which variable should be instantiated to which value. For our experiments, we decided to select decision variables having smallest domain size.
2. **Inference levels** This parameter is used to achieve more domain reduction by changing the inference level of some global constraints, like `All-different`.
3. **Parameters on search**
   - **Search type parameter** This parameter controls the type of (constructive) search applied to a problem. The search type we selected is based on the `Restart` strategy, jointly with two other parameters, namely, `Fail limit`, and `Restart grow factor`.

Interestingly, there has been recent works in identifying a parameter tuning that achieves good performance (cf. [12]). However, the best combinations of parameter settings differ according to problem types, which is of course the reason that such design choices are mainly given as parameters.

Discussion

Due to the slower solving performance of the proposed MIP model, we have investigated another MIP model, where the key idea consists to reformulate the lexicographic ordering method as a *weighted sum* optimization method, providing that the chosen weighting vector prohibits any compensation between the criteria. Clearly, to react as a lexicographic method, we chose a large enough weights to rule out any compensation. Additionally, the number of generated constraints of this MIP model is of $\mathcal{O}(m^2)$ in the worse case (i.e., $m(m-1)/2$ constraints to model the preference relations) against $\mathcal{O}(n \times m^2)$ in the first MIP model. However, even if this reformulation seems to be compact, it may result in numerical instability and slower performance.

8   Related Works

The problem of elicitation in multicriteria decision making has been also tackled by many works in decision making community. The main focus of these works is on preference elicitation [7,4,8,20]. These works encompass both (pro)active and iterative learning procedures, and are advocated as means to restrict the domain of admissible preferences which enables to make a good decision. Here, we suppose that the multicriteria decision making is done in the context of a multicriteria method. For instance, the paper [3] tackles parameter elicitation in the context of OWA methods.

In our previous work related to parameter elicitation [2,1], we have investigated a CP approach, along with two approximate approaches based on statistics. However, the main contribution of the present work, is on the use of two exact approaches. In particular, an improved CP model, and a linear reformulation of this model based on MIP techniques, to handle efficiently the elicitation problem regarding the lexicographic method.

9   Conclusion and Perspectives

In this paper, we have proposed and discussed two exact approaches toward solving the problem of parameter elicitation regarding the lexicographic method. More precisely, we highlighted the effectiveness of an exact approach, since it ensures the optimality of the solution, and offers a concise and expressive problem formulation, especially with CP modeling. We also showed some empirical results, carried on a significant multicriteria problem. These experiments have shown that the CP approach is more suitable than the MIP approach in term of solving performance. Among our future works, we plan to improve the proposed MIP model, particularly by investigating other linearization techniques.

Acknowledgment. We would especially like to thank Olivier Lhomme for his assistance, suggestions and insightful comments.

References

1. Aribi, N., Lebbah, Y.: Approches d'élicitation basées sur le coefficient de Spearman pour l'optimisation multicritère par ordre lexicographique. In: JFPC 2013 - Neuvièmes Journées Francophones de Programmation par Contraintes, Aix-en-Provence, France, pp. 51–60 (2013)
2. Aribi, N., Lebbah, Y.: Statistical and constraint programming approaches for parameter elicitation in lexicographic ordering. In: Amine, A., Mohamed, O.A., Bellatreche, L. (eds.) Modeling Approaches and Algorithms. SCI, vol. 488, pp. 47–56. Springer, Heidelberg (2013)
3. Beliakov, G.: How to build aggregation operators from data. International Journal of Intelligent Systems 18, 903–923 (2003)
4. Boutilier, C., Regan, K., Viappiani, P.: Simultaneous elicitation of preference features and utility. In: Proceedings of the Twenty-fourth AAAI Conference on Artificial Intelligence (AAAI 2010), Atlanta, GA, USA, pp. 1160–1167. AAAI Press (July 2010)
5. Brase, C.H., Brase, C.P.: Understandable Statistics: Concepts and Methods. Cengage Learning (2011)
6. Corder, G.W., Foreman, D.I.: Nonparametric Statistics for Non-Statisticians: A Step-By-Step Approach. Wiley (2009)
7. Delecroix, F., Morge, M., Routier, J.-C.: An algorithm for active learning of lexicographic preferences. In: Pirlot, M., Mousseau, V. (eds.) Proc. of the Workshop from Multiple Criteria Decision Aiding to Preference Learning, pp. 115–122 (November 2012)
8. Escoffier, B., Lang, J., Öztürk, M.: Single-peaked consistency and its complexity. In: Proceedings of the 2008 Conference on ECAI 2008: 18th European Conference on Artificial Intelligence, pp. 366–370. IOS Press, Amsterdam (2008)
9. Figueira, J., Greco, S., Ehrgott, M.: Multiple Criteria Decision Analysis: State of the Art Surveys. Springer, Boston (2005)
10. Van Hentenryck, P.: Constraint satisfaction in logic programming. MIT Press (1989)
11. Hooker, J.: Integrated Methods for Optimization. International series in operations research & management science. Springer, US (2012)
12. Hutter, F., Hoos, H.H., Leyton-Brown, K.: Automated configuration of mixed integer programming solvers. In: Lodi, A., Milano, M., Toth, P. (eds.) CPAIOR 2010. LNCS, vol. 6140, pp. 186–202. Springer, Heidelberg (2010)

13. Kendall, M.G.: A New Measure of Rank Correlation. Biometrika 30(1/2), 81–93 (1938)
14. Marler, R.T., Arora, J.S.: Survey of multi-objective optimization methods for engineering. Structural and Multidisciplinary Optimization 26, 369–395 (2004), doi:10.1007/s00158-003-0368-6
15. Montanari, U.: Networks of constraints: Fundamental properties and applications to picture processing. Information Sciences 7, 95–132 (1974)
16. Refalo, P.: Linear formulation of constraint programming models and hybrid solvers. In: Dechter, R. (ed.) CP 2000. LNCS, vol. 1894, pp. 369–383. Springer, Heidelberg (2000)
17. Régin, J.-C.: A filtering algorithm for constraints of difference in csps. In: Hayes-Roth, B., Korf, R.E. (eds.) AAAI, pp. 362–367. AAAI Press / The MIT Press (1994)
18. Régin, J.-C.: Global Constraints: a survey. In: Milano, M., Van Hentenryck, P. (eds.) Hybrid Optimization. Springer (2011)
19. Rossi, F., van Beek, P., Walsh, T. (eds.): The Handbook of Constraint Programming. Elsevier (2006)
20. Roy, B., Bouyssou, D.: Aiding Decisions with Multiple Criteria: Essays in Honor of Bernard Roy. International Series in Operations Research & Management Science. Springer (2002)
21. Sedgewick, R., Wayne, K.: Algorithms, 4th edn. Addison-Wesley (2011)
22. Tabachnick, B.G., Fidell, L.S.: Using Multivariate Statistics, 5th edn. Allyn & Bacon, Inc., Needham Heights (2006)
23. Wolsey, L.A.: Integer Programming. A Wiley-Interscience Publication, Wiley (1998)
24. Xu, W., Chang, C., Hung, Y.S., Kwan, S.K., Fung, P.C.W.: Order statistics correlation coefficient as a novel association measurement with applications to biosignal analysis. IEEE Transactions on Signal Processing 55(12), 5552–5563 (2007)

Possible Winners in Approval Voting

Nathanaël Barrot[1,2], Laurent Gourvès[2,1], Jérôme Lang[2,1],
Jérôme Monnot[2,1], and Bernard Ries[1,2]

[1] PSL, Université Paris-Dauphine, 75775 Paris Cedex 16, France
[2] CNRS, LAMSADE UMR 7243
{nathanael.barrot,laurent.gourves,lang,monnot,ries}@lamsade.dauphine.fr

Abstract. Given the knowledge of the preferences of a set of voters over a set of candidates, and assuming that voters cast sincere approval ballots, what can we say about the possible (co-)winners? The outcome depends on the number of candidates each voter will approve. Whereas it is easy to know who can be a unique winner, we show that deciding whether a set of at least two candidates can be the set of co-winners is computationally hard. If, in addition, we have a probability distribution over the number of candidates approved by each voter, we obtain a probability distribution over winners; we study the shape of this probability distribution empirically, for the impartial culture assumption. We study variants of the problem where the number of candidates approved by each voter is upper and/or lower bounded. We generalize some of our results to multiwinner approval voting.

Keywords: Computational social choice, Approval voting, Voting under incomplete knowledge, Computational complexity.

1 Introduction

While most voting rules take as input a collection of *rankings* over candidates, approval voting stands as an exception and takes as input a collection of *subsets* of candidates [7]. It is well-known that there is no single sincere approval ballot given a voter's preferences over a set of candidates: for any candidate c, approving the set of all candidates that are preferred to c is a sincere ballot [8]. If the voter's preference relation over a set of m candidates is a linear order, this makes m sincere ballots[1].

Assume that we[2] know the preference relation of every voter (each assumed to be a linear order) but that we cannot predict the *threshold* they will fix, that is, the number of candidates they will approve. For each vector of such thresholds (one for each voter), there will be a winner, or, in case of a tie, a set of co-winners, called a co-winning set. We say that a subset of candidates is a

---

[1] Sometimes, voting for *all* candidates is excluded, which makes only $m - 1$ sincere ballots. See for instance [9].

[2] 'We' is generic, and represents anyone who may reason about the outcome of the vote; the chair, for instance.

P. Perny, M. Pirlot, and A. Tsoukiàs (Eds.): ADT 2013, LNAI 8176, pp. 57–70, 2013.

possible co-winning set if it is the set of co-winners for some vector of thresholds, and candidate x is a possible unique winner if $\{x\}$ is a possible co-winning set. The properties of the set of possible approval winners has been addressed first in [9], with the restriction that voters cannot approve all candidates (nor none). They show that the set of possible approval winners contains the Condorcet winner (if any) and also the winner(s) of many voting rules. Another related work is [28], who gives a geometric interpretation of the set of possible approval winners. None of both works characterizes possible winning sets, nor addresses the computational difficulties of identifying them.

We go further in several respects. First, we consider a more general setting where the number of approved candidates can be anything between a fixed lower bound and a fixed upper bound. In the case where voters are totally free of the number of approvals, that is, when these bounds are respectively 1 and m candidates[3], characterizing the set of candidates that can be a unique winner (without a tie) turns out to be straightforward: x is a possible unique winner if it is Pareto-undominated in the original profile. We give a similar characterization when the bounds are different. Then we consider the problem of recognizing co-winning sets, and show that it is NP-complete, even for sets of size two. Next, we consider a probabilistic version of the problem, starting with a probability distribution over approval vectors; we focus on the uniform distribution, and in this case we first observe that the probability that a candidate is in the co-winning set is proportional to its Borda score; then, assuming impartial culture, we study experimentally the shape of the probability distribution over winners.

This work is related to (at least) four research streams. The first of these is a series of works in social choice theory that relate approval voting to the classical Arrovian model, which considers social choice functions mapping a collection of weak orders into a nonempty subset of candidates (whereas approval voting generates the social outcome by aggregating collections of subsets of candidates). For this, the key notion is that of sincere ballot, already evoked above. Most works in this research stream (with the exception of [9] and [28] cited above) study the conditions under which approval voting can, or cannot, be considered strategyproof, and the extent to which strategic behaviour may lead to an undesirable outcome; see [29,30,23,15,26,24,14].

The second related research stream is the characterization and computation of possible and necessary winners given some incomplete information about the votes. The main difference with our setting is that in all these works (up to one exception, discussed below), the voting rule used takes a classical profile, that is, a collection of rankings, as input, and the incomplete information consists of a collection of *partial orders*: a possible (resp. necessary) winner is then a candidate that wins in some completion (respectively, all completions) of this collection of partial orders [21,32,4,3,33,10,5,1,22,18]. An exception is [33], which, in Section 4, states a characterization of possible winners in approval voting, given an

---

[3] Approving no candidate and approving all of them are equivalent, in the sense that whatever the remaining votes, the outcome will be the same. Therefore, without loss of generality, we exclude the possibility for a voter to approve 0 candidate.

initial approval ballot over an initial set of candidates, and given a number of new candidates to be added; the nature of the incomplete information about approval ballots in their setting and ours (an approval profile over a subset of candidates vs. a ranking profile over all candidates) is totally different, and results cannot easily be compared.

The third related research stream is a series of works that focuses on the computational aspects of strategic behaviour in approval voting; see in particular [16,2]. The reason why it relates to our work is that we also find computationally hard problems in approval voting; but, once again, our problems do not come from any form of strategic behaviour. Lastly, the computational aspect of strategic behaviour in *multiwinner* versions of approval voting was considered in [25].

Lastly, our Section 4, where we study the complexity of identifying possible outcomes in multiwinner approval voting, relates to the computational study of multiwinner election schemes, such as full proportional representation [27,6,31], Condorcet winning sets [13] or other approaches to committee selection [11,12,20] (we discuss [11] in more detail in Section 4).

The paper is organized as follows. In Section 2 we introduce the necessary background. In Section 3 we define possible and necessary (co)winners and give some characterizations as well as some hardness results about the identification of possible winning sets. In Section 4 we consider multiwinner elections, and generalize some of our results of Section 3. In Section 5 we present further research issues.

2   Preliminaries

We are given n *voters* $N = \{1, \ldots, n\}$ and m *candidates* (or *alternatives*) $X = \{x_1, \ldots, x_m\}$. A *ranking profile* $P = (P_i)_{i \in N}$ is a collection of linear orders (also called rankings) over X. P_i is also denoted by $\succ_i$.

An *approval ballot* is a nonempty subset of X. An *approval profile* is a collection $A = \langle A_1, \ldots, A_n \rangle$ where $A_i \subseteq X$ is the set of candidates approved by voter i. Such an approval ballot is called *sincere*, if for every voter i and every candidate x_j approved by i there exists no candidate x not approved by i such that $x \succ_i x_j$. We denote by k_i, for $i = 1, \ldots, n$ and $1 \le k_i \le m$, the number of candidates approved by voter i. Hence, in a sincere approval ballot, each voter i approves its k_i best candidates according to the ranking given by P_i.

Given an approval profile A, the *approval score* of candidate x_j, denoted by $app_A(x_j)$, or, when there is no ambiguity, $app(x_j)$, is the number of voters i such that $x_j \in A_i$, for $i = 1, \ldots, n$ and $j = 1, \ldots, m$. The set of *approval co-winners for A*, denoted by $App(A)$, is the set of candidates with maximal approval score. If $App(A)$ is a singleton $\{a\}$ then a is said to be a *single winner* for A^4.

For a ranking profile P, voter $i \in N$ and candidate $x \in X$, $\text{rk}_P(i, x) \in \{1, \ldots, m\}$ denotes the rank of x in the ranking P_i. For $X' \subset X$, let $P_{X'}$ be the

---

[4] Approval voting is here considered an *irresolute* voting rule; a *resolute* version of approval voting can be defined by applying a tie-breaking priority mechanism.

restriction of P to candidates in X'. We denote by $pl(x, X')$ the plurality score approval score of candidate $x \in X'$ in profile $P_{X'}$, that is, the number of voters in $P_{X'}$ who rank x on top. For $X' \subset X$ and $x \in X \setminus X'$, we write $X' \succ_i x$ if $\forall x' \in X'$, $x' \succ_i x$ and candidates among X' are ranked arbitrarily. Finally, we say that candidate x *dominates* candidate x' according to profile P if $\forall i \in N$, $x \succ_i x'$.

Approval voting can also be used for *multiwinner* elections. Here the goal is to elect a set of alternatives, or a *committee*, of fixed size K. There are several procedures for determining a committee using approval voting, which are reviewed in [19]. The most obvious way consist in choosing the candidates with the K highest approval scores (using some tie-breaking mechanism if necessary).

Sometimes, a further constraint on the number of approvals is added: each voter is only allowed to approve at least d and most k candidates, where $k \geq d \geq 1$; a typical choice, often implemented in real-world elections, consists in fixing d to 1 and k to an arbitrary constant (such as, in multi-winner elections, the number of positions to be filled). The corresponding voting rule, mapping any collection of n subsets of X of cardinality between d and k, is called $[d, k]$-*approval voting*. Notice that approval voting is equivalent to $[1, m]$-approval voting.

3   Single-Winner Approval Voting

3.1   Restriction-Free Approval Voting

We start by defining the set of approval ballots that are compatible with a ranking profile.

Definition 1

- A threshold vector *(for N and X) is a vector* $\boldsymbol{k} = \langle k_1, \ldots, k_n \rangle \in \{1, \ldots, m\}^n$.
- Let $P = \langle P_1, \ldots, P_n \rangle$ *be a ranking profile over X, and $\boldsymbol{k}$ a threshold vector. For all $i \leq n$, let $(P_i)^{1 \to k_i}$ be the subset of X defined by*

$$(P_i)^{1 \to k_i} = \{x \in X \mid \mathrm{rk}(x, P_i) \leq k_i\}$$

The approval profile induced by P and $\boldsymbol{k}$, *denoted by* $A^{P,\boldsymbol{k}}$, *is defined as*

$$A^{P,\boldsymbol{k}} = \langle (P_1)^{1 \to k_1}, \ldots, (P_n)^{1 \to k_n} \rangle$$

- *The set of all approval profiles compatible with P is defined as*

$$CAP(P) = \{A^{P,\boldsymbol{k}} \mid \boldsymbol{k} \in \{1, \ldots, m\}^n\}$$

Example 1. Let $m = n = 3$, $P = \langle x_1 \succ x_2 \succ x_3, x_1 \succ x_2 \succ x_3, x_3 \succ x_1 \succ x_2 \rangle$ and $\boldsymbol{k} = \langle 2, 1, 2 \rangle$; then $A^{P,\boldsymbol{k}} = \langle \{x_1, x_2\}, \{x_1\}, \{x_1, x_3\} \rangle$.

Definition 2. *Let P be a ranking profile P over X. A subset $X' \subseteq X$ is called a* possible co-winner set *for P if there exists a threshold vector $\boldsymbol{k}$ such that $X' = App(A^{P,\boldsymbol{k}})$. The set of all possible co-winner sets for P is denoted by $PCS(P)$. $x \in X$ is a* possible single winner *for P if $\{x\} \in PCS(P)$, a* possible co-winner *if it belongs to some possible co-winner set, a* necessary co-winner *if it belongs to all co-winner sets for P, and a* necessary single winner *if $PCS(P) = \{x\}$.*

Example 1, continued. $\{x_1\}$ is a possible co-winner set (and hence x_1 a possible single winner) for P, obtained for instance for $\boldsymbol{k} = \langle 1, 3, 3 \rangle$, for $\boldsymbol{k} = \langle 1, 2, 2 \rangle$, and for many other threshold vector; and

$$PCS(P) = \{\{x_1, x_2, x_3\}, \{x_1, x_2\}, \{x_1, x_3\}, \{x_1\}, \{x_3\}\}$$

whereas the possible single winners for P are x_1 and x_3.

Without any restriction on the allowed thresholds, the notions of possible co-winner, necessary co-winner and necessary single winner turn out to trivialize: all candidates are possible co-winners, no candidate is a necessary single winner, and a x is a necessary co-winner if and only if it is ranked on top of all votes.

We now consider the following question: given a ranking profile P and a subset of X' of candidates, is X' a possible winner set for P? We call this problem the POSSIBLE CO-WINNER SET PROBLEM FOR APPROVAL VOTING[5]. This problem turns out to be easy in the case where X' is a singleton:

Theorem 1. *x is a possible single winner for P if and only if no candidate in $X \setminus \{x\}$ dominates x in P.*

Proof. Assume no y dominates x in P. Define $\boldsymbol{k}$ by $k_i = \mathrm{rk}_P(i, x)$ for any $i \in N$. x is approved n times in $A^{P, \boldsymbol{k}}$; if $y \neq x$ is also approved n times $A^{P, \boldsymbol{k}}$, then for all i, $\mathrm{rk}_P(i, y) \leq k_i$, i.e., y would dominates x in P; therefore, $App(A^{P, \boldsymbol{k}}) = \{x\}$. Conversely, if y dominates x in P, then for all $\boldsymbol{k}$, y will be approved at least as many times as x in $A^{P, \boldsymbol{k}}$, therefore x cannot be a posssible single winner.

As a consequence, the restriction of the possible co-winner set problem to singletons can be solved in polynomial time. This property does not generalize to subsets of arbitrary size. Indeed, the possible co-winner set problem is computationally hard, even under the restriction to sets of candidates of fixed size $\ell \geq 2$. We first prove the following lemma.

Lemma 1. *If $X' \in PCS(P)$, then there exists a solution $(k_i)_{i \in N}$ satisfying the following properties:*

(a) For any $i \in N$, $k_i \in \{\mathrm{rk}_P(i, x) : x \in X'\}$.
(b) The score of any co-winner is at least $\max_{x \in X'} pl(x, X')$.

Proof. Let $X' \in PCS(P)$. (a): Let $(k_i)_{i \in N}$ be any solution such that the candidates in $X' = \{x_1, \ldots, x_\ell\}$ are exactly the co-winners for profile P. Consider a voter i and, without loss of generality, assume that $x_1 \succ_i \cdots \succ_i x_\ell$. Moreover, assume $k_i \notin \{\mathrm{rk}_P(i, x) : x \in X'\}$. If $\mathrm{rk}_P(i, x_j) < k_i < \mathrm{rk}_P(i, x_{j+1})$ with $j \in \{1, \ldots, \ell - 1\}$, then we replace k_i by $\mathrm{rk}_P(i, x_j)$. If $k_i < \mathrm{rk}_P(i, x_1)$ or $k_i > \mathrm{rk}_P(i, x_\ell)$, we replace k_i by $\mathrm{rk}_P(i, x_\ell)$. It is not difficult to see that X' remains exactly the co-winner set. By repeating this procedure for each voter, we obtain the expected result. (b): Using (a), we know that there is a solution $(k_i)_{i \in N}$ such that the global score of candidate $x_j \in X'$ is at least $pl(x_j, X')$. Since the candidates in X' are the co-winners, we must have that each candidate of X' is approved at least $\max_{x \in X'} pl(x, X')$ times.

---

[5] From now on we will generally omit "for approval voting".

Theorem 2. *Let $\ell \geq 2$. Given a profile P and a subset of candidates X' such that $|X'| = \ell$, determining whether X' is a possible co-winner set for P is **NP**-complete.*

Proof. The problem is clearly in **NP** for all $\ell \geq 2$. Let us give a proof for the case when $\ell = 2$ and explain then how to generalize to all other cases. The proof of the **NP**-completeness is based on a reduction from EXACT 3-SET COVER (X3C in short). In an instance of X3C, we are given a family of m sets $\mathcal{S} = \{S_1, \ldots, S_m\}$ over a ground set $Y = \{y_1, \ldots, y_{3n}\}$ such that $\cup_{i=1}^{m} S_i = Y$ and $|S_i| = 3$, for $i = 1, \ldots, m$. The question is whether there exists a subset $J \subseteq \{1, \ldots, m\}$ of size n such that $\sum_{j \in J} S_j = Y$? This problem is known to be **NP**-complete [17].

Let $I = (\mathcal{S}, Y)$, with $\mathcal{S} = \{S_1, \ldots, S_m\}$ and $Y = \{y_1, \ldots, y_{3n}\}$, be an instance of X3C. We build an instance of POSSIBLE CO-WINNER SET FOR APPROVAL VOTING, with $\ell = 2$, as follows. There are $2m - n$ voters $N = \{1, \ldots, m - n\} \cup \{1', \ldots, m'\}$ and $m + 2n + 2$ candidates $X = E \cup Y \cup \{a, b\}$ where $E = \{e_1, \ldots, e_{m-n}\}$ and we set $X' = \{a, b\}$ as the target candidates. The profile P is given by:

- For $1 \leq i \leq m - n$, $E \setminus \{e_i\} \succ_i Y \succ_i a \succ_i e_i \succ_i b$.
- For $1 \leq j \leq m$, $b \succ_{j'} Y \setminus S_j \succ_{j'} E \succ_{j'} a \succ_{j'} S_j$.

This clearly gives us an instance I' of POSSIBLE CO-WINNER SET. We claim that there exists a subset $J \subseteq \{1, \ldots, m\}$ with $|J| = n$ such that $\sum_{j \in J} S_j = Y$ if and only if $\{a, b\}$ is a possible co-winner set for P.

Suppose that I is a yes-instance of X3C, i.e., there exists $J \subseteq \{1, \ldots, m\}$ with $|J| = n$ such that $\sum_{j \in J} S_j = Y$. We set $k_{j'} = \text{rk}_P(j', a)$ for $j \in J$. For the remaining voters $i \in N \setminus \{j' : j \in J\}$, we set $k_i = \min\{\text{rk}_P(i, a), \text{rk}_P(i, b)\}$. a and b are approved m times while candidates in $E \cup Y$ are approved at most $m - 1$ times. Thus $X' = \{a, b\}$ is a possible co-winner set.

Conversely, assume that I' is a yes-instance of POSSIBLE CO-WINNER SET. Using (a) and (b) of Lemma 1, there exists k with $k_i \in \{\text{rk}_P(i, a); \text{rk}_P(i, b)\}$ for any $i \in N$ and a, b must be approved at least m times. Thus, there exists $J \subseteq \{1, \ldots, m\}$ such that $k_{j'} = \text{rk}_P(j', a)$ for $j \in J$ and $k_{j'} = \text{rk}_P(j', b)$ for $j \notin J$. In particular, we deduce that $\sum_{j \in J} S_j = X$ since otherwise any candidate of $X \setminus (\sum_{j \in J} S_j)$ necessarily dominates a; hence, $|J| \geq n$. Moreover, if $|J| \geq n + 1$, then a gets approved at least $m + 1$ times. Thus, there exists at least one voter $i \in \{1, \ldots, m - n\}$ such that $k_i = \text{rk}_P(i, b)$ (since a and b must get approved the same number of times). But then $app(e_i) \geq app(a)$, a contradiction. So we conclude that $|J| = n$ and $\sum_{j \in J} S_j = Y$: I is a yes-instance of X3C.

This shows the **NP**-completeness of POSSIBLE CO-WINNER SET FOR APPROVAL VOTING, restricted to co-winner sets of size 2. Now it is not difficult to see that, if we proceed exactly the same way and replace everywhere in the previous proof a by $\{a_1, \ldots, a_{\ell-1}\}$ and we set $X' = \{b\} \cup \{a_1, \ldots, a_{\ell-1}\}$, for $\ell \geq 3$, we can show the **NP**-completeness of POSSIBLE CO-WINNER SET FOR APPROVAL VOTING restricted to co-winner sets of size ℓ.

3.2   Approval Voting with Restriction on the Number of Approvals

We now consider, more generally, $[d, k]$-approval voting. The definitions are natural generalizations of those in Section 3.1, with the difference that each k_i should be such that $d \leq k_i \leq k$. The set of all $[d, k]$-approval profiles compatible with P is defined by $CAP_{d,k}(P) = \{A^{P,\boldsymbol{k}} \mid \boldsymbol{k} \in [d, k]^n\}$, and the set of possible $[d, k]$-approval co-winner sets for P is denoted by $PCS_{d,k}(P)$.

Example 1, continued.

– $PCS_{1,2}(P) = \{\{x_1\}, \{x_1, x_2\}\}$;
– $PCS_{2,3}(P) = \{\{x_1\}, \{x_1, x_2\}, \{x_1, x_3\}, \{x_1, x_2, x_3\}\}$

Again, in order to check whether x is a possible single winner, it is enough to check it for a specific choice of $\boldsymbol{k}$, namely, the best possible choice for x.

Theorem 3. $\{x\} \in PCS_{d,k}(P)$ *if* $App(A^{P,\boldsymbol{k}}) = \{x\}$ *for* $\boldsymbol{k}$ *defined by* $k_i = \mathrm{rk}_P(i, x)$ *when* $\mathrm{rk}_P(i, x) \in [d, k]$, *and* $k_i = d$ *otherwise.*

Proof. ($\Leftarrow$) is direct from the definition. For ($\Rightarrow$), suppose $\{x\} \in PCS_{d,k}(P)$. Then there is a vector $(k_i')_{i \in N}$ for which x is the single winner. If $\boldsymbol{k}' = \boldsymbol{k}$ then we are done. Otherwise, take the voter i with minimum index that satisfies $k_i' \neq k_i$. If $k_i' < k_i$ then doing $k_i' \leftarrow k_i$ increases the score of a subset of candidates by one unit, and this subset includes x. If $k_i' \geq k_i$ then by doing $k_i' \leftarrow k_i$ the score of x remains unchanged, while the score of some other candidates decreases. In all, x remains the single winner by the operation $k_i' \leftarrow k_i$. Repeating the operation until $\boldsymbol{k}' = \boldsymbol{k}$ leads to the result. $\qquad\square$

Theorem 3 generalizes Lemma 2 in [9]; for $d = 1$ and $k = m - 1$, we recover their notion of *critical strategy profile for* x: every voter who ranks i as his worst candidate approves only one candidate; the other voters vote for i and all candidates above. Then x is a possible $1, m - 1$- approval winner (called AV outcome in [9]) if x wins at his critical strategy profile.

As a corollary, we get simple characterizations of possible and necessary co-winners and single winners, which we state without proof: let $D_P^+(x, y) = \{i \mid \mathrm{rk}_i(P, x) \leq k, \mathrm{rk}_i(P, y) > d \text{ and } x \succ_i y\}$ and $D_P^-(x, y) = \{i \mid \mathrm{rk}_i(P, x) \leq d \text{ and } \mathrm{rk}_i(P, y) > k\}$. Then x is a possible $[d, k]$-approval co-winner (respectively, possible single winner, necessary co-winner, single winner) for P if and only if for all $y \neq x$, $|D_P^+(x, y)| \geq |D_P^-(y, x)|$ (respectively, $|D_P^+(x, y)| > |D_P^-(y, x)|$, $|D_P^-(x, y)| \geq |D_P^+(y, x)|$, $|D_P^-(x, y)| > |D_P^+(y, x)|$).

Theorem 2 immediately extends to $[1, k]$-approval for $k = m - 2$ because in the proof of Theorem 2 we do not approve more $m - 2$ candidates for each voter.

Theorem 4. *For any integer* $\ell \geq 2$, *the problem of checking whether* X' *is a* $[1, m - 2]$-*approval possible co-winner set is* **NP**-*complete, even under the restriction* $|X'| = \ell$.

Remark: The proof of Theorem 2 can be adapted in such a way that for any integers $\ell \geq 2$ and $d \geq 2$, checking whether X' is a $[d, k]$- approval possible co-winner set, under the restriction $|X'| = \ell$, is **NP**-complete for some k. Moreover, using Algorithm 1 described in Subsection 3.3 we can prove that checking whether X' is a $[d, k]$- approval possible co-winner set is polynomial whenever $k - d$ is upper bounded by a constant.

3.3   The Probability of Possible Co-winner Sets

Definition 3. *Let p be a probability distribution on all threshold vectors. Given a profile P and a subset of candidates $X' \subseteq X$, the probability that X' is the co-winner set (for approval) is equal to $\sum_{\mathbf{k}|App(A^{P,\mathbf{k}})=X'} pr(\mathbf{k})$.*

A simple assumption consists in assuming that $\pi(i, r)$ approves his r most preferred candidates with a given probability $\pi(i, r)$, and that voters' choices are probabilistically independent. Under this assumption, we show how to compute efficiently the probability of each co-winner subset.

We first show how to enumerate all possible scores and their probabilities. Given a voter i and a threshold $k_i \in [d..k]$, we define TRACE(i, k_i) as the m-dimensional 0-1 vector whose coordinate j is equal to 1 if candidate x_j belongs to the k_i most preferred candidates of i, and 0 otherwise. For example, there are 4 candidates and voter i's preference profile is $x_2 \succ_i x_3 \succ_i x_1 \succ_i x_4$; we have TRACE$(i, 1) = (0, 1, 0, 0)$, TRACE$(i, 2) = (0, 1, 1, 0)$, TRACE$(i, 3) = (1, 1, 1, 0)$ and TRACE$(i, 4) = (1, 1, 1, 1)$.

We suppose wlog. that the voters provide their ballots sequentially, by ascending index, and a list L_i contains all possible scores after voter i's turn. Therefore L_i is defined as L_{i-1} to which one adds the possible ballots of voter i.

An element of a list is a couple composed of an m-dimensional vector (a score for each candidate) and a probability. We suppose that a list never contains two elements with the same vector. In addition, a list is sorted by its elements' vectors which are sorted in lexicographic order (e.g. $(1, 3, 4) <_{lex} (1, 4, 0)$). A possible list can be $\langle((1, 3, 4), 0.24), ((1, 4, 0), 0.36), ((2, 0, 1), 0.15)\rangle$.

We use a subroutine MERGE-LISTS(L, L') that merges the lists L and L'. If several elements have the same vector then they are combined in a unique element whose probability is the sum of all condensed elements' probabilities. For example, MERGE-LISTS$(\langle((1, 4, 0), 0.36), ((2, 0, 1), 0.15)\rangle, \langle((1, 2, 6), 0.41), ((1, 4, 0), 0.06)\rangle)$ is equal to $\langle((1, 2, 6), 0.41), ((1, 4, 0), 0.42), ((2, 0, 1), 0.15)\rangle$. MERGE-LISTS$(L, L')$ needs $|L| + |L'|$ operations.

Given a list L, a vector vec and its probability π, $L \oplus (vec, \pi)$ means that we add vec to every vector of L (component by component) and we multiply every probability by π. For example, $\langle((1, 4, 0), 0.3), ((2, 0, 1), 0.1)\rangle \oplus ((1, 0, 1), 0.3)$ gives $\langle((2, 4, 1), 0.09), ((3, 0, 2), 0.03)\rangle$. $L \oplus (vec, \pi)$ requires $|L|$ operations.

Algorithm 1 gives the exhaustive list of outcomes with their probabilities, where $\mathbf{0}$ denotes the m-dimensional vector whose coordinates are all equal to 0.

Then we can retrieve from L_n the winner sets and their probabilities. The size of L_i is at most $(k - d + 1)|L_{i-1}|$, and $|L_0| = 1$ so $|L_n| \leq (k - d + 1)^n \leq m^n$.

Algorithm 1. All possible scores with probabilities

1: $L_0 \leftarrow \langle (\mathbf{0}, 1) \rangle$
2: **for** $i = 1$ to n **do**
3:    $L' \leftarrow \langle \rangle$
4:    **for** $r = d$ to k **do**
5:       $L' \leftarrow$ MERGE-LISTS$(L', L_{i-1} \oplus (\text{TRACE}(i, r), \pi(i, r)))$
6:    **end for**
7:    $L_i \leftarrow L'$
8: **end for**
9: **return** L_n

Meanwhile the final score of any candidate belongs to $[0..n]$ so there are at most $(n+1)^m$ distinct vectors of scores and $|L_n| \leq (n+1)^m$. Thus Algorithm 1 is exponential in the input size but it is polynomial when n or m is a fixed constant.

As a short example, consider an instance with $m = n = 3$, $d = 1$ and $k = 3$. The profiles are $x_1 \succ_1 x_2 \succ_1 x_3$, $x_3 \succ_2 x_1 \succ_2 x_2$ and $x_2 \succ_3 x_3 \succ_3 x_1$. We suppose that for every voter i, the probabilities that the k_i first candidates are approved are 0.3, 0.5 and 0.2 when k_i is equal to 1, 2 and 3 respectively. These probabilities are independent. Hence voter 1 approves $\{x_1, x_2\}$ with probability 0.5. And voter 2 approves $\{x_1, x_2\}$ with probability 0 since it is not a sincere vote. Running Algorithm 1 yields the values given in Table 1.

Table 1. Output of Algorithm 1 on the example

detailed scores & corresponding prob.	winner(s) & total prob.	detailed scores & corresponding prob.	winner(s) & total prob.
(322) (312) (211) 0.082 0.03 0.045	$\{x_1\}$ 0.157	(212) 0.093	$\{x_1, x_3\}$ 0.093
(121) (231) (232) 0.045 0.03 0.062	$\{x_2\}$ 0.137	(122) (233) 0.093 0.02	$\{x_2, x_3\}$ 0.113
(112) (123) (223) 0.045 0.03 0.012	$\{x_3\}$ 0.087	(111) (222) (333) 0.027 0.235 0.008	$\{x_1, x_2, x_3\}$ 0.27
(221) (332) 0.123 0.02	$\{x_1, x_2\}$ 0.143		

3.4   Experimental Analysis

Finally, we provide an experimental analysis of the sensitivity of the winner to the choice of the thresholds. We generate $5 * 10^4$ ranking profiles with an uniform distribution (*impartial culture assumption*). For each profile we generate $5 * 10^4$ threshold vectors with a uniform distribution, for each of these vectors we compute the winner (ties being broken randomly), and we obtain the winning probability of each candidate. We reorder these winning probabilities decreasingly. Then we compute the average, over all generated profiles, of the largest winning probability. The results of these experiments are summarized in Table 2. We observe that the largest winning probability is above 50% with a low number of candidates and any number of voters. This probability decreases when the number of candidates increases.

Table 2. Largest winning probability with uniformly drawn profiles and thresholds

	n = 5	n = 20	n = 50	n = 100
m = 5	55.9	58.5	55.4	54.7
m = 20	32.8	35.0	34.6	35.1
m = 50	23.9	27.1	27.3	27.3
m = 100	18.3	22.4	22.7	22.8

We also compute the average of the second and third largest winning probabilities. Figure 1 shows us the evolution of the largest, second and third largest winning probabilities as a function of the number of candidates, with $n = 5$. Finally, in Figure 2 we represent the largest winning probability as a function of the number of voters. The largest winning probability appears to be independent from the number of voters.

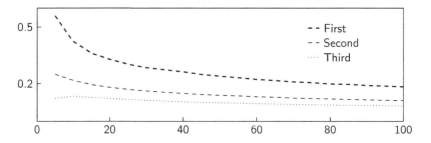

Fig. 1. Largest, second and third largest winning probabilities as a function of the number of candidates, with $n = 5$

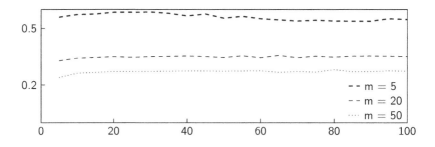

Fig. 2. Largest winning probability as a function of the number of voters

4   Multiwinner Approval Voting

We now briefly reconsider some of the questions addressed in Section 3 in the context of *multiwinner* approval voting. We are now given an integer $K \leq m$, and look for a committee of size K.

Definition 4. *For any approval profile A, $App_K(A)$ is the set of all committees $X' \subseteq X$ of size K such that for any $x \in X$ and $z \notin X'$, $app_A(x) \geq app_A(z)$. Let P a ranking profile over X. $X' \subseteq X$ is a* possible winning K-committee *for P if $X' \in App_K(A^{P,k})$ for some threshold vector $\boldsymbol{k}$. The set of all possible winning K-committees for P is denoted by $PCS_K(P)$. $x \in X$ is possibly (resp. necessarily) elected w.r.t. K and P if it belongs to some (resp. all) possible winning K-committee(s) for P. Let $Poss_K(P)$ and $Nec_K(P)$ be the set of possibly (resp. necessarily) elected candidates w.r.t. K and P. These definitions naturally generalize to $[d, k]$-approval voting.*

The following result generalizes Theorem 1.

Theorem 5. *$x \in Poss_K(P)$ if and only if x is not Pareto-dominated by K candidates or more.*

Proof. Suppose that x is member of a possible winning K-committee, then the candidates that dominate x are also in this winning K-committee, therefore at most $K - 1$ candidates dominate x. Conversely, assume that x is dominated by $K - 1$ candidates or less. For all $i \in N$, let $k_i = \text{rk}_P(i, x)$. Only x and the candidates that dominates x have an approval score equal to n, therefore x belongs to a winning K-committee.

Theorem 6. *$x \in Nec_K(P)$ if and only if x dominates at least $n - K$ candidates.*

The proof is similar to the proof of Theorem 5.

We now consider the following problem: given a ranking profile P over X and a subset $X' \subset X$ of size K, is K a possible winning K-committee for P? We first establish the following lemma, for $K = 2$.

Lemma 2. *Let $X' = \{x'_1, x'_2\}$. If $X' \in PCS_2(P)$, then $X' \in App_2(A^{P,k})$ for some $\boldsymbol{k}$ satisfying $|\{i \in N : \text{rk}_P(i, x'_1) \leq k_i\}| = |\{i \in N : \text{rk}_P(i, x'_2) \leq k_i\}|$.*

Proof. Let $X' \in PCS_2(P)$ and let $\boldsymbol{k}$ such that $X' \in App_2(A^{P,k})$. If $app_{A^{P,k}}(x'_1) = app_{A^{P,k}}(x'_2)$, we are done. Otherwise, assume without loss of generality that $app_{A^{P,k}}(x'_1) > app_{A^{P,k}}(x'_2)$. There exists a subset $N' \subset N$ of size $app_{A^{P,k}}(x'_1) > app_{A^{P,k}}(x'_2)$ such that for voters $i \in N', \text{rk}_P(i, x'_1) \leq k_i < \text{rk}_P(i, x'_2)$. We build a new vector $\boldsymbol{k}$ as follows: (i) for $i \in N', k'_i = 0$; (ii) for $i \in N \setminus N', k'_i = k_i$. We have $app_{A^{P,k'}}(x'_1) = app_{A^{P,k}}(x'_1) - (app_{A^{P,k}}(x'_1) - app_{A^{P,k}}(x'_2))$ and $app_{A^{P,k'}}(x'_2) = app_{A^{P,k}}(x'_2)$, therefore $app_{A^{P,k'}}(x'_2) = app_{A^{P,k'}}(x'_1)$.

Theorem 7. *Determining whether X' is a possible winning 2-committee is* **NP-complete.**

Proof. Hardness is shown by a reduction from the problem of determining whether a set of 2 candidates is a possible winning set in single-winner approval (**NP**-complete, cf. Theorem 2). Let $I = (P, N, X, X')$ be an instance of this problem, with $X' = \{x'_1, x'_2\}$. From I, we build an instance I' of the possible winning 2-committee problem, with the same N, X, P, and X'. We claim that X' is a possible winning set in I if X' is a possible winning 2-committee in I'. Clearly, if $X' = App(A^{P,\boldsymbol{k}})$, then $X' = App_2(A^{P,\boldsymbol{k}})$. Conversely, assume that $X' = App_2(A^{P,\boldsymbol{k}})$. By lemma 2, we know that there exists $\boldsymbol{k}'$ such that $X' = App_2(A^{P,\boldsymbol{k}})$ and $app_{A^{P,\boldsymbol{k}'}}(x'_1) = app_{A^{P,\boldsymbol{k}'}}(x'_2)$, therefore, $X' = App_2(A^{P,\boldsymbol{k}'})$.

Unsurprisingly, this difficulty carries on to committees of larger size (the proof, by reduction from the POSSIBLE WINNING 2-COMMITTEE, is easy and omitted):

Theorem 8. *For any integer $K \geq 2$, determining whether X' is a possible winning K-committee is **NP**-complete.*

This complexity result extends to $[1, k]$-approval:

Theorem 9. *For any integer $K \geq 2$, and $k \geq 3$, determining whether X' is a possible winning K-committee for $[1, k]$-approval is **NP**-complete.*

A related series of results on the complexity of multiwinner elections with approval ballots is in [11] (Theorems 3.4 to Corollary 3.9). There the setting is different from ours: each voter approves *exactly* t candidates; if voter i approves $A_i \subseteq X$ (with $|A_i| = t$), then given two k-committees X and Y, i is assumed to prefer X over Y ($X \gg_i Y$) if $|X \cap A_i| > |Y \cap A_i|$. A k-committee X is a *popular k-committee* if it majority-wise defeats all other k-committees (that is, if it a Condorcet winner in the set of all k-committees for the profile $\langle \gg_1, \ldots, \gg_n \rangle$). Darmann shows that deciding whether a k-committee is a popular committee is NP-hard as soon as $2 \leq t \leq m - 2$ (finding such a committee is probably even harder). Unlike ours, the hardness results in [11] are not due to the uncertainty about the number of approvals and they do not imply, nor are implied by, any of our results.

5   Further Issues

When thresholds vectors are generated with a uniform probability, the winning probability of a candidate for a given profile is proportional to its Borda score; more generally, if the probabilities on the number of approvals for voters are i.i.d., the winning probability of a candidate for a profile is proportional to its score for some positional scoring rule. This connection is worth exploring further.

Another interesting topic that we did not explore is the control of an election by a chair who has the power to fix the lower and upper bounds d and k on the number of approvals. Assume that the chair moreover knows the voters' rankings and has some subjective probability distribution on the number of candidates the voters will approve (to be conditioned by the bounds d and k). Clearly, the choice of d and k has an influence on the winning probability of a candidate; this election control is computationally hard if computing winning probabilities is computationally hard — a question that we have not addressed yet.

References

1. Bachrach, Y., Betzler, N., Faliszewski, P.: Probabilistic possible-winner determination. In: Proc. of AAAI 2010 (2010)
2. Baumeister, D., Erdèlyi, G., Hemaspaandra, E., Hemaspaandra, L., Rothe, J.: Computational aspects of approval voting. In: Laslier, J.-F., Sanver, R. (eds.) Handbook of Approval Voting, pp. 199–251. Springer (2010)
3. Baumeister, D., Rothe, J.: Taking the final step to a full dichotomy of the possible winner problem in pure scoring rules. In: Proceedings of ECAI 2010 (2010)
4. Betzler, N., Dorn, B.: Towards a dichotomy of finding possible winners in elections based on scoring rules. In: Královič, R., Niwiński, D. (eds.) MFCS 2009. LNCS, vol. 5734, pp. 124–136. Springer, Heidelberg (2009)
5. Betzler, N., Hemmann, S., Niedermeier, R.: A multivariate complexity analysis of determining possible winners given incomplete votes. In: Proceedings of IJCAI 2009, pp. 53–58 (2009)
6. Betzler, N., Slinko, A., Uhlmann, J.: On the computation of fully proportional representation. Journal of Artificial Intelligence Research (2013)
7. Brams, S., Fishburn, P.: Approval voting. American Political Review 72(3), 831–847 (1978)
8. Brams, S., Fishburn, P.: Approval Voting, 2nd edn. Birkhäuser (1987)
9. Brams, S., Sanver, R.: Critical strategies under approval voting: Who gets ruled in and ruled out. Electoral Studies 25(2), 287–305 (2006)
10. Chevaleyre, Y., Lang, J., Maudet, N., Monnot, J., Xia, L.: New candidates welcome! possible winners with respect to the addition of new candidates. Mathematical Social Sciences 64(1), 74–88 (2012)
11. Darmann, A.: Popular committees. Mathematical Social Sciences (to appear, 2013)
12. Delort, C., Spanjaard, O., Weng, P.: Committee selection with a weight constraint based on a pairwise dominance relation. In: ADT, pp. 28–41 (2011)
13. Elkind, E., Lang, J., Saffidine, A.: Choosing collectively optimal sets of alternatives based on the condorcet criterion. In: IJCAI 2011, pp. 186–191 (2011)
14. Endriss, U.: Sincerity and manipulation under approval voting. Theory and Decision (2011)
15. Endriss, U., Pini, M.S., Rossi, F., Venable, K.B.: Preference aggregation over restricted ballot languages: Sincerity and strategy-proofness. In: Boutilier, C. (ed.) IJCAI 2009, pp. 122–127 (2009)
16. Erdélyi, G., Nowak, M., Rothe, J.: Sincere-strategy preference-based approval voting broadly resists control. In: Ochmański, E., Tyszkiewicz, J. (eds.) MFCS 2008. LNCS, vol. 5162, pp. 311–322. Springer, Heidelberg (2008)
17. Garey, M.R., Johnson, D.S.: Computers and Intractability: A Guide to the Theory of NP-Completeness. W. H. Freeman & Co., New York (1979)
18. Kalech, M., Kraus, S., Kaminka, G.A., Goldman, C.V.: Practical voting rules with partial information. Autonomous Agents and Multiagent Systems 22(1), 151–182 (2011)
19. Kilgour, M.: Approval balloting for multi-winner elections. In: Laslier, J.-F., Sanver, R. (eds.) Handbook of Approval Voting, pp. 105–124. Springer (2010)

20. Klamler, C., Pferschy, U., Ruzika, S.: Committee selection with a weight constraint based on lexicographic rankings of individuals. In: Rossi, F., Tsoukias, A. (eds.) ADT 2009. LNCS, vol. 5783, pp. 50–61. Springer, Heidelberg (2009)
21. Konczak, K., Lang, J.: Voting procedures with incomplete preferences. In: IJCAI 2005 Multidisciplinary Workshop on Advances in Preference Handling (2005)
22. Lang, J., Pini, M.S., Rossi, F., Salvagnin, D., Venable, K.B., Walsh, T.: Winner determination in voting trees with incomplete preferences and weighted votes. In: Autonomous Agents and Multi-Agent Systems (2011)
23. Laslier, J.-F.: The leader rule – a model of strategic approval voting in a large electorate. Journal of Theoretical Politics 21, 113–136 (2009)
24. Laslier, J.-F., Sanver, R.: The basic approval voting game. In: Laslier, J.-F., Sanver, R. (eds.) Handbook of Approval Voting. Springer (2010)
25. Meir, R., Procaccia, A., Rosenschein, J., Zohar, A.: The complexity of strategic behavior in multi-winner elections. JAIR 33, 149–178 (2008)
26. Nuñez, M.: Condorcet consistency of approval voting: a counter example in large Poisson games. Journal of Theoretical Politics 22, 64–84 (2010)
27. Procaccia, A., Rosenschein, J., Zohar, A.: On the complexity of achieving proportional representation. Social Choice and Welfare 30(3), 353–362 (2008)
28. Saari, D.: Systematic analysis of multiple voting rules. Social Choice and Welfare 34(2), 217–247 (2010)
29. Sertel, M., Yılmaz, B.: The majoritarian compromise is majoritarian-optimal and subgame-perfect implementable. Social Choice and Welfare 16(4), 615–627 (1999)
30. De Sinopoli, F., Dutta, B., Laslier, J.-F.: Approval voting: three examples. International Journal of Game Theory 35(1), 27–38 (2006)
31. Skowron, P., Faliszewski, P., Slinko, A.: Achieving fully proportional representation is easy in practice. In: AAMAS 2013, pp. 399–406 (2013)
32. Xia, L., Conitzer, V.: Determining possible and necessary winners under common voting rules given partial orders. In: Proceedings of AAAI 2008, pp. 196–201 (2008)
33. Xia, L., Lang, J., Monnot, J.: Possible winners when new alternatives join: new results coming up! In: Sonenberg, L., Stone, P., Tumer, K., Yolum, P. (eds.) AAMAS. IFAAMAS, pp. 829–836 (2011)

Computational Aspects of Manipulation and Control in Judgment Aggregation*

Dorothea Baumeister[1], Gábor Erdélyi[2], Olivia Johanna Erdélyi[3], and Jörg Rothe[1]

[1] Institut für Informatik, Heinrich-Heine-Universität Düsseldorf, 40225 Düsseldorf, Germany
[2] School of Economic Disciplines, University of Siegen, 57076 Siegen, Germany
[3] Institut für Versicherungsrecht, Heinrich-Heine-Universität Düsseldorf, 40225 Düsseldorf, Germany

Abstract. We study computational aspects of various forms of manipulation and control in judgment aggregation, with a focus on the premise-based procedure. For manipulation, we in particular consider incomplete judgment sets and the notions of top-respecting and closeness-respecting preferences introduced by Dietrich and List [13]. This complements previous work on the complexity of manipulation in judgment aggregation that focused on Hamming-distance-induced preferences [14,6], which we also study here. Regarding control, we introduce the notion of control by bundling judges and show that the premise-based procedure is resistant to it in terms of NP-hardness.

1 Introduction

Judgment Aggregation is the task of aggregating individual judgment sets of possibly interconnected logical propositions (see the surveys by List and Puppe [21] and by List [20]). Manipulability and (the game-theoretic concept of) strategy-proofness for the formal framework of judgment aggregation was first introduced by Dietrich and List [13]. We focus on their notion of strategy-proofness, since their (non)manipulability condition is not always appropriate in our setting. Manipulation has been studied in a wide variety of settings (voting, mechanism design, game theory, fair division, judgment aggregation, etc.). The incentive of a manipulative attack is always to achieve a "better" result by agents (voters, players, etc.) providing untruthful information. In judgment aggregation, this untruthful information is the manipulator's individual judgment set and the result is the collective outcome of a judgment aggregation procedure. However, it is not at all obvious what a "*better*" result is. To compare two collective judgment sets, a preference over all possible judgment sets would be needed, but such preferences are rarely elicited, and they may be exponentially large in the number of formulas in the agenda (see Section 2 for the notions not defined here). One way to avoid this obstacle, is to derive an order from a given individual judgment set. Based on the notions introduced by Dietrich and List [13], we in particular consider incomplete judgment sets and the notions of top-respecting and closeness-respecting preferences. Since

---

* This work was supported in part by DFG grants RO 1202/15-1 and ER 738/1-1 and NRF (Singapore) grant NRF-RF 2009-08. Work done in part while the second and third author were visiting the University of Rochester.

P. Perny, M. Pirlot, and A. Tsoukiàs (Eds.): ADT 2013, LNAI 8176, pp. 71–85, 2013.

most judgment aggregation rules are not strategy-proof, we study the computational complexity of the corresponding decision problems. This complements previous work on the complexity of manipulation in judgment aggregation (initiated by Endriss et al. [14], see also the work of Baumeister et al. [6]) that focused on Hamming-distance-induced preferences, which we also study here.

Regarding control in judgment aggregation, we extend previous work by Baumeister et al. [3,4] who, inspired by the notion of control in voting (see, e.g., the book chapter [5] and the references cited therein) studied the complexity of control by adding, deleting, or replacing judges. We introduce a new type of control, *control by bundling judges*, which is well-motivated for judgment aggregation by real-world scenarios and is somewhat reminiscent of control by partitioning voters in voting. We show that one specific judgment aggregation procedure, namely the premise-based procedure is resistant to this control type in terms of NP-hardness.

This paper is organized as follows. In Section 2, we provide the basic framework of judgment aggregation and define the relevant notions formally. In Section 3, we study the complexity of manipulation in judgment aggregation, and in Section 4 that of the problem modeling control by bundling judges. Finally, Section 5 summarizes our results and presents a number of interesting open problems for future research.

2   Preliminaries

We adopt the framework on judgment aggregation described by Endriss et al. [14] and used also by Baumeister et al. [6,3]. Let $N = \{1,\ldots,n\}$ be a set of judges who have to judge over the formulas in the agenda Φ. We assume that the agenda is a finite, nonempty subset of the set $\mathscr{L}_{PS}$ of all propositional formulas that are built from the boolean constants 1 and 0 and the propositional variables in PS using the boolean connectives $\vee$, $\wedge$, $\rightarrow$, and $\leftrightarrow$. Further, we assume that the agenda does not contain doubly negated formulas. To this end, we denote by $\sim\alpha$ the complement of α: $\sim\alpha = \neg\alpha$ if α is not negated, and $\sim\alpha = \beta$ if $\alpha = \neg\beta$. We also assume that the agenda is closed under complementation (if $\alpha \in \Phi$ then $\sim\alpha \in \Phi$) and under propositional formulas (every literal that occurs in a formula of the agenda is itself contained in the agenda).

An *(individual or collective) judgment set* is a subset of the agenda Φ, where "*individual*" refers to the judgment set of an individual judge and "*collective*" refers to the outcome of a judgment aggregation procedure. A judgment set is said to be *complete* if it contains α or $\sim\alpha$ for all $\alpha \in \Phi$; it is said to be *consistent* if all its formulas can be satisfied by some truth assignment simultaneously; and it is said to be *complement-free* if it does not contain α and $\sim\alpha$ simultaneously for any $\alpha \in \Phi$. Let $\mathscr{J}(\Phi)$ denote the set of all complete and consistent judgment sets.

The well-known doctrinal paradox says that under the majority rule the collective outcome may be inconsistent even if all underlying individual judgment sets are consistent. To avoid this, we focus on the *premise-based procedure* (*PBP*) for an odd number of judges, which—under the assumptions made below—always guarantees a complete and consistent outcome. For a given profile $\mathbf{T} = (J_1,\ldots,J_n) \in \mathscr{J}(\Phi)^n$ of individual judgment sets, the agenda Φ is divided into the set of premises Φ_p and the set of conclusions Φ_c. $PBP(\mathbf{T})$ first aggregates the individual judgment sets on the premises Φ_p

using the majority rule and then derives the collective outcome for the conclusions Φ_c. Formally, it is a function $PBP : \mathcal{J}(\Phi)^n \to 2^\Phi$ mapping each given profile of individual judgment sets to the collective judgment set

$$PBP(\mathbf{T}) = \triangle \cup \{\varphi \in \Phi_c \mid \triangle \models \varphi\},$$

where

$$\triangle = \left\{ \varphi \in \Phi_p \;\middle|\; \|\{i \mid \varphi \in J_i\}\| > \frac{n}{2} \right\}.$$

To guarantee complete and consistent outcomes, we follow Endriss et al. [14] and identify the premises with the set of literals from the agenda. Furthermore, we will extend *PBP* to work also for an even number of judges by assuming that in case of a tie the negated literal will be contained in the collective judgment set.

3   Various Forms of Manipulation in Judgment Aggregation

3.1   Definitions

As mentioned in the introduction, we apply the notions introduced by Dietrich and List [13] to study various types of preferences. If for two judgment sets $X, Y \in \mathcal{J}(\Phi)$, X is preferred to Y for a given type of preference T and some individual judgment set J, we write $X \succ_T^J Y$.

Definition 1. *Given some individual judgment set J, we define preferences to be (strictly)*

- unrestricted (U) *if there is no restriction on* $\succ_U^J$;
- top-respecting (TR) *if* $J \succ_{TR}^J X$ *for all* $X \in \mathcal{J}(\Phi) \setminus \{J\}$;
- closeness-respecting (CR) *if for all* $X, Y \in \mathcal{J}(\Phi)$, *we have* $X \succ_{CR}^J Y$ *if* $Y \cap J \subset X \cap J$;
- Hamming-distance-induced (HD) *if for all* $X, Y \in \mathcal{J}(\Phi)$, $X \succ_{HD}^J Y$ *if and only if* $HD(X, J) < HD(Y, J)$, *where the Hamming distance* $HD(X, Y)$ *between two (possibly incomplete) judgment sets X and Y is the number of disagreements on propositions that occur in both judgment sets.*

By allowing equalities the Hamming-distance-induced preference is the only complete relation among the above. Intuitively, unrestricted preferences capture the setting where we know nothing about the individual preferences. The slightly more restricted case of top-respecting preferences at least requires the given judgment set to be the most preferred one. This also holds for closeness-respecting preferences, but in addition judgment sets that have additional agreement are preferred. In contrast, the Hamming-distance-induced preferences focus only on the total number of disagreements. Hence, for $X, Y \in \mathcal{J}(\Phi)$, if $X \succ_{TR}^J Y$ then it holds that $X \succ_{CR}^J Y$, and if $X \succ_{CR}^J Y$ then it holds that $X \succ_{HD}^J Y$.

Example 1. For variables a, b, c, and d, let the agenda contain the formulas

$$a, \quad b, \quad c, \quad d, \quad a \vee b, \quad b \vee c, \quad a \vee c, \quad b \vee d,$$

Table 1. Applying the premise-based judgment aggregation procedure

	a	b	c	d		$a \vee b$	$b \vee c$	$a \vee c$	$b \vee d$
Judge 1	1	1	0	0		1	1	1	1
Judge 2	0	0	0	0		0	0	0	0
Judge 3	1	0	1	1		1	1	1	1
PBP	1	0	0	0	$\Rightarrow$	1	0	1	0

and their negations. The individual judgment sets of three judges are shown in Table 1. A 0 indicates that the negation of the formula is in the judgment set, and a 1 indicates that the formula itself is contained in the judgment set.

The result according to the premise-based procedure is also given in the table. Now assume that the third judge is trying to manipulate and reports the untruthful individual judgment set $\{a,b,c,d\}$ and the corresponding conclusions. Then the collective outcome equals the individual judgment set of the first judge.

– If the manipulator has unrestricted preferences, we do not know whether she prefers this new outcome or not.
– If she has closeness-respecting preferences, we again do not know whether she prefers the new outcome, since the agreement on $\neg b$ is no longer given. However, if she is interested only in the conclusions, then she does prefer the new outcome, since the agreement on $a \vee b$ and $a \vee c$ is preserved and there are the two additional agreements on $b \vee c$ and $b \vee d$.
– The same holds for top-respecting preferences: If the manipulator is interested in the whole collective judgment set, we do not know which outcome is better for her, but restricted to the conclusions the new outcome equals her initial individual judgment set and thus is preferred to all other outcomes.
– If the manipulator has Hamming-distance-induced preferences, we know that the new outcome is preferred to the old one, since before the manipulation the Hamming distance was 4, but now it is only 3.

Just as Dietrich and List [13], we study settings where the desired judgment set is incomplete, to also capture their "reason-oriented" and "outcome-oriented" preferences. However, we will not generally restrict the desired judgment set to the premises or the conclusions; rather, we allow arbitrary incomplete desired judgment sets (which still must have a consistent extension to the whole agenda). In this case, we restrict the preferences to the formulas that occur in the desired judgment set. Since we want to compare two preferences with each other, but most of the induced preferences will be incomplete, we distinguish the cases where the relation between them is known or unknown. Let $\mathscr{T} \in \{U, TR, CR\}$ be a type of induced preferences.

– A judge *necessarily prefers X to Y for type* $\mathscr{T}$ if $X >_{\mathscr{T}}^J Y$ for all complete extensions of $\succ_{\mathscr{T}}^J$.
– A judge *possibly prefers X to Y for type* $\mathscr{T}$ if $X >_{\mathscr{T}}^J Y$ for some complete extension of $\succ_{\mathscr{T}}^J$.

Definition 2. *A judgment aggregation rule F is* necessarily/possibly strategy-proof *with respect to induced preferences of type $\mathcal{T} \in \{U, TR, CR\}$ if for all profiles $(J_1, \ldots, J_n)$ and each i, $1 \leq i \leq n$, agent i necessarily/possibly prefers the outcome $F(J_1, \ldots, J_n)$ to the outcome*

$$F(J_1, \ldots, J_{i-1}, J_i^*, J_{i+1}, \ldots, J_n)$$

(with respect to preferences of type $\mathcal{T}$ and the individual judgment set J_i) for any $J_i^* \in \mathcal{J}(\Phi)$.

This definition applies to complete desired judgment sets J_i only. More generally, the definition can easily be extended to incomplete desired judgment sets $J \subseteq J_i$ as well.

These notions are remotely inspired by "possible" vs "necessary winner" in voting theory due to Konczak and Lang [19] (see also the work of Xia and Conitzer [32]), and by "possible" vs "necessary envy-freeness" in fair division due to Bouveret et al. [8] (see also the papers by Brams et al. [9,10]). The stronger notion of *necessary strategy-proofness* corresponds to the "strategy-proofness" condition defined by Dietrich and List [13], whereas the weaker notion of *possible strategy-proofness* is introduced here. Note that since the Hamming-distance-induced preferences are a complete relation, we simply say that F is *strategy-proof with respect to Hamming-distance-induced preferences* if for each individual judge the actual outcome is at least as good as all outcomes obtained by reporting a different individual judgment set.

The result of Dietrich and List [13] says that an aggregation rule that satisfies the "universal domain" condition is necessarily strategy-proof with respect to non-strict closeness-respecting preferences if and only if it is independent and monotonic. *Universal domain* is satisfied if the domain of the aggregation function is the set of all possible profiles from $\mathcal{J}(\Phi)^n$, which obviously is true for *PBP*. *Independence* means that the collective decision on each proposition only relies on the individual judgments of this proposition. Since *PBP* derives the outcome for the conclusions from the outcome of the premises, it is not independent and hence not necessarily strategy-proof with respect to non-strict closeness-respecting preferences. An aggregation function is *monotonic* if additional support for some proposition that is currently accepted may never result in a non-acceptance for this formula, provided everything else remains unchanged. In the case where the agenda contains solely premises, *PBP* is independent and monotonic, and hence necessarily strategy-proof also for the case of strict closeness-respecting preferences.

Endriss et al. [14] initiated the study of the complexity of manipulation in judgment aggregation. Their work (and also the follow-up work of Baumeister et al. [6]) focuses only on preferences induced by the Hamming distance to the complete desired judgment set of the manipulator. We extend this study to the setting where the manipulator may be interested only in parts of the agenda, so her desired judgment set can be an incomplete subset of her true judgment set. For a given type $\mathcal{T} \in \{U, TR, CR\}$ of preference induced by the desired judgment set $J \subseteq J_n$ (i.e., judge n is the manipulator), we define the manipulation problem $\mathcal{T}$-NECESSARY-MANIPULATION as follows:

$\mathscr{T}$-NECESSARY-MANIPULATION

Given: An agenda Φ, a profile $\mathbf{T} = (J_1, \ldots, J_n) \in \mathscr{J}(\Phi)^n$, and the manipulator's desired consistent (possibly incomplete) judgment set $J \subseteq J_n$.

Question: Does there exist a judgment set $J^* \in \mathscr{J}(\Phi)$ such that

$$PBP(J_1, \ldots, J_{n-1}, J^*)|_J >_{\mathscr{T}}^J PBP(J_1, \ldots, J_n)|_J$$

for all extensions $>_{\mathscr{T}}^J$ that are consistent with $\succ_{\mathscr{T}}^J$?

Here, $PBP(J_1, \ldots, J_n)|_J$ denotes the restriction of $PBP(J_1, \ldots, J_n)$ to the formulas that occur, negated or not, in the desired judgment set J. In $\mathscr{T}$-POSSIBLE-MANIPULATION, we ask whether

$$PBP(J_1, \ldots, J_{n-1}, J^*)|_J >_{\mathscr{T}}^J PBP(J_1, \ldots, J_n)|_J$$

for some extension $>_{\mathscr{T}}^J$ that is consistent with $\succ_{\mathscr{T}}^J$. In the case of Hamming-distance-induced preferences we will simply say HD-MANIPULATION. Furthermore, we introduce and study the exact variant, EXACT-MANIPULATION, where the manipulator seeks to achieve not only a better, but a *best* outcome for a given subset of her desired judgment set. Here, the question is whether there is some judgment set $J^* \in \mathscr{J}(\Phi)$ such that

$$J \subseteq PBP(J_1, \ldots, J_{n-1}, J^*).$$

We assume the reader is familiar with complexity classes such as P and NP and the notion of NP-completeness (w.r.t. the polynomial-time many-one reducibility, $\leq_m^p$).

3.2   Results

Theorem 1. EXACT-MANIPULATION *is* NP-*complete, even for only three judges.*

Proof.   The proof is by a reduction from the NP-complete satisfiability problem. Let φ be a given formula in conjunctive normal form, where the clauses are built from the set $A = \{\alpha_1, \ldots, \alpha_m\}$ of variables. The question is whether there is a satisfying assignment for this formula. Without loss of generality, we may assume that neither setting all variables to true, nor setting all variables to false is a satisfying assignment for φ. Now construct an agenda Φ that consists of the variables in A and their negations, an additional variable β and its negation, and the formula $\varphi \vee \beta$ and its negation. The profile $\mathbf{T}$ consists of three judges. The individual judgment set of the first one contains A and $\neg\beta$ and the individual judgment set of the second one contains $\neg\alpha_i$ for each i, $1 \leq i \leq m$, and $\neg\beta$. The third judge is the manipulative one and his individual judgment set contains A and β. The desired outcome he tries to achieve exactly consists of only the conclusion $\varphi \vee \beta$. It holds that

$$PBP(\mathbf{T}) = A \cup \{\neg\beta\} \cup \{\neg(\varphi \vee \beta)\}.$$

Note also that the third judge is decisive for every formula in A, and that independently of the individual judgment set of the manipulator, β is never contained in the collective judgment set. Hence, the only way to obtain the conclusion $\varphi \vee \beta$ in the collective outcome is to evaluate the formula φ to true. This implies that there is a satisfying assignment for φ if and only if the individual judgment set of the third judge can be modified such that $\varphi \vee \beta$ is contained in the collective outcome.   ❑

Theorem 2.   *1.* EXACT-MANIPULATION $\leq_m^p$ $\mathscr{T}$-NECESSARY-MANIPULATION *for each type* $\mathscr{T} \in \{TR,CR\}$.
2. EXACT-MANIPULATION $\leq_m^p$ $\mathscr{T}$-POSSIBLE-MANIPULATION *for each type* $\mathscr{T} \in \{U,TR,CR\}$.
3. EXACT-MANIPULATION $\leq_m^p$ HD-MANIPULATION.

Proof.   For the exact problem, we have an agenda Φ, some profile $\mathbf{T} = (J_1,\dots,J_n)$, and some desired judgment set $J = \{\alpha_1,\dots,\alpha_m\} \subseteq J_n$, and we are looking for a modified judgment set J_n^* such that

$$J \subseteq PBP(J_1,\dots,J_{n-1},J_n^*).$$

In the trivial case that $J \subseteq PBP(\mathbf{T})$, $J_n^* = J_n$ obviously fulfills the requirement, so we can construct an arbitrary yes-instance for the corresponding manipulation problem. We will prove all three assertions via the same reduction, but using different arguments.

Assume that $J \setminus PBP(\mathbf{T}) \neq \emptyset$ and consider the following problem. Fix some $\mathscr{T} \in \{TR,CR,HD\}$, let the agenda Φ' be the union of Φ, the formula $\varphi = \alpha_1 \wedge \cdots \wedge \alpha_m$, and its negation. Let $\mathbf{T}' \in \mathscr{J}(\Phi')^n$ be the consistent extensions of $\mathbf{T}$. In particular, $J_n' = J_n \cup \varphi$. Let the desired judgment set be $J' = \varphi$, and we are looking for a modified judgment set $J_n'^*$ such that for all extensions $>_{\mathscr{T}}^{J'}$ of $\succ_{\mathscr{T}}^{J'}$, we have

$$PBP(J_1',\dots,J_{n-1}',J_n'^*)|_{J'} >_{\mathscr{T}}^{J'} PBP(J_1',\dots,J_{n-1}',J_n')|_{J'}.$$

Since J' consists of the single formula φ, there are only two different collective outcomes when restricted to J'. Since $\varphi \subseteq J_n$, it obviously holds that $\varphi \succ_{\mathscr{T}}^{J'} \neg\varphi$ for all $\mathscr{T} \in \{TR,CR,HD\}$, and since in this case $\succ_{\mathscr{T}}^{J'}$ is complete, there is no difference between the notions of necessary and possible preference. In the case of unrestricted preferences and the possible manipulation problem, we ask whether there is some different outcome, since they all may be possibly preferred. Since there is some J_n^* with

$$J \subseteq PBP(J_1,\dots,J_{n-1},J_n^*)$$

if and only if there is some $J_n'^*$ with

$$\varphi \subseteq PBP(J_1',\dots,J_{n-1}',J_n'^*),$$

the reduction works in all cases.   ❑

This reduction requires a partial desired judgment set for $\mathscr{T}$-NECESSARY-MANIPULATION, $\mathscr{T}$-POSSIBLE-MANIPULATION, and HD-MANIPULATION; together with Theorem 1, this implies NP-completeness of HD-MANIPULATION, $\mathscr{T}$-NECESSARY-MANIPULATION for $\mathscr{T} \in \{TR,CR\}$, and $\mathscr{T}$-POSSIBLE-MANIPULATION for $\mathscr{T} \in \{U,TR,CR\}$ whenever the desired judgment set of the manipulator is incomplete. Alternatively, the reduction given by Endriss et al. [14] in fact shows NP-completeness for HD-MANIPULATION even if the desired judgment set of the manipulator is complete.

Proposition 1. *For $\mathcal{T} \in \{U,TR\}$, $\mathcal{T}$-POSSIBLE-MANIPULATION can be solved in polynomial time if the desired judgment set of the manipulator is complete.*

Proof. This result holds, since a U-POSSIBLE-MANIPULATION instance is positive exactly if there is some premise from the desired judgment set for which the manipulator is decisive, i.e., the collective outcome depends on the decision of the manipulator. For a TR-POSSIBLE-MANIPULATION instance to be positive, it must additionally be required that the desired judgment set is not the actual outcome.   ❑

Proposition 2. *PBP is possibly strategy-proof when closeness-respecting preferences are assumed and the desired judgment set of the manipulator is complete.*

Proof. If closeness-respecting preferences are assumed, a judgment set that is necessarily preferred to the actual collective outcome must preserve all agreements between the desired judgment set and the actual outcome. Now consider a premise α that is contained in the collective judgment set, but $\sim\alpha$ is contained in the desired judgment set. It can obviously never be the case that a switch from the manipulator to α causes $\sim\alpha$ to be in the collective judgment set. Hence there can be no additional agreement among the premises. Since the desired judgment set is complete and the outcome for the conclusions depends solely on the outcome of the premises, *PBP* is possibly strategy-proof in this case.   ❑

Note that this does not contradict the results of Dietrich and List [13], since they impose different conditions on nonmanipulability and non-strict preferences.

4   Control by Bundling Judges

Previous work on control in judgment aggregation (see [3,4]) considered the problems of control by adding, deleting, or replacing judges. Although adding and deleting judges is inspired by the corresponding control problems in voting, explicit examples for such control actions in judgment aggregation are given, and the third type, control by replacing judges, was motivated by real-world examples from international arbitration. We here introduce another type of control motivated by real-world scenarios, *control by bundling judges*, which is remotely akin to control by partitioning voters in voting. A prominent natural example for control by bundling judges can be found in European legislation. Certain European legislative acts, such as Directives, give considerable freedom to Member States regarding the concrete implementation of these acts. Yet, in some cases uniform implementation is crucial, so the basic act confers implementing powers on the European Commission or the Council of the European Union to adopt the required implementing acts.[1] The exercise of implementing powers through the Commission and Council is controlled by the member states through so-called comitology committees in accordance with previously specified rules.[2] The committees are set

---

[1] Article 291 of the Treaty on the Functioning of the European Union.

[2] Regulation (EU) No 182/2011 of the European Parliament and of the Council of 16 February 2011 laying down the rules and general principles concerning mechanisms for control by Member States of the Commission's exercise of implementing powers (Implementing Acts Regulation).

up by the basic act in question.[3] Some of these committees are concerned with such a broad range of issues that they are divided into subcommittees, each of which is dealing with different issues. When preparing implementing acts covering several issues, each subcommittee votes on the issues assigned to it, and the implementing act is shaped according to the decisions of the different subcommittees.[4]

4.1 Definitions

The problem EXACT CONTROL BY ADDING JUDGES asks, given an agenda Φ, two complete profiles $\mathbf{T} \in \mathcal{J}(\Phi)^n$ and $\mathbf{S} \in \mathcal{J}(\Phi)^{\|S\|}$, a positive integer k, and a desired judgment set J (which may be incomplete, i.e., $J \subseteq J'$ for some $J' \in \mathcal{J}(\Phi)$), whether there is a subset $\mathbf{S}' \subseteq \mathbf{S}$ of the potential new judges of size at most k, which can be added such that $J \subseteq PBP(\mathbf{T} \cup \mathbf{S}')$. The variant of this problem asking for a preferred outcome when Hamming-distance-induced preferences are assumed will be denoted by CONTROL BY ADDING JUDGES. The problem EXACT CONTROL BY DELETING JUDGES asks, given an agenda Φ, a complete profile $\mathbf{T} \in \mathcal{J}(\Phi)^n$, a positive integer k, and a desired (possibly incomplete) judgment set J, whether it is possible to delete at most k judges from $\mathbf{T}$ such that J is a collective outcome, and the corresponding problem CONTROL BY DELETING JUDGES asks, for the same input, whether there is a preferred outcome when Hamming-distance-induced preferences are assumed.

When analyzing the complexity of these problems, Baumeister et al. [3,4] follow the terminology introduced by Bartholdi, Tovey, and Trick [2] for control problems in voting. For a given judgment aggregation procedure F (such as PBP) and a given control type $\mathscr{C}$ (such as those defined above), F is said to be

- *immune to control by* $\mathscr{C}$ if it is never possible to successfully exert this type of control,
- *susceptible to control by* $\mathscr{C}$ if F is not immune,
- *vulnerable to control by* $\mathscr{C}$ if F is susceptible to control by $\mathscr{C}$ and the corresponding decision problem is in P, and
- *resistant to control by* $\mathscr{C}$ if F is susceptible to control by $\mathscr{C}$ and the corresponding decision problem is NP-hard.

Baumeister et al. [3,4] have shown that the premise-based procedure is resistant to control by adding judges, to control by deleting judges, and to control by replacing judges (which in some sense combines control by deleting with control by adding judges) when preferences are assumed to be Hamming-distance-induced and in the exact variant. We will study the new problem of CONTROL BY BUNDLING JUDGES also in these two variants for the premise-based procedure. The formal definition for the Hamming-distance-induced version is as follows. In the problem definition below, we will use the notation

$$\Delta = \bigcup_{1 \leq i \leq k} PBP(\mathbf{T}|_{\Phi_p^i, N_i}),$$

---

[3] Recital 6 of the Preamble of Implementing Acts Regulation.

[4] One example is the Customs Code Committee, see Articles 1 (1) and 5 (7) (8) of the Rules of procedure for the Customs Code Committee.

where $PBP(\mathbf{T}|_{\Phi_p^i, N_i})$ is the collective judgment set obtained by restricting the agenda to Φ_p^i and the set of judges to $N_i \subseteq N$.

CONTROL BY BUNDLING JUDGES

Given: An agenda Φ, where the premises are partitioned into k subsets $\Phi_p^1, \ldots, \Phi_p^k$, a complete profile $\mathbf{T} \in \mathcal{J}(\Phi)^n$, and a consistent and complement-free judgment set J (not necessarily complete).

Question: Is there a partition $N_1, \ldots, N_k$ of the n judges such that

$$H(J, \Delta \cup \{\varphi \in \Phi_c \mid \Delta \models \varphi\}) < H(J, PBP(\mathbf{T}))?$$

In EXACT CONTROL BY BUNDLING JUDGES we ask, for the same input, whether there is a partition $N_1, \ldots, N_k$ of the n judges such that

$$J \subseteq \Delta \cup \{\varphi \in \Phi_c \mid \Delta \models \varphi\}).$$

Example 2. Consider the same variables a, b, c, and d and the same individual judgment sets as in Example 1. Assume that the set of premises is partitioned into $\Phi_1^p = \{a, b\}$ and $\Phi_2^p = \{c, d\}$, and that the desired judgment set J contains

$$a \vee b, \quad b \vee c, \quad \neg(a \vee c), \quad \text{and} \quad b \vee d.$$

Note that this is a consistent judgment set, since it can be reached by accepting b and the negation of all other variables. The Hamming distance between the current collective outcome and J is 3. But if we partition the set of judges into two groups, where the first judge forms the first group and the last two judges are in the second group, the outcome is as shown in Table 2, where the individual judgments for a single variable not belonging to the group who decides over this variable are marked with $\mathbb{1}$ or $\emptyset$. Recall that the negative literal is contained in the collective judgment set in case of a tie by convention.

Table 2. Example for CONTROL BY BUNDLING JUDGES

	a	b	c	d		$a \vee b$	$b \vee c$	$a \vee c$	$b \vee d$
Judge 1	1	1	$\emptyset$	$\emptyset$		1	1	1	1
Judge 2	$\emptyset$	$\emptyset$	0	0		0	0	0	0
Judge 3	$\mathbb{1}$	$\emptyset$	1	1		1	1	1	1
PBP	1	1	0	0	$\Rightarrow$	1	1	1	1

After bundling the judges, the Hamming distance between the collective outcome and J has decreased to 1. Hence, this is a positive instance of CONTROL BY BUNDLING JUDGES. However, since it is not possible to bundle the jugdes into two groups to obtain exactly J as a subset of the collective outcome, it is a negative instance of EXACT CONTROL BY BUNDLING JUDGES.

Remotely related bundling problems in judgment aggregation have recently been studied by Alon et al. [1]. However, their setting is different from ours. They consider judgment aggregation over independent variables, and only the variables are bundled in their bundling attacks. It is assumed that then all judges decide over all bundles by deciding uniformly for all variables contained in the same bundle. Furthermore, the goal in their model is to always accept all positive variables, that is, a complete desired judgment set. This setting in fact covers a restriction of judgment aggregation known as optimal lobbying (see, e.g., the papers by Christian et al. [12], Erdélyi et al. [16], and Bredereck et al. [11]).

4.2 Results

We show that the exact variant and the Hamming-distance-induced variant defined above are closely related. In fact, the proof of Lemma 1 below applies to all the control problems in judgment aggregation studied in the literature (control by adding, deleting, replacing, or bundling judges; for the formal definition of control by replacing judges, see [3,4]).

Lemma 1. *Let $\mathscr{C}$ be a control type.* EXACT CONTROL BY $\mathscr{C} \leq_m^p$ CONTROL BY $\mathscr{C}$.

Proof.   In the exact problem variant, we have an agenda Φ, some profile $\mathbf{T}$, and some desired judgment set $J = \{\alpha_1, \ldots, \alpha_m\}$, and we are looking for a modified profile $\mathbf{U}$ such that $PBP(\mathbf{U}) = J$. Now consider the following problem. Let the agenda Φ' be the union of Φ, the formula $\varphi = \alpha_1 \wedge \cdots \wedge \alpha_m$, and its negation. Let $\mathbf{T}'$ and $\mathbf{U}'$ (both in $\mathscr{J}(\Phi')$) be the consistent extensions of, respectively, $\mathbf{T}$ and $\mathbf{U}$, and let $J' = \varphi$. In the trivial case that $PBP(\mathbf{T}) = J$, we have $H(J, PBP(\mathbf{T}')) = 0$. In the nontrivial case that $PBP(\mathbf{T}) \neq J$, we have $H(J, PBP(\mathbf{T}')) = 1$. This implies

$$H(J, PBP(\mathbf{U}')) < H(J, PBP(\mathbf{T}')) \text{ if and only if } PBP(\mathbf{U}') = J',$$

so $H(J, PBP(\mathbf{U}')) < H(J, PBP(\mathbf{T}'))$ is equivalent to $PBP(\mathbf{U}) = J$.   ❑

Note that the above proof requires the desired judgment set of the Hamming-distance-induced variant to be incomplete. Note further that Lemma 1 implies that NP-hardness of a Hamming-distance-induced variant is inherited from NP-hardness of the corresponding exact problem variant.

The problem CONTROL BY BUNDLING JUDGES is somewhat similar to the problem of CONTROL BY DELETING JUDGES. We will exploit this in the following proof.

Theorem 3. *PBP is resistant to* EXACT CONTROL BY BUNDLING JUDGES *and to* CONTROL BY BUNDLING JUDGES.

Proof.   The proof will be by a reduction from the related problem EXACT CONTROL BY DELETING JUDGES. Given an agenda $\Phi = \Phi_p \cup \Phi_c$, a complete profile $\mathbf{T} \in \mathscr{J}(\Phi)^n$, and a positive integer k that is the bound on the number of judges that may be deleted. We assume that the individual judgment set of the manipulator is J_n, and $J \subseteq J_n$ is the desired judgment set. Now, we construct an instance of EXACT CONTROL BY BUNDLING JUDGES, resistance for CONTROL BY BUNDLING JUDGES then follows from Lemma 1. Without loss of generality, we assume that $n \geq k + 2$. The agenda is $\Phi' = \Phi \cup \{\alpha, \neg\alpha\}$, and is divided into two subsets. The first one consists of Φ_p, and the second one is $\{\alpha, \neg\alpha\}$. The profile $\mathbf{S} \in \mathscr{J}(\Phi')^{n+k+1}$ contains the individual judgment sets from $\mathbf{T}$, each extended by $\neg\alpha$. Furthermore, there are $k + 1$ new individual judgment sets that each contain $\varphi \in \Phi_p$ if and only if $\sim\varphi \in J$, they each contain α, and the conclusions are evaluated accordingly. These $k + 1$ new judges will be denoted by N'. The desired judgment set is $J' = J \cup \{\alpha\}$. We show that it is possible to obtain the desired judgment set J by deleting at most k judges from $\mathbf{T}$ if and only if the judges from $\mathbf{S}$ can be bundled into two groups such that the desired outcome is J'.

For the direction from left to right, assume that there is a subset $\mathbf{T}' \subseteq \mathbf{T}$, $\|\mathbf{T}'\| \leq k$, such that $PBP(\mathbf{T} \setminus \mathbf{T}') = J$. Then the judges can be bundled as follows. The $k + 1$ new judges and the judges corresponding to $\mathbf{T}'$ decide over α. Then obviously α is contained in the collective outcome, hence the constructed instance is a positive one for EXACT CONTROL BY BUNDLING JUDGES.

For the direction from right to left, assume that the judges can be bundled into N_1 and N_2 such that the collective outcome is J'. Hence, it holds that $PBP(\mathbf{S}|_{\Phi, N_1}) = J$. We will show that $\|N_2 \setminus N'\| \leq k$ and $PBP(\mathbf{S}|_{\Phi, N_1 \setminus N'}) = J$. Since α is contained in the collective judgment set and since there are only $k + 1$ judges having α in their individual judgment set, at most k of the initial judges can be in N_2. Due to the premise-based procedure, it is enough to show that $PBP(\mathbf{S}|_{\Phi_p, N_1}) = PBP(\mathbf{S}|_{\Phi_p, N_1 \setminus N'})$. This holds trivially, since for all judges from N_1 it holds that $\varphi \in \Phi_p$ is contained in the individual judgment set if and only if $\sim\varphi \in J$.   ☐

5   Conclusions and Future Work

To conclude, we investigated various forms of manipulation in judgment aggregation that originate from different assumptions on the incentives and the type of preferences of the manipulator. Our results show that whether one considers a judgment aggregation rule to be (necessarily or possibly) strategy-proof or manipulable crucially depends on the given setting. Table 3 summarizes our results for the various manipulation problems. The last two columns consider the Hamming-distance-induced preferences and the exact variant; note that there is no distinction between the possible and necessary manipulation problem for these preference types. The first two rows concern the general problem with an incomplete desired judgment set (abbreviated by DJS in the table), whereas the last two rows show the results for the restricted problem where the desired judgment set is required to be complete. We abbreviate "NP-complete" by "NP-c." All results stated in the table are new to this paper, except for the one for Hamming-distance-induced preferences with a complete desired judgment set, which is due to Endriss et al. [14].

We propose to launch a systematic study of the computational aspects of manipulation in judgment aggregation for complete and incomplete desired judgment sets, in particular by solving the open problems indicated by question marks in Table 3. Furthermore, the concepts studied here for manipulation can be transferred to other forms of interference as well, such as bribery and control [3,4,6]. Regarding the latter, we have proposed a new control type, *control by bundling judges*, which in some

Table 3. Overview of results for various manipulation problems

$\mathscr{T}$	U	TR	CR	HD	EXACT
$\mathscr{T}$-POSSIBLE-MANIPULATION for incomplete DJS	NP-c	NP-c	NP-c	NP-c	NP-c
$\mathscr{T}$-NECESSARY-MANIPULATION for incomplete DJS	?	NP-c	NP-c		
$\mathscr{T}$-POSSIBLE-MANIPULATION for complete DJS	in P	in P	?	NP-c [14]	strategy-proof
$\mathscr{T}$-NECESSARY-MANIPULATION for complete DJS	?	?	possibly strategy-proof		

sense corresponds to control by partitioning voters in voting. We have argued why this control type models a natural real-world scenario and showed that the premise-based procedure is resistant to it. It would be interesting to complement such worst-case complexity results by typical-case studies, or with respect to parameterized complexity, as has been done successfully in voting (see, e.g, the papers by Betzler and Uhlmann [7], Erdélyi et al. [15], Liu et al. [22,23], and Rothe and Schend [25,26,24] for control and the papers by Isaksson et al. [18], Friedgut et al. [17], Walsh [27,28,29], and Xia and Conitzer [30,31] for manipulation).

References

1. Alon, N., Falik, D., Meir, R., Tennenholtz, M.: Bundling attacks in judgment aggregation. In: Proceedings of the 27th AAAI Conference on Artificial Intelligence. AAAI Press (to appear, 2013)
2. Bartholdi III, J., Tovey, C., Trick, M.: How hard is it to control an election? Mathematical and Computer Modelling 16(8/9), 27–40 (1992)
3. Baumeister, D., Erdélyi, G., Erdélyi, O., Rothe, J.: Bribery and control in judgment aggregation. In: Brandt, F., Faliszewski, P. (eds.) Proceedings of the 4th International Workshop on Computational Social Choice, AGH University of Science and Technology, Kraków, Poland, pp. 37–48 (September 2012)
4. Baumeister, D., Erdélyi, G., Erdélyi, O., Rothe, J.: Control in judgment aggregation. In: Proceedings of the 6th European Starting AI Researcher Symposium, pp. 23–34. IOS Press (August 2012)
5. Baumeister, D., Erdélyi, G., Hemaspaandra, E., Hemaspaandra, L., Rothe, J.: Computational aspects of approval voting. In: Laslier, J., Sanver, R. (eds.) Handbook on Approval Voting, ch. 10, pp. 199–251. Springer (2010)
6. Baumeister, D., Erdélyi, G., Rothe, J.: How hard is it to bribe the judges? A study of the complexity of bribery in judgment aggregation. In: Brafman, R. (ed.) ADT 2011. LNCS, vol. 6992, pp. 1–15. Springer, Heidelberg (2011)
7. Betzler, N., Uhlmann, J.: Parameterized complexity of candidate control in elections and related digraph problems. Theoretical Comput. Sci. 410(52), 5425–5442 (2009)
8. Bouveret, S., Endriss, U., Lang, J.: Fair division under ordinal preferences: Computing envy-free allocations of indivisible goods. In: Proceedings of the 19th European Conference on Artificial Intelligence, pp. 387–392. IOS Press (August 2010)

9. Brams, S., Edelman, P., Fishburn, P.: Fair division of indivisible items. Theory and Decision 5(2), 147–180 (2004)
10. Brams, S., King, D.: Efiicient fair division—help the worst off or avoid envy? Rationality and Society 17(4), 387–421 (2005)
11. Bredereck, R., Chen, J., Hartung, S., Kratsch, S., Niedermeier, R., Suchý, O.: A multivariate complexity analysis of lobbying in multiple referenda. In: Proceedings of the 26th AAAI Conference on Artificial Intelligence, pp. 1292–1298. AAAI Press (2012)
12. Christian, R., Fellows, M., Rosamond, F., Slinko, A.: On complexity of lobbying in multiple referenda. Review of Economic Design 11(3), 217–224 (2007)
13. Dietrich, F., List, C.: Strategy-proof judgment aggregation. Economics and Philosophy 23(3), 269–300 (2007)
14. Endriss, U., Grandi, U., Porello, D.: Complexity of judgment aggregation. Journal of Artificial Intelligence Research 45, 481–514 (2012)
15. Erdélyi, G., Fellows, M., Rothe, J., Schend, L.: Control complexity in Bucklin and fallback voting. Technical Report arXiv:1103.2230 [cs.CC], Computing Research Repository, arXiv.org/corr/ (March 2012) (March 2011) (revised August 2012)
16. Erdélyi, G., Fernau, H., Goldsmith, J., Mattei, N., Raible, D., Rothe, J.: The complexity of probabilistic lobbying. In: Rossi, F., Tsoukias, A. (eds.) ADT 2009. LNCS, vol. 5783, pp. 86–97. Springer, Heidelberg (2009)
17. Friedgut, E., Keller, N., Kalai, G., Nisan, N.: A quantitative version of the Gibbard–Satterthwaite theorem for three alternatives. SIAM Journal on Computing 40(3), 934–952 (2011)
18. Isaksson, M., Kindler, G., Mossel, E.: The geometry of manipulation – A quantitative proof of the Gibbard-Satterthwaite theorem. Combinatorica 32(2), 221–250 (2012)
19. Konczak, K., Lang, J.: Voting procedures with incomplete preferences. In: Proceedings of the Multidisciplinary IJCAI 2005 Workshop on Advances in Preference Handling, pp. 124–129 (July/August 2005)
20. List, C.: The theory of judgment aggregation: An introductory review. Synthese 187(1), 179–207 (2012)
21. List, C., Puppe, C.: Judgment aggregation: A survey. In: Anand, P., Pattanaik, P., Puppe, C. (eds.) Oxford Handbook of Rational and Social Choice, ch. 19. Oxford University Press (2009)
22. Liu, H., Feng, H., Zhu, D., Luan, J.: Parameterized computational complexity of control problems in voting systems. Theoretical Comput. Sci. 410(27-29), 2746–2753 (2009)
23. Liu, H., Zhu, D.: Parameterized complexity of control problems in maximin election. Inf. Process. Lett. 110(10), 383–388 (2010)
24. Rothe, J., Schend, L.: Challenges to complexity shields that are supposed to protect elections against manipulation and control: A survey. Annals of Mathematics and Artificial Intelligence (to appear)
25. Rothe, J., Schend, L.: Control complexity in bucklin, fallback, and plurality voting: An experimental approach. In: Klasing, R. (ed.) SEA 2012. LNCS, vol. 7276, pp. 356–368. Springer, Heidelberg (2012)
26. Rothe J., Schend, L.: Control complexity in bucklin, fallback, and plurality voting: An experimental approach. Technical Report arXiv:1203.3967 [cs.GT], Computing Research Repository, arXiv.org/corr/ (March 2012) (revised August 2012)
27. Walsh, T.: Where are the really hard manipulation problems? The phase transition in manipulating the veto rule. In: Proceedings of the 21st International Joint Conference on Artificial Intelligence, IJCAI, pp. 324–329 (July 2009)
28. Walsh, T.: An empirical study of the manipulability of single transferable voting. In: Proceedings of the 19th European Conference on Artificial Intelligence, pp. 257–262. IOS Press (August 2010)

29. Walsh, T.: Is computational complexity a barrier to manipulation? Annals of Mathematics and Artificial Intelligence 62(1-2), 7–26 (2011)
30. Xia, L., Conitzer, V.: Generalized scoring rules and the frequency of coalitional manipulability. In: Proceedings of the 9th ACM Conference on Electronic Commerce, pp. 109–118. ACM Press (June 2008)
31. Xia, L., Conitzer, V.: A sufficient condition for voting rules to be frequently manipulable. In: Proceedings of the 9th ACM Conference on Electronic Commerce, pp. 99–108. ACM Press (June 2008)
32. Xia, L., Conitzer, V.: Determining possible and necessary winners given partial orders. Journal of Artificial Intelligence Research 41, 25–67 (2011)

Property-Based Preferences
in Abstract Argumentation

Richard Booth[1], Souhila Kaci[2], and Tjitze Rienstra[1,2]

[1] Université du Luxembourg
6 rue Richard Coudenhove-Kalergi, Luxembourg
{richard.booth,tjitze.rienstra,leon.vandertorre}@uni.lu
[2] LIRMM (CNRS/Université Montpellier 2)
161 rue Ada, Montpellier, France
souhila.kaci@lirmm.fr

Abstract. Many works have studied preferences in Dung-style argumentation. Preferences over arguments may be derived, e.g., from their relative specificity, relative strength or from values promoted by the arguments. An underexposed aspect in these models is change of preferences. We present a dynamic model of preferences in argumentation, centering on what we call property-based AFs. It is based on Dietrich and List's model of property-based preference and it provides an account of how and why preferences in argumentation may change. The idea is that preferences over arguments are derived from preferences over properties of arguments, and change as the result of moving to different motivational states. We also provide a dialogical proof theory that establishes whether there exists some motivational state in which an argument is accepted.

Keywords: argumentation, preferences, property-based, dialogue, Dung.

1 Introduction

Dung's theory of abstract argumentation [1] plays a central role in many approaches to reasoning and decision making in AI. It is based on the concept of an *argumentation framework* (AF, for short), i.e., a set of abstract *arguments* and a binary *attack relation* encoding conflict between arguments. The outcome of an AF is a set of justifiable points of view on the acceptability of its arguments, represented by extensions and computed under a given semantics, different semantics corresponding to different degrees of skepticism or credulousness.

Many works have recognized the importance of *preferences* in this setting. Preferences over arguments may be derived, e.g., from their relative specificity or from the relative strength of the beliefs with which they are built. On the abstract level preferences can be represented by *preference-based* AFs, which instantiate AFs with a preference relation over the set of arguments [2,3]. An attack of an argument x on y then *succeeds* only if y is not strictly preferred over x. *Value-based* AFs provide yet another account of how preferences are derived [4]. The idea here is

P. Perny, M. Pirlot, and A. Tsoukiàs (Eds.): ADT 2013, LNAI 8176, pp. 86–100, 2013.

that arguments promote certain *values* and that different *audiences* have different preferences over values, from which the preferences over arguments are derived.

An underexposed aspect in these models is change of preferences [5,6]. Preferences are usually assumed to be fixed and no account is provided of how or why they may change. We address this aspect by applying Dietrich and List's recently introduced model of *property-based preference* [7,8]. In this model, preferences over alternatives are derived from preferences over sets of properties satisfied by the alternatives. Furthermore, agents are assumed to have a motivational state, consisting of the properties on which the agent focuses in a given situation, when forming preferences over alternatives. The authors present an axiomatic characterization of their model, in terms of a number of reasonable constraints on the relationship between motivational states and preferences.

Our contribution is a new, dynamic model of preferences in argumentation, centering on what we call *property-based* AFs. It is based on the model of Dietrich and List and provides an account of how and why preferences in argumentation may change. Our model generalizes preference-based AFs as well as value-based AFs, if properties are used to represent values. We look at two types of acceptance, called *weak* and *strong* acceptance (i.e., acceptance in *some* or *all* motivational states). We also provide a dialogical proof theory that establishes whether an argument is weakly accepted. It is based on the grounded game [9] and extends it with dialogue moves consisting of properties.

The outline of this paper is as follows. We start in section 2 with some preliminaries concerning abstract argumentation theory. In section 3 we first give a brief outline of preference-based and value-based abstract argumentation. Then we give in section 4 an overview of the relevant parts of Dietrich and List's model of property-based preferences. We move on to our own work in section 5, where we present our model of property-based AFs, followed by a dialogical proof procedure for weak acceptance in section 6. We discuss some related work in section 7 and we conclude in section 8.

2   Preliminaries

We start out with some preliminaries concerning Dung's model of abstract argumentation [1]. We assume that argumentation frameworks are finite.

Definition 1. *An* argumentation framework *(AF for short) is a pair* $AF = (A, \rightarrow)$ *where* A *is a finite set of* arguments *and* $\rightarrow \subseteq A \times A$ *an* attack relation.

Given an AF $(A, \rightarrow)$ we say that x *attacks* y and also write $x \rightarrow y$ instead of $(x, y) \in \rightarrow$. The outcome of an AF consists of possible sets of arguments, called *extensions*. A *semantics* embodies a set of conditions that an extension must satisfy. The most studied ones are defined as follows:

Definition 2. *Let* $AF = (A, \rightarrow)$. *An* extension *of* AF *is a set* $E \subseteq A$. *We say that* E *is* conflict-free *iff* $\nexists x, y \in E$ *s.t.* $x \rightarrow y$; *that it* defends *an argument* $x \in A$ *iff* $\forall y \in A$ *s.t.* $y \rightarrow x$, $\exists z \in E$ *s.t.* $z \rightarrow y$; *and we define* $Def(E)$ *by*

$Def(E) = \{x \in A \mid E \text{ defends } x\}$. *An extension $E \subseteq A$ is said to be:* admissible *iff E is conflict free and $E \subseteq Def(E)$,* complete *iff E is conflict free and $E = Def(E)$,* stable *iff E is admissible and $\forall x \in A \setminus E, \exists y \in E$ s.t. $y \to x$,* preferred *iff E is maximal (w.r.t. set inclusion) among the set of admissible extensions of AF and* grounded *iff E is minimal (w.r.t. set inclusion) among the set of complete extensions of AF.*

Note that the grounded extension is unique and always exists, and represents the most skeptical viewpoint on the acceptability of the arguments in the AF. Although the concepts we introduce in this paper can be applied generally to all semantics, we will focus in this paper on the grounded semantics.

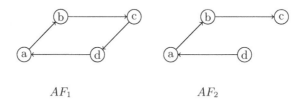

AF_1 AF_2

Fig. 1. Two argumentation frameworks

Example 1. Consider the AF AF_1 shown in figure 1 (nodes represent arguments and arrows represent attacks). The AF has three complete extensions, namely $\emptyset$, $\{a, c\}$ and $\{b, d\}$. The extension $\emptyset$ is also the grounded extension, while $\{a, c\}$ and $\{b, d\}$ are also stable and preferred extension. The AF AF_2 has a single complete extension namely $\{d, b\}$. This extension is thus also a grounded, stable and preferred extension.

3   Preferences and Values in Argumentation

Preference-based AFs [2] extend AFs with a preference relation over arguments, used to represent the relative strength of arguments. The idea is that an attack of an argument x on y succeeds only if y is not strictly preferred over (i.e., not stronger than) x. A preference-based AF *represents* a unique AF $(A, \to)$, where the attack relation $\to$ consists only of the attacks that succeed [10]. The extensions of a preference-based AF are those of the AF that it represents. Formally:

Definition 3. *A preference-based AF (PAF for short) is a triple $PAF = (A, \rightsquigarrow, \preceq)$ where A is a finite set of arguments, $\rightsquigarrow$ an attack relation and $\preceq$ a partial pre-order (i.e., a reflexive and transitive relation) or a total pre-order (i.e., a reflexive, transitive and complete relation) over A. A PAF $(A, \rightsquigarrow, \preceq)$ represents the AF $(A, \to)$ where $\to$ is defined by $\forall x, y \in A, x \to y$ iff $x \rightsquigarrow y$ and not $(x \prec y)$.*

Example 2. Consider the *PAF* $(A, \rightsquigarrow, \preceq)$ where A and $\rightsquigarrow$ are as in AF_1 in example 1 and $\preceq$ is a total pre-order defined by $x \preceq y$ iff $x \in \{b, c\}$ or $y \in \{a, d\}$. We have that $(A, \rightsquigarrow, \preceq)$ represents AF_2, shown in figure 1. This AF has one complete, grounded, stable and preferred extension, namely $\{d, b\}$.

Preference-based AFs give—at least at the abstract level—no account of how preferences over arguments are formed. Bench-Capon's [4] model of *value-based* AFs does. In a value-based AF, the idea is that arguments may promote certain *values* and that different *audiences* have different preferences over values, from which the preferences over arguments are derived. An *audience specific* value-based AF encodes a single audience's preferences over values.

Definition 4. *A value-based AF (VAF for short) is a 5-tuple $(A, \rightsquigarrow, V, val, U)$, where A is a set of arguments, $\rightsquigarrow$ an attack relation, V a set of values, $val : A \rightarrow V$ a mapping from arguments to values and U a set of audiences. An audience specific value-based AF (aVAF for short) is a 5-tuple $(A, \rightsquigarrow, V, val, <_a)$ where $a \in U$ is an audience and $<_a$ a partial order (i.e. an irreflexive and transitive relation) over V.*

An *aVAF* represents a unique *PAF* [10]:

Definition 5. *An aVAF $(A, \rightsquigarrow, V, val, <_a)$ represents the PAF $(A, \rightsquigarrow, \preceq)$, where $\preceq$ is defined by $\forall x, y \in A, x \preceq y$ iff $val(x) <_a val(y)$ or $val(x) = val(y)$.*

Since a *PAF* represents a unique AF, an *aVAF* also represents a unique AF. The extensions of an *aVAF* are the extensions of this AF.

Example 3. Consider the *aVAF* $(A, \rightsquigarrow, V, val, <_a)$ where A and $\rightsquigarrow$ are as in example 1, $V = \{blue, red\}$, $val(a) = val(d) = blue, val(b) = val(c) = red$ and $<_a$ is defined by $x <_a y$ iff $x = red$ and $y = blue$. It can be checked that this *aVAF* represents the *PAF* from example 2 and thus the AF AF_2 shown in figure 1.

4   Dietrich and List's Model of Property-Based Preference

Dietrich and List's model of *property-based preference* [7,8] aims at giving an account of rational choice that explains how preferences are formed and how they may change. This is opposed to traditional models that assume an agent's preferences over alternatives to be given and fixed. In this model, every alternative $x \in X$ is associated with a set $P(x)$ of *properties* satisfied by x, each $P(x)$ being a subset of a set $\mathcal{P}$ of possible properties. Furthermore, a set $\mathcal{M} \subseteq 2^{\mathcal{P}}$ of *motivational states* encodes sets of properties on which an agent may focus in a given situation. That is, if $M \in \mathcal{M}$ is the agent's state then only the properties in M matter to the agent when forming preferences over X. Change of preferences can then be understood as being caused by moving from one motivational state to another. Note that $\mathcal{M}$ may coincide with $2^{\mathcal{P}}$ but in general this need not be the case, as certain combinations of properties may be deemed inconsistent.

Every state $M \in \mathcal{M}$ gives rise to a preference order (i.e., a total pre-order) $\preceq_M$ over X representing the agent's preferences in the state M. There is thus a family $(\preceq_M)_{M \in \mathcal{M}}$ of preference orders over X. Strict and indifference relations $\prec_M$ and $\sim_M$ are defined as usual.

According to the model of property-based preference, preferences over X are formed using an underlying *weighing relation* $\leq$ over combinations of properties. This relation can be thought of as a 'betterness' relation, i.e., if $S \leq S'$ then the set of properties S' is at least as good as the set of properties S.

Definition 6. *A family* $(\preceq_M)_{M \in \mathcal{M}}$ *of preference orders is called* property-based *if there is a* weighing relation $\leq \subseteq 2^{\mathcal{P}} \times 2^{\mathcal{P}}$ *such that, for every* $M \in \mathcal{M}$ *and* $x, y \in X$, $x \preceq_M y$ *iff* $P(x) \cap M \leq P(y) \cap M$.

The authors present an axiomatic characterization of their model, in terms of two constraints on the relationship between motivational states and preferences.

Theorem 1. *[An axiomatic characterization [7]] Let* $(\preceq_M)_{M \in \mathcal{M}}$ *be a family of preference orders. Consider the following axioms:*

Axiom 1 $\forall x, y \in X$, $\forall M \in \mathcal{M}$, *if* $P(x) \cap M = P(y) \cap M$, *then* $x \sim_M y$.

Axiom 2 $\forall x, y \in X$, $\forall M, M' \in \mathcal{M}$ *s.t.* $M \subseteq M'$, *if* $P(x) \cap (M' \setminus M) = P(y) \cap (M' \setminus M) = \emptyset$ *then* $x \preceq_M y \leftrightarrow x \preceq_{M'} y$.

It holds that if $\mathcal{M}$ *is intersection-closed (i.e.* $M, M' \in \mathcal{M}$ *implies* $M \cap M' \in \mathcal{M})$ *then a family of preference orders* $(\preceq_M)_{M \in \mathcal{M}}$ *satisfies axioms 1 and 2 iff it is property-based.*

Axiom 1 says that the preference relation is indifferent on pairs of alternatives that have the same properties that are at the same time motivational, while axiom 2 says that preferences on pairs of alternatives change only if additional properties become motivational that are satisfied by at least one of the alternatives. A third axiom, strengthening the second and concerned with the class of *separable* weighing relations may be considered as well. The reader is referred to Dietrich and List [7] for details.

5 Property Based AFs

The value-based AF model gives an account of where an agent's (or audience's) preferences over arguments come from, namely the relative importance of the values they promote. However, it gives no account of how or why they may change. This motivates us to apply the model of property-based preference in argumentation, giving rise to what we call *property-based AFs*. In a property-based AF, each argument is associated with a set of properties that it satisfies. Among the types of properties we may consider are values promoted by the argument.

Furthermore, a property-based AF consists of a set of motivational states $\mathcal{M}$ and a weighing relation $\leq$ over sets of properties. The idea is as before: $\leq$ encodes the agent's preferences over sets of properties but only properties in the agent's state $M \in \mathcal{M}$ matter when forming preferences over arguments.

Definition 7. *A property-based AF is a 6-tuple $(A, \leadsto, \mathcal{P}, P, \mathcal{M}, \leq)$ where A is a set of arguments, $\leadsto$ an attack relation, $\mathcal{P}$ is a set of properties, $P : A \to 2^{\mathcal{P}}$ a mapping of arguments to sets of properties, $\mathcal{M} \subseteq 2^{\mathcal{P}}$ is an intersection-closed set of motivational states and $\leq \subseteq 2^{\mathcal{P}} \times 2^{\mathcal{P}}$ a reflexive, transitive and complete weighing relation.*

Note that there are cases where $\leq$ does not need to be transitive and complete over all sets of properties. For simplicity, however, we assume that it is. The reader is referred to Dietrich and List [7, Remark 1] for details.

If we focus on values as properties then the weighing relation can be understood as encoding the relative importance that an agent associates with different combinations of values, and the motivational state as consisting of the values of which an agent is aware in a given situation.

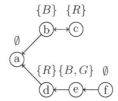

Fig. 2. An argumentation framework

Given a property-based AF, each motivational state $M \in \mathcal{M}$ represents a unique *PAF* which we denote by PAF_M. Preferences in PAF_M are formed by comparing sets of properties satisfied by the arguments, that are at the same time motivational. The AF according to which the agent determines the extensions in the motivational state M, denoted by AF_M, is the AF represented by PAF_M.

Definition 8. *Given a property-based AF $(A, \leadsto, \mathcal{P}, P, \mathcal{M}, \leq)$ and a motivational state $M \in \mathcal{M}$ we say that:*
- *M represents the $PAF_M = (A, \leadsto, \preceq)$, where $\preceq$ is defined by $\forall x, y \in A, x \preceq y$ iff $P(x) \cap M \leq P(y) \cap M$.*
- *M represents the AF $AF_M = (A, \to_M)$, which is the AF represented by PAF_M.*

Given an attack $x \leadsto y$ and state $M \in \mathcal{M}$, we say that $x \leadsto y$ is enabled *(otherwise* disabled*) in M iff $x \to_M y$.*

Let us illustrate the definitions with an example.

Example 4. Consider the property-based AF $(A, \rightsquigarrow, \mathcal{P}, P, \mathcal{M}, \leq)$ where A and $\rightsquigarrow$ and the properties assigned by P to the arguments are as shown in figure 2. Furthermore, $\mathcal{P} = \{R, G, B\}$, $\mathcal{M} = 2^{\mathcal{P}}$ and $\leq$ is defined via a weight function $w : \mathcal{P} \to \mathbb{Z}$ with $w(R) = w(G) = 1$ and $w(B) = -2$ as follows: $X \leq X'$ iff $\sum_{x \in X} w(x) \leq \sum_{x \in X'} w(x)$. This gives rise to the weighing relation $\{B\} < \{R, B\} = \{G, B\} < \{R, G, B\} = \emptyset < \{R\} = \{G\} < \{R, G\}$, where $<$ is the strict counterpart of $\leq$.

Figure 3 shows the AFs represented by all possible motivational states. We have, e.g., that in $PAF_{\{G\}}$ the argument e is strictly preferred over f, so that the attack from f to e is disabled $AF_{\{G\}}$. On the other hand, in $PAF_{\emptyset}$ and $PAF_{\{B,G\}}$ the argument e is not preferred over f. Here, the attack from f to e succeeds and is therefore enabled in $AF_{\emptyset}$ and $AF_{\{B,G\}}$.

Arguments in the AFs in figure 3 that are a member of the grounded extension of the respective AFs are shown black. We can see, e.g., that a is accepted only in the motivational state $\{R, G\}$.

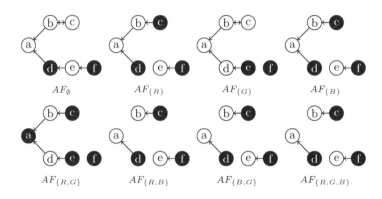

Fig. 3. AFs represented by all motivational states in example 4

We should remark that in many systems of argumentation, arguments have (in)formal 'logical content'. As a result, conflicts between arguments cannot generally be disregarded, on pain of inconsistency of the AF's outcome. This can be taken into account by requiring, for example, the relation $\rightsquigarrow$ to be symmetric, representing a conflict relation over two arguments, i.e. both arguments cannot be accepted together. In this way one attack between a pair of arguments always remains enabled.

Apart from looking at acceptance of arguments in a given motivational state, we can look at acceptance of arguments in some or all possible states. We will say that an argument is *weakly* (resp. *strongly*) accepted iff it is a member of the grounded extension given some (resp. all) motivational states. Weak acceptance thus means that the agent may accept an argument, namely when she moves to the right motivational state, whereas strong acceptance means that an agent accepts an argument regardless of her motivational state.

Definition 9. *Let* $(A, \leadsto, \mathcal{P}, P, \mathcal{M}, \leq)$ *be a property-based AF and* $x \in A$ *an argument. We say that* x *is* weakly accepted (*resp.* strongly accepted) *iff* x *is a member of the grounded extension of* AF_M *for some (resp. all)* $M \in \mathcal{M}$.

Example 5 (Continued from example 4). All arguments except b are weakly accepted. Only f is strongly accepted.

The following properties follow directly from theorem 1.

Proposition 1. *Let* $(A, \leadsto, \mathcal{P}, P, \mathcal{M}, \leq)$ *be a property-based AF. We have:*

Property 1 $\forall x, y \in A$ *s.t.* $x \leadsto y$, $\forall M \in \mathcal{M}$ *s.t.* $P(x) \cap M = P(y) \cap M, x \rightarrow_M y.$

Property 2 $\forall x, y \in A$, $\forall M, M' \in \mathcal{M}$ *s.t.* $M \subseteq M'$, *if* $P(x) \cap (M' \setminus M) = P(y) \cap (M' \setminus M) = \emptyset$ *then* $x \rightarrow_M y$ *iff* $x \rightarrow_{M'} y.$

Property 1 states that an attack $x \leadsto y$ is enabled in a motivational state M if x and y have the same set of properties that are also motivational in M, while property 2 states that an attack between two arguments x and y changes only if additional properties become motivational that are satisfied either by x or by y.

6 A Dialogical Proof Theory for Weak Acceptance

In this section we present a proof procedure to establish weak acceptance of an argument in a property-based AF. It is a dialogical proof procedure because it is based on generating dialogues where two players (PRO and OPP) take alternating turns in putting forward attacks according to a certain set of rules. This is similar in spirit to the *grounded game*, a dialogical proof procedure that establishes an argument's membership of the grounded extension [9]. In the grounded game, PRO repeatedly puts forward arguments (either as an initial claim or in defence against OPP's attacks) and OPP can initiate different disputes by putting forward possible attacks on the arguments put forward by PRO. PRO wins iff it can end every dispute in its favor according to a "last-word" principle.

By contrast, the proof procedure we present simply generates dialogues won by PRO. Such dialogues represent proofs that the initial argument is weakly accepted, and are structured as single sequences of moves where PRO and OPP put forward attacks and, in addition, PRO puts forward properties. If the procedure generates no dialogues then the argument is not weakly accepted.

Dialogical proof procedures make it possible to relate a semantics to a stereotypic pattern of dialogue. It has been shown, e.g., that the grounded and preferred credulous semantics can be related to persuasion and socratic style dialogue [11,12]. Dialogues generated by our procedure can also be thought of as persuasion dialogues, where PRO has the additional freedom to change the motivational state of the players by putting forward properties. Intuitively, this may benefit PRO in two ways: PRO can enable attacks necessary to put up a successful line of defence, and disable attacks put forward by the opponent from which PRO cannot defend its own arguments. PRO thus persuades OPP to accept an argument, where PRO decides which properties become motivational. Dialogues are structured as follows.

Definition 10. *Let* $(A, \rightsquigarrow, \mathcal{P}, P, \mathcal{M}, \leq)$ *be a property-based AF. A dialogue is a sequence* $S = (m_1, \ldots, m_n)$, *where each* m_i *is either:*
- *an attack move* "***OPP:*** $x \rightsquigarrow y$", *where* $x, y \in A$ *and* $x \rightsquigarrow y$,
- *a defence move* "***PRO:*** $x \rightsquigarrow y$", *where* $x, y \in A$ *and* $x \rightsquigarrow y$,
- *an enabling property move* "***PRO:*** $P+$", *where* $P \subseteq \mathcal{P}$,
- *a disabling property move* "***PRO:*** $P-$", *where* $P \subseteq \mathcal{P}$,
- *a conceding move* "***OPP: ok***",
- *a success claim move* "***PRO: win***".

We denote by $S \cdot S'$ *the concatenation of* S *and* S' *and we say that* S *is a subsequence of* S' *iff* $S' = S'' \cdot S \cdot S'''$ *for some* S'', S''', *and that* S *is a* proper *subsequence of* S' *iff* $S' = S'' \cdot S \cdot S'''$ *for nonempty* S'' *or* S'''.

Definition 11. *Let* $S = (m_1, \ldots, m_n)$ *be a dialogue. We denote the motivational state in* S *at index* i *by* M_i^S, *defined recursively by:*

$$M_i^S = \begin{cases} \emptyset & \text{if } i = 0, \\ M_{i-1}^S \cup P & \text{if } m_i = \textbf{\textit{PRO:}} \, P+ \text{ or } m_i = \textbf{\textit{PRO:}} \, P-, \\ M_{i-1}^S & \text{otherwise.} \end{cases}$$

We now define a set of production rules that generate *weak x-acceptance dialogues*. Note that AFs containing cycles may generate infinite sequences of moves. We prevent this by requiring dialogues to be finite.

Definition 12 (Weak acceptance dialogue). *Let* $(A, \rightsquigarrow, \mathcal{P}, P, \mathcal{M}, \leq)$ *be a property-based AF and let* $x \in A$.
- *A weak x-acceptance dialogue is a finite sequence*

$$S_1 \cdot (\textbf{\textit{PRO: win}})$$

   where S_1 *is an x-attack sequence.*
- *An x-attack sequence is a sequence*

$$(\textbf{\textit{OPP:}} \, y_1 \rightsquigarrow x) \cdot S_1 \cdot \ldots \cdot (\textbf{\textit{OPP:}} \, y_n \rightsquigarrow x) \cdot S_n \cdot (\textbf{\textit{OPP: ok}})$$

   where $\{y_1, \ldots, y_n\} = \{y \mid y \rightsquigarrow x\}$ *and each* S_i *is a* y_i-*defence sequence.*
- *An x-defence sequence is either:*
  - *a regular x-defence sequence*

$$(\textbf{\textit{PRO:}} \, y \rightsquigarrow x) \cdot S_1$$

     for some $y \in A$ *s.t.* $y \rightsquigarrow x$, *where* S_1 *is a* y-*attack sequence,*
  - *an enabling property defence sequence*

$$(\textbf{\textit{PRO:}} \, P+) \cdot S_1$$

     for some $P \subseteq \mathcal{P}$, *where* S_1 *is a regular x-defence sequence,*

- *a* disabling property defence sequence

$$(\textbf{PRO:}\, P-)$$

for some $P \subseteq \mathcal{P}$.

Intuitively, a disabling property move can be interpreted as saying "the preceding move is invalid considering the properties P." An enabling property move, on the other hand, says "the following move is valid considering the properties P." Not every weak x-acceptance dialogue, generated by the production rules in definition 12, will follow this interpretation. We need to impose a number of additional constraints to ensure that property moves make sense.

Definition 13 (Property-consistency). Let $(A, \rightsquigarrow, \mathcal{P}, P, \mathcal{M}, \leq)$ be a property-based AF and $S = (m_1, \ldots, m_n)$ a sequence. We say that S is property-consistent iff for all $i \in [1, \ldots, n]$, we have:

1. $M_i^S \in \mathcal{M}$
2. If $m_i = \textbf{PRO:}\, x \rightsquigarrow y$ then for all $j \in [i, \ldots, n]$, $x \rightarrow_{M_j^S} y$,
3. If $m_i = \textbf{PRO:}\, P-$ and $m_{i-1} = \textbf{OPP:}\, x \rightsquigarrow y$ then for all $j \in [i, \ldots, n]$, $x \not\rightarrow_{M_j^S} y$.

Condition 1 ensures that property moves are valid in the sense that they actually lead to a new motivational state $M \in \mathcal{M}$. Conditions 2 and 3 ensure that a property move does not undermine preceding property moves. That is, condition 2 ensures that attacks put forward by PRO remain enabled in subsequent states and condition 3 ensures that disabled attacks remain disabled.

Example 6 (Continued from example 4). Consider the following two property-consistent weak acceptance dialogues for the argument *a*.

Index	Move	State		Index	Move	State
1	**OPP:** $b \rightsquigarrow a$	$\emptyset$		1	**OPP:** $b \rightsquigarrow a$	$\emptyset$
2	**PRO:** $c \rightsquigarrow b$	$\emptyset$		2	**PRO:** $c \rightsquigarrow b$	$\emptyset$
3	**OPP:** $b \rightsquigarrow c$	$\emptyset$		3	**OPP:** $b \rightsquigarrow c$	$\emptyset$
4	**PRO:** $\{R\}-$	$\{R\}$		4	**PRO:** $\{R, G\}-$	$\{R, G\}$
5	**OPP:** ok	$\{R\}$		5	**OPP:** ok	$\{R, G\}$
6	**OPP:** $d \rightsquigarrow a$	$\{R\}$		6	**OPP:** $d \rightsquigarrow a$	$\{R, G\}$
7	**PRO:** $\{G\}+$	$\{R, G\}$		7	**PRO:** $e \rightsquigarrow d$	$\{R, G\}$
8	**PRO:** $e \rightsquigarrow d$	$\{R, G\}$		8	**OPP:** $f \rightsquigarrow e$	$\{R, G\}$
9	**OPP:** $f \rightsquigarrow e$	$\{R, G\}$		9	**PRO:** $\emptyset-$	$\{R, G\}$
10	**PRO:** $\emptyset-$	$\{R, G\}$		10	**OPP:** ok	$\{R, G\}$
11	**OPP:** ok	$\{R, G\}$		11	**OPP:** ok	$\{R, G\}$
12	**OPP:** ok	$\{R, G\}$		12	**PRO:** win	$\{R, G\}$
13	**PRO:** win	$\{R, G\}$				

Explanation: In the dialogue shown on the left, the initial exchange of attacks consists of $b \rightsquigarrow a, c \rightsquigarrow b$ and $b \rightsquigarrow c$. PRO must end this line of argument by making a disabling property to disable the attack $b \rightsquigarrow c$. PRO moves **PRO:** $\{R\}-$

and as a result, the motivational state of the dialogue becomes $\{R\}$. OPP's next attack is $d \rightsquigarrow a$. PRO cannot move $e \rightsquigarrow d$ because this attack is disabled in the current motivational state. PRO moves **PRO:** $\{G\}+$, changing the motivational state of the dialogue to $\{R, G\}$, so that $e \rightsquigarrow d$ is enabled. To OPP's attack $f \rightsquigarrow e$ PRO responds with an empty disabling move, as $f \rightsquigarrow e$ is already disabled in the current motivational state. The dialogue on the right is similar with the exception that PRO immediately moves both R and G when making a disabling property move on line 4. As a result, no enabling property move is needed on line 7 because the attack $d \rightsquigarrow e$ is already enabled.

The existence of a property-consistent weak x-acceptance dialogue implies weak acceptance of x, i.e., it is a sound proof procedure:

Lemma 1 (Soundness). *Let $(A, \rightsquigarrow, \mathcal{P}, P, \mathcal{M}, \leq)$ be a property-based AF and $x \in A$. If there exists a property-consistent weak x-acceptance dialogue $S = (m_1, \ldots, m_n)$ then x is a member of the grounded extension of the AF represented by M_n^S. Hence x is weakly accepted.*

Proof (of lemma 1). Let $(A, \rightsquigarrow, \mathcal{P}, P, \mathcal{M}, \leq)$ be a property based AF, $x \in A$ and S a property-consistent weak x-acceptance dialogue. A subsequence S' of S that is a y-attack sequence (for some $y \in A$) will be called a *y-attack subsequence.* We denote the *depth* of an attack subsequence S' by $D(S')$ and define it by $D(S') = 0$, if $S' = (\mathbf{OPP: ok})$ and $1 + k$ otherwise, where $k = max(\{D(S'') \mid S'' \in T\})$, where T is the set of attack sequences that are proper subsequences of S'. Furthermore from hereon we denote the grounded extension of $(A, \rightarrow_{M_n^S})$ by G. We show that for every y-attack subsequence S' it holds that $y \in G$. We prove this by strong induction on the depth of S'. Let the induction hypothesis $H(k)$ stand for "if S' is a y-attack subsequence with depth k then $y \in G$."
- Base case ($H(0)$): Here $S' = (\mathbf{OPP: ok})$, thus y has no attackers in $(A, \rightsquigarrow)$, hence no attackers in $(A, \rightarrow_{M_n^S})$. It follows that $y \in G$.
- Induction step: Assume $H(0), \ldots, H(k-1)$ holds. We need to prove $H(k)$. It can be checked that for every z s.t. $z \rightsquigarrow y$, either:
  - There is a z'-attack sequence S'' that is a proper subsequence of S'. Thus $D(S'') < k$ and $z' \rightsquigarrow z$. From $H(D(S''))$ and the fact that S is property-consistent it follows that z is attacked by G.
  - S' contains a disabling property move. Hence $z \not\rightarrow_{M_n^S} y$.

This means that for every z such that $z \rightarrow_{M_n^S} y$, G attacks z, hence $y \in G$. By the principle of strong induction it follows that if there is a y-attack subsequence then $y \in G$. Thus we have $x \in G$, hence x is weakly accepted.   □

Conversely, if x is weakly accepted then a property-consistent weak x-acceptance dialogue exists:

Lemma 2 (Completeness). *Let $(A, \rightsquigarrow, \mathcal{P}, P, \mathcal{M}, \leq)$ be a property-based AF and $x \in A$ be weakly accepted. There exists a weak x-acceptance dialogue S that is property-consistent.*

Proof. Let $(A, \rightsquigarrow, \mathcal{P}, P, \mathcal{M}, \leq)$ be a property-based AF and $x \in A$ be weakly accepted. Then there is some $M \in \mathcal{M}$ s.t. x is a member of the grounded extension of $(A, \rightarrow_M)$. From hereon we use M to refer to any such motivational state and G to refer to the grounded extension of $(A, \rightarrow_M)$.

First some notation: The *characteristic function* $C : 2^A \rightarrow 2^A$ of an AF $(A, \rightarrow)$ is defined by $C(X) = \{x \in A \mid x \text{ is defended by } X\}$. It is well known that G coincides with the least fixed point of C [1]. We define the *degree* $Deg(x)$ of any $x \in G$ as the smallest positive integer s.t. $x \in C^n(\emptyset)$.

We now prove, by strong induction over the degree of an argument $y \in G$ that there exists a property consistent weak y-acceptance dialogue. Let $H(k)$ stand for "If $y \in G$ and $Deg(y) = k$ then there exists a property consistent weak y-acceptance dialogue."

- Base case $(H(0))$: If $y \in G$ and $Deg(y) = 0$ then there is no $z \in A$ s.t. $z \rightarrow_M y$ and we can define S by (**OPP:** $z_1 \rightsquigarrow y$) $\cdot$ S' $\cdot \ldots \cdot$ (**OPP:** $z_n \rightsquigarrow y$) $\cdot S' \cdot$ (**OPP: ok**) $\cdot$ (**PRO: win**), where $\{z_1, \ldots, z_n\} = \{z' \mid z' \rightsquigarrow y\}$ and $S' = $ (**PRO:** $M-$). It can be checked that S is a property consistent weak y-acceptance dialogue.
- Induction step: Assume $H(0), \ldots, H(k-1)$ holds. Thus if $y' \in G$ and $Deg(y') < k$ then there exists a property consistent weak y'-acceptance dialogue. We denote this dialogue by $S(y')$. We need to prove $H(k)$.

  Assume that $y \in G$ and $Deg(y) = k$. It follows that for every $z \in A$ s.t. $z \rightarrow_M y$, there exists an argument which we denote by $def(z, y)$ such that $def(z, y) \in G$ and $def(z, y) \rightarrow_M z$. Furthermore from the fixpoint construction it follows that $Deg(def(z, y)) < k$, so that $S(def(z, y))$ is well defined. Now, for every $z \in A$ s.t. $z \rightsquigarrow y$ we define $T_y(z)$ by (1) $T_y(z) = $ (**OPP:** $z \rightsquigarrow y$) $\cdot$ (**PRO:** $M-$), if $z \not\rightarrow_M y$ and (2) $T_y(z) = $ (**OPP:** $z \rightsquigarrow y$) $\cdot$ (**PRO:** $M+$) $\cdot S'$, if $z \rightarrow_M y$—where S' is defined by $S(def(z, y)) = S' \cdot$ (**PRO: win**). It can be checked that $T_y(z_1) \cdot \ldots \cdot T_y(z_i) \cdot$ (**OPP: ok**) $\cdot$ (**PRO: win**) (where $\{z_1, \ldots, z_i\} = \{z' \mid z' \rightsquigarrow y\}$) is a property consistent weak y-acceptance dialogue.

By the principle of strong induction it follows that for every $y \in G$, there exists a property consistent weak y-acceptance dialogue. Hence, there exists a property consistent weak x-acceptance dialogue. □

Notice that in the fourth move of in the second dialogue in example 6, PRO puts forward both R and G in a disabling property move. However, it suffices to put forward just R, as in the first dialogue, because G is not relevant with respect to disabling the attack $b \rightsquigarrow c$. We call a dialogue in which property moves are relevant a *property-relevant* dialogue. Property moves in a property-relevant dialogue consist only of properties satisfied by one of the arguments involved in the attack that is enabled or disabled.

Definition 14 (Property-relevance). *Let* $(A, \rightsquigarrow, \mathcal{P}, P, \mathcal{M}, \leq)$ *be a property-based AF and* $S = (m_1, \ldots, m_n)$ *a weak acceptance dialogue. We say that* S *is property-relevant iff for all* $i, j \in [1, \ldots, n]$ *s.t.* $j = i + 1$, *we have:*

1. If $m_i = \textbf{\textit{OPP:}}\, x \rightsquigarrow y$ and $m_j = \textbf{\textit{PRO:}}\, P-$ then $P \subseteq P(x) \cup P(y)$.
2. If $m_i = \textbf{\textit{PRO:}}\, P+$ and $m_j = \textbf{\textit{PRO:}}\, x \rightsquigarrow y$ then $P \subseteq P(x) \cup P(y)$.

Note that in example 6 the first dialogue is property-relevant, whereas the second one is not. Focusing on property-relevant dialogues can be used to optimize the algorithm. Furthermore, it makes sense intuitively: when persuading an opponent to accept an argument, one does not refer to properties not relevant to this objective.

As a final result we show that weak acceptance of an argument implies the existence of a property-consistent weak x-acceptance dialogue that is, in addition, property relevant. However, this requires that $\mathcal{M}$ is sufficiently rich to ensure that PRO is not forced to put forward irrelevant properties. This can be achieved by assuming that $\mathcal{M} = 2^{\mathcal{P}}$, but note that there are cases where a weaker assumption is sufficient.

Lemma 3 (Property-relevant completeness). *Let $(A, \rightsquigarrow, \mathcal{P}, P, \mathcal{M}, \leq)$ be a property-based AF where $\mathcal{M} = 2^{\mathcal{P}}$, and let $x \in A$ be weakly accepted. There exists a weak x-acceptance dialogue S that is property-consistent and property-relevant.*

Proof. Let $(A, \rightsquigarrow, \mathcal{P}, P, \mathcal{M}, \leq)$ be a property-based AF and $x \in A$ be weakly accepted. Let $S = (m_1, \ldots, m_n)$ be the property-consistent weak x-acceptance dialogue (for x a member of the grounded extension of $(A, \rightarrow_M)$) as constructed in the proof of lemma 2. That is, every property move in S is either of the form $\textbf{\textit{PRO:}}\, M+$ or $\textbf{\textit{PRO:}}\, M-$. Using property 1 (2) it can be checked that the dialogue S' formed by

- replacing every move $m_i = \textbf{\textit{PRO:}}\, M+$ in S by $\textbf{\textit{PRO:}}\, M'+$, where $M' = M \cap P(x) \cup P(y)$ where x, y are defined by $m_{i+1} = \textbf{\textit{PRO:}}\, x \rightsquigarrow y$, and
- replacing every move $m_i = \textbf{\textit{PRO:}}\, M-$ in S by $\textbf{\textit{PRO:}}\, M'-$, where $M' = M \cap P(x) \cup P(y)$ where x, y are defined by $m_{i-1} = \textbf{\textit{OPP:}}\, x \rightsquigarrow y$,

is also a property-consistent weak x-acceptance dialogue, that is in addition property-relevant. $\qquad\square$

Summarizing, we have the following result.

Theorem 2. *Let $(A, \rightsquigarrow, \mathcal{P}, P, \mathcal{M}, \leq)$ be a property-based AF.*
- *An argument $x \in A$ is weakly accepted iff there exists a weak x-acceptance dialogue that is property-consistent.*
- *If $\mathcal{M} = 2^{\mathcal{P}}$ then an argument $x \in A$ is weakly accepted iff there exists a weak x-acceptance dialogue that is property-consistent and property-relevant.*

Proof. Follows from lemmas 1, 2 and 3. $\qquad\square$

7   Related Work

We already mentioned the relation of our model with that of preference and value-based AFs [2,4]. Also related is a study of value-based AFs where arguments promote multiple values [10], concerned mainly with the problem of deriving a unique preference order over arguments from a preference relation over individual values. Note that in our approach, a property-based AF together with a motivational state already defines a unique preference order over arguments. Furthermore, Bench-Capon et al. have considered dialogues in which a proponent can make moves consisting of value preferences [13]. In this approach, the outcome of a winning dialogue corresponds to the specification of an audience (i.e., a preference order over values) such that some initial set of arguments is accepted in the corresponding $aVAF$.

Also related are Modgil's model of *extended AFs*, in which arguments attack and disable attacks between other arguments [14]. Such arguments can be seen as meta-level arguments expressing preferences over object level arguments. Whereas we take the agent's state (which determines whether individual attacks are enabled) to be external to the AF, here it is part of AF itself. That is, whether an attack is enabled depends on the status of a metalevel argument.

Our work shares methodological similarities with work of Kontarinis et al. [15], who present a goal-oriented procedure to determine which attacks to disable or enable in order to make an argument accepted under a given semantics. While the procedure that they present is designed to be implemented as a term rewriting system, our procedure is defined simply by a set of production rules, amenable to implementation using e.g. PROLOG.

8 Conclusion and Future Work

We presented a dynamic model of preferences in argumentation, based on Dietrich and List's model of property-based preference. It provides an account of how and why preferences in argumentation may change and generalizes both preference-based AFs and value-based AFs, if properties are taken to be values. We consider a number of directions for future work. First, we plan to complete the proof-theoretic picture by looking at the problem of deciding whether an argument is strongly accepted. In addition, we will consider other semantics in addition to grounded. Second, we plan to investigate the possibility of axiomatizing property-based AFs, in the spirit of Dietrich and List's axiomatization as presented in section 4. Finally, we intend to look at connections between property-based AFs and Modgil's model of extended AFs.

Acknowledgements. Richard Booth is supported by the FNR (National Research Fund) DYNGBaT project.

References

1. Dung, P.M.: On the acceptability of arguments and its fundamental role in non-monotonic reasoning, logic programming and n-person games. Artif. Intell. 77(2), 321–358 (1995)
2. Amgoud, L., Cayrol, C.: A reasoning model based on the production of acceptable arguments. Ann. Math. Artif. Intell. 34(1-3), 197–215 (2002)

3. Simari, G., Loui, R.: A mathematical treatment of defeasible reasoning and its implementation. Artificial Intelligence 53, 125–157 (1992)
4. Bench-Capon, T.J.M.: Persuasion in practical argument using value-based argumentation frameworks. J. Log. Comput. 13(3), 429–448 (2003)
5. Liu, F.: Reasoning about preference dynamics. Springer (2011)
6. Grüne-Yanoff, T., Hansson, S.O. (eds.): Preference change: Approaches from philosophy, economics and psychology. Springer (2009)
7. Dietrich, F., List, C.: Where do preferences come from? International Journal of Game Theory, 1–25 (2012)
8. Dietrich, F., List, C.: A reason-based theory of rational choice. Nous 47(1), 104–134 (2013)
9. Modgil, S., Caminada, M.: Proof theories and algorithms for abstract argumentation frameworks. In: Argumentation in Artificial Intelligence, pp. 105–129. Springer (2009)
10. Kaci, S., van der Torre, L.: Preference-based argumentation: Arguments supporting multiple values. Int. J. Approx. Reasoning 48(3), 730–751 (2008)
11. Caminada, M., Podlaszewski, M.: Grounded semantics as persuasion dialogue. In: Proceedings of COMMA 2012, pp. 478–485 (2012)
12. Caminada, M.: Preferred semantics as socratic discussion. In: Proceedings of the 11th AI* IA Symposium on Artificial Intelligence, pp. 209–216 (2010)
13. Bench-Capon, T.J.M., Doutre, S., Dunne, P.E.: Audiences in argumentation frameworks. Artif. Intell. 171(1), 42–71 (2007)
14. Modgil, S.: Reasoning about preferences in argumentation frameworks. Artif. Intell. 173(9-10), 901–934 (2009)
15. Kontarinis, D., Bonzon, E., Maudet, N., Perotti, A., van der Torre, L., Villata, S.: Rewriting rules for the computation of goal-oriented changes in an argumentation system. In: Kontarinis, D., Bonzon, E., Maudet, N., Perotti, A., van der Torre, L., Villata, S. (eds.) CLIMA. LNCS. Springer (forthcoming 2013)

Learning Multicriteria Utility Functions with Random Utility Models

Géraldine Bous[1,2] and Marc Pirlot[2]

[1] BIT Advanced Development, SAP, Sophia Antipolis, France
geraldine.bous@sap.com
[2] Dept. of Mathematics & Operations Research, Faculté Polytechnique,
Université de Mons, Belgium

Abstract. In traditional multicriteria decision analysis, decision maker evaluations or comparisons are considered to be error-free. In particular, algorithms like UTA*, ACUTA or UTA-GMS for learning utility functions to rank a set of alternatives assume that decision maker(s) are able to provide fully reliable training data in the form of e.g. pairwise preferences. In this paper we relax this assumption by attaching a likelihood degree to each ordered pair in the training set; this likelihood degree can be interpreted as a choice probability (group decision making perspective) or, alternatively, as a degree of confidence about pairwise preferences (single decision maker perspective). Since binary choice probabilities reflect order relations, the former can be used to train algorithms for learning utility functions. We specifically address the learning of piecewise linear additive utility functions through a logistic distribution; we conclude with examples and use-cases to illustrate the validity and relevance of our proposal.

1 Introduction

Preference learning consists in determining a model that reflects the subjective value, i.e. as perceived by a decision maker (DM), of alternatives or items belonging to a set S (the reader can refer to Fürnkranz & Hüllermeier, 2011, for a general introduction on the topic). In artificial intelligence and decision theory, this problem is frequently solved by learning a *value* or *utility function* u such that the order obtained by ranking the alternatives by order of decreasing utility corresponds to the order induced on S by the preferences of the DM. Typically, the preference relation on S is not entirely known; therefore, it is common practice to obtain a sample of the preference relation on a subset $S_L \subset S$ – the learning set – to train the utility model. The thereby obtained utility function can then be used to evaluate alternatives in $S \setminus S_L$ and to obtain an estimate of the preference relation on S as a whole, thereby allowing to solve ranking or choice problems on S.

In multicriteria decision theory, alternatives are characterized by their performances of several criteria; in this context, the preference learning problem aims at producing a utility function that evaluates items as a function of their 'scores'

P. Perny, M. Pirlot, and A. Tsoukiàs (Eds.): ADT 2013, LNAI 8176, pp. 101–115, 2013.
© Springer-Verlag Berlin Heidelberg 2013

on the criteria *and* that reflects the preference relation of the DM. In other terms, the 'global utility' $u(i)$ of an item $i \in S$ is an aggregate of its scores on all criteria and the order induced by u on S (or S_L) is the same as the one induced by the preferences. The most common aggregation model in decision theoretic literature is the additive value model, which, starting with the contribution of Jacquet-Lagrèze & Siskos (1982), has led to a variety of preference learning approaches known as the UTA (for 'UTilités Additives') family of methods. For instance, UTA (Jacquet-Lagrèze & Siskos, 1982), UTA* (Siskos & Yannacopoulos, 1985), UTAMP I & II (Beuthe & Scannella, 1996, 2001) and ACUTA (Bous et al., 2010) are methods based on piecewise linear utility functions that are adjusted to the preference sample through different optimization methods. As opposed to these methods, which produce a unique utility function, other recent proposals (see, in particular, Greco et al., 2008; Figueira et al., 2009; Kadziński et al., 2012) seek to determine the set of all preference-compatible utility functions (not necessarily piecewise), often requiring a post-processing step to prune the results (see Kadziński et al., 2012).

Information regarding preferences can be expressed in different manners, the most common being direct evaluations and binary (i.e. pairwise) comparisons of items, which may take the form of a ranking or any other order relation. In traditional multicriteria decision analysis (including previously cited references), such preference information is typically assumed to be entirely reliable, i.e. error-free. There is however longstanding evidence that this is not necessarily true. The pioneering work of Fechner (1860) and Thurstone (1927a,b) showed that binary comparisons on two items $i, j \in S$ may be subject to unintentional (perception) errors, which translate as random fluctuations in the outcome of the comparison. Such fluctuations have experimentally been shown to be magnified by a variety of factors, including the similarity or strong dissimilarity of items, the cardinality of S and the existence of multiple attributes (see, e.g., Kahneman, 1973; Shugan, 1980). For the multiattribute preference learning problem, this implies that the DM preference sample, i.e. the order on S_L, may not be entirely accurate. Instead, pairwise comparisons have a likelihood degree that takes the form of a binary choice probability.

In addition to problems involving a single 'inaccurate' DM, we may also consider group decision making problems. That is, learning a utility function as compatible as possible with the preference samples of all DM's on S. Unless all DM's have the same preference relation, there will be some who, for any two $i, j \in S$, prefer i, while others prefer j. The proportion of DM's preferring one or the other alternative thus allows to deduce the frequency with which one item is chosen, i.e. it allows to deduce the binary choice probability regarding the pair i, j.

From this probabilistic perspective, the (multicriteria) preference learning paradigm thus becomes a problem of determining a utility function as compatible as possible with a preference relation on S_L taking into account that ordered pairs have a likelihood degree. For instance, if some binary preferences are contradicting or cannot be satisfied, priority should be given to more likely pairs. However, 'learning' is in this case more than just a problem of preference prioritization: in addition to deriving a utility function to evaluate alternatives S, it is necessary to learn a probability distribution that reflects the binary choice probabilities of the items in S_L *and* that predicts the choice probabilities of the items in S as whole. In other terms, the preference learning problem is about adjusting a *random utility model* (see, e.g., Block & Marschak, 1960; Luce & Suppes, 1965; Fishburn, 1998) as a function of multiattribute utility functions such that it reflects both the order relation and the binary choice probabilities of the items in S_L.

In this paper we consider a random utility model based on the *multinomial logistic distribution*, which is the most widely applied random utility model (McFadden, 1986), in combination with the piecewise linear additive (PLA) model to characterize the utilities of multiattribute items. We measure the 'adjustment' of these models through the relative entropy between target probabilities (provided by the DM) and calculated probabilities (obtained by fitting the model). Altogether, this constitutes the CEUTA method, a nonlinear optimization problem to learn utility functions subject to both preference-related constraints and PLA model constraints. The paper is structured as follows: after introducing the problem, we present the PLA model and then discuss the problem of learning random utility models. Next, we present the CEUTA method and finally conclude with an experimental analysis and directions for future research.

2 Utility Functions with the Piecewise Linear Additive Model

2.1 Preferences and Utility Functions

The standard approach to modeling preferences is to model them as a binary relation over the items of a set, i.e. as an *order* (see e.g. Luce & Suppes, 1965; Fishburn, 1970; Aleskerov et al., 2007). Throughout the paper, we focus on preference relations that are strict linear orders, i.e. a binary relation $\succ$ on a nonempty set S that is complete, asymmetric and transitive. In other terms, whenever $i \succ j$ holds for any two $i, j \in S$, $i \neq j$, it means that i is strictly preferred to j. Preference relations can be represented by utility (or value) functions provided the latter are 'order preserving'. More formally, for a finite set $S = \{1, 2, ..., N\}$ of N items and a strict linear order $(S, \succ)$, a utility function is a numeric representation $u : S \to \mathbb{R}^+$ of $(S, \succ)$, such that $i \succ j \Leftrightarrow u(i) > u(j)$. It has been shown (see e.g. Fishburn, 1970, ch. 2) that, provided S is finite, there always exists an order-preserving function u for strict linear orders on S.

2.2 Multicriteria Decision Problems and Additive Utility

In the multidimensional case, items are characterized by several measurable criteria, i.e. they are described by m-dimensional vectors with each dimension corresponding to one criterion. We write this $x(i) = (x_1(i), x_2(i), ..., x_m(i))$, where $x_k(i)$ corresponds to the actual score of item i on criterion k. In decision theoretic literature, items are usually called *alternatives* and are denoted $\mathcal{A} = \{a_1, a_2, ..., a_N\}$; for simplicity, we retain the 'label notation' whenever possible, i.e. $i \equiv a_i$, and use S to denote the set of items, i.e. $S \equiv \mathcal{A}$.

As opposed to the real score of an item, we are interested in its *utility*, which reflects the (subjective) value that a person would assign to this item. It is of use to distinguish between the overall value of an item, known as total or *global utility*, and criterion-specific value, which is known as *marginal* or *partial utility*. The marginal utility of an item i on criterion k, denoted $u_k(i)$, thus reflects the perceived value of i on k. It is a function of $x_k(i)$, i.e. $u_k(i)$ is a short notation for $u_k(x_k(i))$. A convenient model to aggregate marginal utilities into a global utility is the additive model, i.e.

$$u(i) = \sum_{k=1}^{m} u_k(i) \qquad \forall\, i \in S, \tag{1}$$

where S denotes the set of alternatives. Note that the additive utility model requires criteria to be (mutually) preferentially independent of each other (see e.g. Fishburn, 1970; Dyer, 2005, for details).

2.3 Modeling Marginal Utility with Piecewise Linear Functions

The representation of marginal utilities is usually done using functions $u_k(x_k)$ whose domain cover the scores of all items considered in the decision problem, i.e.

$$x_k \in \left[\min_{i \in S} x_k(i),\ \max_{i \in S} x_k(i) \right], \tag{2}$$

where $S = \{1, 2, ..., N\}$. In the piecewise linear utility function model (Jacquet-Lagrèze & Siskos, 1982), interval x_k is divided into s_k sub-intervals of equal length. This leads to a discretization with $s_k + 1$ score-utility pairs (u_k^l, u_k^{l+1}), for $l = 0, 1..., s_k - 1$; the rest of the values of the function are obtained through linear interpolation. On this basis, the marginal utility of an item i at an arbitrary point $x_k(i) \in x_k$ is given by:

$$u_k(i) = \begin{cases} u_k^l & \text{if } x_k(i) = x_k^l \\ \left(\frac{x_k(i) - x_k^l}{x_k^{l+1} - x_k^l} \right) u_k^{l+1} + \left(\frac{x_k^{l+1} - x_k(i)}{x_k^{l+1} - x_k^l} \right) u_k^l & \text{if } x_k(i) \in (x_k^l; x_k^{l+1}) \end{cases} \tag{3}$$

There are two important 'types' of (marginal) utility functions: those corresponding to the case of criteria that have to be maximized, modeled with monotonously increasing functions, and those that have to be minimized, modeled with monotonously decreasing functions. In the forthcoming formulations, we focus on the maximization case (the minimization case is straightforward to derive from the former; for details, the reader can refer to e.g. Siskos et al., 2005). Obtaining a monotonously increasing function is achieved by constraining the values of the variables $u_k^0, u_k^1, ..., u_k^{s_k}$ as follows:

$$u_k^l \leq u_k^{l+1} \quad \text{for } l = 0, 1, ..., s_k - 1 \text{ and } k = 1, ..., m. \tag{4}$$

Let $\underline{x}_k = x_k^0$; the piecewise linear model imposes a utility of zero to the worst score $\underline{u}_k = u_k(\underline{x}_k)$ in the x_k-scale, i.e.

$$\underline{u}_k = 0 \quad k = 1, 2, ..., m. \tag{5}$$

Moreover, utilities of items are limited to at most one, a constraint known as the *normalization constraint*. Let $\overline{x}_k = x_k^{s_k}$ so that $\overline{u}_k = u_k(\overline{x}_k)$ is the *weight* of criterion k; the normalization constraint is then given by

$$\sum_{k=1}^{m} \overline{u}_k = 1. \tag{6}$$

Equations (1) - (6) constitute the PLA utility model (Jacquet-Lagrèze & Siskos, 1982). We take one more step into rewriting the global utility of an item as a linear combination of the variables u_k^l. To this end, we rewrite the marginal utility as

$$u_k(i) = \sum_{l=0}^{s_k} \alpha_k^l(i) \cdot u_k^l , \tag{7}$$

where the $\alpha_k^l(i)$ are coefficients derived from (3), so that the global utility of item i may be written as

$$u(i) = \sum_{k=1}^{m} \sum_{l=0}^{s_k} \alpha_k^l(i) \cdot u_k^l . \tag{8}$$

Methods for learning PLA utility functions include UTA (Jacquet-Lagrèze & Siskos, 1982), UTA* (Siskos & Yannacopoulos, 1985), UTAMP I & II (Beuthe & Scannella, 1996, 2001) and ACUTA (Bous et al., 2010). All of these methods require preference information in the form of rankings (possibly non strict) on S_L. ACUTA is a nonlinear optimization approach that determines the analytic center (Huard, 1967; Sonnevend, 1985) of the constraints, while the remaining methods are formulated as linear programming problems.

3 Learning Random Utility Models

3.1 Random Utility Models

Preferences expressed in the form of (strict or non-strict rankings) can be broken down to ordered pairs. For example, the ranking $i \succ j \succ k$ corresponds to $i \succ j$, $i \succ k$ and $j \succ k$. To provide a ranking on a set S, the DM must thus be able to *choose* and state the most preferred item in every pair $i, j \in S$, $i \neq j$. As previously mentioned, there is evidence that performing such choices is difficult and may consequently lead to unintentional errors, like e.g. occasionally stating $j \succ i$ when actually $i \succ j$. Alternatively, we may consider a group of non-consensual DMs, some stating $i \succ j$ and others declaring $j \succ i$, which also expresses uncertainty on whether i or j is *overall* preferred. Uncertainty in choice behavior led to the development of *probabilistic choice theories* (see Luce & Suppes, 1965, for a general overview) in which choice – and hence also binary comparisons – are governed by choice probabilities $p_{ij} = \mathbb{P}(i \succ j)$, $\forall\, i, j \in S$, that describe the probability of choosing (i.e. preferring) i over j.

Random utility models are a special type of probabilistic choice models in which choice probabilities are a function of item utilities. More precisely, $p_{ij} = \mathbb{P}(u(i) \geq u(j))$, i.e. the probability of choosing i over j is the probability that the utility of i is greater than that of j. While many random utility models exist, the most common are the normal distribution (as originally proposed by Thurstone, 1927a) and the *multinomial logit choice model*, which is based on the *logistic distribution* and is defined as (see McFadden, 1986)

$$p_{ij} = \frac{e^{(u(i)-u(j))}}{1 + e^{(u(i)-u(j))}}. \tag{9}$$

In what follows (section 4), we use this model to learn random utility functions based on the previously introduced PLA model.

3.2   Learning Functions

For a given learning set S_L of alternatives for which an order and binary choice probabilities are given, learning a random utility model such as (9) is thus a problem of fitting a utility function u such that

1. the order obtained by sorting the alternatives of S_L by decreasing utility corresponds to the order provided by the DM on S_L
2. the binary choice probabilities, as given by (9), also correspond to those given as input to the problem.

Fortunately, there exists a correspondence between binary choice probabilities and orders: under certain conditions, binary choice probabilities induce different types of order relations (Fishburn, 1973). In particular, binary choice probabilities are equivalent to a nested family of strict linear orders if, for any three items $i, j, k \in S_L$ with $i \succ j \succ k$, they satisfy *partial stochastic transitivity* (see Fishburn, 1973, for a detailed proof):

$$p_{ij} > \frac{1}{2} \;\&\; p_{jk} > \frac{1}{2} \Rightarrow p_{ik} \geq \min(p_{ij}, p_{jk}). \tag{10}$$

As a consequence, to learn a utility function that reflects a strict linear order *and* binary choice probabilities (satisfying partial stochastic transitivity) on the items of S_L, it is sufficient to adjust u to have $p_{ij} = p_{ij}^\star$, $\forall\, i, j \in S_L$, where p_{ij} denotes the 'calculated probability' given by (9) and $p_{ij}^\star$ denotes the target probability provided as input to the decision problem. The utility function learning problem thus becomes a problem of learning a probability model which 'implicitly' produces a utility function that can be used to evaluate and rank the alternatives.

Several types of learning (i.e. fitting) functions can be considered to adjust a probability model to target probabilities. Least squares minimization remains a very popular fitting approach in the learning domain in general (see e.g. Bishop, 2006; Hastie et al., 2009, for applications and references in the machine and statistical learning areas); another popular option for probability distribution learning is *relative entropy minimization* (or cross-entropy minimization), which has been successfully applied in the machine learning community in the context of neural networks (Baum & Wilczek, 1987) as well as in the web ranking community to adjust logistic distributions (Burges et al., 2005; Burges, 2006). Relative entropy (or *Kullback-Leibler distance*) is a distance-like measure of the resemblance of two probability distributions. For two probability mass functions $p(x)$ and $q(x)$ it is defined as (see Cover & Thomas, 2006, p. 19)

$$D(p\|q) = \sum_{x \in X} p(x) \log \frac{p(x)}{q(x)} \qquad (11)$$

with the conventions $0 \log \frac{0}{0} = 0$, $0 \log \frac{0}{q} = 0$ and $p \log \frac{p}{0} = \infty$. While relative entropy is not a true distance, it still is a measure of the inefficiency of assuming that the probability distribution is $q(x)$ when it actually is $p(x)$ (Cover & Thomas, 2006). Relative entropy satisfies $D(p\|q) \geq 0$ for any $p(x)$ and $q(x)$; the closer $q(x)$ to $p(x)$, the smaller $D(p\|q)$ with $D(p\|q) = 0$ if and only if $p(x) = q(x)$ (Cover & Thomas, 2006, p. 28). In our context, the minimization of the relative entropy between $p_{ij}^\star$ and p_{ij}, as given by (9), thus means adjusting p_{ij} such that it is 'as close as possible' to $p_{ij}^\star$.

4   The CEUTA Method

4.1   Utility and Probability Models

The CEUTA method is based on the PLA utility model, as defined by equations (1) - (8). To slightly simplify notations, we let

$$\alpha_k^l(i,j) = \alpha_k^l(i) - \alpha_k^l(j) \qquad \forall\, i,j \in S, \qquad (12)$$

so that the utility difference between any two items $i, j \in S$ is given by

$$u(i) - u(j) = \sum_{k=1}^{m} \sum_{l=0}^{s_k} \alpha_k^l(i,j) \cdot u_k^l. \qquad (13)$$

Regarding the probability model, CEUTA relies on a parametric logistic function. The 'traditional' logistic distribution (9) is a probability distribution on $[-\infty, +\infty]$, which means that it cannot take the scaling of the value (or utility) function into account. With the PLA model, utilities are however defined on the $[0,1]$ interval and utility differences $u(i) - u(j)$ are *at most* equal to one; this scaling should however not forbid to obtain binary choice probabilities that are close to 1 or 0 for certain pairs of alternatives. We therefore scale (9) as follows

$$p_{ij} = \frac{e^{\mu(u(i)-u(j))}}{1 + e^{\mu(u(i)-u(j))}}, \qquad (14)$$

where μ is a strictly positive parameter that influences the steepness of the probability function at the inflexion point $u(i) = u(j)$. Equation (14) has an upper asymptote at $p_{ij} = 1$ for $u(i) - u(j) >> 0$ and a lower asymptote at $p_{ij} = 0$ for $u(i) - u(j) << 0$. As a consequence, if μ is sufficiently large, the probability to choose i over j satisfies $p_{ij} \approx 1$ for items whose utility difference $u(i) - u(j) > 0$ is large (on a [0,1] scale). In practice, μ can be chosen such that the choice probability between the ideal alternative (with $u(i) = 1$) and the anti-ideal alternative (with $u(j) = 0$) is very close to one.

Combining the PLA model with the parametric logistic distribution leads to

$$
p_{ij} = \frac{e^{\mu\left(\sum\limits_{k=1}^{m}\sum\limits_{l=0}^{s_k} \alpha_k^l(i,j)u_k^l\right)}}{1 + e^{\mu\left(\sum\limits_{k=1}^{m}\sum\limits_{l=0}^{s_k} \alpha_k^l(i,j)u_k^l\right)}},
\tag{15}
$$

which is obtained by inserting (13) into (14); this equation will be used shortly to calculate the gradient and Hessian for the ranking learning problem.

4.2   Objective Function

Consider a learning set $S_L \subset S$ of cardinality n with known binary choice probabilities $p_{ij}^\star$ for each pair of items $i, j \in S_L$, $i \neq j$. It is assumed that binary choice probabilities satisfy partial stochastic transitivity, as defined in (10), so that cutting the $p_{ij}^\star$ at the level $\frac{1}{2}$ induces a strict linear order on S_L; we call this order the *reference ranking* and denote it π_0. We use the convention $\pi_0(i) = i$ for all $i = 1, ..., n$, i.e. items are labeled according to their rank in the reference ranking, which means that $i < j \Leftrightarrow i \succ j, \forall i, j \in S_L$. With this labeling convention, the reference ranking thus corresponds to the strict linear order $1 \succ 2 \succ ... \succ n$. The relative entropy of one pair i, j of π_0 with the logistic distribution p_{ij} and the goal probability $p_{ij}^\star$ is

$$
p_{ij}^\star \log \frac{p_{ij}^\star}{p_{ij}} + (1 - p_{ij}^\star) \log \frac{(1 - p_{ij}^\star)}{(1 - p_{ij})},
\tag{16}
$$

where the logarithm is base two. For the ranking π_0, we consider the sum of the relative entropy of all pairs, i.e.

$$
\sum_{\substack{i=1,...,n-1 \\ j=i+1,...,n}} p_{ij}^\star \log \frac{p_{ij}^\star}{p_{ij}} + (1 - p_{ij}^\star) \log \frac{(1 - p_{ij}^\star)}{(1 - p_{ij})}.
\tag{17}
$$

As the $p_{ij}^\star$ are actually constants, the expression above can be expanded and all logarithms involving constant values can be eliminated, leading to

$$
f = - \sum_{\substack{i=1,...,n-1 \\ j=i+1,...,n}} p_{ij}^\star \log p_{ij} + (1 - p_{ij}^\star) \log(1 - p_{ij}).
\tag{18}
$$

4.3   Optimization Problem

The CEUTA (Cross Entropy UTA) method combines the PLA model and its constraints, the parametric logistic distribution (15) and objective function (18) to determine a utility function and probability model that reflect the preference relation and the binary choice probabilities of the items in S_L. Following the relative entropy minimization principle, all this combined leads to the following nonlinear optimization problem:

$$
\begin{cases}
\min f = - \displaystyle\sum_{\substack{i=1,\ldots,n-1 \\ j=i+1,\ldots,n}} p_{ij}^{\star} \log p_{ij} + (1 - p_{ji}^{\star}) \log(1 - p_{ij}) \\
\text{s.t. PLA model constraints,}
\end{cases}
\tag{19}
$$

where p_{ij} is given by (15) and the PLA model constraints are (4), (5) and (6).

CEUTA is solved by applying Newton's method for linearly constrained optimization (see, e.g. Gill et al., 1981; Fletcher, 1987; Antoniou & Lu, 2007). Our implementation uses nullspace projection (see Gill et al., 1981) to handle equality constraints (5)-(6) and a logarithmic barrier function for inequality constraints (4), which allows to introduce the latter into the objective function. To this end, let $B \cdot u \leq 0$ be the matrix notation for the monotonicity constraints in the form $u_k^l - u_k^{l+1} \leq 0$ (B is the coefficient matrix and u is the vector containing the u_k^l variables). Also, let $\sigma_t = -b_{t \cdot} \cdot u$ be the tth slack vector and $b_{t \cdot}$ the tth line of B. The *logarithmic barrier* is defined as (see, e.g., Nesterov & Nemirovsky, 1994; Ye, 1997)

$$
\mathcal{B}_\lambda(u) = \lambda \cdot \sum_{t=1}^{n_v} \ln(\sigma_t),
\tag{20}
$$

where $\lambda > 0$ is a small positive constant known as the *barrier parameter* and n_v is the number of monotonicity constraints. Introducing (20) into the objective function leads to

$$
\begin{cases}
\min f_\lambda = - \displaystyle\sum_{\substack{i=1,\ldots,n-1 \\ j=i+1,\ldots,n}} p_{ij}^{\star} \log p_{ij} + (1 - p_{ji}^{\star}) \log(1 - p_{ij}) - \lambda \cdot \sum_{t=1}^{n_v} \ln(\sigma_t) \\
\text{s.t. PLA model equality constraints}
\end{cases}
\tag{21}
$$

To solve this problem with Newton's method, it is required to calculate the gradient and the Hessian of f_λ. For the barrier function, they respectively are (details are given in Bous et al., 2010)

$$
\nabla(-\mathcal{B}_\lambda(u)) = \lambda \sum_{t=1}^{n_v} \frac{b_{t \cdot}^T}{\sigma_t} = \lambda B^T S^{-1} e,
\tag{22}
$$

where $S = \text{diag}(\sigma)$ and e is a column vector of ones, and

$$
\nabla^2(-\mathcal{B}_\lambda(u)) = \lambda \sum_{t=1}^{n_v} \frac{b_{t \cdot} \cdot b_{t \cdot}^T}{\sigma_t^2} = \lambda B^T S^{-2} B.
\tag{23}
$$

Regarding the relative entropy contribution f in f_λ, its partial derivative is given by the following expression:

$$\frac{\partial f}{\partial u_k^l} = -\frac{\mu}{\ln(2)} \sum_{\substack{i=1,\ldots,n-1 \\ j=i+1,\ldots,n}} \alpha_k^l(i,j)\left(p_{ij}^\star - p_{ij}\right). \tag{24}$$

Similarly, the second order partial derivative with respect to the PLA model variables, here generically denoted u_k^l and u_s^t, is given by:

$$\frac{\partial^2 f}{\partial u_k^l \partial u_s^t} = \frac{\mu^2}{\ln(2)} \sum_{\substack{i=1,\ldots,n-1 \\ j=i+1,\ldots,n}} \alpha_k^l(i,j)\alpha_s^t(i,j)\, p_{ij}(1 - p_{ij}). \tag{25}$$

In practice, it is more useful to have these formulas in matrix form, like for the barrier function. To this purpose, let p and p* be column vectors containing pairwise probabilities (model and goal probabilities, respectively) and A be the matrix containing the $\alpha_k^l(i,j)$ coefficients. This leads to the simple expression

$$\nabla f = -\frac{\mu}{\ln(2)} A^T(p^\star - p). \tag{26}$$

Finally, if we let P $=$ diag(p), the Hessian is given by

$$\nabla^2 f = \frac{\mu^2}{\ln(2)} A^T P(I - P)A, \tag{27}$$

where I denotes the identity matrix. Overall, we thus have

$$\nabla f_\lambda = -\frac{\mu}{\ln(2)} A^T(p^\star - p) + \lambda B^T S^{-1} e \tag{28}$$

and

$$\nabla^2 f_\lambda = \frac{\mu^2}{\ln(2)} A^T P(I - P)A + \lambda B^T S^{-2} B. \tag{29}$$

Optimization problem (21) is easily solved with Newton's method using nullspace projection for the equality constraints as done by Bous et al. (2010) in the ACUTA method. Newton's method consists in computing iterates $u_{i+1} = u_i + \gamma_i \triangle_i$, where index i here denotes the ith iteration, γ_i the step-size and $\triangle_i$ the Newton direction. In unconstrained problems, the latter is obtained at each step by solving

$$\nabla^2 f_\lambda(u_i)\, \triangle_i = -\nabla f_\lambda(u_i). \tag{30}$$

Let $B_e \cdot u = c_e$ be the matrix notation of inequality constraints (5)-(6); taking these into account in Newton's method can be achieved by identifying a basis of the null-space of B_e, i.e. a non-null matrix M such that $B_e M = 0$, and projecting both the gradient and the Hessian onto the null-space. The Newton-step is then given by $\triangle_{M,i} = M \triangle_i$ and (30) becomes (Gill et al., 1989)

$$M^T \nabla^2 f_\lambda(u_i) M\, \triangle_{M,i} = -M^T \nabla f_\lambda(u_i). \tag{31}$$

Starting from an initial point u_0 such that all slack variables are strictly positive, iterates are computed until the stopping condition $||M^T \nabla f_\lambda(u_i) = 0||$ is met with a certain precision level ε. The step-size γ_i should be as large as possible under the constraint that slack variables remain strictly positive at each iteration; in practice, it is frequently computed with simple line-search procedures like, for example, starting with $\gamma_i = 1$ and multiplying with 0.9 until $\sigma_{t,i} > 0$, $\forall t$. The interested reader can refer to (Bous et al., 2010) for further details on the algorithm.

5   Experimental Analysis

To validate the CEUTA method, a series of experiments using simulated data were performed. Due to the large number of parameter combinations to analyze (number of criteria, number of segments, size of the learning set S_L, size of $S \setminus S_L$ and μ), we chose to set $\mu = 20$, the number criteria to $m = 10$ and the number of alternatives to evaluate to $|S \setminus S_L| = 1000$, thereby focusing our attention on the number of segments and the size n of S_L. The number of segments s was evaluated for $s = \{1, 2, 3\}$ and equal for all marginal utility functions, i.e. $s_k = s$, $\forall k$; regarding n, we considered the cases $n = \{10, 25, 50, 100\}$. For each combination of values of these two parameters, 100 tests were run. For each test, with a given value of s and n, the following procedure was executed:

1. generate marginal utility functions of s segments randomly;
2. generate $n + |S \setminus S_L|$ alternatives $x(i) = (x_1(i), x_2(i), ..., x_10(i))$ randomly with the uniform distribution;
3. evaluate the global utility u^* of the generated data with the utility functions of step 1;
4. build a learning set by clustering the alternatives into n clusters according to their global utility, and then extracting one alternative per cluster (the one closest to the average utility of the cluster);
5. evaluate the probabilities p_{ij}^* of each alternative with (15) and u^*;
6. deduce the order of the learning set from the probabilities computed at the previous step;
7. run the CEUTA method (with s segments and $\mu = 20$ as well) on the learning set to obtain a 'learned utility function';
8. evaluate the remaining alternatives, i.e. $S \setminus S_L$, with the learned utility function to obtain their global utility u;
9. compute the Kendall rank correlation coefficient (i.e. Kendall's τ) between the order induced by u^* and u on $S \setminus S_L$ and, finally,
10. compute the relative entropy between p_{ij}^* and p_{ij} for the entire set S.

CEUTA converged for all tests performed, requiring on average 5.4083 iterations across all tests (the minimum and maximum being 3 and 44 iterations, respectively). Fig. 1 shows the results for Kendall's τ, grouped by s and n. It is clearly visible that the greater n, the better the results, while $n = 10$ is insufficient to train the model efficiently. Moreover, a larger number of segments negatively influences Kendall's τ, requiring larger learning sets to 'compensate' for an increased number of variables. In general, however, CEUTA produces good results and the correlation between the orders induced by $u^\star$ and u is high, even with learning sets that are quite small in comparison to the number of alternatives to evaluate.

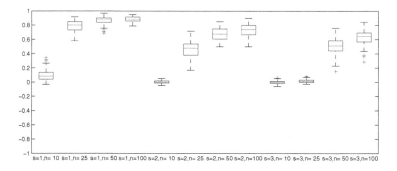

Fig. 1. Kendall's τ as a function of the number of segments s and the size of the learning set n for 1000 alternatives evaluated with CEUTA

Fig. 2 shows the distribution of the relative entropy, again grouped by s and n. Confirming the results of Fig. 1, relative entropy shows high dispersion and very large values for $n = 10$, an indication that the learning set is too small and fails to train the model. For other values of n and s however, relative entropy is low, which shows that the learning algorithm is efficient in fitting the PLA model and the parametric logistic distribution to the data set as whole.

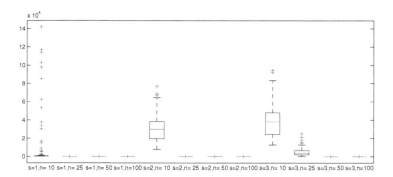

Fig. 2. Relative entropy as a function of the number of segments s and the size of the learning set n for $n + 1000$ alternatives evaluated with CEUTA

Besides evaluating CEUTA with real data-sets, other interesting experiments to perform in future work are to analyze the robustness of the method with respect to different number of segments, as well as different values of μ in the generated and learned functions. In addition, it is of interest to evaluate the impact of imprecise information regarding the target probabilities $p_{ij}^{\star}$.

6 Conclusion

In traditional multicriteria preference learning algorithms, decision maker evaluations or comparisons are considered to be error-free despite longstanding evidence that this is not necessarily the case. We have therefore relaxed this assumption by attaching a likelihood degree to the preference relation of the DM(s) and developed a new method, CEUTA, to learn PLA utility functions by taking into account the probabilities. CEUTA is based on the multinomial logistic distribution to model probabilities and relies on relative entropy to fit the model to the training set. Preliminary experimental results show that CEUTA is very efficient in that it is both fast and accurate, even with relatively small training sets for large data-sets.

Besides further experimentation, we consider two main directions for future research. First, the development of a variant of CEUTA in which μ is not a parameter, but instead a variable of the optimization problem. We expect this to improve the robustness of the method by making it 'independent' of a parameter which currently is determined approximately. Second, we wish to pursue in the direction of group decision making analysis. While CEUTA can be applied to this type of problems, it produces a unique utility function as compatible as possible with the preferences of all DM's, which implies that it may be incompatible with the preferences of one or more subgroups of DM's. It is therefore of interest to investigate the possibility of extending CEUTA as to identify subgroups of DM's who have similar preferences and produce possibly several utility functions (i.e. one per subgroup) if the preferences between subgroups show little correlation.

References

Aleskerov, F., Bouyssou, D., Monjardet, B.: Utility Maximization, Choice and Preference. Springer, Berlin (2007)

Antoniou, A., Lu, W.S.: Practical Optimization: Algorithms and Engineering Applications. Springer, Berlin (2007)

Baum, E.B., Wilczek, F.: Supervised learning of probability distributions by neural networks. In: Anderson, D. (ed.) Neural Information Processing Systems, American Institute of Physics, pp. 52–61 (1987)

Beuthe, M., Scannella, G.: Applications comparées des méthodes d'analyse multicritère UTA. RAIRO Operations Research 30, 293–315 (1996)

Beuthe, M., Scannella, G.: Comparative analysis of UTA multicriteria methods. European Journal of Operational Research 130, 246–262 (2001)

Bishop, C.M.: Pattern Recognition and Machine Learning. Information Science and Statistics. Springer, Berlin (2006)

Block, H.D., Marschak, J.: Random orderings and stochastic theories of responses. In: Olkin, I., Ghurye, S.G., Hoeffding, W., Madow, W.G., Mann, H.B. (eds.) Contributions to Probability and Statistics, pp. 97–132. Stanford University Press, Stanford (1960)

Bous, G., Fortemps, P., Glineur, F., Pirlot, M.: ACUTA: A novel method for eliciting additive value functions on the basis of holistic preference statements. European Journal of Operational Research 206, 435–444 (2010)

Burges, C.: Ranking as function approximation. In: Iske, A., Levesley, J. (eds.) Proceedings of the 5th International Conference on Algorithms for Approximation, pp. 3–18. Springer, Heidelberg (2006)

Burges, C., Shaked, T., Renshaw, E., Lazier, A., Deeds, M., Hamilton, N., Hullender, G.: Learning to rank using gradient descent. In: Proceedings of the 22nd International Conference on Machine Learning, pp. 89–96. ACM Press, New York (2005)

Cover, T.M., Thomas, J.A.: Elements of Information Theory. Wiley, Hoboken (2006)

Dyer, J.S.: MAUT – Multiattribute Utility Theory. In: Figueira, J., Greco, S., Ehrgott, M. (eds.) Multiple Criteria Decision Analysis: State of the Art Surveys. International Series in Operations Research & Management Science, vol. 78, pp. 265–292. Springer, New York (2005)

Fechner, G.T.: Elemente der Psychophysik. Breitkopf and Härtel, Leipzig (1860)

Figueira, J., Greco, S., Słowiński, R.: Building a set of additive value functions representing a reference preorder and intensities of preference: GRIP method. European Journal of Operational Research 195, 460–486 (2009)

Fishburn, P.C.: Utility Theory for Decision Making. Wiley, New York (1970)

Fishburn, P.C.: Binary choice probabilities: on the varieties of stochastic transitivity. Journal of Mathematical Psychology 10, 327–352 (1973)

Fishburn, P.C.: Stochastic utility. In: Barberà, S., Hammond, P.J., Seidl, C. (eds.) Handbook of Utility Theory, vol. 1, pp. 273–319. Kluwer, Dordrecht (1998)

Fletcher, R.: Practical Methods of Optimization. Wiley, Chichester (1987)

Fürnkranz, J., Hüllermeier, E.: Preference Learning: An Introduction. In: Fürnkranz, J., Hüllermeier, E. (eds.) Preference Learning, pp. 1–17. Springer, Berlin (2011)

Gill, P.E., Murray, W., Saunders, M.A., Wright, M.H.: Constrained nonlinear programming. In: Nemhauser, G.L., Rinnooy Kan, A.H.G., Todd, M.J. (eds.) Optimization - Handbooks in Operations Research and Management Science, pp. 171–210. North Holland, Amsterdam (1989)

Gill, P.E., Murray, W., Wright, M.H.: Practical Optimization. Academic Press, London (1981)

Greco, S., Mousseau, V., Słowiński, R.: Ordinal regression revisited: Multiple criteria ranking using a set of additive value functions. European Journal of Operational Research 191, 416–436 (2008)

Hastie, T., Tibshirani, R., Friedman, J.: The Elements of Statistical Learning: Data Mining, Inference and Prediction. Statistics. Springer, Berlin (2009)

Huard, P.: Resolution of mathematical programming with nonlinear constraints by the method of centers. In: Abadie, J. (ed.) Nonlinear Programming, pp. 209–219. Wiley, New York (1967)

Jacquet-Lagrèze, E., Siskos, Y.: Assessing a set of additive utility functions to multicriteria decision-making: the UTA method. European Journal of Operational Research 10, 151–164 (1982)

Kadziński, M., Greco, S., Słowiński, R.: Selection of a representative value function in robust multiple criteria ranking and choice. European Journal of Operational Research 217, 541–553 (2012)

Kahneman, D.: Attention and effort. Prentice-Hall, Englewood Cliffs (1973)

Luce, R.D., Suppes, P.: Preference, utility and subjective probability. In: Luce, R.D., Bush, R.R., Galanter, E. (eds.) Handbook of Mathematical Psychology, vol. III, pp. 249–410. Wiley, New York (1965)

McFadden, D.: The choice theory approach to market research. Marketing Science 5, 275–297 (1986)

Nesterov, Y., Nemirovsky, A.: Interior-point polynomial algorithms in convex programming. SIAM, Philadelphia (1994)

Shugan, S.M.: The cost of thinking. Journal of Consumer Research 7, 99–111 (1980)

Siskos, Y., Grigoroudis, E., Matsatsinis, N.: UTA methods. In: Figueira, J., Greco, S., Ehrgott, M. (eds.) Multiple Criteria Decision Analysis: State of the Art Surveys, pp. 297–344. Springer, Berlin (2005)

Siskos, Y., Yannacopoulos, D.: UTASTAR: an ordinal regression method for building additive value functions. Investigaçao Operacional 5, 39–53 (1985)

Sonnevend, G.: An 'analytical centre' for polyhedrons and new classes of global algorithms for linear (smooth, convex) programming. In: Prekopa, A., Szelezsan, J., Strazicky, B. (eds.) System Modelling and Optimization. LNCIS, vol. 84, pp. 866–876. Springer, Heidelberg (1985)

Thurstone, L.L.: A law of comparative judgment. Psychological Review 34, 278–286 (1927a)

Thurstone, L.L.: Psychophysical analysis. American Journal of Psychology 38, 368–389 (1927b)

Ye, Y.: Interior point algorithms: theory and analysis. Wiley, New York (1997)

An Evolutionary Algorithm for the Biobjective Capacitated m-Ring Star Problem*

Herminia I. Calvete[1,**], Carmen Galé[2], and José A. Iranzo[1]

[1] Dpto. de Métodos Estadísticos, IUMA, Universidad de Zaragoza, Pedro Cerbuna, 12, 50009 Zaragoza, Spain
{herminia,joseani}@unizar.es
[2] Dpto. de Métodos Estadísticos, IUMA, Universidad de Zaragoza, María de Luna 3, 50018 Zaragoza, Spain
cgale@unizar.es

Abstract. This paper addresses the biobjective capacitated m-ring star problem. The problem consists of finding a set of m simple cycles (rings) through a subset of nodes of a network. The network consists of a distinguished node called the depot and two different kinds of nodes, the customers and the transition points. Each ring contains the depot, a number of customers and some transition points. The customers not in any ring are directly connected to nodes in the rings. The rings must be node-disjoint and the total number of customers in a ring or connected to a ring is limited by the capacity of the ring. The aim is to minimize two objective functions, one referring to the cost due to the links of the rings and the other referring to the cost of allocating customers to nodes in the ring. An evolutionary algorithm is developed to approximate the Pareto front. The algorithm combines standard characteristics of evolutionary algorithms with the use of a heuristic to construct feasible solutions to the problem. A computational experiment is carried out using benchmark instances to show the performance of the algorithm.

1 Introduction

The capacitated m-ring star problem (CmRSP) aims to design a set of m cycles (rings) through a subset of nodes of a network (see Fig.1). The network consists of a distinguished node called the depot and two different kinds of nodes, the customers and the transition points. Each ring contains the depot, a number of customers and some transition points. The customers not in any ring are directly connected to nodes in the rings. The rings must be node-disjoint (except for the depot) and the total number of customers in a ring or connected to a ring is limited by the capacity of the ring. The relevant costs are due to the links of the rings (ring cost) and to the connections of customers with the nodes in the ring (allocation cost). The CmRSP was first proposed by Baldacci et al. [1]

---

* This research work has been funded by the Gobierno de Aragón under grant E58 (FSE) and by UZ-Santander under grant UZ2012-CIE-07.
** Corresponding author.

P. Perny, M. Pirlot, and A. Tsoukiàs (Eds.): ADT 2013, LNAI 8176, pp. 116–129, 2013.

in the context of designing telecommunication networks. The objective was to minimize the sum of the ring costs and the allocation costs. They present two integer programming formulations of the problem and propose a branch-and-cut approach to solve it. Hoshino and Souza [4] propose a branch-and-cut-and-price approach that outperforms in many instances the algorithm introduced in [1]. Mauttone et al. [6] develop a hybrid metaheuristic algorithm based on GRASP and Tabu Search. Naji et al. [7,8] propose two heuristics based on the use of different local search procedures and the use of the general Variable Neighborhood Search scheme together with exact optimization. These algorithms show a good performance in terms of computing time involved and closeness to the optimal solution.

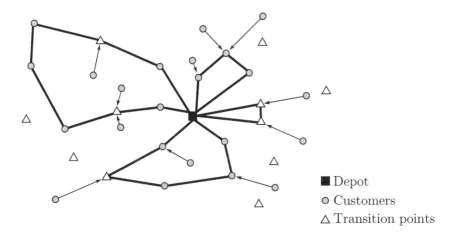

Fig. 1. A feasible solution of the CmRSP with $m = 4$

In this paper we address the biobjective capacitated m-ring star problem (B-CmRSP) in which both the ring cost and the allocation cost are considered individually instead of jointly. Liefooghe et al. [5] point out the importance of considering the ring cost and the allocation cost separately when dealing with the ring star problem, a particular case of the CmRSP in which there are no transitions points and only a ring is constructed without capacity bounds. They remark that both objective functions are comparable only if it is assumed that they are proportional one to another, which is rarely the case in practice. The CmRSP as formulated in [1] can be considered a scalar approach of the B-CmRSP in which the ring cost and the allocation cost are added. The aim of the paper is to propose an evolutionary algorithm to find a good approximation of the Pareto front. The paper is organized as follows. Section 2 states the biobjective capacitated m-ring star problem. Section 3 presents the main characteristics of the evolutionary algorithm developed to tackle the problem. In Section 4 the computational performance of the algorithm is evaluated using the benchmark instances dealt with in the literature. Finally, Section 5 concludes the paper with some final remarks and the main lines for future work.

2   The Biobjective Capacitated m-Ring Star Problem

Let $G = (V, E \cup A)$ be a mixed graph, where V is the node set, E is the edge set and A is the arc set. The node set is defined as $V = \{0\} \cup U \cup W$, where node 0 represents the depot, U is the set of customers and W is the set of transition points (Steiner nodes). The set of edges is $E = \{[i,j] : i,j \in V, i \neq j\}$. Edges represent undirected arcs used to link the nodes of the ring. The set of arcs is $A = \{(i,j) : i \in U, j \in V\}$. Arcs are directed links which are used to connect the customers to the ring. We assume that there is a nonnegative ring cost c_{ij} associated with each edge $[i,j]$, representing the cost of connecting nodes i and j, and a nonnegative allocation cost d_{ij} associated with each arc (i,j), referring to the cost of customer i being allocated (connected) to node j.

A ring R is a simple cycle visiting a subset of nodes including the depot. Each customer i is assigned to a ring R, meaning that either the ring R passes through the node i or the node i is allocated to a node of the ring R. Let m be the number of rings to be constructed and Q the capacity of each ring, i.e. the maximum number of customers which can be assigned to it. We assume that $mQ \geqslant |U|$, where $|U|$ stands for the cardinal of U, i.e. the number of customers.

A feasible solution $\mathcal{R} = \{R_1, \ldots, R_m, A_{\mathcal{R}}\}$ of the B-CmRSP consists of m rings $R_1, \ldots, R_m$ and a set of index pairs $A_{\mathcal{R}}$ so that each customer is in exactly one ring or is allocated to exactly one node of a ring, each transition point is visited at most once and the total number of customers assigned to each ring is less than or equal to Q. The nodes in the rings R_k, $k = 1, \ldots, m$, will be called ring nodes. Let V_k be the ring node set of the ring R_k. For each customer $i \notin V_1 \cup \cdots \cup V_m$, let j_i be the index of the node to which he is allocated. Hence, $A_{\mathcal{R}} = \{(i, j_i) : i \notin V_1 \cup \cdots \cup V_m\}$. Let $\mathcal{X}$ be the set of feasible solutions in the decision space.

The ring cost of the feasible solution $\mathcal{R}$ is defined as:

$$Z_1(\mathcal{R}) = \sum_{k=1}^{m} \sum_{[i,j] \in R_k} c_{ij} \tag{1}$$

and the allocation cost as:

$$Z_2(\mathcal{R}) = \sum_{(i,j_i) \in A_{\mathcal{R}}} d_{ij_i} \tag{2}$$

The B-CmRSP aims to minimize the objective functions (1)-(2) in the set of feasible solutions:

$$\min \quad [Z_1(\mathcal{R}), Z_2(\mathcal{R})]$$
$$\text{subject to} \ \ \mathcal{R} \in \mathcal{X}$$

For every $\mathcal{R} \in \mathcal{X}$, let $Z(\mathcal{R}) = [Z_1(\mathcal{R}), Z_2(\mathcal{R})]$ and let $\mathcal{Z} = \{Z(\mathcal{R}) : \mathcal{R} \in \mathcal{X}\}$ be the set of feasible points in the objective space. According to the theory of multiobjective optimization [3], a feasible solution $\mathcal{R}$ is efficient if and only if there is no other feasible solution $\widetilde{\mathcal{R}}$ so that $Z_s(\widetilde{\mathcal{R}}) \leqslant Z_s(\mathcal{R})$, $s = 1, 2$ and there exists $t \in \{1, 2\}$ such that $Z_t(\widetilde{\mathcal{R}}) < Z_t(\mathcal{R})$. A point $Z \in \mathcal{Z}$ is a nondominated outcome vector if there exists at least one efficient solution $\mathcal{R}$ so that $Z = Z(\mathcal{R})$. The set of all nondominated outcome vectors is the Pareto front.

3 An Evolutionary Algorithm

In order to approximate the Pareto front of the B-CmRSP we develop an evolutionary algorithm. One of its distinctive aspects is that the chromosome contains information on the ring nodes but neither on the rings themselves nor on the allocation of the remaining nodes.

3.1 Chromosome Encoding

We encode the chromosome as a binary $|V|$-dimensional vector $C \in \{0, 1\}^{|V|}$, so that for each $i \in V \setminus \{0\}$

$$C_i = \begin{cases} 1, & \text{if } i \text{ is a ring node} \\ 0, & \text{otherwise} \end{cases}$$

We define $C_0 = 1$, representing that node 0 is always in all the rings.

A chromosome is associated with a feasible solution of the B-CmRSP. Note that the chromosome C gives information on the ring nodes but gives no indication of how the rings are constructed and how the customers which are no ring nodes are allocated to the rings. In order to build the rings, we propose to use the main ideas of the heuristic algorithm proposed in [7] to construct an initial solution of the CmRSP.

For this purpose, we select the ring node i_1 which is furthest from the node 0. The first ring consists of the nodes 0 and i_1. Next the ring node i_2 is selected which is furthest from nodes 0 and i_1. The second ring consists of the nodes 0 and i_2. The process is continued until the ring node i_m is selected which is furthest from nodes $0, i_1, \ldots, i_{m-1}$ and the m-th ring is formed with nodes 0 and i_m. Let S be the set of ring nodes of the chromosome and $\widetilde{S} \subseteq S$. The furthest ring node from $\widetilde{S}$ is defined as $\arg\max\{\min\{c_{ij} : j \in \widetilde{S}\} : i \in S \setminus \widetilde{S}\}$.

From these initial m rings, the remaining nodes are randomly taken and the following procedure is applied:

- If $i \in U$ and $C_i = 1$, it is inserted in the ring which provides the minimum insertion cost and is not full of customers.
- If $i \in U$ and $C_i = 0$, it is allocated to the ring node which provides the minimum allocation cost, bearing in mind the constraint on the capacity of the corresponding ring.
- If $i \in W$ and $C_i = 1$, it is inserted in the ring which provides the minimum insertion cost.
- If $i \in W$ and $C_i = 0$, it is discarded.

Finally, two local search procedures are applied. In order to improve the rings, 2-opt local search is applied to each of the m rings. Next, to improve the allocation of the customers which are not ring nodes, these nodes are analyzed looking for a better feasible allocation. Therefore, each of them is selected in a random order and is allocated to the node ring with the lowest allocation cost belonging to a ring with available capacity. At the end of this process, we have a feasible solution of the B-CmRSP associated with the chromosome C.

3.2 Initial Population

The initial population is formed by randomly generated P chromosomes, where P represents the population size. In order to encourage diversification, for each chromosome, a number $p \in [0, 1]$ is generated. Then, each node is selected to be a ring node with probability p. Given the chromosome C, if

$$\sum_{r=1}^{|V|-1} C_r < m$$

the chromosome is 'repaired' by switching the allele of $m - \sum_{r=1}^{|V|-1} C_r$ nodes randomly selected amongst the current no ring nodes.

3.3 Crossover, Mutation and Local Search

We apply a uniform crossover operator which enables the parent chromosomes to contribute the gene level. For each population, the crossover operator randomly selects P pairs of parents and generates one offspring from each pair. Each gene of the offspring is selected from one of the parents with a probability of 0.5.

Next, the mutation operator is applied to the offspring. The mutation probability is 0.5. After a chromosome has been selected, a gene is randomly selected and its allele value is switched.

At the end of this process, if a chromosome has fewer than m ring nodes (excluding the depot) the chromosome is repaired as indicated above.

3.4 Fitness Evaluation

We define the *cost* of a chromosome as the pair formed by the ring cost and the allocation cost of its associated feasible solution. Based on the chromosome cost, the NSGA-II procedure [2] is applied to the incumbent population (current population plus offspring). First, the chromosomes are ranked into several nondominated fronts. The first nondominated front involves all nondominated points of the incumbent population. After discarding these points, the second nondominated front involves all nondominated points amongst the remaining ones. The process is repeated until all chromosomes of the incumbent population are classified and are assigned a nondomination rank. Moreover, for each chromosome a second value called crowding distance is computed which gives an estimation of the density of solutions surrounding the solution associated to the chromosome in the population. Then the new population is generated by selecting the best P individuals (without repetition) in accordance with the nondomination rank or, in case of a tie, according to the crowding distance. Moreover, an archive is maintained which contains potentially efficient solutions already found.

4   Computational Experiment

In order to analyze the performance of the algorithm, a computational experiment has been carried out. The numerical experiments have been performed on a PC Intel® Core™ I7-3820 CPU at 3.6 GHz having 32 GB of RAM under Ubuntu Linux 12.04 LTS. The code has been written in C++, GCC 4.6.3.

The algorithm has been tested on test problems generated following the ideas proposed in [1] to test the C*m*RSP. The problems were generated from TSPLIB instances [9] named *eil51*, *eil76* and *eil101*, whose node set consists of 51, 76 and 101 points. Moreover, an instance *eil26* with 26 nodes was created which consists of the first 26 nodes of *eil51*. The first point of the TSPLIB instance defines the depot, the following $|U|$ points are the customers and the remaining points are the transition points. For each TSPLIB instance, several B-C*m*RSP instances are generated by varying m and $|U|$. The number of rings is set to $m = 3, 4, 5$. The number of customers is given by $|U| = \lfloor \alpha(n-1) \rfloor$ where n is the number of nodes and $\alpha = 0.25, 0.5, 0.75, 1.0$. The ring capacity Q is given by $Q = \lceil \frac{|U|}{0.9m} \rceil$. The ring cost c_{ij} and the allocation cost d_{ij} of a pair of nodes i, j are equal to the Euclidean distance according to the TSPLIB standard, i.e. both costs are integer. Unlike instances in [1], no constraint is set on the nodes to which a customer can be allocated.

The aim of the study was to assess the influence of the population size P and the computing time involved on the performance of the algorithm. We have taken $P = 100$ and $P = 200$ and the total computing time has been established at 5 minutes, recording the results every 30 seconds. A set of 20 runs per instance has been executed. To evaluate the performance of the algorithm we have used the additive ϵ indicator $I^1_{\epsilon+}$ and the hypervolume indicator I^-_H. They have been computed using the tool suite provided in PISA (available at http://www.tik.ee.ethz.ch/pisa/?page=assessment.php). For each instance, the reference set has been computed as the Pareto front of the union of the outputs obtained throughout the whole experiment.

From the study we conclude that there are no significant differences between the values of both indices obtained when varying the population size. Hence, Tables 1 and 2 display only the results of the experiment when $P = 100$. Both tables are similar, except for the indicator shown. The first column gives the name of the problem. The second, third and fourth columns show the number of rings, the number of customers and the capacity value. The fifth to fourteenth columns show the average of the corresponding indicator values obtained in the 20 runs (in order to obtain the real value, the number in the table must be multiplied by 10^{-3}). These results are displayed in Figs. 2-5. These tables and

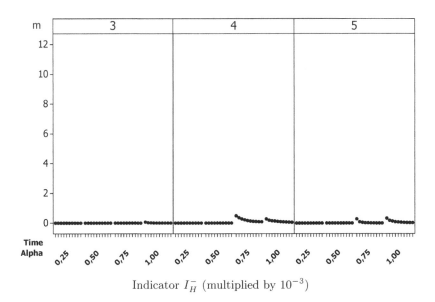

Indicator I_H^- (multiplied by 10^{-3})

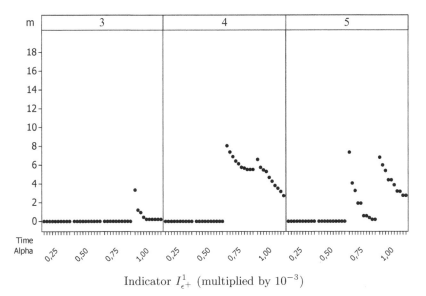

Indicator $I_{\epsilon+}^1$ (multiplied by 10^{-3})

Fig. 2. Problem *eil26*

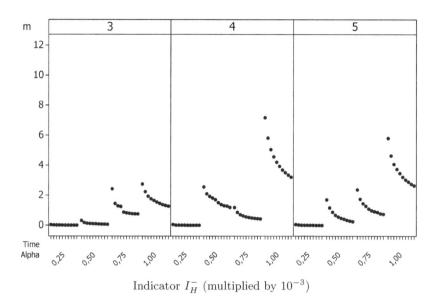

Indicator I_H^- (multiplied by 10^{-3})

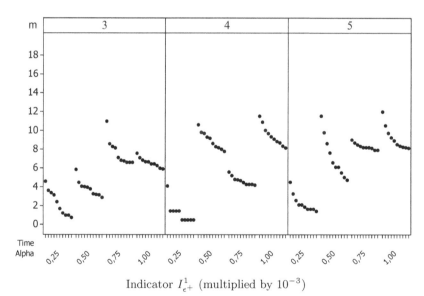

Indicator $I_{\epsilon+}^1$ (multiplied by 10^{-3})

Fig. 3. Problem *eil51*

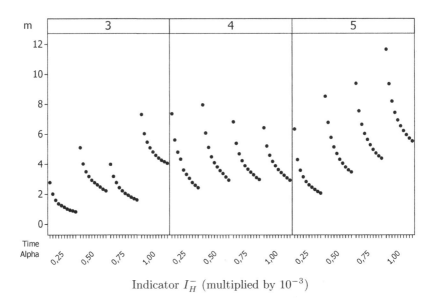

Indicator I_H^- (multiplied by 10^{-3})

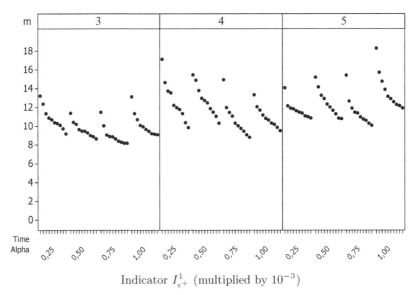

Indicator $I_{\epsilon^+}^1$ (multiplied by 10^{-3})

Fig. 4. Problem *eil76*

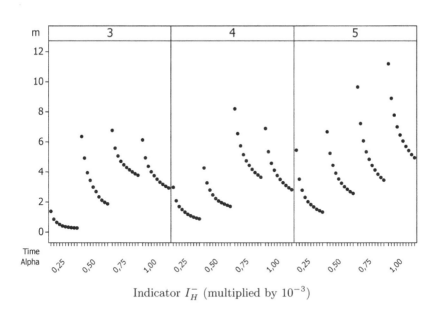

Indicator I_H^- (multiplied by 10^{-3})

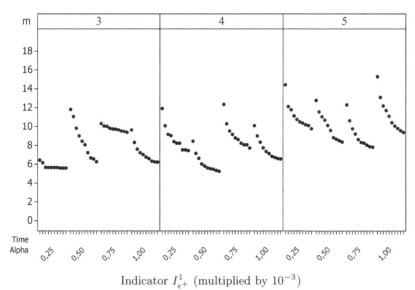

Indicator $I_{\epsilon+}^1$ (multiplied by 10^{-3})

Fig. 5. Problem *eil101*

Table 1. Results of the I_H^- indicator, multiplied by 10^{-3} ($P = 100$)

Name	m	$|U|$	Q	Computing time in seconds									
				30	60	90	120	150	180	210	240	270	300
eil26	3	6	3	0.00	0.00	0.00	0.00	0.00	0.00	0.00	0.00	0.00	0.00
eil26	3	12	5	0.00	0.00	0.00	0.00	0.00	0.00	0.00	0.00	0.00	0.00
eil26	3	18	7	0.00	0.00	0.00	0.00	0.00	0.00	0.00	0.00	0.00	0.00
eil26	3	25	10	0.07	0.03	0.02	0.01	0.00	0.00	0.00	0.00	0.00	0.00
eil26	4	6	2	0.00	0.00	0.00	0.00	0.00	0.00	0.00	0.00	0.00	0.00
eil26	4	12	4	0.00	0.00	0.00	0.00	0.00	0.00	0.00	0.00	0.00	0.00
eil26	4	18	5	0.49	0.36	0.28	0.22	0.17	0.14	0.12	0.11	0.09	0.09
eil26	4	25	7	0.28	0.19	0.16	0.14	0.12	0.09	0.09	0.08	0.07	0.06
eil26	5	6	2	0.00	0.00	0.00	0.00	0.00	0.00	0.00	0.00	0.00	0.00
eil26	5	12	3	0.00	0.00	0.00	0.00	0.00	0.00	0.00	0.00	0.00	0.00
eil26	5	18	4	0.27	0.08	0.04	0.02	0.02	0.01	0.01	0.00	0.00	0.00
eil26	5	25	6	0.32	0.18	0.13	0.08	0.07	0.06	0.04	0.04	0.03	0.03
eil51	3	12	5	0.04	0.02	0.02	0.02	0.01	0.01	0.01	0.00	0.00	0.00
eil51	3	25	10	0.32	0.18	0.14	0.12	0.11	0.10	0.09	0.08	0.07	0.07
eil51	3	37	14	2.42	1.45	1.29	1.25	0.88	0.83	0.80	0.78	0.77	0.76
eil51	3	50	19	2.75	2.24	1.93	1.76	1.65	1.55	1.45	1.38	1.32	1.28
eil51	4	12	4	0.06	0.01	0.01	0.01	0.01	0.00	0.00	0.00	0.00	0.00
eil51	4	25	7	2.56	2.10	1.93	1.82	1.71	1.51	1.38	1.31	1.28	1.19
eil51	4	37	11	1.18	0.86	0.71	0.63	0.57	0.53	0.49	0.47	0.45	0.43
eil51	4	50	14	7.17	5.82	5.06	4.57	4.22	3.93	3.68	3.51	3.35	3.21
eil51	5	12	3	0.06	0.03	0.02	0.01	0.01	0.01	0.01	0.01	0.01	0.01
eil51	5	25	6	1.71	1.18	0.89	0.69	0.57	0.48	0.43	0.35	0.30	0.26
eil51	5	37	9	2.38	1.75	1.46	1.29	1.10	0.99	0.93	0.88	0.78	0.75
eil51	5	50	12	5.81	4.64	4.08	3.74	3.47	3.22	3.03	2.90	2.75	2.65
eil76	3	18	7	2.78	2.02	1.60	1.37	1.25	1.14	1.02	0.95	0.90	0.83
eil76	3	37	14	5.11	4.02	3.50	3.18	2.93	2.77	2.63	2.48	2.33	2.23
eil76	3	56	21	4.00	3.18	2.77	2.44	2.23	2.05	1.93	1.80	1.70	1.62
eil76	3	75	28	7.32	6.04	5.48	5.11	4.83	4.60	4.42	4.28	4.17	4.07
eil76	4	18	5	7.37	5.62	4.81	4.34	3.61	3.32	3.06	2.78	2.60	2.44
eil76	4	37	11	7.96	6.08	5.13	4.50	4.11	3.82	3.57	3.38	3.14	2.94
eil76	4	56	16	6.83	5.40	4.70	4.24	3.90	3.66	3.46	3.29	3.10	2.97
eil76	4	75	21	6.43	5.22	4.60	4.19	3.90	3.64	3.44	3.26	3.08	2.93
eil76	5	18	4	6.34	4.30	3.58	3.15	2.83	2.62	2.46	2.31	2.17	2.06
eil76	5	37	9	8.53	6.78	5.78	5.15	4.71	4.33	4.07	3.81	3.61	3.47
eil76	5	56	13	9.39	7.55	6.65	6.04	5.64	5.29	4.98	4.74	4.55	4.40
eil76	5	75	17	11.66	9.36	8.19	7.44	6.94	6.55	6.23	5.96	5.73	5.54
eil101	3	25	10	1.38	0.86	0.64	0.51	0.41	0.36	0.33	0.30	0.28	0.27
eil101	3	50	19	6.36	4.93	3.94	3.43	2.98	2.69	2.34	2.12	1.98	1.88
eil101	3	75	28	6.76	5.58	5.07	4.71	4.48	4.29	4.14	4.00	3.87	3.77
eil101	3	100	38	6.13	4.94	4.38	4.01	3.75	3.51	3.32	3.17	3.04	2.93
eil101	4	25	7	2.98	2.08	1.70	1.50	1.33	1.18	1.09	1.00	0.93	0.88
eil101	4	50	14	4.26	3.27	2.79	2.47	2.23	2.07	1.96	1.87	1.78	1.71
eil101	4	75	21	8.20	6.55	5.74	5.16	4.75	4.44	4.16	3.96	3.79	3.64
eil101	4	100	28	6.89	5.35	4.58	4.11	3.76	3.50	3.28	3.10	2.94	2.81
eil101	5	25	6	5.45	3.51	2.78	2.30	2.02	1.83	1.67	1.52	1.42	1.33
eil101	5	50	12	6.67	5.24	4.44	3.91	3.53	3.24	3.01	2.84	2.68	2.57
eil101	5	75	17	9.65	7.22	6.07	5.34	4.84	4.43	4.12	3.84	3.63	3.45
eil101	5	100	23	11.19	8.89	7.78	7.00	6.46	6.05	5.70	5.43	5.16	4.95

Table 2. Results of the $I_{\epsilon+}^1$ indicator, multiplied by 10^{-3} ($P = 100$)

Name	m	$|U|$	Q	Computing time in seconds									
				30	60	90	120	150	180	210	240	270	300
eil26	3	6	3	0.00	0.00	0.00	0.00	0.00	0.00	0.00	0.00	0.00	0.00
eil26	3	12	5	0.00	0.00	0.00	0.00	0.00	0.00	0.00	0.00	0.00	0.00
eil26	3	18	7	0.00	0.00	0.00	0.00	0.00	0.00	0.00	0.00	0.00	0.00
eil26	3	25	10	3.35	1.21	0.95	0.44	0.22	0.22	0.22	0.22	0.22	0.22
eil26	4	6	2	0.00	0.00	0.00	0.00	0.00	0.00	0.00	0.00	0.00	0.00
eil26	4	12	4	0.00	0.00	0.00	0.00	0.00	0.00	0.00	0.00	0.00	0.00
eil26	4	18	5	8.05	7.38	6.88	6.42	6.13	5.76	5.67	5.52	5.52	5.52
eil26	4	25	7	6.59	5.75	5.47	5.31	4.68	4.27	3.82	3.52	3.18	2.73
eil26	5	6	2	0.00	0.00	0.00	0.00	0.00	0.00	0.00	0.00	0.00	0.00
eil26	5	12	3	0.00	0.00	0.00	0.00	0.00	0.00	0.00	0.00	0.00	0.00
eil26	5	18	4	7.35	4.07	3.28	1.92	1.92	0.59	0.59	0.39	0.20	0.20
eil26	5	25	6	6.81	6.00	5.40	4.42	4.42	3.89	3.22	3.19	2.77	2.77
eil51	3	12	5	4.59	3.62	3.38	3.14	2.42	1.69	1.21	0.97	0.97	0.73
eil51	3	25	10	5.84	4.47	4.06	4.02	3.96	3.79	3.26	3.20	3.16	2.88
eil51	3	37	14	10.98	8.58	8.28	8.14	7.11	6.82	6.75	6.60	6.60	6.60
eil51	3	50	19	7.57	7.09	6.83	6.67	6.65	6.44	6.44	6.26	5.98	5.91
eil51	4	12	4	4.10	1.44	1.44	1.44	1.44	0.48	0.48	0.48	0.48	0.48
eil51	4	25	7	10.62	9.81	9.70	9.30	9.18	8.61	8.28	8.17	8.00	7.77
eil51	4	37	11	5.59	5.21	4.80	4.76	4.67	4.47	4.27	4.27	4.27	4.20
eil51	4	50	14	11.53	10.90	10.03	9.69	9.35	9.08	8.83	8.68	8.34	8.15
eil51	5	12	3	4.50	3.28	2.57	2.11	2.11	1.87	1.64	1.64	1.64	1.41
eil51	5	25	6	11.54	9.79	8.61	7.63	6.58	6.11	6.11	5.51	5.02	4.74
eil51	5	37	9	9.00	8.69	8.48	8.33	8.20	8.19	8.19	8.12	7.92	7.92
eil51	5	50	12	11.99	10.53	9.73	9.24	8.91	8.52	8.35	8.24	8.19	8.13
eil76	3	18	7	13.22	12.37	11.34	10.89	10.70	10.39	10.29	10.11	9.73	9.17
eil76	3	37	14	11.38	10.42	10.21	9.64	9.49	9.48	9.29	9.00	8.87	8.64
eil76	3	56	21	11.48	10.05	9.05	8.89	8.85	8.63	8.39	8.28	8.18	8.18
eil76	3	75	28	13.12	11.34	10.70	10.09	9.94	9.66	9.46	9.20	9.14	9.09
eil76	4	18	5	17.13	14.63	13.73	13.54	12.21	11.96	11.77	11.33	10.37	9.85
eil76	4	37	11	15.45	14.89	13.78	12.97	12.74	12.47	11.87	11.49	11.05	10.32
eil76	4	56	16	14.93	11.97	11.45	11.07	10.32	10.04	9.76	9.46	9.07	8.81
eil76	4	75	21	13.33	12.06	11.70	11.17	10.83	10.66	10.33	10.19	9.87	9.51
eil76	5	18	4	14.07	12.15	11.91	11.82	11.62	11.47	11.38	11.07	11.00	10.85
eil76	5	37	9	15.19	14.16	13.27	12.92	12.33	12.03	11.67	11.29	10.84	10.78
eil76	5	56	13	15.41	12.66	11.90	11.47	11.38	10.97	10.75	10.60	10.30	10.08
eil76	5	75	17	18.29	15.71	14.76	13.89	13.15	12.91	12.58	12.28	12.17	11.92
eil101	3	25	10	6.43	6.16	5.67	5.66	5.66	5.66	5.66	5.60	5.60	5.60
eil101	3	50	19	11.82	11.06	9.83	8.99	8.44	8.06	7.22	6.66	6.56	6.27
eil101	3	75	28	10.32	10.06	10.02	9.81	9.75	9.71	9.64	9.52	9.48	9.38
eil101	3	100	38	9.63	8.30	7.59	7.22	7.02	6.76	6.59	6.33	6.25	6.23
eil101	4	25	7	11.92	10.09	9.16	9.03	8.39	8.23	8.23	7.52	7.52	7.44
eil101	4	50	14	8.44	7.15	6.62	6.03	5.81	5.62	5.53	5.47	5.35	5.25
eil101	4	75	21	12.34	10.30	9.52	9.15	8.78	8.62	8.23	8.07	8.06	7.71
eil101	4	100	28	10.10	8.99	8.33	7.74	7.35	7.13	6.83	6.71	6.60	6.56
eil101	5	25	6	14.43	12.12	11.78	11.13	10.76	10.50	10.37	10.20	10.11	9.77
eil101	5	50	12	12.77	11.56	11.04	10.69	10.13	9.58	8.82	8.65	8.50	8.37
eil101	5	75	17	12.30	10.62	9.76	9.20	8.62	8.31	8.25	8.03	7.86	7.81
eil101	5	100	23	15.29	13.10	12.20	11.71	11.10	10.42	10.05	9.80	9.55	9.36

figures allow us to assess the evolution of the corresponding indicator when the computing time varies. For the smaller problems, these indices are zero or close to zero from the smaller computing time values. For the remaining problems the indices decrease as the computing time increases. However, both indices have a different behavior. For instance, for the problem $eil26$, $m = 3$, $|U| = 25$, the indicator I_H^- ranges from 0.07×10^{-3} to 0.00 whereas the indicator $I_{\epsilon+}^1$ ranges from 3.35×10^{-3} to 0.22×10^{-3}. Moreover, as we might expect, the characteristics of the problem, i.e. the number of nodes, rings and customers, greatly affect the value of the indicators and hence the slope of the curve.

5 Conclusions and Future Work

This paper addresses for the first time the biobjective capacitated m ring star problem and proposes an evolutionary algorithm for approaching the Pareto front. The chromosomes of the algorithm contain information on the ring nodes but neither on the rings themselves nor on the allocation of the remaining nodes. A feasible solution of the problem associated to the chromosome is constructed by applying a heuristic algorithm. The evolutionary algorithm has been tested on a set of benchmark problems. Our work can be extended in several ways. First, the proposed algorithm could be enriched by considering local search procedures which allow for the interchange of ring nodes which belong to different rings. Moreover, other techniques to construct the rings from the information provided by the chromosome should be assessed. Second, it is worthwhile to explore different approaches to approximate the Pareto front based on local search. We hope to explore these extensions in future work.

References

1. Baldacci, R., Dell'Amico, M., Salazar González, J.J.: The capacitated m-ring-star problem. Operations Research 55(6), 1147–1162 (2007)
2. Deb, K., Pratap, A., Agrawal, S., Meyarivan, T.: A fast and elitist multiobjective genetic algorithm: NSGA-II. IEEE Transactions on Evolutionary Computation 6(2), 182–197 (2002)
3. Ehrgott, M.: Multicriteria Optimization, 2nd edn. Springer, Berlin (2005)
4. Hoshino, E.A., de Souza, C.C.: A branch-and-cut-and-price approach for the capacitated m-ring-star problem. Discrete Applied Mathematics 160(18), 2728–2741 (2012)
5. Liefooghe, A., Jourdan, L., Talbi, E.-G.: Metaheuristics and cooperative approaches for the bi-objective ring star problem. Computers and Operations Research 37(6), 1033–1044 (2010)
6. Mauttone, A., Nesmachnow, S., Olivera, A., Robledo, F.: A hybrid metaheuristic algorithm to solve the capacitated m-ring star problem. In: International Network Optimization Conference (2007)

7. Naji-Azimi, Z., Salari, M., Toth, P.: A heuristic procedure for the capacitated m-ring-star problem. European Journal of Operational Research 207(3), 1227–1234 (2010)
8. Naji-Azimi, Z., Salari, M., Toth, P.: An integer linear programming based heuristic for the capacitated m-ring-star problem. European Journal of Operational Research 217(1), 17–25 (2012)
9. Reinelt, G.: TSPLIB–a traveling salesman problem library. Journal of Computing 3(4), 376–384 (1991)

Planning System for Emergency Services

Adil Chennaoui[1] and Marc Paquet[2]

[1] département de génie des technologies de l'information,
[2] département de génie de la production automatisée,
École de technologie supérieure, 1100, rue Notre-Dame Ouest, Montréal (Québec) H3C 1K3
{adil.chennaoui.1,marc.paquet}@etsmtl.ca

Abstract. In this paper, we are dealing with the problem of paramedics scheduling, we propose a new system sufficiently flexible for its resolution. The optimization method used to solve this problem has been selected based mainly on the time available to solve an instance of the problem and the required level for the quality of the solution. Our goal was to produce a usable system on desktop computers as well as being used frequently for evaluating a multiple scenarios. Our approach is an adaptation of a linear programming model that solves a Pydxp6g4roblem of coverage, also called "set covering", by selecting the optimal mix of elements from the bank of the available scheduling patterns.

Keywords: Planning, scheduling, optimization, decision support, simulation.

1   Introduction

The "scheduling" is a very wide range of research. [Ernst-1] has identified more than 600 scientific articles in this field. Making scheduling staff is just one of the various branches of "scheduling". However, very few articles of scheduling deal with the ambulance services or even emergency services in general (for example: police, fire, etc.).

A quick literature reviews show that very few articles resume the same modeling solutions that have been already used. This still true even if a very large number of articles deal with problems having a lot issues in common. For instance, there are more than 100 articles that deal with nurses only as identified in [Ernst-1]. This is an indication of the specificity of each problem making schedules, and the difficulty of adapting an existing model to a new problem. The models are generally-specific to each situation.

The classification of issues making schedules made by [Cheang] is structured, detailed, and logical and it is made by many of the same authors who designed the annotated bibliography [Ernst-1]. The classification of [Ernst-2] (that concerns the Schedule process) is made by modules; a problem can be solved by using one or more of these modules. These modules are to some extent the steps of a process of scheduling. And to our knowledge, only one article studies the paramedics scheduling. This is the [Ernst-0]. More than a dozen other articles discuss "ambulance scheduling", but they focus on other issues such as ambulance positioning of vehicles in the territory.

In this article, we focus on the problem of making schedule paramedics in "*Urgences-santé de Québec*" (U.S.) that has the responsibility to provide pre-hospital

P. Perny, M. Pirlot, and A. Tsoukiàs (Eds.): ADT 2013, LNAI 8176, pp. 130–138, 2013.

services to the islands of Montreal and Laval. Paramedics are working continuously round the clock, 7 days a week. Some of them work full-time (FT), other part-time (PT). Some have regular jobs with strict working conditions, others do not. To cover a large territory, the vehicles have only three deposits. Paramedics do not have the same skill levels. Work rules are governed by a relatively strict and complex collective agreement, but also by several work habits deeply rooted in the culture of the company and therefore difficult to change. In addition, time must be made in order to take into account implicitly vehicle management.

The system proposed in this paper enable us to optimize and reduce considerably the time effectively and making schedules paramedics.

This system consists of two main parts. The first part is an algorithmic component incorporating the "set covering" and the constraint programming approach. The second part is a human-machine interface (HMI) facilitating to the user the intervention in the resolution process and visualization of the results.

Our system has been tested in "Urgences-santé de Québec". We present the results of tests on real data to scheduling paramedics for the year 2013.

2   Definition of Paramedic Scheduling Problem

The problem of scheduling is part of a class of problems that are very complex to model and solve. In fact, this problem is of nature combinatorial in which several constraints are conflicting.

Solve a paramedics scheduling problem equivalent to specifying the number of paramedics, vehicles, work days, days off and duration of the working day and lunch break for each employee. A permanent service 24 hours a day and 7 days a week, must be assured.

Paramedics are grouped into pairs on each vehicle. Each pair of paramedics has the same schedule for the full duration of a quarter. Pairs of paramedics are always assigned to a vehicle. So the problem is to solve a problem of scheduling pairs of paramedics and this leads implicitly to take into account the management of vehicles. This allows us to simplify the representation of the problem by simply combining two paramedics to the same line of the schedule.

Considering the large number of paramedics to which we must make a schedule and also the number of shifts possible for one pair of paramedics, the problem to resolve becomes very complex. Indeed, in a real context such as that of U.S., there are over 600 paramedics, so more than 300 pairs, and the shifts of each of them can start almost any time of day, any day of the week. The size of the space of possible solutions to a problem instance is huge.

Since we have only few minutes to resolve the problem (to be usable in a real context), a heuristic (and even a meta-heuristic) could explore a very small part of the solution space. We therefore concluded that the problem must be broken.

The conflicting nature of the constraints of the problem has naturally led us to divide them into two types: hard and flexible constraints that can be violated but we minimize the violation.

3   Solving Approach

Following an analysis of the structure of the problem and its characteristics, our choice fell on a decomposition of the problem in two steps:

3.1   The First Step: The Bank of Schedules Patterns

It allows creating a database of schedules patterns from a merger of two distinct elements: the cycle and day. After the creation of cycles and days, the module developed at this level enables to generate schedules patterns bank and calculate the annual cost (in hours) for each zone based on the following process:

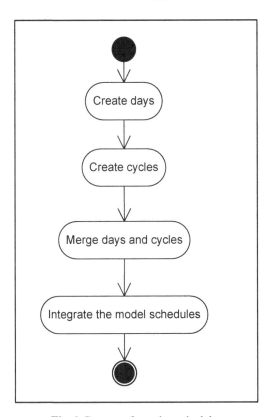

Fig. 1. Process of creating schedules

In the figure above, the schedules that are bank scheduling patterns are obtained with the fusion of cycles and days via the GUI of our system in which we have developed algorithmic components integrating constraint programming. This bank will be supplied as required by new combinations cycle / day and can be operated in whole or in part in making different scenarios. This module also includes a calculation function of ergonomic note per schedule.

3.2   The Second Step: The Model Resolution

Like any optimization problem, our problem requires the optimization of a function objective respecting certain constraints. Our approach is the application of a linear programming model that solves a problem of coverage, also called "set covering". This problem is resolved by selecting time from the bank scheduling patterns available in order to meet the real demand accurately and optimize management of the fleet while maintaining the best possible operational and financial constraints and ergonomic criteria.

All data in our optimization system are stored in a database. The majority of data constitutes the constraints of the linear program schedules, columns and the results of previously optimized time models.

The first step to perform in our approach is to assess the demand for ambulance transport. This application is assessed using historical data such as the number of interventions each time of the day and the duration of these interventions. The results of the analysis are expressed in our system in terms of number of vehicles required on the entire territory and discretized into half-hour for the seven different days of the week.

The objective function is to minimize the sum of costs associated with time and the amount of costs associated with the management of the fleet. It should be noted that the costs will not consist of monetary costs, but the cost in terms of penalties in violations of ergonomic constraints or the number of hours each year by that time. The more violations are great, the greater the penalty in the objective function will be. This aims to:

Minimize:
Summation (associated with each hourly cost * amount each time) + (cost for each additional vehicle * number of additional vehicles) - (associated with a unit gap between demand and offer service cost)

The objective defined in the objective function is to minimize the violations of soft constraints. This is due to the approach that is more focused on ensuring that employees have a certain number of hours of work over a definite period.

It should there fore be highlighted that all variables can be set via the HMI of the proposed system.

4   Methodology

In our model, we selected periods of 30 minutes (half an hour) because we did not seem to have more accurate discretization and half-hour discretization is generally acceptable for many making scheduling problems. However, given the rapidity of the resolution of our system obtained during the tests, it would be quite possible to solve it with a more finite discretization if the actual application context required it.

The translation in discretization per periods of half an hour has to go along with a selection of solutions shifts from the bank of schedules patterns so that the beginning of shifts would be the best possible spread in the resolution horizon. This spread is necessary to optimize the management of the fleet to limit the number of the required vehicles, but also to optimize the coverage of demand that varies greatly from half an hour to another of a day.

The first phase of our model (the creation of schedules patterns bank) is solved by a constraint programming and a set of algorithms translated into computer modules. The second phase of the production schedule is determined by a mixed linear program. The relatively simple structure of the modeling problem is the main motivation for this choice. Indeed, given that this problem can be solved optimally within a very short time by this method, at this stage, it is not necessary to consider other possibilities of solving methods.

Our model optimizes the location of each quarter of the solution and locates their breaks in order to maintain the best possible coverage of the application.

Specifically, the objective function minimizes the sum of violations of   short-comings in offer, surplus in offer, the maximum number of employees in full and part-time, the maximum number of vehicles, of non-respected holiday requests, of soft ergonomic constraints and preference.

Our goal is to get a workable solution in a real environment with a consideration of breaks. The quarters of our model are variables; duration shift can be 8 hours with a 30 minute break, 10 hours with a 45 minute break or 12 hours with 60 minutes break (rules of the collective agreement). For each day, there are three possible types of quarters which are: night (N), day (D) and evening (E). Shifts can start at the earliest at 5:00 and at the latest at 23:30. Every paramedic must have a weekend off every two weeks.

5   Analysis of Numerical Results

In this section, we will present the preliminary results that we have obtained using our system called SYSCONF. The evaluation of the quality of the results is simple to make for certain elements like cover the demand. However, with regard to other elements such as the quality of individual schedules, evaluation is less simple to do.

We created a list of measures allowing the user to assess the quality of a schedule as a whole. These measures are:

— Hours of vehicles
— Number of man hours
— Number of times.
— Ergonomic Cost
— Number of vehicles
— Number of additional vehicles
— Cover demand
— Percentage of shifts starting less than 12 hours after the previous quarter.
— Execution Time
— Number of hours by type of cycle / day.

We considered that these statistics are those necessary and / or useful to be able to assess the quality of a schedule, in terms of respecting the demand and also the compliance with the available number of vehicles, ergonomic, but most of all in terms of its potential implementation in U.S.

SYSCONF contains several parameters. The vast majority of these are included in the of schedules patterns bank. Making the schedules model requires very little setup.

The size of the problem is a parameter of high importance because it affects significantly over the duration of resolution. Of course, the user has little control over this parameter (it can identify the amount and nature of schedules to include in the model).

The size of the problem has three dimensions: the number of employees, the schedule selection and the horizon of resolution. A normal instance has an horizon of two weeks (14 days) and between 300 to 400 pairs of paramedics. Under these conditions, it is possible to obtain a satisfactory solution generally between 1 and 30 minutes of computing time on a desktop Intel Core Duo CPU E6850 at 3.00 GHz and 3.48 RAM. The time resolution is highly variable because it depends on the number of constraints and integrated active parameters of the model we are trying to resolve. Generally, the more the number of constraints and parameters are activated, the longer the time of resolution will be.

The tests in this section are of a size of 298 schedules, a schedule for each pair of paramedics (278 schedules full-time and 20 part-time) over a period of two weeks as it is illustrated in the following table:

Table 1. Number of schedules to make by SYSCONF for each test

Number of schedules	Test 1	Test 2	Test 3
Full time	278	278	278
Part time	20	20	20
Total	298	298	298

This corresponds to a time model for the employees in the U.S.; for the 596 paramedics 556 are FT and 40 are TP.

For the three tests, SYSCONF responded fully to the request of the weekend with 596 employees.

The constraint of the covered demand pushed the offer to be greater than or equal to the demand for all periods of the year where it is feasible. If the request cannot be fully covered for a given period, the objective function is penalized. The surplus in offer is not taken into account since they have no direct negative impact on the quality of the solution. The indirect consequences of a surplus in offer are the improper distribution on the horizon of the early shift (causing bottlenecks in vehicle warehouse) and the need to add additional vehicles to the fleet when we have a surplus in offer during peak periods.

The constraint of respecting the number of available vehicles in the fleet is very important for two following reasons: first, an increase in the number of the required vehicles usually occurs in parallel with an increase in traffic while storing vehicles, which conducts to bottlenecks during the go in/out of vehicles. These bottlenecks increase the complexity of managing the fleet and reduce the availability of vehicles. Second, the cost of adding an additional vehicle is very expensive and is around $ 200,000 over five years. This cost does not include additional products by managing problems caused by the increase in fleet size and the increase in traffic while storing the additional vehicles to the fleet.

Ergonomics is the less penalized criterion in the phase of the generation of schedules patterns bank.

The criteria for judgment used in our tests were always stated in terms of the number of iterations without improving the best known solution.

The main parameters selected in each test are:

Table 2. The main parameters of our time models for the 3 tests

	Test 1	Test 2	Test 3
Maximum number of vehicles	110	115	123
Maximum number of additional vehicles	2	2	2
Part-time	20	20	20
Full time	278	278	278
Cycle	1	1	1
Cost of additional vehicle	1000	1000	1000
Surplus gap	200	200	200
Type of cycle (for each test)	• 6 days on 14 • 7 days on 14 • 8 days on 14 • 10 days on 14		
Type of day (for each test)	• 8 hours • 10 hours • 12 hours		
Start time shift	From 5h00 at the earliest to 23h30 at the latest		
Schedules (for each test)	• All schedules are 10 working days on 14 with a weekend off every two weeks and 8 hours with 30 minutes of breaks. • All schedules are 8 working days on 14, with a weekend off every two weeks and 10 hours with 45 minutes breaks. • All schedules are 7 working days on 14, with a weekend off every two weeks and 12 hours with 60 minutes of breaks. • All schedules are 6 working days on 14, with a weekend off every two weeks and 10 hours with 45 minutes breaks.		

The table above summarizes the main parameters of three tests we performed in SYSCONF when making three scenarios schedules models.

It should be highlighted that we have not used the constraints relative to schedules groupings or constraints related to the limitations of a certain type of schedules.

The following table shows the duration of resolution in each test:

Table 3. Time resolution

	Test 1	Test 2	Test 3
Time	00:10:18.63	00:10:18.78	00:33:39.80

By changing a few parameters, the running time differs from one test to another including the maximum number of vehicles.

The following table shows the distribution of results for the three tests based on the type of schedules:

Table 4. Distribution of schedules by type for the three tests

Type	Test 1	Test 2	Test 3
614 / 10h	20	20	20
714 / 12h	205	202	206
814 / 10h	13	10	10
1014 / 8h	60	66	62
Total	298	298	298

These results are presented in detail in SYSCONF as a report description schedules for solution.

The report description schedules gives a detailed presentation of all the schedules of the solution indicating the cycle type and the schedule type, the start and end of the shift, the schedule number, the required number of schedule per combination to meet the demand, the hour break, and a graphical presentation of working days and holiday on 14 days cycle.

Summary of other results is presented in the following table:

Table 5. Summary of other test results

Item	Test 1	Test 2	Test 3
Hours-vehicles	568.483	568.750	568.769
Man-hours	1.136.967	1.137.500	1.137.539
Schedules	596	596	596

The strength of our method is the rapid resolution of the linear program since it (the linear program) must select the schedules among a sample to be included in the solution and take into consideration several types of shift. In the short term, we see SYSCONF as an excellent tool for scenarios simulation, for modifications and for improvements of existing structures and processes of schedules in U.S. To conclude, the following section will discuss the avenues for improving our SYSCONF system.

6   Conclusion and Future Application

The objective of SYSCONF was to produce simple and flexible schedules with best quality that can responses to the majority of organisational and operational constraints of U.S. We believe this is achieved quickly and our model provides good solutions on desktop machines and its flexibility allows you to test and evaluate a large number of scenarios without ever having to create manually schedules. Schedules created are ergonomic and easily meet the demand as the number of employees is sufficient. In addition, fleet management is done efficiently, thereby reducing the high costs associated with the acquisition of additional vehicles and several instances have achieved savings of 15% in terms of number of vehicles required in relation to the schedules model of U.S. SYSCONF is as effective in terms of time needed for the preparation of a schedule.

Future research should be conducted in order to add components to optimize the distribution shifts in operational center. Finally, the addition of a replacements management system in real time could further improve SYSCONF and this is another focus which would be interesting to study.

References

1. Blais, M., Guertin, Y., Morel, C.: « Horaires de travail des techniciens ambu-lanciers: État de la situation ». Service de statistiques et recherche opération-nelle, Urgences-Santé de Québec, 20 pages (2011)
2. Cheang, B., Li, H., Lim, A.: Rodrigues: "Nurse rostering problems - A bibliographic survey". European Journal of Operational Research - Amsterdam 151(3), 447–460 (2003)
3. Ernst, A.T., Hourigan, P., Krishnamoorthy, M., Mills, G., Nott, H., Sier, D.: "Rostering Ambulance Officers". In: Proceedings of the 15th National Conference of the Australian Society for Operations Research, pp. 470–481 (1999)
4. Ernst, A.T., Jiang, H., Krishnamoorthy, M., Owens, B., Sier, D.: An Anno-tated Bibliography of Personnel Scheduling and Rostering. Annals of Operations Research - Basel 127(1:4), 21–144 (2004)
5. Ernst, A.T., Jiang, H., Krishnamoorthy, M., Sier, D.: Staff scheduling and rostering: A review of applications, methods and models. European Journal of Operational Research 153, 3–27 (2004)
6. Guertin, Y.: « Projet d'amélioration des horaires des paramédics et des superviseurs, Description préliminaire des besoins ». Direction des services préhospitaliers, 7 pages (2010)

What Is a Decision Problem?
Preliminary Statements

Alberto Colorni[1] and Alexis Tsoukiàs[2]

[1] Design Department, Politecnico di Milano
[2] LAMSADE-CNRS, Université Paris Dauphine

Abstract. This paper presents a general framework about what is a decision problem. The aim is to provide a theory under which the existing methods and algorithms can be characterised, designed, chosen or justified. The framework shows that 5 features are necessary and sufficient in order to completely describe the whole set of existing methods. It also explains why optimisation remains the general approach under which decision problems are algorithmically considered.

1 Introduction

The reader should be aware that this paper does not address the title question in a comprehensive way. The problem of what is a decision and what is a decision problem has been addressed in philosophy, psychology and the cognitive sciences, economy, political science etc.. We are not going to make a survey of this literature which is out of the scope of the paper. The reader interested in these aspects can have a look to a number of fundamental texts such as [10], [12], [16], [24], [26], [28], [30], [35].

Our proposition is instead pretty technical and formal. Operational Research and Decision Analysis are seen as part of a more general Decision Aiding Methodology (see [32]) aiming to help real decision makers to understand, formulate and model their problems and possibly reach a reasonable solution (if any). We are concerned by that type of activities occurring in a decision aiding situation where a "client" (very broadly defined) asks for some advice or help to an "analyst", such an advice being expected to come under form of a formal model allowing some form of rationality. We call such activities a "decision aiding process" (see [31]). At a certain point of that process the analyst will have to formulate a "decision problem" requiring some computing to be performed by some algorithms providing a result which is expected to be used in order to present a recommendation relevant to the decision maker's "decision problem".

Our focus is exactly here: what is a decision problem for the analyst? The proposal of the paper is to suggest a general framework under within which it is possible to identify all possible models, algorithms, procedures which routinely analysts use in their job as well as to allow to invent ones (if possible). The paper introduces two hypotheses:

- It is possible to establish a common framework under which any formal decision problem can be formulated, enabling to construct wide classes of methods characterised by common features.
- From an algorithmic point of view any decision problem can be reduced to an optimisation problem.

P. Perny, M. Pirlot, and A. Tsoukiàs (Eds.): ADT 2013, LNAI 8176, pp. 139–153, 2013.

In the following section we introduce notation. Then in section 3 we show what the primitives of a decision problem are. Then in section 4 we describe the five characteristic features under which methods can be described. Section 5 introduces some methodology principles, while section 6 discusses the two running examples of the paper. Further research challenges are introduced within the conclusion.

2   Concepts and Notation

In the following A will always represent the set of "alternatives" considered either within a model or by a method. Although in practice such a set is never readily available, but constructed, for the purpose of this paper we are going to consider it as "given".

Along the paper we are going to use extensively preference relations. The basic relation we will adopt will be $\succeq$ (possibly indexed $\succeq_i$) which will read as "at least as good as" ($\succ$ will represent the asymmetric part of $\succeq$, while $\sim$ will represent the symmetric part). We will only make the hypothesis that this is a reflexive binary relation. The interested reader can see more about preference structures in [21] or [27] from which we adopt definitions and notation. We are now able to make our first claim.

Claim 1. *A decision problem for the analyst consists in finding an appropriate partitioning of the set A, relevant for the decision maker's concerns.*

The presentation of any algorithm or method discussed in this paper will be based on separating the "primitives" (what is the strictly necessary information required to be provided by the decision maker in order to allow some reasonable advice) and the "output" (what is the information the algorithm or method provides to the user). The reader should note that we distinguish between primitives and "input" to the algorithm. The reason is that the input to a precise existing algorithm or method is in reality constructed out of the primitives. Let's summarise. Our hypothesis is that the modelling process, that is the dialogue between the client and the analyst, follows (roughly) a sequence starting with the client providing *ground information*, which through *learning protocols* is transformed in *primitives* and these through *modelling tools* are transformed to the input to some method. For the description of these concepts see figure 1.

Fig. 1. The modelling process

Ground Information contains the problem description and for the purposes of this paper we will focus to what we call "preference statements": pieces of client's statements (in his own language), expressing values, opinions and likelihoods. In other terms it is how the client see his/her problem. *Learning Protocols* are procedures allowing to identify preference statements within the client's discourse and to translate them in ordering relations. In order to do so we need to establish the sets on which such relations apply. As it will become clear in section 4 such protocols are aimed at establishing the set A, the problem statement and the preference relations upon A. *Primitives* are the ordering relations "learned" using the protocols and we will discuss them extensively in the next section. *Modelling Tools* are the usual analytic tools an analyst uses in order to transform primitives in decision aiding models. Examples include the procedures allowing to construct a value function, a set of constraints, a probability distribution etc.. The *Input* is the information modelled in such a way that a decision aiding method can be applied. For instance in a linear programming method the input are the decision variables, the constraints and the objective function. In the following we present two running examples explaining some of the concepts introduced in each section.

Example 1.1 *Ground information.* The client is a horse races gambler. He is considering the next bet to make. In order to assess the "value" of each possible bet the client considers three different information: the quality of the horse, the quality of the jockey who runs it and the weather conditions. The client wants to rank all possible bets.

Example 2.1 *Ground information.* A hospital is considering the recrutement of nurses for three of their departments: General Medicine (GM), Oncology (ON), Children (CH). The hirings are managed by two: the general manager and the surgeon general. Candidates fill an application form and go through an interview. Practically the result is a report where the two managers consider three information: the age, the specialisation (if any) and the motivations of the candidate.

In both cases the learning protocols are procedures through which the analyst will try to gather the preferences of the client(s). Which horses (s)he prefers? With which jockey? Under which weather conditions? What is a good nurse for a given department? How specialisations compare with respect to the requirements of each department? How age influences the fitting of a candidate to a given department?

3 Primitives and Problems

What type of information can generally affect a decision? Since the origins (see for instance [7], [23]) most of the decision analysis literature will classify such information in three categories: values, opinions and likelihoods.

1. Values (related to attributes). Values should represent "what matters for the decision maker" (for a nice discussion see [13]). Under a more formal perspective we consider that the set A can be described against a set of attributes D, each attribute being equipped with a scale from a set of scales E. Following measurement theory (see [25]) such scales can be nominal, ordinal, ratio or interval ones. However, this is just descriptive information about A (x is 10cm long, y is yellow etc.). In order to be able to talk about values affecting decisions we need further information coming under form of preferential statements ("I prefer long tables to short ones", " I do not like yellow shoes"), possibly of more complex content ("I prefer a train travel to Paris to a flight at Amsterdam", "my preference of apples against oranges is stronger than my preference of peaches against apricots"). We distinguish two types of sentences:
- *comparative ones*, where elements of A are compared among them (under one or more attributes) in order to express a preference;

- *absolute ones*, where an element of A is directly assessed against some "value struc-ture" (under one ore more attributes).

2. Opinions (related to stakeholders). Decisions can be affected by the judgements and opinions of many stakeholders. In this case preference statements are going to be associated to "opinions". It is reasonable however, to distinguish once again among:
- *comparative opinions* (stakeholder i prefers x to y), where preferences are expressed among elements of the set A;
- *absolute opinions* (stakeholder i considers x as "worthy"), where preferences are ex-pressed under form of value assessments.

3. Likelihoods (related to scenarios). When we express preferences it is likely that these depend from uncertain future conditions. Although the intuitive temptation is to use estimates (it is likely to rain) or quantifications of uncertainty (the probability of raining is p), if we focus on decision situations the primitives we need to consider will once again be preference statements of the type "under scenario j, I prefer x to y" or of the type "under scenario j, x is unworthy".

The reader will note that we do not include among the primitives the concept of rela-tive importance of the dimensions under which preferences are expressed. The reason is simple: relative importance is a derivable information. Consider the case where $x \succ_I y$ and $y \succ_J x$ for some $I, J \subset H$ (x, y being elements of A and H being the set of crite-ria). If we add the information that "globally" $x \succ y$ (which is once again a primitive) we can derive that $I \gg J$ ($\gg$ representing an ordering relation upon the power set of H: $\gg \subset 2^H \times 2^H$). This will be true if I and J are sets of values, but also if they are opinions or likelihoods. Of course a decision maker may wish to make direct state-ments comparing two dimensions (I is more likely to occur than J), but there are two reasons for which it is better to avoid it. The first has to do with the fact that there is no general model for "relative importance", this depending on how primitive preferences are considered at the global level (see [17]). The second is that these are second order comparisons allowing for more cognitive biases and potential inconsistencies. Indeed more often than less decision makers are ready to change their statements about relative importance as soon as they realise the impact they may have on first order preference statements (that is comparing elements of A).

From the above discussion and in order to model "absolute statements" we need to introduce, besides the set A (of potential decisions), a set B being a collection of "norms" or "standards" or "thresholds" representing an external (with respect to A) value structure. In other terms if we want to claim that x is "nice" (under a certain point of view) we need to establish somewhere (not in A) what "nice" means and compare x to that norm. Under such a perspective:

Definition 1.
- *Comparative preference statements come under form of $x \succeq_i y$ $x, y \in A$, i being any among attributes, stakeholders or scenarios (thus $\forall i \succeq_i \subseteq A \times A$).*
- *Absolute preference statements come under form of $x \succeq_i b$ $x \in A$ $b \in B$, i being any among attributes, stakeholders or scenarios (thus $\forall i \succeq_i \subseteq A \times B \cup B \times A$).*

From the above presentation we can establish our second general claim.

Claim 2. *Decisions are based on primitives which always come under form of comparative or absolute preference statements.*

Remark 1. We use the term preference statement in a very broad way. However, the formalism adopted should not conduct to confusion. Preference statements can be modelled either under form of asymmetric ordering relations (x is strictly before y) or under form of symmetric ordering relations such as similarities or nearness relations (x is similar or near to y) and these can be learned directly from the client. The use of the $\succeq$ relation is a comfortable way to combine such relations in a unique definition.

We continue with our running examples.

Example 1.2 *Primitives.* The alternatives considered by the client are the "bets" (a combination of a horse with its jockey). This is a finite enumeration of the participants to the next race. The client provides different types of preference statements (examples):
- horse x is better than horse y;
- jockey i is better than jockey j;
- horse w run by jockey j is better than horse z run by jockey i;
- if it rains horse y is better than horse w.

Such sentences need to be interpreted. For instance should we understand the first sentence as "horse x being better than horse y independently from the jockeys they run them and the weather conditions?" Or should we understand it as "considering the same jockey and the same weather conditions then horse x is better than horse y?" The difference can be important. In the first case we consider that $\forall j \in J, t \in T \langle xjt \rangle \succeq \langle yjt \rangle$ where J is the set of jockeys and T is the set of weather conditions. This will imply for instance that $\langle x, \text{Paul}, \text{rain} \rangle \succeq \langle y, \text{John}, \text{dry} \rangle$. In the second case such a comparison is not allowed. We can only write formulas of the type $\langle x, \text{Paul}, \text{rain} \rangle \succeq \langle y, \text{Paul}, \text{rain} \rangle$ (ceteris paribus comparisons).

Example 2.2 *Primitives.* The set A is composed by those candidates who filled the application form and got the interview. However, this is an assignment problem where we also need to specify the classes where the elements of A are assigned. These are four: the three hospital departments and the rejected. The type of preference statements we need here are of the type: "being young is ideal for the Childrens' Department", "not having a specialisation is very bad for Oncology", "being motivated and specialised are the candidates we are looking for", these being more or less nuanced among the two managers (who may possibly disagree on some of them).

However, once again these need to be better understood. For instance we need to establish what "young" means (we need a threshold for that). Perhaps too young is not that ideal, that implying measuring an absolute distance from some ideal "young nurse age". We also need to understand if not having a specialisation is an eliminating handicap for the candidate or if it is a general negative assessment becoming more important for the special case of Oncology. On the other hand the reader will note that at this stage we do not need to compare candidates among them.

The reader will note that in the case of the horse races we are using comparative preference statements (a horse is better than another), while in the case of the nurses we are using absolute preference statements (a nurse feature fits, more or less, the requirements of a given department).

Using claim 2 and definition 1 we can summarise the possible primitive information used in a decision problem as in figure 2. As we move away the origin along any of the axes we start considering multiple values (opinions, likelihoods).

Definition 2. *We call optimisation problem any decision problem considering a unique dimension under which primitives are expressed (point $\mathcal{O}$ of figure 2).*

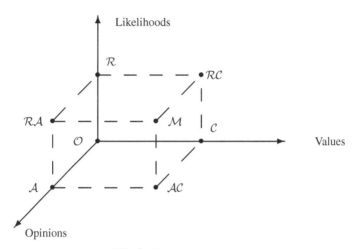

Fig. 2. The archetype problems

Figure 2 establishes eight archetype "problems" represented by the eight points: $\mathcal{O}, \mathcal{A}, \mathcal{C}, \mathcal{R}, \mathcal{AC}, \mathcal{RC}, \mathcal{RA}$ and $\mathcal{M}$. We call:

$\mathcal{O}$: an optimisation problem (see more details in section 4);
$\mathcal{A}$: an agreement problem, since different opinions need to be taken into account;
$\mathcal{C}$: a compromise problem, since different values need to be considered;
$\mathcal{R}$: a robustness problem, since a solution needs to be considered worthy under different likelihoods;
$\mathcal{AC}$: an agreed compromise (a combination of $\mathcal{A}$ and $\mathcal{C}$);
$\mathcal{RC}$: a robust compromise (a combination of $\mathcal{R}$ and $\mathcal{C}$);
$\mathcal{RA}$: a robust agreement (a combination of $\mathcal{R}$ and $\mathcal{A}$);
$\mathcal{M}$: a "mess", because the problem starts to become really messy ...

However, the above eight archetype problems do not stand alone. Behind a compromise problem other compromises may need to be considered in a hierarchy of criteria. Behind an agreement problem other agreement problems may have to be solved along a hierarchy of delegates, community representatives and other organisational structures. Behind a robustness problem many states of the nature may have to be considered in a hierarchy of likelihoods establishing complex scenarios. And, any combination of the above may in reality occur as complex as possible (see figure 3).

What can we observe in analysing these archetypal decision problems? Despite the different semantics behind values, opinions and likelihoods, the underlying formal structure is always the same. We have one common primitive: preference statements which we represent through preference relations. And we have one principal task: move along the hierarchy, from the leaves of the most elementary preference statements up to the root where "$x \succeq y$ all relevant information being considered" (x, y being either both in A or one in A and the other in B). This allows to introduce our third claim.

Claim 3. *A decision problem can be represented as a sequence of preference aggregations along an hierarchy of opinions, values and likelihoods, combined arbitrarily.*

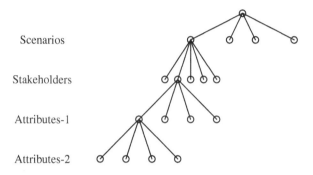

Fig. 3. Many decision problems hierarchically related

Example 3.1 Consider the case where a committee is assessing a number of development projects for an urban area. The outcomes of these projects depend on a number of uncertain issues due to the unstable economic situation of the whole region. At this stage we have many levels of the dimensions hierarchy: the different members of the committee and the different scenarios considered as "realistic" for the region, the set of attributes describing the projects. We consider as first level of our hierarchy the scenarios and as second level the committee members. We can expect that each committee members will assess the projects on a number of attributes (such as cost, sustainability, environmental impact etc.). It is reasonable to consider that some of such attributes decompose further in other attributes (such as direct costs, maintenance costs, financial costs etc.). This situation is captured by figure 3.

4 Main Features

In the following we present the 5 features which constitute the key parameters designing the whole set of conceivable formal decision problems. For the time being only one feature (the problem statement) will be discussed in an extensive way, the other four being essentially sketched.

4.1 The Set of Alternatives

A can be of different types:
- a countable enumeration of objects, $A = \{a_1 \cdots a_n\}$;

- a subset of all possible combinations of the attribute scales in the attributes space, $A \subseteq E_1 \times \cdots E_m$;
- a combinatorial structure resulting from the product of a set of discrete decision variables (possibly 0 or 1), $A \subseteq X_1 \times \cdots X_m$, $X_i \subseteq \mathbb{Z}$ or $X_i = \{0, 1\}$;
- a vector space resulting from the product of a set of real valued decision variables, $A \subseteq X_1 \times \cdots X_m$, $X_i \subseteq \mathbb{R}$.

4.2 The Problem Statements

Partitioning a set A consists in establishing a set of "equivalence classes" to which associate the elements of A. We can distinguish two different cases:

1. The first case concerns the fact that such classes can be ordered or not. Typical examples in the first case are classes of merit, the equivalence classes of a weak order etc.. Typical examples of the second case are problems of medical diagnosis, failure detection, pattern recognition etc..

2. The second case concerns the fact that such classes can be pre-defined with respect to some norm, standard, profile etc. or not. Typical examples of the first case include assigning elements of A to given ratings or patterns. Typical examples of the second case are clustering a population for some attribute or ranking it.

We summarise the above cases in table 1.

Table 1. Basic Problem Statements

	Pre-defined wrt some external standard	NOT pre-defined
Ordered	Rating	Ranking
Not Ordered	Assignment	Clustering

A special cases within the above problem statements is the one where the number of classes are just two, one being the complement of the other.

Let's discuss more in details the above problem statements.

1. Ranking. The primitive in this case will be a binary relation on A: $\succeq \subseteq A \times A$ to be read "at least as good as". The expected result is a partitioning of A in equivalence classes $[A]_1, \cdots [A]_n$ such that:

- $\exists \succcurlyeq \subseteq A \times A$
- $\succcurlyeq \ = \ \succ \cup \approx$
- $[A]$ is the set of equivalence classes constructed by $\approx$
- $\succ \subseteq [A] \times [A]$, $\succ$ being a strict partial order such that:
- $[A]_j \succ [A]_i \Leftrightarrow j > i$ and
- $\forall x \in [A]_j, y \in [A]_i : x \succ y$ and
- $\forall x, y \in [A]_j \ x \approx y$

Discussion. The reader will note that the ordering relation among the equivalence classes is not the primitive relation comparing the elements of A. Generally speaking $\succeq$ is not an ordering relation since preferences can be partial and or inconsistent. If we have to proceed with some operational procedure we need to transform the preference relation $\succeq$ to an ordering relation $\succcurlyeq$. We may impose that $\succeq \subset \succcurlyeq$, but this is not mandatory (for instance in case of inconsistent preference statements we may want to drop some primitive comparisons). As already mentioned $\succ$ is not necessary a complete relation. In case we impose completeness ($\succ$ becoming a total order) we get that $[A]_1 = \sup_A(\succ)$ (n being the number of equivalence classes in which A is partitioned).

What is a choice problem? We consider as choice the particular case where A is partitioned in two classes $[A]_1 \succ [A]_2$. We can generalise this observation claiming that any ranking problem partitions A in a set of classes, the first being the "optimal elements", then the second one being the "second optimal ones" and so on. The current literature considers as optimisation the special case where:
$\succeq = \succcurlyeq$ and $\succcurlyeq$ is a weak order on A such that $\exists f : A \mapsto \mathbb{R} : x \succcurlyeq y \Leftrightarrow f(x) \geq f(y)$.
Under such a hypothesis it is clear that $[A]_1 = \max_A f(x)$. However, we do not really need the function f in order to "optimise". Generalising the concept of optimisation we can always construct an algorithm such that $[A]_1 = \sup_A(\succcurlyeq)$. Extending further this reasoning the whole set of equivalence classes can be constructed as result of some optimisation: $[A]_{n+1} = \sup_{A \setminus [A]_n}(\succcurlyeq)$ etc..

2. Clustering The primitive in this case is a binary relations on A: $\approx \subseteq A \times A$ to be read "similar to". The expected result is a partitioning of A in $[A]_1, \cdots [A]_n$ such that: $\forall x, y \in [A]_j \; x \approx y$ and $\forall x \in [A]_j, \; y \in [A]_i : \neg(x \approx y)$.

Discussion. In case $\approx$ is an equivalence relations then the partitioning of A results in constructing the indiscernibility relation on A ([22]). However, this is not generally the case. Elements of A are more or less similar between them (or differently similar, see [34]). Under such a perspective we consider that instead of a single similarity relation we have a set of nested similarity relations $\approx_l$ and $[A]_j = \sup_A(\approx_l)$. In other terms we try to maximise similarity within classes (clusters) and minimise similarity among classes (clusters). If $\approx_l$ are nested similarity relations with nice properties then we can establish metrics (see [11]:

- $s(x, y)$: how similar is x to y?
- $d(x, y)$: how distant is x from y?

Then, establishing the equivalence class of any element $y \in A$,
$[A]_y = \{x | \max_A F(s(x, y))\}$, F being a measure (a fitting function) of the overall similarity of the elements of $[A]$ with respect to y, we can construct the clusters $[A]_j$. Meyer and Olteanu generalised this idea (see [18]) for general preference structures.

Remark 2. The reader should note that both ranking and clustering problem statements boil down in solving some mathematical optimisation problem. This should not be surprising: in absence of any external information and being allowed only to compare elements of A among them, the only mathematical notion we have, in order to clearly separate classes between them, is the one of "optimality".

3. Rating The primitive here is a binary relation from the set A to the set B: $\succeq \subseteq A \times B \cup \overline{B \times A}$ to be read "at least as good as", B being the set of external "norms" characterising the ordered classes $C_1 \triangleright \cdots \triangleright C_n$. The expected result is to assign each element of A in a C_j such that: $x \in C_j \Leftrightarrow x \succcurlyeq p_j, p_{j+1}, \cdots p_n$ and $p_1 \cdots p_{j-1} \succcurlyeq x$.

Discussion. As in the ranking problem statement we need to differentiate between the primitive preference relation $\succeq$ and the operational result represented by the ordering relation $\succcurlyeq$. By transitivity of $\succcurlyeq$ it is clear that if element x is in C_j and element y is in

C_{j+1}, $x \succ y$, while if both elements are assigned in the same class $x \sim y$ ($\succ \cup \sim = \succsim$). However, the reader should remember that classes are pre-established.

Suppose now that the relation $\succsim$ is a weak order such that we can establish a function $f : A \cup B \mapsto \mathbb{R}$ such that $x \succsim p_j \Leftrightarrow f(x) \geq f(p_j)$. The problem of assigning the elements of A to the ordered classes represented by the "norms" p_j turns to be a classical constraint satisfaction problem: $C_j = \{x : f(x) \geq f(p_j) \text{ and } f(p_{j-1}) \geq f(x)\}$. We can generalise this concept dropping the function and claim that the rating problem can be considered a generalised constraint satisfaction problem.

4. Assignment The primitive is a binary relation on A: $\approx \subseteq A \times B \cup B \times A$ to be read "similar to", B being the set of external "norms" characterising the classes $C_1 \cdots C_n$ (the difference with the rating problem statement being the fact that these classes are not ordered among them). The result is to assign each element of A in a C_j such that: $x \in C_j \Leftrightarrow x \approx p_j$, where p_j is the norm characterising class C_j.

Discussion. Assigning objects to unordered classes could be seen as a constraint satisfaction problem where constraints are expressed as equalities ($C_j = \{x : f(x) = f(p_j)\}$), where f is a function representing a metric of similarity.

Concluding: the problem statements we present here can be handled either as an optimisation problem or as a constraint satisfaction one. Considering that any constraint satisfaction problem can be transformed in an optimisation one we can state one of our principal claims, based on the hypothesis that our problem statements are exhaustive of all possible decision problems (partitionings).

Claim 4. *From an algorithmic point of view any decision problem is an optimisation problem.*

Remark 3. The reader should not make confusion with the notion of decision problem typical in the algorithmic complexity literature ([8]). On the other hand we want to emphasise that what we are talking here concerns how algorithmically primitives get transformed in ordering relations such that can be used for recommending something or being used for further aggregations.

4.3 Independence

As already mentioned primitives come under form of statements of the type "x is at least as good as y, under I", x, y being mono or multi-dimensional objects and I being a subset among values, likelihoods and opinions. However, despite its intuitive meaning, such a sentence can still be interpreted in different ways. We are going to distinguish two principal interpretations:

- "x is at least as good as y, under I", independently on what happens to $H \setminus I$ (H being the set of criteria);
- "x is at least as good as y, under I", provided a condition holds in some $J \subseteq H \setminus I$.

These two interpretations lead to completely different problem formulations and consequently to different methods and resolution algorithms. Preferential independence (the first interpretation) allows to envisage a linear (additive) model representing preferences. Conditional preferences lead to more complex preference structures implying non linear aggregation functions ([9], [14], [29]) or specialised algorithms ([3], [4]).

4.4   Differences of Preferences

Let's recall once again our primitives and let's consider the sentence "x is strictly better than y and these are both better than z (under I)". We know we can represent this sentence giving numerical values to x, y, z (for instance $x = 3, y = 2, z = 1$ and adopt the natural ordering of the numbers. However we could choose the numerical representation $x = 100, y = 10, z = 1$ and it would be the same. Preferences are orders and the numbers we use only carry ordinal information.

The point is that in many cases we could either have richer information (we know for instance that x is twice more heavy than y) or we would like to have richer information of the type "x is much more better than y". We need to reason in terms of "differences of preferences" and their representation. In other terms we need primitives of the type: "xy is not less than zw" where xy (zw) represents the difference of preference between x and y (z and w). It is interesting to note that primitives of this type can be used also in order to express ordinal preferences, while the opposite is not true. Under such a perspective we can claim that primitives should always be considered as sentences about differences of preferences, the ordinal case being a special one. The interested reader can see more in the literature about conjoint measurement (see [15]) and how this helped in reframing multidimensional preferences (see [1], [2], [17]).

4.5   Positive and Negative Reasons

Consider a preference statement of the type: "I do not like x", or "any candidate, but not x". Such statements can be considered as explicit "negative preferential statements" to be considered independently from the "positive ones" (which are the usual ones). The idea here is that there are cases where decision makers need to express negative judgements and values which are not complementary to the positive ones (such as a veto on a specific dimension). Such statements have been explicitly considered in the literature both in decision theory (see [6], [20], [33]) and in argumentation theory (see [19]). When such situations occur we need to develop specific procedures adding thus a further dimension of characterisation of the decision problem at hand.

5   Methodology

Let's summarise in order to outline how our framework can be used for methodological purposes. We have a set A, information describing the set A against a number of attributes and preference statements (these being values, opinions or likelihoods) comparing either the elements of A among them or the elements of A to elements of a set B (the set of norms). We aim at partitioning A appropriately.

The first problem we have is reducing the problem to an "optimisation problem": that is, obtaining one-dimensional preference statements. In other terms we are trying to aggregate preference statements expressed on several different dimensions to a single one. For the time we consider that transforming some attributes to "constraints" (thus bounding the space of feasible solutions) has already been considered in establishing the set A. How do the different parameters described in section 4 influence the design (or the choice) of an appropriate solution method?

Allowing to have explicit measures of differences of preferences allows to handle richer preferential information, such that we can consider to obtain at the aggregated level preference statements sufficiently rich to satisfy nice properties (for instance obtaining directly an ordering relation). In case we need to work with purely ordinal information we should expect the negative consequences of Arrow's impossibility theorem (see the discussion in [5]). In case preferential independence holds we are in the "easy case": given a set of primitives holding at the same level of the hierarchy of dimensions (see figure 3) these can be aggregated (possibly through a linear model) to the parent node. In case independence does not hold we have two options: either we need explicitly non linear models accounting for the observed dependencies or we need to reformulate the modelling dimensions (these options being not exclusive). This second case may result in aggregating more levels of the hierarchy in a single step.

The presence of explicit negative preference statements (not complementary to the positive ones) will result in duplicating the decision model creating an hierarchy of "negative reasons" (to be associated to the hierarchy of "positive reasons"). It will also require to establish how and when these two sources of information should merge.

The second problem we need to handle is to obtain, out of the one-dimension primitives computed in the previous step, an ordering relation allowing the partitioning of the set A (the recommendation to hand to the decision maker). It is clear that the type of problem statement adopted strongly influences how this step will be considered since it establishes both the type of primitives we need to construct and the type of algorithm to be used. Before concluding this discussion we note that the properties of the set A will also influence the design or the choice of the method for obvious algorithmic reasons.

At this stage is easy to show that the five features we introduced in this paper (properties of the set A (and B), problem statement, preferential independence, difference of preferences, explicit negative preference statements) are necessary in order to clearly establish, design and axiomatically characterise any model, algorithm and method aiming at handling a decision problem (on set A). Our claim, not demonstrated here, is that these features are also sufficient. We thus get:

Claim 5. *The properties of the set A, the type of problem statement, the holding or not of preferential independence, the explicit use of differences of preferences, the explicit use of negative preference statements, are the necessary and sufficient features for choosing, designing, justifying and axiomatically characterising any decision problem and the associated resolution methods and algorithms.*

Discussion. The reader should recall that the use of the "learning protocols" provides the analyst with some first basic information: the set A, the problem statement and the preferences about A. Without these information we cannot really make any tentative to formalise a decision problem of a client. However, there are many methods through which this information can be handled. In order to choose among them, to explain and justify them besides maintaining an axiomatic coherence) we need to know more about the preferences: whether preferential independence is met, whether differences of preferences hold or not and whether explicit negative statements are done. For instance the use of linear programming will make sense in a situation where the set A is a subset of a vector space, the problem statement is a ranking one (and more precisely a choice) and the preferences among the elements of A are such to allow a numerical representation which is also an interval scale of "profits" (or costs), which means that differences of preferences hold, and this is additive, which means that preferential independence also holds, while there is no use of explicit negative statements.

6 Horse Races and Nurses

Let's finalise the discussion of the two running examples.

Example 1.3 *Modelling*. Having established the correct primitives representing the client's preference statements we can check whether preferential independence is met and how differences of preferences are measured. If the preferences among horses are independent from the preferences among jockeys and among weather conditions then we can clearly look for a linear aggregation of such values (in case differences of preferences are meaningful) and even allow to compute a relative importance for each weather condition under form of probabilities. This will allow to use expected utility theory.

Discussion. The problem presented is a "robust compromise". A compromise because we need to take into account two different type of values (the quality of the horse and the quality of the jockey) and a robustness problem because we need to take into account three different likelihoods: raining, humid weather, dry weather. The expected utility problem formulation will hold under very specific conditions. In case these do not hold we can look for alternative ones (using CP nets [3], [4] or fuzzy integrals [9]).

Example 2.3 *Modelling*. Each candidate is described against three attributes which are very different: age is a continuous numerical scale (probably discretised), specialisation is a nominal scale (which specialisation, if any, has the candidate), motivation is (probably) an ordinal scale reporting the judgement of the expert who did the interview. On the other hand the classes also need to be described against the same attributes either providing the "ideal" values for each class (for instance the "ideal nurse" for the Childrens' Department is 30 year old) or the "minimal" ones (not more than 35 years old). Then we need to establish how the candidates compare to such classes. For instance we need to value the distance between two ages or between two specialisations. The reader should remember that such models might be different among the two managers. Should we be able to measure these differences and should these be commensurable among them, then we can envisage to write a value model (possibly additive) assessing the "fitting" of each candidate for each category. Otherwise we could opt for some ordinal model using some majority principle (of the type: if two criteria agree that candidate x fits to class a then assign him/her there). However, the presence of negative assessments play an important role here, since they may exclude a candidate from a certain class independently from any other assessment (if the candidate does not have a specialisation in oncology cannot work in Oncology).

Discussion. The situation is a an Agreed Compromise since we need to find a compromise among the three fitness criteria and an agreement among the two managers. There are two paths leading to the final assignment. The first (more cooperative) consists in finding an agreement between the two managers for each criterion separately and then find a compromise using the agreed assessments. The second (more negotiation oriented) consists in finding the assignments for each manager and then try to find an agreement for the cases where these are different (possibly all of them). The reader will note that there is no reason for which these two paths lead to the same result.

7 Discussion and Conclusion

In this paper we introduced a general framework describing the whole set of methods and decision problems an analyst may have to design in order a problem provided by a client. We have shown that decision problems can be generally seen as optimisation problems after a complex hierarchy of values, opinions and likelihoods gets transformed to a single dimension through a sequence of "preference aggregations". In order to construct this general framework we make use of what we call primitives (the strictly necessary information to be provided in order to model meaningfully the decision problem). Then we described five main features which characterises any method aiming at modelling and solving the decision problem. Our general claim is that these five features (the type of the set A, the problem statement, the holding of preferential independence, how differences of preferences are considered, the presence of explicit negative reasons) are necessary and sufficient in order to choose, design, justify and characterise any decision support procedure. Further research includes on the one hand showing how formal argumentation theory can help in explaining and justifying methods design and their outcomes and on the other hand exploring how formal model reformulation techniques can help finding the most appropriate algorithm to adopt for resolution purposes.

References

1. Bouyssou, D., Marchant, T., Pirlot, M., Tsoukiàs, A., Vincke, P.: Evaluation and decision models with multiple criteria: Stepping stones for the analyst, 1st edn. Springer, Boston (2006)
2. Bouyssou, D., Pirlot, M.: Conjoint measurement models for preference relations. In: Bouyssou, D., Dubois, D., Pirlot, M., Prade, H. (eds.) Decision Making Process, pp. 617–672. J. Wiley, Chichester (2009)
3. Brafman, R., Dimopoulos, Y.: Extended semantics and optimization algorithms for CP-networks. Computational Intelligence 20(2), 219–245 (2004)
4. Brafman, R., Domshlak, C.: Preference handling: An introductory tutorial. AI Magazine 30(1), 58–86 (2008)
5. Dubois, D., Fargier, H., Perny, P., Prade, H.: Qualitative decision theory: from Savage's axioms to non-monotonic reasoning. Journal of the ACM 49, 455–495 (2002)
6. Dubois, D., Prade, H.: An introduction to bipolar representations of information and preference. International Journal of Intelligent Systems 23, 866–877 (2008)
7. French, S.: Decision theory - An introduction to the mathematics of rationality. Ellis Horwood, Chichester (1988)
8. Garey, M., Johnson, D.: Computers and Intractability. Freeman and Co., New York (1979)
9. Grabisch, M., Labreuche, C.: Fuzzy measures and integrals in MCDA. In: Figueira, J., Greco, S., Ehrgott, M. (eds.) Multiple Criteria Decision Analysis: State of the Art Surveys, pp. 563–608. Springer, Boston (2005)
10. Habermas, J.: Theorie des kommunikativen Handelns. In: Suhrkamp, Frankfurt am Main, 1981. engl. version: The Theory of Communicative Action. McCarthy, T. (trans.). Polity, Cambridge (1984-1987)

11. Janowitz, M.F.: Ordinal and Relational Clustering. World Scientific, Singapore (2010)
12. Kaplan, M.: Decision Theory as Philosophy. The Cambridge University Press, Cambridge (1996)
13. Keeney, R.L.: Value-Focused Thinking. A Path to Creative Decision Making. Harvard University Press, Cambridge (1992)
14. Keeney, R.L., Raiffa, H.: Decisions with multiple objectives: Preferences and value tradeoffs. J. Wiley, New York (1976)
15. Krantz, D.H., Luce, R.D., Suppes, P., Tversky, A.: Foundations of measurement. Academic Press, New York (1971)
16. Luce, R.D., Raiffa, H.: Games and Decisions. J. Wiley, New York (1957)
17. Marchant, T.: Towards a theory of MCDM: stepping away from social choice theory. Mathematical Social Sciences 45, 343–363 (2003)
18. Meyer, P., Olteanu, A.-L.: Formalizing and solving the problem of clustering in mcda. European Journal of Operational Research 227, 494–502 (2013)
19. Modgil, S., Prakken, H.: A general account of argumentation with preferences. Artificial Intelligence 195, 361–397 (2013)
20. Öztürk, M., Tsoukiàs, A.: Bipolar preference modelling and aggregation in decision support. International Journal of Intelligent Systems 23, 970–984 (2008)
21. Öztürk, M., Tsoukiàs, A., Vincke, P.: Preference modelling. In: Ehrgott, M., Greco, S., Figueira, J. (eds.) State of the Art in Multiple Criteria Decision Analysis, pp. 27–72. Springer, Berlin (2005)
22. Pawlak, Z.: Rough Sets - Theoretical Aspects of Reasoning about Data. Kluwer Academic, Dordrecht (1991)
23. Raiffa, H.: Decision Analysis – Introductory Lectures on Choices under Uncertainty. Addison-Wesley, Reading (1968)
24. Rescher, N.: Introduction to Value Theory. Prentice Hall, Englewood Cliffs (1969)
25. Roberts, F.S.: Measurement theory, with applications to Decision Making, Utility and the Social Sciences. Addison-Wesley, Boston (1979)
26. Rosenhead, J.: Rational analysis of a problematic world, 2nd revised edn. J. Wiley, New York (2001)
27. Roubens, M., Vincke, P.: Preference Modeling. In: LNEMS, vol. 250, Springer, Berlin (1985)
28. Simon, H.A.: Administrative behaviour: a study of Decision Making Processes in Administrative Organizations. Mac Millan, New York (1947)
29. Suppes, P., Krantz, D.H., Luce, R.D., Tversky, A.: Foundations of Measurement. In: Geometrical, Threshold and Probabilistic Representations, vol. 2. Academic Press, New York (1989)
30. Toulmin, S.: The Uses of Argument. Cambridge University Press, Cambridge (1958)
31. Tsoukiàs, A.: On the concept of decision aiding process. Annals of Operations Research 154, 3–27 (2007)
32. Tsoukiàs, A.: From decision theory to decision aiding methodology. European Journal of Operational Research 187, 138–161 (2008)
33. Tsoukiàs, A., Perny, P., Vincke, P.: From concordance/discordance to the modelling of positive and negative reasons in decision aiding. In: Bouyssou, D., Jacquet-Lagrèze, E., Perny, P., Slowinski, R., Vanderpooten, D., Vincke, P. (eds.) Aiding Decisions with Multiple Criteria: Essays in Honour of Bernard Roy, pp. 147–174. Kluwer Academic, Dordrecht (2002)
34. Tversky, A.: Features of similarity. Psychological Review 84, 327–352 (1977)
35. Watzlawick, P., Weakland, J.H., Fisch, R.: Change; principles of problem formation and problem resolution. Norton, New York (1974)

Risk Information Extraction and Aggregation

Experimenting on Medline Abstracts

Léa Deleris[1], Stéphane Deparis[1], Bogdan Sacaleanu[1], and Lamia Tounsi[2]

[1] IBM Research – Ireland, Dublin
{lea.deleris,stephane.deparis,bogdan.sacaleanu}@ie.ibm.com
[2] Dublin City University
lamia.tounsi@computing.dcu.ie

Abstract. By exploiting advances in natural language processing, we believe that information contained in unstructured texts can be leveraged to facilitate risk modeling and decision support in healthcare. In this paper, we present our initial investigations into dependence relation extraction and aggregation into a Bayesian Belief Network structure. Our results are based on a corpus composed of MEDLINE® abstracts dealing with breast cancer risk factors.

Keywords: Probability, influence, Natural Language Processing, risk analysis, information extraction, expert aggregation, Bayesian networks.

1 Introduction

Bayesian belief networks (BBNs) [1] have been applied within a variety of contexts, including engineering, computer science, medicine and bioinformatics. They are a popular tool for risk analysis modeling and decision support systems [2, 3]. While BBNs can be constructed automatically from structured data, the amount of human effort required to construct a BBN manually, either from expert opinion or based on a literature review, may be impractical on a large scale.

However, there is a wealth and growing amount of information in unstructured format that could be leveraged to facilitate the creation of BBNs (Medical academic knowledge being one example). The challenges associated with unstructured texts are many. Relevant information is (i) sparse, (ii) scattered among a large amount of irrelevant sentences and (iii) ambiguous due to the richness of the human language. Beyond the extraction of that information, synthesizing it in a coherent model presents its own challenges, akin to those associated with the construction of risk models from multiple experts, in particular diverging opinions and partial information.

Our goal is nevertheless to investigate how current advances in natural language processing techniques could be used to address the scalability problem of extracting relevant pieces of information to build BBN-based risk models from unstructured texts. Note that our objective is not to replace experts but rather to facilitate their task. Therefore the focus is on extracting the information and creating initial versions of BBNs which are then to be edited by domain experts.

P. Perny, M. Pirlot, and A. Tsoukiàs (Eds.): ADT 2013, LNAI 8176, pp. 154–166, 2013.
© Springer-Verlag Berlin Heidelberg 2013

In this paper we describe our overarching approach to construct BBN from text followed by a detailed description of extraction and aggregation of structural information. The extraction and aggregation of the quantitative information is not addressed here.

2   Description of the Research Problem

2.1   Overarching Approach

This section presents the general approach that we have defined to construct BBNs from unstructured texts. BBNs are composed of a graph, whose nodes (or vertices) represent random variables and whose arcs (or edges) capture dependence statements (independence, the stronger statements, is derived from the absence of arcs). Each node is associated with a conditional probability table which provides numerical information about the strength of the dependence. People commonly refer to the graph (nodes and arcs) as the graphical layer or structure of the BBN while the term quantitative layer pertains to the conditional probability tables (CPTs) of the nodes. The construction of BBN *based on expert knowledge* typically follows three main steps:

1. Identification of variables (nodes), possibly definition of the states (although this does not need to be done till before the third step)
2. Definition of the network structure, i.e., the dependence and independence relations among the variables (arcs)
3. Specification of the parameters of the CPTs.

How does this process translate into text-based extraction of expertise? Because the elements of the quantitative layer depend on the graphical layer (the structures of the conditional probability tables depend on the parents of each node) it is reasonable when obtaining information from experts to determine the structure of the BBN before populating it with quantitative information. This constraint does not apply when information is extracted from a set of unstructured texts because the extraction step is not tailored to a given structure. Indeed it is unlikely that the probability statements that can be extracted will perfectly match the required CPT inputs. Limiting the search for quantitative information solely to the CPTs would result in discarding relevant information. We assume therefore that the dependence relation extraction and the probability statement extraction can be done in parallel as independent tasks.

By contrast, the output of the transformation of the extracted information into a BBN can vary depending on the sequence of subtasks. Rather than constructing a BBN for each source and then combine those together, we chose to first aggregate all dependence information into a network structure and thereafter to populate the associated CPTs based on all the probability statements extracted. We think that pooling information together as early as possible will reduce aggregation challenges: When probability and dependence statements are few at each source level but aggregated over many sources, the individual BBNs for each source are likely to be dissimilar.

Our suggested methodology is summarized on Fig. 1. We start by building the information base through the extraction of (i) structural information in the form of dependence and independence relations among risk factors (e.g. Age at Menarche → Breast Cancer Incidence) and (ii) quantitative information in the form of probability statements (e.g., P(Lifetime Breast Cancer | Malaysian Women) = 0.05). The content of the information base is then processed to generate variables, states, aggregated network structure and finally parameters for the CPTs.

In this paper we describe in more details the dependence extraction and the construction of the aggregated network structure. Despite not having achieved full automation of the extraction process, we are able to shed some light on the feasibility of both extraction and aggregation and on some of the related challenges. Regarding variable and state identification, we intend to address them in future research and resort to simple heuristics in the meantime. Specifically, different terms are considered as different variables and variables are considered as binary so that state identification is not required. In the future, we will leverage knowledge bases such as UMLS along with machine learning methods to cluster terms together into variables. Regarding quantitative information, we have defined and evaluated algorithms for extraction and evaluation but chose not to present them due to space constraints.

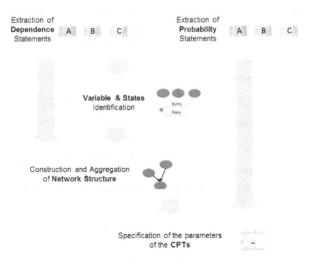

Fig. 1. Steps involved in building BBN from Text

2.2   Related Research

To the best of our knowledge, there is limited research that focuses on building BBNs from unstructured text. There has been some early work related to building BBN from structured text information. [4] presents a graph grammar which takes as input structured information (disease, treatment, test) and constructs an influence diagram (BBN augmented by decision nodes) from it. Similarly, [5] leverages the information contained in Medical Subject Headings (MESH) and subheadings in MEDLINE citations to build the structure of an influence diagram. Both approaches are specific to the medical domain as they rely on pre-determined keywords to guide the structuring of the network. [6] appears to be the first paper to specifically focus on presenting an NLP framework for building BBN from unstructured texts. The authors, coming from the natural language processing community, restrict their search to causal relation extraction to determine the edges of the network. While they leverage semantic information extracted from WordNet for variable identification, they provide a limited discussion on the construction of the graph, ignoring challenges related to aggregation of all the causal information into a coherent BBN such as the creation of cycles and the need to distinguish between direct and indirect relations. Our approach, while similar in spirit, builds more significantly upon the decision science literature. In that domain, there has been a significant amount of work in combining expert opinions when those opinions are probabilities, for instance [7], yet the combination of Bayesian network structures provided by experts is seldom considered. Notable exceptions are [8-10] which provide practical and theoretical perspectives on the fusion of Bayesian networks. One important requirement of these research efforts is that input are composed of true Bayesian networks, meaning directed dependence relations among random variables, while our input would be better described as degenerate versions of Bayesian network in the form of undirected dependence and independence statements among pairs of random variables. The works of [11-12], which we will discuss in section 4, are better aligned with our task as relying on looser input requirements.

2.3   Focus on Healthcare

So far, we have concentrated our efforts in applying the above framework in the healthcare domain. Indeed, BBNs have long been advocated as a useful decision and risk modeling framework in medicine [13] and used in several decision support systems [14-16] yet the difficulty of building BBNs from scratch has limited widespread adoption [17]. In fact, [17] reports that the process of building the structure of model for therapy selection for the treatment of the cancer of the esophagus required 11 sessions with 2 experts and one knowledge engineer, each session lasting 2-4 hours and requiring about 20 hours of preparation. Our focus on automating the extraction and aggregation of relevant information seeks to address this practical challenge.

This paper focuses specifically on breast cancer, simply motivated by the large amount of academic papers published on the topic each year and the fact that it is a disease that is well understood. We created a *reference information base* (called gold standard in the natural language processing community) from about 300 MEDLINE® abstracts selected according to the query: "KW[1]:breast cancer AND parity" over the past 5 years. This reference information base consists of the manually generated output of the two extraction tasks. It is used to understand the characteristics of the elements to be extracted, to develop machine learning algorithms for extraction and, in this paper, as proxy for automatically extracted information (for aggregation).

---

[1] KW stands for Keyword(s).

3   Structural Information Extraction

3.1   Defining Dependence and Independence

Structural information extraction involves identifying in texts information indicating dependence or independence among variables. Dependence and independence here are to be understood as defined by probability theory where A depends on B iff $P(A) \neq P(A|B)$. As independence statements are seldom, we focus predominantly on dependence in the remaining of this paper and define independence as the negation of dependence. Future research will focus on understanding the specificities of independence statements to increase the chances of extracting such statements.

Our objective in the extraction step is to transform a sentence into set of dependence relation entries structured as follows:

- One Variable A and One Variable B (compulsory)
- Influence terms (compulsory)
- A modifier (optional)
- A context variable (Optional)
- A negation (Optional)

While dependence and independence of variables are symmetric relations, we make a loose distinction between Variable A and Variable B, where Variable A captures the causing factor and Variable B the influenced factor. We make influence terms compulsory because from a language perspective, they serve as the cornerstone of the dependence relation. Examples of influence terms include: "associated with", "reduction", "correlated", "higher" or "likely". The modifier can provide nuances about the strength of the statement (e.g., "significantly", "may", and "positively"), the context variable limits the population to which the statement applies, and the negation element enables to capture independence. Other optional variables could be added, typically meta-data such as authorship, venue, publication year and original language. As an example, from the sentence "For endometrial cancer, body mass index represents a major modifiable risk factor; about half of all cases in postmenopausal women are attributable to overweight or obesity.", we extract two structured relations:

- Variable A: body mass index – Variable B: endometrial cancer – Influence Term: risk factor – Modifier: major
- Variable A: overweight or obesity – Variable B: cases – Influence Term: attributable – Context: postmenopausal women.

Our first step is to characterize the elements to be extracted (so far variables and influence terms) and then to develop machine learning approaches using features derived from the characteristics uncovered. Dependence information extraction has many similarities with relation extraction from clinical texts whose focus has been on protein-protein interactions or protein-gene interactions, with less attention being paid to the extraction of other types of relationship [18-20]. Our goal here is different as we approach a new type of relation (risk) and do not have any predefined restrictions for the variables beside the medical domain.

3.2   Characterizing Elements

Using 30 abstracts[2] from our information base, we analyzed the syntactic and semantic characteristics of variables. Syntactically, variables are noun phrases (NP) with different levels of complexity: Variables A are mostly specific displaying context information (e.g. *age at menarche, duration of breast feeding, personal history of breast cancer*), while Variables B are rather concise (e.g. *breast carcinoma, mammographic density*). Semantically, using UMLS Metathesaurus semantic types[3] referred to in biomedical texts, we observed the following patterns:

* Variables A: [Finding] / [Therapeutic or Preventive Procedure] / [Organism Function] / [Disease or Syndrome] / [Gene or Genome] / [Organism Attribute]
* Variables B: [Neoplastic Process] / [Qualitative Concept] / [Disease or Syndrome]

Regarding influence terms, they appear to belong to a limited controlled language. From our full information base, we identified 1000 relations representing only 117 unique terms out of a vocabulary of about 10000 words. We report in Table 1 frequency counts of the 10 most frequent roots of those influence terms along with ratio of occurrences in a dependence relation by the total occurrences. Note that the five most common roots account for about 50% of the influence terms occurrences and the 10 most frequent for almost two third. Note also that for many roots, their occurrence is highly indicative of the presence of a dependence relation (fourth column).

Table 1. Frequency Counts of Most Frequent Influence Terms

Root	Occurrences in relations	Frequency among identified influence terms	Occurrences in relation / Total occurences
associate	333	34.5%	64.8%
increase	76	7.9%	37.4%
risk	62	6.4%	9.2%
relate	42	4.3%	36.5%
reduce	33	3.4%	55.0%
influence	26	2.7%	54.2%
likely	23	2.4%	71.9%
effect	22	2.3%	27.2%
correlate	20	2.1%	62.5%
high	17	1.8%	22.7%

---

[2] Those corresponds to the first 30 abstracts that were returned from Medline with our query.
[3] UMLS Metathesaurus semantic types enable to categorize the concepts listed in the Metathesaurus. There are 133 semantic types which are grouped around the following broad categories: organism, anatomical structure, biologic function, chemical, physical object, idea or concept. See [22] for a more in-depth description.

3.3   Extraction Algorithm

Consequently, using our information base we will build a dictionary of influence terms. The main challenge that we currently face is to determine for each occurrence of such terms whether it does correspond to a dependence relation and to extract the associated variables. As commonly done in natural language processing, we rely on machine learning for this task. Each term in a text will represent a candidate to be classified using features ranging from lexical ones (e.g., distance to influence term), to syntactic ones (e.g., part-of-speech, grammatical function, constituent type), to semantic ones (e.g., UMLS type). We plan to experiment with different classification algorithms for each variable type (A and B). Since the training set is small, generative classifiers like Naive Bayes are preferred initially as they are less subject to overfitting. Afterward, we plan to use additional unlabeled data with the Naive Bayes classifier and combine them in a semi-supervised learning approach. The resulting extended set of labeled data will be considered for learning with ensemble methods for decision trees that tend to outperform other algorithms for classification problems.

However, there are several steps that have to be undertaken before using the current information base for learning. Medical abstracts have to be preprocessed to deal with two linguistic challenges highly present in this domain, namely anaphoric reference and ellipsis. Referencing happens when the authors refer back to previously mentioned entities to avoid repetition (e.g., it referencing BRCA[4]), while ellipsis occurs when, after a more specific mention, words are partially or completely omitted when the phrase needs to be repeated.

Similarly we need to post-process the variables extracted to handle terms referring to an underdefined set of variables (e.g., physician characteristics, personal factors) and to cluster those referring to the same risk factor (e.g., smokers, smoking, cigarettes, tobacco consumption) into a single variable. This corresponds to our variable identification step which we will undertake using general lexical resources like the Roget Metathesaurus or specialised ones as the UMLS Metathesaurus and borrowing from [6]. The final output of the extraction and variable identification steps would be a set of cleaned evidence statements representing dependence and independence between variables rather than terms. Those are the input for the information aggregation described hereafter.

4   Structural Information Aggregation

4.1   General Problem Description

One way to aggregate a cleaned set of evidence statements into a BBN structure is:

1. to reconcile statements involving the same pair of variables A and B. The simplest approach being a majority vote, which can be modulated by adding weights to each statement, based on modifiers, publication year of the paper (higher weights for more recent papers) or any other relevant information.

---

[4] BRCA is a mutation of the genes BRCA1 and BRCA2 associated with higher risk of developing breast cancer.

2. to aggregate reconciled statements into a BBN structure, i.e., into a directed acyclic graph. One process to do so is to first create the undirected graph associated with the reconciled statements, assuming lack of information is understood for instance as independence (but the reverse convention is as valid, if not more) and second to orient the edges by assigning a random order to the variables thus ensuring that no cycle is created in the process.

The specific methods described in each step above are only the most straightforward solutions provided for the sake of illustration. More sophisticated approaches on the majority vote of the adjacency matrix of the graph, adapted for instance from [11] could also be applied. In addition, a Bayesian approach as described by Richardson and Domingos [12] is a valid alternative where, instead of having as input BBN structures from multiple experts, we have evidence statements from multiple papers. Finally yet another approach can be used based on an adaptation of the PC Algorithm [21]. While the PC Algorithm is based on independence tests, it can be adapted by replacing those tests with the independence statements extracted from the evidence statement extraction after reconciliation for conflicting opinions.

4.2   Our Method

We adapt Richardson and Domingos (RD hereafter) method for aggregation of multiple BBNs provided by human experts (and data, but that is not relevant in this paper). In their approach, each expert gives a description of the dependencies among a fixed set of variables, which is used to compute the most probable structure doing Bayesian updating. Some assumptions are made to simplify the computations, in particular that experts statements are independent from one another both across experts and across pairs of variables. As will become clear in the following paragraph, we make similar independence assumptions. However, the main difference in our method is that we assume that the information provided by sources (abstracts for us, experts for RD) is *not* directed. This simplifies the specification of the likelihood function yet implies that the output of the model is non-directed. We thus need to orient the edges to obtain a full BBN.

We now describe our approach, starting with some notation:

- Let $X = \{x_1, \cdots, x_N\}$ denote the set of variables in the information base, arbitrarily ordered.
- Let G denote a non-directed graph over A, we choose to represent $G = (g_1, \cdots, g_j, \cdots, g_{N'})$ as a vector of size $N' = N \cdot (N-1)/2$ where g_j corresponds to the jth variable pair for some arbitrary ordering of all possible pairs. $g_j = 1$ if there is an edge between the variables forming the jth pair and $g_j = 0$ otherwise.
- Let M denote the number of sources (texts) in our information base and $E_m = (e_{m1}, \cdots, e_{mj}, \cdots, e_{mN'})$ represent the evidence statements from the mth source where $e_{mj} = \emptyset$ means that no statement has been made about the jth pair, $e_{mj} = D$, means a dependence statement has been made and $e_{mj} = I$ means an independence statement has been made. Finally we denote by $E = (E_1, \cdots, E_m, \cdots, E_M)$ the full content of the information base.

Our objective is to evaluate the posterior distribution of the underlying non-directed graph P(G|E), which by Bayes theorem is P(G|E) = αP(G)P(E|G). We need to specify the prior P(G) and the likelihood function P(E|G). For simplicity, we assume that the presence of an arc between a pair of variables is independent from the other pairs, thus $P(G) = \prod_{j=1}^{N'} P(g_j)$. In addition, we assume that sources are conditionally independent given graph G so that $P(E|G) = P(E_1, \cdots, E_m, \cdots, E_M) |G) = \prod_{m=1}^{M} P(E_m|G)$. Finally, we assume that evidence statement from source m about pair j is conditionally independent from the other evidence statements from source m and depends only on g_j and not $\{g_l\}_{l \neq j}$. This leads to

$$P(G|E) = \alpha P(G)P(E|G) = \alpha \prod_{m=1}^{M} \left(\prod_{j=1}^{N'} P(g_j) \cdot P(e_{mj}|g_j) \right) \qquad (1)$$

We acknowledge that we are making strong independence assumptions. We will revisit those assumptions in future research yet the current model has value in the sense that it provides a benchmark for more sophisticated models. Moreover, in several other situations, it has been found that naïve models such as ours provide a useful first approximation (e.g., naïve Bayes classification, Markov chains modeling of complex systems, aggregation of expert opinions about probabilities).

Applying (1) enables us to derive the most likely non directed structure G* associated with the information base E. To orient the arcs, we rely on a user-defined order of the variables, which we obtained by clustering the diseases (and associated UMLS concepts) at the end of the list so that arcs would typically go from factors/symptoms to disease. This is consistent with our findings in section 3.2 related to Variables A and B.

4.3   Experimental Results

Data. From the reference information base of about 1000 relations, we selected 198 cleaned evidence statements, limiting ourselves to simple concepts that were unambiguous to non-medically trained researchers. *"Her2-overexpressing cases"*, *"basal-like molecular subtype"* were thus discarded, as were references to subtypes of breast cancer (Triple-negative, ER+/GR+). To reduce the number of variables to a manageable size, we selected the 18 most frequent which led us to keep 96 evidence statements. As expected given the limited size of our text data, we only have very partial information about the relationships among those variables. In our example, out of the 153 possible pairs, we obtained

- 18 pairs associated with only one evidence statement (11 pairs with one dependence statement and 7 pairs with one independence statement),

- 10 pairs associated with multiple concurring dependence statements (none for independence statements),
- 5 pairs associated with conflicting evidence statements (interactions between breast cancer and respectively age at menarche, breastfeeding, hormone replacement therapy, fertility, and single-nucleotide polymorphism).

We believe however that expanding the corpus may not resolve the lack of full coverage. Medical research focuses mostly on interactions between factors and diseases rather than among factors.

Parameters. In our experiment, we assume that each pair of variables independently presents a probability p_0 of being connected and set p_0 to 0.05, similarly to RD. We similarly assume that all likelihood parameters $P(e_{jm}|g_j)$ are identically distributed according to Table 2. Through this parameterization, lack of mention of a relation $(\emptyset)$ is neutral in the sense that observation of "no mention" leaves the prior probability unchanged. While it has no effect on the computations, for the sake of completeness we set $p_{\emptyset} = 0.995$ (based on observed value in our information base). We set $p_D = p_I = 0.9$, a somewhat arbitrary estimation of the reliability of medical research on which we perform sensitivity analysis.

Table 2. Likelihood Model

		e_{mj}		
		$\emptyset$	D	I
g_j	1	$p_{\emptyset}$	$(1 - p_{\emptyset})p_D$	$(1 - p_{\emptyset})(1 - p_D)$
	0	$p_{\emptyset}$	$(1 - p_{\emptyset})(1 - p_I)$	$(1 - p_{\emptyset})p_I$

Output. Applying the method described above, the BBN structure that we would obtain is presented on Fig. 2a. where labels indicate the associated posterior probabilities that there is an arc between the pair of variables j i.e. $P(g_j = 1|E)$. In fact, as our goal is not to provide a definite model but rather an easily understandable synthesis of the extracted information that tends toward a BBN representation, we deviate from providing the most likely structure. Instead, we report two graphs derived from the posterior probabilities as follows: If $P(g_j|E) > \beta P(g_j)$ then we show an arc on Fig. 2a, if $P(g_j|E) < P(g_j)/\beta$ then we show an arc on Fig. 2b (with $\beta = 3$ in this illustration). Any pair of variables that is not connected in either graphs is therefore in the unknown category (which could be represented graphically). In this perspective, the order of the variable used for arc orientation is not critical; its role is mostly to avoid cycles and to generate an intuitive output.

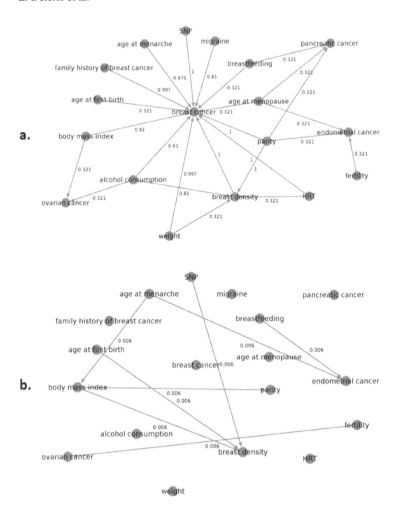

Fig. 2. Aggregation Output: Dependence Relations (top) and Independence Relations (bottom)

Overall, the information presented on Fig. 2 is designed to be challenged by a human expert. Variable order could be modified; logical relations (such as an arc between fertility and parity) could be added. In a real implementation, the expert would also be able to select an arc and view the underlying information for validation. In addition to the graphs, pairs of variables having strong support (multiple concurring relations) or high conflict (multiple conflicting relations) would be highlighted. Furthermore, we envision querying the user for validation of non-consensual statements and indirect influences (i.e. when we have both A→B→C and A→C), asking whether the second influence really exists or was generated from a simplification of the indirect one. For the BBN presented in Fig. 2 for instance, we would question whether the (protective) effect of parity on breast cancer is limited to breastfeeding.

One question that can be raised is about the benefits of the sophisticated Bayesian approach which seems to behave very much like majority vote. This is in part due to our strong independence assumptions. In future research, we will investigate how social networks about authorship (co-authors / referred and referring authors) could be used to build a simple dependence model about text sources so as to make the benefits of the Bayesian updating method more salient.

A Note on Sensitivity Analysis. We performed sensitivity analysis on the parameters, using graph edit distance to measure the effect of parameters changes. If p_0 stays within [0.01-0.2], then the output graph remains exactly the same, thus displaying reasonable robustness. In addition, the values of p_D and p_I only influence the arcs supported by multiple *conflicting* statements, which would nonetheless be singled out for appraisal by a human expert. More care, however, needs to be spent in defining the default values of those parameters.

5   Conclusion

This paper outlines our general framework for facilitating the construction of Bayesian networks from unstructured texts, focusing initially on the medical domain. Our objective is to reduce the need for human intervention in just the same way that machine translation of a text provides a human translator with an imperfect but labor-saving first draft. We discuss more specifically the problem of extracting and aggregating dependence information for which we propose, if not fully defined algorithms, a detailed description of the steps to follow. The main shortcomings of our current analyses are (i) the lack of final algorithm for extraction and (ii) the limited size of our reference set for aggregation. Therefore we cannot reach definite conclusions. In particular, we can expect that the input to aggregation will be several orders of magnitude larger once the extraction step is fully automated. We feel nonetheless that the findings discussed in this paper are encouraging in that they confirm the feasibility of our endeavor and illustrate its usefulness.

Acknowledgements. The authors gratefully acknowledge the financial support provided by the Irish Industrial Development Agency under reference 199954.

References

1. Pearl, J.: Probabilistic Reasoning in Intelligent Systems: Networks of Plausible Inference. Morgan Kauffman, San Francisco (1988)
2. Pietzsch, J.B., Paté-Cornell, E.: Early technology assessment of new medical devices. International Journal of Technology Assessment in Health Care 24, 37–45 (2008)
3. Deleris, L.A., Yeo, G., Siever, A., Paté-Cornell, E.: Engineering risk analysis of a hospital oxygen supply system. Medical Decision Making 26, 162–172 (2006)
4. Egar, J.W., Musen, M.A.: Automated modeling of medical decisions. In: Proceedings of the Annual Symposium on Computer Application in Medical Care, pp. 424–428 (1993)

5. Zhu, A., Li, J., Leong, T.: Automated knowledge extraction for decision model construction: A data mining approach. In: AMIA 2003 Symposium Proceedings, pp. 758–762 (2003)
6. Sanchez-Graillet, O., Poesio, M.: Acquiring Bayesian networks from text. In: Proceedings of LREC (2004)
7. Clemen, R.T., Winkler, R.L.: Aggregating probability distributions. In: Edwards, W., Miles, R., von Winterfeldt, D. (eds.) Advances in Decision Analysis, pp. 154–176. Cambridge University Press, Cambridge (2007)
8. Matzkevitch, I., Abramson, B.: The topological fusion of Bayes Nets. In: Dubois, D., Wellman, M.P., Ambrosio, B., Smets, P. (eds.) Proceedings of the Eighth Conference on Uncertainty in Artificial Intelligence. Morgan Kaufmann, San Francisco (1992)
9. Sagrado, J., Moral, S.: Qualitative combination of Bayesian networks. International Journal of Intelligent Systems 18, 237–249 (2003)
10. Zhang, Y., Yue, K., Yue, M., Liu, W.: An approach for fusing Bayesian networks. Journal of Information and Computational Science 8, 194–201 (2011)
11. Rush, R., Wallace, W.A.: Elicitation of knowledge from multiple experts using network inference. IEEE Transactions on Knowledge and Data Engineering 9, 688–696 (1997)
12. Richardson, M., Domingos, P.: Learning with knowledge from multiple experts. In: Fawcett, T., Mishra, N. (eds.) Proceedings of the Twentieth International Conference on Machine Learning. Morgan Kaufmann, Washington (2003)
13. Pauker, S.G., Wong, J.B.: The influence of influence diagrams in medicine. Decision Analysis 2, 238–244 (2005)
14. Warner, H.R., Haug, P., Bouhaddou, O., Lincoln, M., Sorenson, D., Williamson, J.W., Fan, C.: Iliad as an expert consultant to teach differential diagnosis. In: Proceedings of the Annual Symposium on Computer Applications in Medical Care, vol. 154, pp. 371–376 (1988)
15. Hoffer, E., Feldman, M., Kim, R., Famiglietti, K., Barnett, G.: Dxplain: patterns of use of a mature expert system. In: AMIA 2005 Symposium Proceedings, pp. 321–325 (2005)
16. Fuller, G.: Simulconsult. Journal of Neurology, Neurosurgery and Psychiatry 76, 10 (2005), http://www.simulconsult.com
17. van der Gaag, L.C., Renooij, S., Witteman, C.L.M., Aleman, B., Taal, B.F.: How to elicit many probabilities. In: Laskey, K.B., Prade, H. (eds.) Proceedings of the Fifteenth Conference on Uncertainty, pp. 647–654. Morgan Kaufmann Publishers, San Francisco (1999)
18. Chowdhary, R., Zhang, J., Liu, J.: Bayesian inference of protein-protein interactions from biological literature. Bioinformatics 25, 1536–1542 (2009)
19. Bui, Q.C., Katrenko, S., Sloot, P.M.A.: A hybrid approach to extract protein-protein interactions. Bioinformatics 27, 259–265 (2011)
20. Miwa, M., Sætre, R., Kim, J.-D., Tsujii, J.: Event extraction with complex event classification using rich features. Journal of Bioinformatics and Computational Biology 8, 131–146 (2010)
21. Spirtes, P., Glymour, C., Scheines, R.: Causation, prediction, and search. Lecture Notes in Statistics. Springer (1993)
22. UMLS® Reference Manual [Internet]. Bethesda (MD): National Library of Medicine (US); Semantic Network, http://www.ncbi.nlm.nih.gov/books/NBK9679/

Voting on Actions with Uncertain Outcomes

Ulle Endriss

Institute for Logic, Language, and Computation,
University of Amsterdam

Abstract. We introduce a model for voting under uncertainty where a
group of voters have to decide on a joint action to take, but the individual
voters are uncertain about the current state of the world and thus about
the effect that the chosen action would have. Each voter has preferences
about what state they would like to see reached once the action has been
executed. That is, we need to integrate two kinds of aggregation: beliefs
regarding the current state and preferences regarding the next state.

1 Introduction

Imagine a group of *agents* who have to make a collective decision about what
(joint) *action* to take. This action has to be chosen from a set of available actions.
Each agent has her own *preferences* over the effects of actions, i.e., over the state
of the world after a given action has been executed. Unfortunately, our agents
are *uncertain* about the *current state* of the world and thus about the precise
effects (or *outcomes*) of actions. Each agent only has a (possibly different) *set of
states* she considers plausible (but they all agree on what the effect of executing
a given action in a given state would be). What action should they take?

This is a collective decision making problem that involves the aggregation
of two kinds of information: *social* information regarding the preferences over
states of the world (effects of actions) and *epistemic* information regarding the
plausibility of certain states being the actual current state of the world. This
combination of concerns is what we should expect to encounter in a variety of
application domains, e.g., when devising mechanisms for teams of autonomous
software agents to interact and agree on actions to pursue collectively. However,
while social choice theory has, rightly, been argued to be relevant to multiagent
systems [17,7], the standard model of social choice only deals with the aggregation of preferences [10]. In this paper we put forward a model for voting on
actions with uncertain effects that integrates this standard perspective with a
simple notion of uncertainty.

We model uncertainty in the simplest possible way: a voter can only distinguish between states she considers plausible and those she does not consider
plausible. This model of uncertainty is sometimes called *strict uncertainty* or
complete ignorance [5]. It is different from much work on reasoning under uncertainty, which often assumes the availability of a probability distribution over
the set of possible states [13]. Our chosen representation of voter preferences is
also simpler than what is typically assumed in the literature on decision-making

P. Perny, M. Pirlot, and A. Tsoukiàs (Eds.): ADT 2013, LNAI 8176, pp. 167–180, 2013.

under uncertainty: rather than endowing each voter with a utility function mapping possible outcomes to numerical values, we only assume that they are able rank the outcomes in terms of their relative desirability. There are two reasons for choosing such simple representations of uncertainty and preference. First, it arguably is often not realistic to expect a decision-maker to be able to provide either a fully specified probability distribution (e.g., how should she judge whether a given state has probability 15% or 20% of being the actual state of the world?) or a precise utility function (what does it mean to assign utility 20 rather than 19 to a given outcome?). Second, we want the individual components of our model to be as simple as possible, so as to be able to better focus on the analysis of their interplay.

The remainder of this paper is organised as follows. We begin, in Section 2, with a detailed discussion of the challenges associated with voting under uncertainty and present three paradoxes one might encounter in this context. After presenting our formal model in Section 3, we then focus on two specific aspects of the general problem: aggregating information regarding the current state only, without considering preferences (in Section 4); and integrating uncertainty and preference information for the special case of a single agent (in Section 5). We conclude in Section 6 by using the insights gained to make a first tentative proposal for best practices for voting on actions with uncertain outcomes.

2 Three Paradoxes

In this section we introduce three paradoxes—scenarios that show how a seemingly reasonable approach to voting on actions with uncertain effects can lead to suboptimal outcomes. Our presentation is organised in terms of the point in the aggregation process at which the uncertainty about the current state of the world is being resolved.

For each paradox we also briefly discuss similarities to related phenomena from different strands of the literature on social choice theory, namely judgment aggregation, preference aggregation and voting, and ranking sets of objects.

2.1 The Paradox of Individual Uncertainty Resolution

Suppose there are two possible states of the world, A and B. Whichever state we are in, we can execute one of two actions: to *change* to the other state or to *stay* in the current state. There are three agents, who are all uncertain about the current state. Agents 1 and 2 believe A is most plausible; agent 3 believes B is most plausible. They have to agree on one of the two actions to be executed, and they each have their own preferences about what state they would like to end up in afterwards. Agent 1 prefers A; agents 2 and 3 both prefer B.

If each agent is left to resolve their uncertainty regarding the current state on their own, then each of them will infer their "most preferred action" based on their preferences over states and based on what they believe to be the most plausible current state. That is, agents 1 and 3 will choose to *stay*, and only

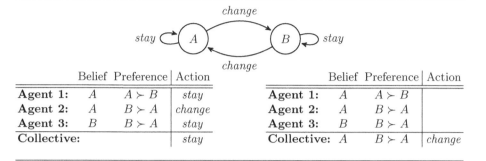

	Belief	Preference	Action
Agent 1:	A	$A \succ B$	stay
Agent 2:	A	$B \succ A$	change
Agent 3:	B	$B \succ A$	stay
Collective:			stay

	Belief	Preference	Action
Agent 1:	A	$A \succ B$	
Agent 2:	A	$B \succ A$	
Agent 3:	B	$B \succ A$	
Collective:	A	$B \succ A$	change

Fig. 1. The Paradox of Individual Uncertainty Resolution

agent 2 will choose to *change*. If we use the majority rule to decide between these two options, we will thus decide that the best move for the collective is to *stay* in the current state (see lefthand side of Fig. 1).

But this, arguably, is not the best course of action for this group. If instead the agents first use the majority rule to aggregate their information regarding the current state (finding A to be most plausible) and then to aggregate their preferences (finding B to be socially preferred), then the appropriate action to take would be to *change* the state (see righthand side of Fig. 1).

That is, resolving uncertainties regarding the current state individually before aggregation can lead to outcomes that, arguably, are suboptimal.

This paradox is related to the *discursive dilemma* familiar from judgment aggregation [15]. In judgment aggregation, we are asked to aggregate the views of several individuals regarding the truth or falsity of a number of formulas of propositional logic, and the discursive dilemma is a family of paradoxical situations that we may encounter in this framework. We can model our paradox as a problem of judgment aggregation as follows. Let p stand for "A is the most plausible current state" (and thus $\neg p$ for "B is the most plausible current state"); let q stand for "A is the most preferred next state" (and thus $\neg q$ for "B is the most preferred next state"). Then we should choose the action *stay* if and only if $p \leftrightarrow q$ is true (i.e., we should choose *change* if it is false). Then the following situation corresponds to the scenario described earlier:

	p	q	$p \leftrightarrow q$
Agent 1:	Yes	Yes	Yes
Agent 2:	Yes	No	No
Agent 3:	No	No	Yes
Majority:	Yes	No	?

The righthand side of Fig. 1 corresponds to what is known as the *premise-based procedure* in judgment aggregation: we use the majority rule to obtain a collective judgment on p and q and then use logical inference to decide that $p \leftrightarrow q$ must be false. The lefthand side of Fig. 1 corresponds to the *conclusion-based procedure:* to obtain a collective judgment on $p \leftrightarrow q$ we only consider the

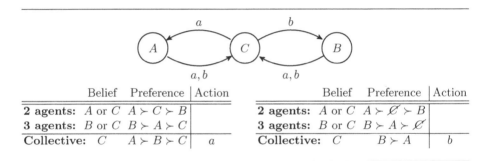

Belief	Preference	Action		Belief	Preference	Action
2 agents: A or C	$A \succ C \succ B$		**2 agents:** A or C	$A \succ \cancel{C} \succ B$	
3 agents: B or C	$B \succ A \succ C$		**3 agents:** B or C	$B \succ A \succ \cancel{C}$	
Collective: C	$A \succ B \succ C$	a	**Collective:** C	$B \succ A$	b

Fig. 2. The Paradox of Late Collective Uncertainty Resolution

individual judgments on the same formula and find that a majority is in favour of labelling it as being true. Thus, two seemingly reasonable forms of aggregation yield contradictory advice, i.e., we are facing a discursive dilemma.

2.2   The Paradox of Late Collective Uncertainty Resolution

Suppose there are three possible states, A, B and C, and two available actions, a and b. Executing a in state C takes us to A; executing b in C takes us to B; and executing either a or b in either state A or B will take us to C. We have two agents who are uncertain whether the current state is A or C, and for the next state they both prefer A over C over B; and we have three agents who are uncertain between B and C, and they all prefer B over A over C.

Let us use the *Borda rule* to aggregate preferences: for each agent a state gets as many points as that agent ranks other states below the state in question (and we then order the states in terms of the points received). That is, A gets $2 \cdot 2 + 3 \cdot 1 = 7$ points, B gets $2 \cdot 0 + 3 \cdot 2 = 6$ points, and C gets $2 \cdot 1 + 3 \cdot 0 = 2$ points, i.e., we obtain the collective preference order $A \succ B \succ C$. To resolve the uncertainty regarding the current state, there really is only one natural choice, namely to take the state considered plausible by the highest number of agents, which is C. A, the collectively most preferred state, is reachable from C, namely by executing action a. Hence, we should execute a (see lefthand side of Fig. 2).

But now consider this: if C is the current state, which is the collectively most plausible assumption to make here, then C cannot be the *next* state, whichever action we execute (if we execute a, we end up in A; if we execute b, we end up in B). So, arguably, our agents' preferences regarding C are not relevant. Then we have two agents who prefer A over B, and three agents who prefer B over A. Hence, the collectively most preferred state is B (under the Borda rule, as well as under any other reasonable aggregator) and we should execute action b to reach it (see righthand side of Fig. 2).

That is, by having postponed the collective uncertainty resolution until after the step of preference aggregation, we have missed the opportunity to eliminate irrelevant information from our aggregation problem which, arguably, has led to a suboptimal outcome.

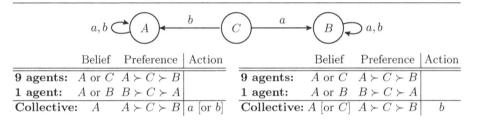

	Belief	Preference	Action		Belief	Preference	Action
9 agents:	A or C	$A \succ C \succ B$		**9 agents:**	A or C	$A \succ C \succ B$	
1 agent:	A or B	$B \succ C \succ A$		**1 agent:**	A or B	$B \succ C \succ A$	
Collective:	A	$A \succ C \succ B$	a [or b]	**Collective:** A [or C]		$A \succ C \succ B$	b

Fig. 3. The Paradox of Early Collective Uncertainty Resolution

This paradox is closely related to Arrow's *independence of irrelevant alternatives* [2]: Arrow postulated that for any reasonable form of preference aggregation the relative collective ranking of two alternatives should only depend on their relative rankings provided by the individuals, and not on any third ("irrelevant") alternative. Our example demonstrates that the Borda rule, when used as an aggregator for preference orders, violates this desideratum: the relative ranking of A and B adopted by the collective *does* depend on C. Our paradox also has close connections to the topic of election control by means of adding (or deleting) candidates, widely studied in computational social choice [4,9], which in turn relies on violations of Arrow's independence axiom: our example demonstrates how adding candidate C to a Borda election in which B was winning can result in A becoming the new winner.

2.3   The Paradox of Early Collective Uncertainty Resolution

Suppose again that there are three possible states, A, B and C, and two available actions, a and b. When in state C, action a will take us to state B, while b will take us to A. If the current state is either A or B, then neither action will change the current state. There are ten agents. Nine of them consider A and C plausible states and prefer A over C over B. The remaining agent considers A and B plausible and prefers B over C over A.

If we aggregate the information regarding the plausibility of different states, we find that A is the most plausible state (but C comes a close second). Now suppose we insist on aggressively exploiting this information to pinpoint the most plausible current state: that is, we take A to be that state. In state A it does not matter what action we execute; the next state will always be A again. Suppose that in such a case, by default, the lexicographically first action is chosen, i.e., a in this case (see the lefthand side of Fig. 3).

But this clearly is a suboptimal choice: C is almost as likely to be the current state as A. If it really is C, then executing (the chosen) action a will result in state B, while executing (the dismissed) action b will result in state A. Considering that the collective preference order is clearly $A \succ C \succ B$ (for any reasonable preference aggregation rule), executing b therefore would be a much better choice. We could obtain this outcome by delaying uncertainty resolution:

if we consider both A and C plausible current states and compute the best action for either case, we find that in the first case it makes no difference which action is chosen, while in the second case b is better (see the righthand side of Fig. 3).

That is, resolving the uncertainty regarding the current state too early, even if we take the information supplied by all agents into account, can lead to a suboptimal outcome.

There are connections here to problems analysed in the literature on *ranking sets of objects* [14,3,11]. The question discussed in that literature is how to extend a preference order over individual objects to a preference order over nonempty sets of such objects (with the most common interpretation of those sets being that we will eventually obtain one of the objects in the set in question, but cannot control which). One of the most basic axioms formulated in this literature postulates that, if we strictly prefer A over B, then we should (at least weakly) prefer $\{A\}$ over $\{A, B\}$—because, whatever choice rule may get used to select from the second set, the outcome can never be worse than A. This is precisely why, in our scenario above, we should prefer action b (which can only result in A when A and C are considered the only plausible states) over action a (which might result in either A or B).

3   A Formal Model

Fix a finite set $\mathcal{Q}$ of *states* and a finite set Σ of *actions*. A *transition function* $\delta : \mathcal{Q} \times \Sigma \to \mathcal{Q}$ determines for any given state $q \in \mathcal{Q}$ and any given action $\sigma \in \Sigma$ the state q' that will be reached after executing σ in q. In other words, δ defines the *effect* of an action for every possible state. Note that δ is a total function, i.e., every action is executable from every state—but it may well be the case that an action σ has no effect in a given state q in the sense that $\delta(q, \sigma) = q$. We use $\delta(\sigma, Q) = \{\delta(q, \sigma) \mid q \in Q\}$ to refer to the set of states that are *reachable* from Q via σ, i.e., the set of states we might reach by executing action σ in a situation where any of the states in $Q \subseteq \mathcal{Q}$ is a plausible current state.

Fix a finite set $\mathcal{N} = \{1, \ldots, n\}$ of agents. We assume that each agent has complete knowledge of δ. On the other hand, an agent does not necessarily know the identity of the current state. We express an agent's uncertainty regarding the current state in terms of a (nonempty) subset of $\mathcal{Q}$: for each agent $i \in \mathcal{N}$, let $Q_i \subseteq \mathcal{Q}$ denote the set of states she considers possible. This is a very minimalist approach towards modelling uncertainty; other options include defining a probability distribution or a plausibility ranking on $\mathcal{Q}$ for each agent.

Agents have preferences over states, which we model in terms of linear orders (i.e., binary relations that are irreflexive, transitive, and complete). We write $\succ_i$ for the preference order of agent $i \in \mathcal{N}$. Let $\mathcal{L}(\mathcal{Q})$ denote the set of all linear orders on $\mathcal{Q}$ and let $\Pi(\mathcal{Q})$ denote the set of nonempty subsets of $\mathcal{Q}$. That is, each agent provides us with an element of $\Pi(\mathcal{Q}) \times \mathcal{L}(\mathcal{Q})$, i.e., a pair consisting of an uncertainty set and a preference order. A *profile* is a vector of n such pairs, one for each agent, i.e., it is an element of $[\Pi(\mathcal{Q}) \times \mathcal{L}(\mathcal{Q})]^n$.

We are now ready to define our main concept: an aggregation mechanism that accepts a profile, i.e., the beliefs and preferences of each agent, and then returns a single action to be executed by the group. Our definition is relative to a given transition function δ (while $\mathcal{Q}$, Σ and $\mathcal{N}$ are taken to be fixed throughout).

Definition 1. *Let δ be a transition function. Then a **social action choice function** for δ, $F_\delta : [\Pi(\mathcal{Q}) \times \mathcal{L}(\mathcal{Q})]^n \to \Sigma$, is a mapping from profiles to actions.*

Note that we insist on F_δ returning a *single* action. Alternatively, we could have defined F_δ as an *irresolute* aggregator returning a nonempty set of most preferred actions and considered the problem of tie-breaking, i.e., of choosing one best action to actually execute, as a separate problem.

A closely related problem is to derive a *weak order* (i.e., a binary relation that is reflexive, transitive, and complete) on the available actions from a given profile. This is useful, for instance, when we are not yet certain which actions will become available and we want to prepare for executing the collectively best available action. We write $\mathcal{W}(\Sigma)$ for the set of all weak orders on Σ.

Definition 2. *Let δ be a transition function. Then a **social action ranking function** for δ, $F_\delta : [\Pi(\mathcal{Q}) \times \mathcal{L}(\mathcal{Q})]^n \to \mathcal{W}(\Sigma)$, is a mapping from profiles to weak orders on actions.*

When agents are certain about the identity of the current state of the world, then social action choice functions essentially correspond to what are known as (resolute) *social choice functions* in the literature, and *social action ranking functions* correspond to *social welfare functions* [10].

4   Uncertainty Resolution in Isolation

Our model deals with problems that require both the aggregation of preferences and the aggregation of beliefs regarding the identity of the current state. The former, when viewed in isolation, is the main problem studied in classical social choice theory [10]. In this section we want to study the latter in isolation, i.e., we want to study the problem of uncertainty resolution on the basis of the reports of the individual agents. Suppose each agent reports a nonempty sets of states (a subset of $\mathcal{Q}$). How should we aggregate such an *uncertainty profile* into a single collective set of states that appropriately reflects the beliefs of the group?

Definition 3. *An **uncertainty resolution rule** $F : \Pi(\mathcal{Q})^n \to \Pi(\mathcal{Q})$ is a mapping from an uncertainty profile to a single nonempty set of states.*

We use $\mathbf{Q} = (Q_1, \ldots, Q_n)$ to refer to a profile of sets of states, with Q_i being the set reported by agent i. Let $N_q^{\mathbf{Q}} = \{i \in \mathcal{N} \mid q \in Q_i\}$ denote the set of agents who include state q in their set under profile $\mathbf{Q}$.

Structurally, an uncertainty resolution rule has the same form as a voting rule based on *approval ballots* [16], of which *approval voting* is the main representative [6]: such a voting rule takes as input a set of candidates from each voter and returns a winning candidate as output (or possibly a set of tied winners). Indeed, the most natural choices for an uncertainty resolution rule all correspond to rules discussed in the literature on voting:

- Under *approval voting* we return the states with maximal support [6]:

$$F(\boldsymbol{Q}) = \operatorname*{argmax}_{q \in \mathcal{Q}} |N_q^{\boldsymbol{Q}}|$$

- Under *even-and-equal cumulative voting* each agent evenly distributes a total of weight 1 over the states they report as plausible and the state(s) with the maximal sum of weights are being returned [1]:

$$F(\boldsymbol{Q}) = \operatorname*{argmax}_{q \in \mathcal{Q}} \sum_{i \in N_q^{\boldsymbol{Q}}} \frac{1}{|Q_i|}$$

- Under the *mean-based rule* we return all those states that receive at least an average amount of support [8]:

$$F(\boldsymbol{Q}) = \left\{ q \in \mathcal{Q} \mid |N_q^{\boldsymbol{Q}}| \geqslant \frac{|Q_1| + \cdots + |Q_n|}{|\mathcal{Q}|} \right\}$$

To see that above equality correctly formalises the mean-based rule, note that $|Q_1| + \cdots + |Q_n| = \sum_{q \in \mathcal{Q}} |N_q^{\boldsymbol{Q}}|$.

Observe that in case there is at least one state that is reported by every agent, approval voting yields the same result as taking the intersection of all individual sets of states. Let us call this the *intersection rule*.

We now want to consider uncertainty resolution rules from an axiomatic point of view and formulate desiderata to characterise appropriate rules. The simplest such desideratum is *neutrality*, a standard concept in social choice theory, which asks that the chosen rule should treat all states symmetrically. Our formulation below is closest to how neutrality has been formalised in the literature on judgment aggregation [12].

Definition 4. *An uncertainty resolution rule F is called **neutral** if, for all profiles $\boldsymbol{Q}$ and all states q and q', $N_q^{\boldsymbol{Q}} = N_{q'}^{\boldsymbol{Q}}$ implies $q \in F(\boldsymbol{Q}) \Leftrightarrow q' \in F(\boldsymbol{Q})$.*

Before we formulate further desiderata, we need to clarify the semantics of the set of states reported by an agent. Do agents report what they *know* about the identity of the current state or do they merely report their *beliefs*?

4.1 Uncertainty Resolution When Agents Report Knowledge

Assuming that *agents report knowledge* means assuming that, for each agent, the true state of the world is an element of the set reported by that agent. In particular, this means that $Q_1 \cap \cdots \cap Q_n \neq \emptyset$. This is not as unrealistic an assumption as it may seem: after all, each agent has the option to report the full set $\mathcal{Q}$, i.e., even a completely ignorant agent can report true (albeit vacuous) information. If we *trust* in our assumption of agents reporting knowledge, then we can exclude any state not reported by all agents.

Definition 5. *Suppose agents report knowledge. An uncertainty resolution rule is called* **trustful** *if, for all profiles* $\boldsymbol{Q}$, *we have that* $F(\boldsymbol{Q}) \subseteq Q_1 \cap \cdots \cap Q_n$.

Proposition 1. *Suppose agents report knowledge. Then an uncertainty resolution rule is both trustful and neutral if and only if it is the intersection rule.*

Proof. Clearly, the intersection rule is both trustful and neutral. For the other direction, suppose F is an uncertainty resolution rule that is trustful and neutral. By virtue of F being trustful, we have $F(\boldsymbol{Q}) \subseteq Q_1 \cap \cdots \cap Q_n$. Now, for the sake of contradiction, assume that $F(\boldsymbol{Q})$ is a *proper* subset of $Q_1 \cap \cdots \cap Q_n$. Then there exist a state $q \in F(\boldsymbol{Q})$ and another state $q' \in Q_1 \cap \cdots \cap Q_n \setminus F(\boldsymbol{Q})$. That is, despite having $N_q^{\boldsymbol{Q}} = N_{q'}^{\boldsymbol{Q}}$ (both are equal to the full set $\mathcal{N}$), we do not have $q \in F(\boldsymbol{Q}) \Leftrightarrow q' \in F(\boldsymbol{Q})$, i.e., we have observed a violation of neutrality. □

4.2 Uncertainty Resolution When Agents Report Mere Beliefs

If agents merely report beliefs regarding the identity of the current state, then the intersection rule ceases to be a viable option (as it may then return the empty set). In this case, approval voting is maybe the most natural choice. Even-and-equal cumulative voting is appropriate if we have reason to assume that agents who report small sets do so because they possess more accurate information. The mean-based rule is an attractive choice if we do not want to exclude too many alternatives before also considering preference information. Axiomatisations of all these rules are available in the literature [18,1,8].

5   The Single-Agent Case

In this section we discuss the special case where there is just a single agent. Even in this severely simplified scenario we still face the challenge of integrating belief and preference information. As we shall see, this special case is interesting in its own right, but it is also relevant to the more general aggregation problem discussed in this paper. The reason is that one natural approach to voting under uncertainty is to first aggregate belief and preference information independently and to then decide what action to take based on the thus obtained collective belief and collective preference order.

For the case of $n = 1$, Definitions 1 and 2 simplify to what we shall call *action choice functions* and *action ranking functions*, respectively. In the sequel, we focus on the latter, so as to be able to discuss connections to the literature on ranking sets of objects [14,3,11].

For any $Q \subseteq \mathcal{Q}$ and any linear order $\succ$ on $\mathcal{Q}$, let $\succcurlyeq_Q$ represent the weak order returned by F_δ (i.e., $\succcurlyeq_Q$ denotes $F_\delta(Q, \succ)$). That is, we write $\sigma \succcurlyeq_Q \sigma'$ to say that σ is *at least as good an action to take as* σ'. We furthermore write $\sigma \succ_Q \sigma'$ in case σ is the strictly better action to take and $\sigma \sim_Q \sigma'$ in case both actions are equally good choices, i.e., $\succ_Q$ is the *strict part* of $\succcurlyeq_Q$ and $\sim_Q$ is the *indifference part* of $\succcurlyeq_Q$. Finally, $\succcurlyeq$ (without a subscript) denotes the reflexive closure of $\succ$ (our agent's strict preference order over outcomes).

5.1   Desiderata for Action Choice Functions

We now want to formulate desiderata (or *axioms*, in the terminology of social choice theory) for an action ranking function F_δ. Recall that, given a set Q of plausible current states, every action σ gives rise to a set of plausible outcomes $\delta(\sigma, Q)$. There are two approaches to formulating desiderata:

- *Outcome-based desiderata:* First, we may choose to formulate desiderata regarding the relative ranking of actions purely in terms of the sets of plausible outcomes they correspond to.
- *Case-based desiderata:* Alternatively, we may also take into account which of the plausible current states would give rise to what outcomes and formulate our desiderata case by case.

We begin with the former approach:

Definition 6. *An action ranking function satisfies* **outcome-dominance** *if, for all $\sigma, \sigma' \in \Sigma$, we have $\sigma \succ_Q \sigma'$ whenever the following conditions hold:*

(i) $q \succ q'$ *for all $q \in \delta(\sigma, Q) \setminus \delta(\sigma', Q)$ and all $q' \in \delta(\sigma', Q)$*
(ii) $q \succ q'$ *for all $q \in \delta(\sigma, Q)$ and all $q' \in \delta(\sigma', Q) \setminus \delta(\sigma, Q)$*
(iii) $\delta(\sigma, Q) \neq \delta(\sigma', Q)$

Outcome-dominance is known as the *Gärdenfors principle* in the literature on ranking sets of objects [3]. It is a natural requirement to impose: it classifies any change of action as an improvement if it results either in a new best possible outcome to be added or in the currently worst possible outcome to be eliminated.

An example for an action ranking function that satisfies outcome-dominance is the *max-min ordering:* Prefer σ over σ' (with Q being the plausible states) if and only if either $\max(\delta(\sigma, Q)) \succ \max(\delta(\sigma', Q))$ or both $\max(\delta(\sigma, Q)) = \max(\delta(\sigma', Q))$ and $\min(\delta(\sigma, Q)) \succ \min(\delta(\sigma', Q))$.[1] That is, under the max-min ordering we first check for which action the best possible outcome is better and in case that is not enough to differentiate the two actions, we go by the worst possible outcome for either action.

Definition 7. *An action ranking function satisfies* **casewise-dominance** *if $\sigma \succ_Q \sigma'$ whenever $\delta(q, \sigma) \succcurlyeq \delta(q, \sigma')$ for all states $q \in Q$, and that preference is strict in at least one case.*

That is, casewise-dominance says that we should prefer an action if it does at least as well or better than a competing action for every state we consider possible. This certainly is a property we would like to see satisfied.

An example for an action ranking function that satisfies it is what we shall call a *casewise-lexicographic ordering:* Fix any strict linear order $\gg$ on Q. We want to rank σ and σ' (with Q being the plausible states). First consider the state q_1 that is maximal in Q with respect to $\gg$. If $\delta(q_1, \sigma) \succ \delta(q_1, \sigma')$, then prefer σ; if $\delta(q_1, \sigma') \succ \delta(q_1, \sigma)$, then prefer σ'; and if $\delta(q_1, \sigma) = \delta(q_1, \sigma')$, then

---

[1] Here max and min are defined with respect to the given order $\succ$ on states.

postpone the decision and compare σ and σ' for state q_2, the next best state with respect to $\gg$. Continue until one action is found to be preferred or until all states in Q have been exhausted (in which case you declare indifference).

Proposition 2. *Outcome-dominance does not imply casewise-dominance, nor does casewise-dominance imply outcome-dominance.*

Proof (sketch). To see this, check that casewise-lexicographic orderings violate outcome-dominance and the max-min ordering violates casewise-dominance. □

Definition 8. *An action ranking function satisfies* **outcome-relevance** *if $\sigma \sim_Q \sigma'$ whenever $\delta(\sigma, Q) = \delta(\sigma', Q)$.*

That is, outcome-relevance says that we should be indifferent between any two actions with identical sets of reachable states (given our beliefs regarding the current state). Intuitively speaking, outcome-relevance is much weaker a condition than outcome-dominance, but we note that technically the latter does not imply the former. The reason is that outcome-dominance does not prescribe how to rank two actions that produce the same outcome set.

5.2 Outcome-Based Desiderata and Ranking Sets of Objects

Our next result shows that we are dealing with a framework that is at least as expressive as the classical framework for *ranking sets of objects* [3].[2] Consider a problem of the latter kind, with m objects. To embed this into an action-ranking problem, we will show how to construct a scenario with a set Q of m states (corresponding to the m objects) and a set Σ of 2^m-1 actions (corresponding to the 2^m-1 nonempty sets of objects), where for every possible nonempty set $Q \subseteq \mathcal{Q}$ of states there exists an action $\sigma \in \Sigma$ such that Q is equal to the set of plausible outcomes associated with σ.[3]

Proposition 3. *For any set $\mathcal{Q}$ of states, we can construct a set Σ of actions and a state transition function δ such that for every nonempty set $Q \subseteq \mathcal{Q}$ there exists exactly one action $\sigma \in \Sigma$ with $\delta(\sigma, \mathcal{Q}) = Q$.*

Proof. Let $\mathcal{Q}$ be an arbitrary set of states with $|\mathcal{Q}| = m$. Define $\Sigma := \Pi(\mathcal{Q})$, i.e., every action is associated with a nonempty set of states. Fix an arbitrary order on $\mathcal{Q}$ so that $\min(Q)$ is well-defined for every set $Q \in \Pi(\mathcal{Q})$. That is, $\min(\sigma) \in \sigma$ for every action σ. Now define a transition function δ as follows:

$$\delta(q, \sigma) = \begin{cases} q & \text{if } q \in \sigma \\ \min(\sigma) & \text{otherwise} \end{cases}$$

Now suppose our agent considers the entire set $\mathcal{Q}$ to be the set of plausible states. Then, given our definition of δ, after executing action σ the states she

---

[2] At this point we shall assume familiarity with the standard model used in the relevant literature [3].

[3] For our specific construction, this will be the case when the set of states considered to be plausible as current states is the full set $\mathcal{Q}$.

will consider possible outcomes are precisely the states in the set corresponding to σ. That is, $\delta(\sigma, \mathcal{Q}) = \sigma$. As every nonempty set $Q \subseteq \mathcal{Q}$ is in fact equal to some action σ, we are done.   □

The interest of this result stems from the fact that any of the impossibility results about satisfying certain combinations of desiderata known for ranking sets of objects [14,3,11] are now inherited by our model. To be sure, this is true only if we allow for arbitrary transition functions. An interesting direction for future work would be to investigate to what extent more positive results are attainable when we limit, for instance, the number of distinct follow-up states any given action may give rise to.

5.3   Impossibility of Satisfying Both Kinds of Desiderata

Our final result shows that, maybe somewhat surprisingly, casewise-dominance (which we suggested to be highly desirable) is in fact incompatible with the weakest possible outcome-related requirement:

Proposition 4. *There exists no action ranking function that satisfies both casewise-dominance and outcome-relevance.*

Proof. Consider a scenario with three sates and three actions: $\mathcal{Q} = \{A, B, C\}$ and $\Sigma = \{a, b, c\}$. Suppose our agent consider all three states plausible and has the preference order $A \succ B \succ C$. Furthermore, suppose the definition of the transition function δ is given by the following figure:

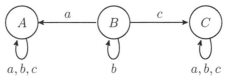

Now, for the sake of contradiction, let us try to construct an action ranking function producing the weak order $\succeq_{\mathcal{Q}}$ for this scenario. We can derive the following relative rankings of actions:

- $a \succ_{\mathcal{Q}} b$. This follows from casewise-dominance: if the current state is A or C, then actions a and b have the same effect; if the current state is B, then a will result in A, while b will only result in B.
- $b \succ_{\mathcal{Q}} c$. This also follows from casewise-dominance: if the current state is A or C, then actions b and c have the same effect; if the current state is B, then b will result in B, while c will only result in C.
- $a \sim_{\mathcal{Q}} c$. This, finally, follows from outcome-relevance, given that we have $\delta(a, \mathcal{Q}) = \delta(c, \mathcal{Q}) = \{A, C\}$.

But there can be no weak order $\succeq_{\mathcal{Q}}$ that satisfies all three constraints, as the first two constraints entail that a should be strictly preferred to c, while the third requires that they should be equally preferred.   □

It is not too hard to see that this impossibility result extends to the natural adaptations of our desiderata to action *choice* functions.

6   Conclusion

We have introduced a simple model for voting under uncertainty and demonstrated its interest through three paradoxes, i.e., situations that show how certain seemingly reasonable choices in that model can lead to unexpected and undesirable outcomes. We have also seen that certain components of the model already give rise to interesting questions. Specifically, we have discussed the problem of aggregating several sets of plausible states into a single such set and we have discussed how to rank available actions when given a single set of plausible states and a single preference order. Finally, while we believe that the main contribution of this paper is of a conceptual nature, we have also established some basic technical results regarding the aforementioned components of the model.

But we have not yet answered our initial question: how should we make a decision when voting on actions with uncertain effects?

Of course, as for standard voting (without uncertainty), we cannot expect a definitive answer to this question: no method of aggregation will satisfy all desiderata that one might wish to impose. Still, our results and discussion above suggest at least some tentative guidelines for best practices. First, while full integration of uncertainty resolution and preference aggregation is desirable in principle, given our current understanding of the problem domain, the best pragmatic approach we can recommend is to first aggregate beliefs and preferences in isolation, and to then integrate the collective beliefs and the collective preferences thus obtained into a decision regarding the action to take. Second, for each of the three phases of this process, we can make some recommendations:

- *Uncertainty resolution:* To avoid the *Paradox of Early Collective Uncertainty Resolution*, uncertainty resolution should not be overly aggressive, i.e., we should not aim at excluding too many possible states. In case the agents are known to report *knowledge* (rather than mere belief) regarding the current state, the intersection rule is the only reasonable choice. Otherwise, the mean-based rule promises to offer a good compromise.

- *Preference aggregation:* It is advisable to use a social welfare function (returning a collective preference order) rather than just a voting rule (returning one or several top states) at this stage, so as to be able to report more information to the next stage. Preference aggregation is subject to the well-understood challenges of social choice, but, e.g., the Kemeny rule is often regarded as an aggregator that makes a good trade-off between desiderata [10], and it may also be a good compromise in our context.

- *Integration:* Finally, we have to integrate the collective beliefs and the collective preference order into a final decision regarding the action to choose. This corresponds to the single-agent case analysed in Section 5. Here a case-based approach appears superior to an outcome-based approach. For instance, we could use a casewise-lexicographic rule. The "salience order" $\gg$ used to initialise this rule could even refer back to the full profile and favour states proposed by larger numbers of individuals.

References

1. Alcalde-Unzu, J., Vorsatz, M.: Size approval voting. Journal of Economic Theory 144(3), 1187–1210 (2009)
2. Arrow, K.J.: Social Choice and Individual Values, 2nd edn. John Wiley & Sons (1963)
3. Barberà, S., Bossert, W., Pattanaik, P.: Ranking sets of objects. In: Handbook of Utility Theory, vol. 2. Kluwer Academic Publishers (2004)
4. Bartholdi III, J.J., Tovey, C.A., Trick, M.A.: How hard is it to control an election? Mathematical and Computer Modelling 16(8–9), 27–40 (1992)
5. Ben Larbi, R., Konieczny, S., Marquis, P.: A characterization of optimality criteria for decision making under complete ignorance. In: Proceedings of the 12th International Conference on Principles of Knowledge Representation and Reasoning (KR-2010), pp. 172–181. AAAI Press (2010)
6. Brams, S.J., Fishburn, P.C.: Approval Voting, 2nd edn. Springer (2007)
7. Brandt, F., Conitzer, V., Endriss, U.: Computational social choice. In: Weiss, G. (ed.) Multiagent Systems, pp. 213–283. MIT Press (2013)
8. Duddy, C., Piggins, A.: Collective approval. Mathematical Social Sciences 65(3), 190–194 (2013)
9. Faliszewski, P., Hemaspaandra, E., Hemaspaandra, L.: Using complexity to protect elections. Communications of the ACM 53(11), 74–82 (2010)
10. Gaertner, W.: A Primer in Social Choice Theory. LSE Perspectives in Economic Analysis. Oxford University Press (2006)
11. Geist, C., Endriss, U.: Automated search for impossibility theorems in social choice theory: Ranking sets of objects. Journal of Artificial Intelligence Research 40, 143–174 (2011)
12. Grossi, D., Pigozzi, G.: Introduction to judgment aggregation. In: Bezhanishvili, N., Goranko, V. (eds.) ESSLLI 2010/2011. LNCS, vol. 7388, pp. 160–209. Springer, Heidelberg (2012)
13. Halpern, J.Y.: Reasoning about Uncertainty. MIT Press (2003)
14. Kannai, Y., Peleg, B.: A note on the extension of an order on a set to the power set. Journal of Economic Theory 32(1), 172–175 (1984)
15. List, C., Pettit, P.: Aggregating sets of judgments: An impossibility result. Economics and Philosophy 18(1), 89–110 (2002)
16. Merrill III, S., Nagel, J.: The effect of approval balloting on strategic voting under alternative decision rules. The American Political Science Review 81(2), 509–524 (1987)
17. Shoham, Y., Leyton-Brown, K.: Multiagent Systems: Algorithmic, Game-Theoretic, and Logical Foundations. Cambridge University Press (2009)
18. Xu, Y.: Axiomatizations of approval voting. In: Sanver, M.R., Laslier, J.F. (eds.) Handbook on Approval Voting, pp. 91–102. Springer (2010)

Restricted Manipulation in Iterative Voting: Condorcet Efficiency and Borda Score

Umberto Grandi[1], Andrea Loreggia[1], Francesca Rossi[1],
Kristen Brent Venable[2], and Toby Walsh[3]

[1] University of Padova
{umberto.uni,andrea.loreggia}@gmail.com, frossi@math.unipd.it
[2] Tulane University and IHMC
kvenabl@tulane.edu
[3] NICTA and UNSW
toby.walsh@nicta.com.au

Abstract. In collective decision making, where a voting rule is used to take a collective decision among a group of agents, manipulation by one or more agents is usually considered negative behavior to be avoided, or at least to be made computationally difficult for the agents to perform. However, there are scenarios in which a restricted form of manipulation can instead be beneficial. In this paper we consider the iterative version of several voting rules, where at each step one agent is allowed to manipulate by modifying his ballot according to a set of restricted manipulation moves which are computationally easy and require little information to be performed. We prove convergence of iterative voting rules when restricted manipulation is allowed, and we present experiments showing that iterative restricted manipulation yields a positive increase in the Condorcet efficiency and Borda score for a number of standard voting rules.

1 Introduction

In multi-agent systems often agents need to take a collective decision. A voting rule can be used to decide which decision to take, mapping the agents' preferences over the possible candidate decisions into a winning decision for the collection of agents. In these kind of scenarios, it may be desirable that agents do not have any incentive to manipulate, that is, to misreport their preferences in order to influence the result of the voting rule in their favor. Indeed, manipulation is usually seen as bad behavior from an agent, to be avoided or at least to be made computationally difficult to accomplish. While we know that every voting rule is manipulable when no domain restriction is imposed on the agents' preferences (Gibbard, 1973; Satterthwaite, 1975), we can at least choose a voting rule that is computationally difficult to manipulate for single agents or coalitions.

In this paper we consider a different setting, in which instead manipulation is allowed in a fair way. More precisely, agents express their preferences over the set of possible decisions and the voting rule selects the current winner as in the usual case. However, this is just a temporary winner, since at this point a single agent may decide to manipulate, i.e., to change her preference if by doing so the result changes in her favor. The process repeats with a new agent manipulating until we eventually reach a convergence state, i.e., a profile where no single agent can get a better result by manipulating. We call

P. Perny, M. Pirlot, and A. Tsoukiàs (Eds.): ADT 2013, LNAI 8176, pp. 181–192, 2013.

such a process *iterative voting*. In this scenario, manipulation can be seen as a way to achieve consensus, to give every agent a chance to vote strategically (a sort of fairness), and to account for inter-agent influence over time.

There are two prototypical situations in which iterative manipulation takes place. The first example is represented by the response of an electorate to a series of information polls about the result of a political election. At each step individuals may realize that their favorite candidate does not have chances to win and report a different preference in the subsequent poll. The second example is Doodle,[1] a very popular on-line system to select a time slot for a meeting. In Doodle, each participant can approve as many time slots as she wants, and the winning time slot is the one with the largest number of approvals. At any point, each participant can modify her vote in order to get a better result, and this can go on for several steps.

Iterative voting has been the subject of numerous publications in recent years. Previous work has focused on iterating the plurality rule (Meir *et al.*, 2010), on the problem of convergence for several voting rules (Lev and Rosenschein, 2012), and on the convergence of plurality decisions between multiple agents (Airiau and Endriss, 2009). Lev and Rosenschein (2012) showed that, if we allow agents to manipulate in any way they want (i.e., to provide their best response to the current profile), then the iterative version of most voting rules do not converge. Therefore, an interesting problem is to seek restrictions on the manipulation moves to guarantee convergence of the associated iterative rule. Restricted manipulation moves are good not only for convergence, but also because they can be easier to accomplish for the manipulating agent. In fact, contrarily to what we aim for in classical voting scenarios, here we do not want manipulation to be computationally difficult to achieve. It is actually desirable that the manipulation move be easy to compute while not requiring too much information for its computation.

An example of a restricted manipulation move is called k-pragmatists in Reijngoud and Endriss (2012): a k-pragmatist just needs to know the top k candidates in the collective candidate order, and will move the most preferred of those candidates to the top position of her preference. To compute this move, a k-pragmatist needs very little information (just the top k current candidates), and with this information it is computationally easy to perform the move. This move assures convergence of all positional scoring rules, Copeland, and Maximin, with linear tie-breaking. Note that each agent can apply this manipulation rule only once (since the top k candidates are always the same), and this is the main reason for convergence.

In this paper we introduce two restricted manipulation moves within the scenario of iterative voting and we analyze some of their theoretical and practical properties. Both manipulation moves we consider are polynomial to compute and require little information to be used. We show that convergence is guaranteed under both moves for those rules we consider, except for STV for which we only have experimental evidence of convergence. Moreover, we show that if a voting rule satisfies some axiomatic properties, such as Condorcet consistency or unanimity, then its iterative version will also satisfy the same properties as well. We then perform an experimental analysis of four restrictions on the set of manipulation strategies. For voting rules that are not Condorcet consistent, we test whether their Condorcet efficiency (that is, the probability to elect

---

[1] http://doodle.com/

the Condorcet winner) improves by adopting the iterative versions. The second parameter that we test is the Borda score of the winner in the truthful profile. Our results show that, with the exception of the Borda rule, both parameters never decrease in iteration, and a significant increase can be observed when the number of candidates is higher than the number of voters, as it is the case in a typical Doodle poll.

The paper is organized as follows. In the first section we introduce the basic definitions of iterative voting and we define two new restricted manipulation moves. The second section contains theoretical results on convergence and preservation of axiomatic properties, and in the third section we present our experimental evaluation of restricted iterative voting. The last section contains our conclusions and points at some directions for future research.

2 Background Notions

In this section we recall the basic notions of social choice theory that we shall use in this paper, we present the setting of iterative voting, and we define three notions of restricted manipulation moves that agents can perform.

2.1 Voting Rules

Let $\mathcal{X}$ be a finite set of m candidates and $\mathcal{I}$ be a finite set of n individuals. We assume individuals have preferences p_i over candidates in $\mathcal{X}$ in the form of *strict linear orders*, i.e., transitive, anti-symmetric and complete binary relations. Individuals express their preferences in form of a *ballot* b_i, which we also assume is a linear order over $\mathcal{X}$, and a profile $\mathbf{b} = (b_1, \ldots, b_n)$ is defined by the choice of a ballot for each of the individuals. A (non-resolute) *voting rule* F associates with every profile $\mathbf{b} = (b_1, \ldots, b_n)$ a non-empty subset of winning candidates $F(\mathbf{b}) \in 2^{\mathcal{X}} \setminus \emptyset$. There is a wide collection of voting rules that have been defined in the literature (see, e.g., Brams and Fishburn, 2002) and in this paper we focus on the following definitions:

Positional scoring rules (PSR): Let $(s_1, \ldots, s_m)$ be a scoring vector such that $s_1 > s_m$ and $s_1 \geq \cdots \geq s_m$. If a voter ranks candidate c at j-th position in her ballot, this gives s_j points to the candidate. The candidates with the highest score win. We focus on four particular PSRs: *Plurality* with scoring vector $(1, 0, \ldots, 0)$, *veto* with vector $(1, \ldots, 1, 0)$, *2-approval* with vector $(1, 1, 0, \ldots, 0)$, *3-approval* with vector $(1, 1, 1, 0, \ldots, 0)$, and *Borda* with vector $(m - 1, m - 2, \ldots, 0)$.

Copeland: The score of candidate c is the number of pairwise comparisons she wins (i.e., contests between c and another candidate a such that there is a majority of voters preferring c to a) minus the number of pairwise comparisons she loses. The candidates with the highest score win.

Maximin: The score of a candidate c is the smallest number of voters preferring it in any pairwise comparison. The candidates with the highest score win.

Single Transferable Vote (STV): At the first round the candidate that is ranked first by the fewest number of voters gets eliminated (ties are broken following a predetermined order). Votes initially given to the eliminated candidate are then transferred to the candidate that comes immediately after in the individual preferences. This process is iterated until one alternative is ranked first by a majority of voters.

With the exception of STV, all rules considered thus far are non-resolute, i.e., they associate a set of winning candidates with every profile of preferences. A tie-breaking rule is then used to eliminate ties in the outcome. In this paper we focus on *linear tie-breaking*: the set $\mathcal{X}$ of candidates is ordered by $\prec_{\mathcal{X}}$, and in case of ties the alternative ranked highest by $\prec_{\mathcal{X}}$ is chosen as the unique outcome. Other forms of tie-breaking are possible, e.g., a random choice of a candidate from the winning set. The issue of tie-breaking has been shown to be crucial to ensure convergence of the iterative version of a voting rule (Lev and Rosenschein, 2012).

2.2 Strategic Manipulation and Iterative Voting

A classical problem studied in voting theory is that of manipulation: do individuals have incentive to misreport their truthful preferences, in order to force a candidate they prefer as winner of the election? The celebrated Gibbard-Satterthwaite Theorem (Gibbard, 1973; Satterthwaite, 1975) showed that under very natural conditions all voting rules can be manipulated. Following this finding, a considerable amount of work has been spent on devising conditions to avoid manipulation, e.g., in the form of restrictions on individual preferences or in the form of computational barriers that make the calculation of manipulation strategies too hard for the agents (Bartholdi and Orlin, 1991; Faliszewski and Procaccia, 2010). In this paper we take a different stance on manipulation: we consider the fact that individuals are allowed to change their preferences as a positive aspect of the voting process, that may eventually lead to a better result after a sufficient number of steps.

We consider a sequence of repeated elections with individuals manipulating one at a time at each step. The iteration process starts at $\mathbf{b}^0$ (which we shall refer to as the *truthful profile*) and continues to $\mathbf{b}^1, \ldots, \mathbf{b}^k, \ldots$. A turn function τ identifies one single individual i_k that is allowed to manipulate at step k, while all other individual ballots remain unchanged (e.g., τ follows the order in which individuals are given)[2]. The individual manipulator uses the *best response* strategy: she changes her full ballot by selecting the linear order which results in the best possible outcome based on her truthful preference. In case the result of the election cannot be changed to a better candidate for the manipulator, we say that the individual does not have incentive to manipulate. The iterative process *converges* if there exists a k_0 such that no individual has incentive to manipulate after k_0 steps of iteration.

The setting of iterative voting was first introduced and studied by Meir *et al.* (2010) for the case of the plurality rule, and expanded by Lev and Rosenschein (2012). In their work, the authors describe the iterated election process as a *voting game*, in which convergence of the iterative process corresponds to reaching a Nash equilibria of the game. They show that convergence is rarely guaranteed with most voting rules under consideration, and that this property is highly dependent on the tie-breaking rule under consideration. For instance, iterative plurality always converges with any tie-breaking rule, while the iterative version of PSRs and Maximin do not always converge. The following

---

[2] Formally, a turn function takes as input the history of moves that have been played at step k, i.e., the sequence of profiles $\mathbf{b}^0 \ldots \mathbf{b}^{k-1}$ and outputs an individual in $\mathcal{I}$. A turn function is strongly related to the notion of scheduler in weakly acyclic games (Apt and Simon, 2012).

example shows that the iterative version of the Copeland rule does not converge even if we choose a linear tie-breaking rule:

Example 1. Let there be two voters and three candidates, with $a \succ b \succ c$ as tie-breaking order. The initial profile $\mathbf{b}^0$ has $c > b > a$ as the preference of voter 1 and $a > c > b$ for the voter 2. Candidate c wins in pairwise comparison with b, and no other candidate win any other pairwise comparison. Thus, the winner using the Copeland rule is c. Voter 2 has now an incentive to change her preferences to $a > b > c$, in which case by the tie-breaking order the winner is a, which is preferred by voter 2 in her truthful preference. Now voter 1 is unhappy, and changes her ballot to $b > c > a$ to force candidate b as the winner. This results in an incentive for voter 2 to change her ballot to $a > c > b$, again forcing a as winner. Finally, voter 1 changes again her ballot to $c > b > a$ to obtain c as winner, moving back to the initial profile and creating a cycle of iterated manipulation.

2.3 Restricted Manipulation Moves

Given that convergence is not guaranteed when voters manipulate choosing their best response, an interesting problem is to devise suitable restrictions on the set of manipulation strategies available to the agents in order to obtain convergence. Initial work on this topic was done by Reijngoud and Endriss (2012). In this section we review their definition and we introduce two novel notions of restricted manipulation. Let $\mathbf{b}^k$ be the current profile at step k, $\mathbf{b}^0$ be the initial truthful profile, and F be a voting rule. Assume that $\tau(k) = i$.

k-pragmatist (Reijngoud and Endriss, 2012): the manipulator i moves to the top of her reported ballot the most preferred candidate following b_i^0 among those that scored in the top k positions.[3]

M1: the manipulator i moves the second-best candidate in b_i^0 to the top of her reported ballot b_i^{k+1}, unless the current winner $w = F(\mathbf{b}^k)$ is already her best or second-best candidate in b_i^0.

M2: the manipulator i moves the most preferred candidate in b_i^0 which is above $w = F(\mathbf{b}^k)$ in b_i^k to the top of her reported ballot b_i^{k+1}, among those that can become the new winner of the election.

Under the *k-pragmatist* restriction voters are allowed to move to the top of their reported ballot the individual which they prefer amongst the top k candidates ranked by F. *M1* allows only a very simple move: switch the first and second candidate in the manipulator's ballot unless she is already satisfied, i.e., in case the current winner is ranked first or second in her truthful ballot. *M2* is more complex: the manipulator selects those candidates that she prefers to the current winner in her current ballot at step k; then, starting from the most preferred one in the truthful ballot, she tries to put such candidate on the top of her current ballot and computes the outcome of the election; the first candidate which succeeds in becoming the new winner of the manipulated election is the one chosen for the top position of her reported ballot.

---

[3] Note that all voting rules considered in this paper can be easily extended to output a ranking of the candidates rather than just a single winner.

While the choice of these restrictions may at first seem arbitrary, we believe they represent three basic prototypes of simple manipulation strategies for agents with bounded computational capabilities and limited access to information. Indeed, when evaluating restrictions on the set of manipulation moves we followed three criteria: (i) the convergence of the iterative voting rule associated with the restriction, (ii) the information to be provided to voters for computing their strategy, and (iii) the computational complexity of computing the manipulation move at every step. An ideal restriction always guarantees convergence, requires as little information as possible, and is computationally easy to compute. Reijngoud and Endriss (2012) show convergence for PSRs using the k-pragmatist restriction, and we shall investigate convergence results for *M1* and *M2* in the following section. Let us move to the other two parameters: on the one hand, *M1* requires as little information as possible to be computed, i.e., only the winner of the current election, and is also very easy to compute. The k-pragmatist restriction has good properties: it is easy to compute, and the information required to compute the best strategy is just the set of candidates which are ranked in the top k positions. *M2* also requires little information: the candidates' final score in case of scoring rules, the majority graph for Copeland and Maximin. Instead, in the case of STV the full profile is required. From the point of view of the manipulator, *M2* is computationally easy (i.e., polynomial) to perform.

We conclude by defining the iterative version of a voting rule depending on the different assumptions we can make on the set of manipulation moves:

Definition 1. *Let F be a voting rule and M a restriction on manipulation moves. $F^{M,\tau}$ associates with every profile* **b** *the outcome of the iteration of F using turn function τ and manipulation moves in M if this converges, and $\uparrow$ otherwise.*

3   Convergence and Axiomatic Properties

In this section we prove that the iterative versions of PSR, Maximin and Copeland converge when using our two new restrictions on the manipulation moves. We also analyze, for a number of axiomatic properties, the behavior of the iterative version of a voting rule with respect to the properties of the non-iterative version.

Theorem 1. $F^{M1,\tau}$ *converges for every voting rule F and turn function τ.*

Proof. The proof of this statement is straightforward from our definitions. The iteration process starts at the truthful profile b_0, and each agent is allowed to switch the top candidate with the one in second position. The iteration process stops after at most n steps.

Theorem 2. $F^{M2,\tau}$ *converges for every turn function τ if F is a PSR, the Copeland rule or the Maximin rule.*

Proof. The winner of an election using a PSR, Copeland or Maximin is defined as the candidate maximizing a certain score (or with maximal score and higher rank in the tie-breaking order). Since the maximal score of a candidate is bounded, it is sufficient to show that the score of the winner increases at every iteration step (or, in case the score remains constant that the position of the winner in the tie-breaking order increases) to show that the iterative process converges. Let us start with PSR. Recall that the score of a candidate c under PSR is $\sum_i s_i$ where s_i is the score given by the position of c in ballot b_i. Using $M2$, the manipulator moves to the top a candidate which lies above the current winner c. Thus, the position – and hence the score – of c remains unchanged, and the new winner must have a strictly higher score (or a better position in the tie-breaking order) than the previous one. The case of Copeland and Maximin can be solved in a similar fashion: it is sufficient to observe that the relative position of the current winner c with all other candidates (and thus also its score) remain unchanged when ballots are manipulated using $M2$. Thus, the Copeland score and the Maximin score of a new winner must by higher than that of c (or the new winner must be placed higher in the tie-breaking order).

This proof generalizes to show the convergence of the $M2$-iterative version of any voting rule which outputs as winners those candidates maximizing a given notion of score. While currently we do not have a proof of convergence for STV, we observed experimentally that its iteration always terminates on profiles with a Condorcet winner when a suitable turn function, which is described in the following section, is used.

Voting rules are traditionally studied using axiomatic properties, and we can inquire whether these properties extend from a voting rule to its iterative version. We refer to the literature for an explanation of these properties (see, e.g., Taylor, 2005). Let us call F_t^M the iterative version of voting rule F after t iteration steps (we omit the superscript τ, indicating that these results hold for every turn function). We say that a restricted manipulation move M *preserves* a given axiom if whenever a voting rule F satisfies the axiom then also F_t^M does satisfy it for all t.

Theorem 3. *M1 and M2 preserve unanimity.*

Proof. Assume that the iteration process starts at a unanimous profile b in which candidate c is at top position of all individual preferences. If F is unanimous, then $F(\mathbf{b}) = c$, and no individual has incentives to manipulate either using $M1$ or $M2$. Thus, iteration stops at step 1 and $F_t^{M1}(\mathbf{b}) = c$ and $F_t^{M2}(\mathbf{b}) = c$, satisfying the axiom of unanimity.

Theorem 4. *M1 and M2 preserve Condorcet consistency.*

Proof. Let c be the Condorcet winner of a profile b. If F is Condorcet-consistent then $F(\mathbf{b}) = c$. As previously observed, when individuals manipulate using either $M1$ or $M2$ the relative position of the current winner with all other candidates does not change, since the manipulation only involves candidates that lie above the current winner in the individual preferences. Thus c remains the Condorcet winner in all iteration steps $\mathbf{b}^k$. Since $F_k^{M1}(\mathbf{b}) = F(\mathbf{b}^k)$ and F is Condorcet-consistent, we have that $F_k^{M1}(\mathbf{b}) = c$ and thus F_k^{M1} is Condorcet consistent. Similarly for $M2$.

Other properties that transfer from a voting rule to its iterative version are neutrality and anonymity (supposing the turn function satisfies an appropriate version of neutrality and anonymity). The Pareto-condition does not transfer to the iterative version, as can be shown by adapting an example by Reijngoud and Endriss (2012, Example 3).

4 Experimental Evaluation of Restricted Manipulation Moves

In this section we evaluate four restrictions on the set of manipulation moves, namely 2 and 3-*pragmatists*, *M1* and *M2*, under two important aspects. First, we measure whether the restricted iterative version of a voting rule has a higher Condorcet efficiency than the initial voting rule, i.e., whether the probability that a Condorcet winner (if it exists) gets elected is higher for the iterative rather than non-iterative rule. Second, we observe the variation of the Borda score of the winner, i.e., we compare the average position of the current winner in the initial truthful profile at convergence with the value of the same parameter in the initial profile. We focus on four voting rules: plurality, STV, Borda, 2 and 3-approval. Our findings show that both parameters never decrease by allowing iterated restricted manipulation, and that a substantial increase can be observed in case the number of candidates is higher than the number of voters (e.g., in a Doodle poll). We conclude the section by reporting on some initial experiments with real-world datasets.

4.1 Experimental Setting

We generated profiles using the Polya-Eggenberger urn model (see, e.g., Berg, 1985). Individual ballots are extracted from an urn initially containing all $m!$ possible ballots, i.e., all linear orders over m candidates, and each time we draw a vote from the urn we put it back with a additional copies of the same vote. In this way we generate profiles with correlated preferences and we control the correlation ratio with the parameter a. In our experiment we tested three different settings: the *impartial culture assumption* (IC) when $a = 0$, the UM10 with 10%-correlation when $a = \frac{m!}{9}$, and the UM50 with 50%-correlation when $a = m!$. We generated 10.000 profiles for each experiment, restricting to profiles with a Condorcet winner when testing the Condorcet efficiency.

Our results are obtained using a program implemented in Java ver.1.6.0. We model two prototypical examples of iterative voting: in the *electoral simulation* we set the number of candidates to $m = 5$ and the number of voters to $n = 500$ to model situations in which a large population needs to decide on a small set of candidates. In the *Doodle simulation* we set the number of candidates to $m = 25$ and the number of voters to $n = 10$ to model a small group of people deciding over a number of time slots. In both cases we performed additional experiments varying the number of voters and candidates but keeping their ratio fixed without observing significant variation in the results.

The turn function used in our experiments associates with each voter i a dissatisfaction index $d_i(k)$, which increases by one point for each iteration step in which the individual has an incentive to manipulate but is not allowed to do so by the turn function. At iteration step k the individual that has the highest dissatisfaction index is allowed to move (in the first step, and in case of ties, the turn follows the initial order in which voters are given). We also performed initial experiments using a sequential turn function which assigns turns depending on the order in which individuals are given obtaining similar results to those shown below (with the exception of STV, which does not converge using the sequential turn function).

It is interesting to observe that the higher the correlation in the profile the smaller the number of profiles in which iteration takes place (with the notable exception of the

Borda rule). In Figure 1 it is shown, for the Doodle simulation, the percentage of profiles in which iteration takes place for the three different correlation ratios considered. In the case of plurality, convergence is reached after an average of 3 steps and a maximal of 9 steps. The figures for the other voting rules are similar.

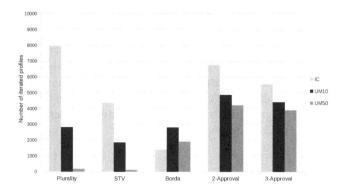

Fig. 1. Number of profiles with iteration compared to the correlation ratio

4.2   Condorcet Efficiency

Figures 2 and 3 compare the four restrictions on manipulation moves with respect to the Condorcet efficiency of the iterative version of the five voting rules under consideration, respectively for the Doodle simulation and the electoral simulation. In both experiments the correlation ratio is set at 10%.

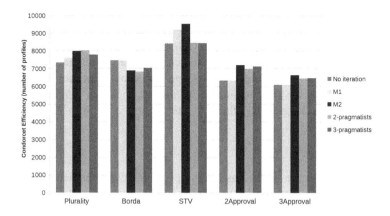

Fig. 2. Doodle experiment with UM10: Condorcet efficiency

Except for the case of the Borda rule, the Condorcet efficiency of the iterative version of a voting rule always improves with respect to the non-iterative version, and the growth is consistently higher when voters manipulate the election using *M2* rather than *M1*. Let us also stress that while the increase in Condorcet efficiency using *M1* is minimal, it is still surprising that such a simple move can result in a better performance than the original version of the voting rule. The 2-*pragmatist* restriction performs quite well with the plurality rule in both experiments. STV has the highest performance of all voting rules considered thus far with respect to Condorcet efficiency and this performance is amplified by the use of iterated manipulation, resulting in the election of a Condorcet winner in almost 95 percent of the cases. As remarked earlier, we observed convergence in all profiles considered. The increase in Condorcet efficiency is more noticeable in the Doodle simulation rather than in the electoral situation. Thus, when the number of individuals is considerably higher than the number of alternatives the iterative process leads to a minimal increase in Condorcet efficiency.

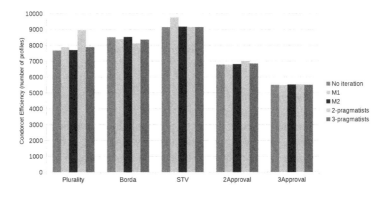

Fig. 3. Electoral experiment with UM10: Condorcet efficiency

We also run the same two experiments with different correlation ratios: Using the IC assumption the increase in Condorcet efficiency is more significant, while with the UM50 assumption the results are much less perturbed by iteration. This should not come as a surprise, given that the amount of profiles in which iteration takes place decrease rapidly with the growth of the correlation ratio.

4.3 Borda Score

The second parameter we used to assess the performance of restricted manipulation moves is known in the literature as the Borda score. Given a candidate c, let p_i be the position of c in the initial preference b_i^0 of voter i (from bottom to top, i.e., if a candidate is ranked first she gets $m - 1$ points, while if she is ranked last she gets 0 points). We compute the Borda score of c as $\sum_{i=1}^{n} p_i$.

For each voting rule and each restriction on the set of manipulation moves we compared the score of the winner of the non-iterative version with that of the winner of the iterative version after convergence. Since the Borda rule elects by definition those candidates with the highest Borda score, we did not evaluate the iterative version of Borda with respect to this parameter. Our results showed that in both the Doodle and the electoral simulation with UM10 the Borda score increases minimally if we allow for iterated restricted manipulation, resulting in a chart similar to that in Figure 3. The best results are in this case obtained by using *M2* and 2-*pragmatists* restriction with 2 and 3-approval. As in the previous section, by decreasing the correlation of the generated preferences we obtain a more significant increase in the Borda score after iteration.

4.4   Real-World Datasets

We performed initial experiments using data from *Preflib* (Mattei and Walsh, 2013), a library of preference datasets collected from various sources. In order to mimic the original preference distribution, we generated 10.000 profile with 5 candidates and 500 voters drawing votes with impartial culture assumption from two original datasets: the Netflix Prize Data (Bennett and Lanning, 2007) and the Skating Data. What we observed is that preferences contained in such datasets are quite correlated, with iteration taking place in just a handful of profiles (in the order of 5–10 per 10.000 profiles). Experiments run on data from political elections may have a chance to lead to more significant results, once our setting has been adapted to the case of partial orders over candidates, as required by the electoral datasets available at present state.

5   Conclusions and Future Work

This paper studies the effect of iteration on classical voting rules by allowing individuals to manipulate the outcome of the election using a restricted set of manipulation moves. We provided two new definitions of manipulation moves *M1* and *M2* and showed that they lead to convergence for all voting rules considered (cf. Theorem 1 and 2) except for STV. We showed that a number of axiomatic properties, such as unanimity and Condorcet consistency, are preserved in the iteration process. We evaluated experimentally the performance of our restricted manipulation moves with respect to the Condorcet efficiency of the iterative version of a voting rule as well as the Borda score of the winner in the initial truthful profile. We performed two simulations based on prototypical examples of iterated manipulation: the first simulation with the number of candidates smaller but comparable to the number of voters to model scheduling with Doodle, and the second with the number of voters much higher than that of candidates to model iterated polls before a political election. With the exception of the Borda rule, we showed that restricted manipulation in iterative voting yields a positive increase in both the Condorcet efficiency and Borda score and that the best performance is obtained when the number of candidates is higher than the number of individuals.

In future work we plan to analyze different versions of manipulation moves, and to compare their performance with that of the existing definitions. By adapting the framework defined in this paper to account for preferences expressed as partial orders it will also be possible to exploit preference data on real-world elections, to assess with more accuracy the effects of iterated restricted manipulation on more realistic distributions of preferences.

References

Airiau, S., Endriss, U.: Iterated majority voting. In: Rossi, F., Tsoukias, A. (eds.) ADT 2009. LNCS, vol. 5783, pp. 38–49. Springer, Heidelberg (2009)

Apt, K.R., Simon, S.: A classification of weakly acyclic games. In: 5th International Symposium on Algorithmic Decision theory (SAGT 2012) (2012)

Bartholdi, J.J., Orlin, J.B.: Single transferable vote resists strategic voting. Social Choice and Welfare 8, 341–354 (1991)

Bennett, J., Lanning, S.: The Netflix prize. In: Proceedings of the KDD Cup and Workshop (2007)

Berg, S.: Paradox of voting under an urn model: The effect of homogeneity. Public Choice 47(2), 377–387 (1985)

Brams, S.J., Fishburn, P.C.: Voting procedures. In: Arrow, K., Sen, A., Suzumura, K. (eds.) Handbook of Social Choice and Welfare, Elsevier (2002)

Faliszewski, P., Procaccia, A.D.: AI's war on manipulation: Are we winning? AI Magazine 31(4), 53–64 (2010)

Gibbard, A.: Manipulation of voting schemes: A general result. Econometrica 41(4), 587–601 (1973)

Lev, O., Rosenschein, J.S.: Convergence of iterative voting. In: Proceedings of the 11th International Conference on Autonomous Agents and Multiagent Systems, AAMAS-2012 (2012)

Mattei, N., Walsh, T.: Preflib: A library of preference data (2013), http://www.preflib.org

Meir, R., Polukarov, M., Rosenschein, J.S., Jennings, N.R.: Convergence to equilibria in plurality voting. In: Proceedings of the Twenty-Fourth Conference on Artificial Intelligence, AAAI 2010 (2010)

Reijngoud, A., Endriss, U.: Voter response to iterated poll information. In: Proceedings of the 11th International Joint Conference on Autonomous Agents and Multiagent Systems, AAMAS-2012 (June 2012)

Satterthwaite, M.A.: Strategy-proofness and Arrow's conditions: Existence and correspondence theorems for voting procedures and social welfare functions. Journal of Economic Theory 10(2), 187–217 (1975)

Taylor, A.D.: Social choice and the mathematics of manipulation. Cambridge University Press (2005)

Controller Compilation and Compression for Resource Constrained Applications

Marek Grześ, Pascal Poupart, and Jesse Hoey

David R. Cheriton School of Computer Science, University of Waterloo,
200 University Avenue West, Waterloo, Ontario N2L 3G1, Canada
{mgrzes,ppoupart,jhoey}@uwaterloo.ca

Abstract. Recent advances in planning techniques for partially observable Markov decision processes have focused on online search techniques and offline point-based value iteration. While these techniques allow practitioners to obtain policies for fairly large problems, they assume that a non-negligible amount of computation can be done between each decision point. In contrast, the recent proliferation of mobile and embedded devices has lead to a surge of applications that could benefit from state of the art planning techniques if they can operate under severe constraints on computational resources. To that effect, we describe two techniques to compile policies into controllers that can be executed by a mere table lookup at each decision point. The first approach compiles policies induced by a set of alpha vectors (such as those obtained by point-based techniques) into approximately equivalent controllers, while the second approach performs a simulation to compile arbitrary policies into approximately equivalent controllers. We also describe an approach to compress controllers by removing redundant and dominated nodes, often yielding smaller and yet better controllers. The compilation and compression techniques are demonstrated on benchmark problems as well as a mobile application to help Alzheimer patients to way-find.

Keywords: Energy-efficiency, Finite-state Controllers, Knowledge compilation, Markov decision processes, Mobile Applications, POMDPs.

1 Introduction

Partially observable Markov decision processes (POMDPs) provide a natural framework for sequential decision making in partially observable domains. Tremendous progress has been made in recent years to develop scalable planning techniques for POMDPs. Point-based value iteration methods for factored and continuous domains can compute good value policies for a wide range of real-world problems [1,2]. In addition, online resources can be used to perform a search at run time to directly select the next action or refine a precomputed policy [3,4].

In this work, we are motivated by an emerging class of applications that pose new challenges for POMDP solvers. We consider monitoring and assistive applications that run on smart-phones, wearable systems or other mobile devices.

P. Perny, M. Pirlot, and A. Tsoukiàs (Eds.): ADT 2013, LNAI 8176, pp. 193–207, 2013.
© Springer-Verlag Berlin Heidelberg 2013

While computational resources are rapidly increasing, energy consumption remains an important bottleneck due to limited battery life. This is especially important in monitoring and assistive applications that need to be continuously running, but should be as power efficient as possible. For such applications, online planning is not an option due to the high computational costs. Computed policies that require online belief monitoring at execution time also consume too much energy. While it is sometimes possible to offload computation through cloud solutions, this requires a data connection, which may not always be available or stable, and which has a high battery consumption.

An effective solution can be found by noting that a POMDP policy can be represented very simply using a finite state controller (FSC) [5], which only requires simple table look-ups during execution. However, controller optimization is notoriously difficult. The non-convex nature of the optimization makes it difficult for many approaches (e.g., gradient ascent [6], quadratically constrained optimization [7], bounded policy iteration [8], expectation maximization [9]) to reliably find the global optimum. An exhaustive search of the space of controllers can avoid local optima, but is clearly intractable [10,11].

In this paper, we describe two novel techniques for compiling an existing POMDP policy (as generated by a point-based method, for example) into a finite state controller (Sec. 3). The first method requires a policy specified as a set of α-vectors and witness belief points and constructs a FSC directly that approximates the given policy. The second method needs only a simulation of the policy, and builds a controller incrementally by building a policy tree and then detecting equivalent conditional plans. We also describe a novel method for compressing a FSC into an equivalent, but smaller, FSC by removing redundant nodes (Sec. 4). We demonstrate our techniques on a set of large benchmark POMDP problems (Sec. 5), and we use policies generated by two state-of-the-art point-based techniques, namely GapMin [12] and SARSOP [13]. We show how we can construct very compact controllers that are equivalent, and sometimes better, than the policies they are derived from. We also demonstrate our methods on a set of POMDPs that are used to provide mobile assistance for persons with Alzheimer's disease for way-finding.

2   Background

A partially observable Markov decision process (POMDP) is formally defined by a tuple $\langle S, A, O, T, Z, R, b_0, \gamma \rangle$ which includes a set S of states s, a set A of actions a, a set O of observations o, a transition function $T(s', s, a) = \Pr(s'|s, a)$, an observation function $Z(o, a, s') = \Pr(o|s', a)$, a reward function $R(s, a) \in \Re$, an initial belief $b_0(s) = \Pr(s)$ and a discount factor $0 \leq \gamma \leq 1$. We assume that the planning horizon is infinite, although the proposed algorithms can be modified easily for finite horizon problems. The goal is to find an optimal policy that maximizes the discounted sum of rewards. A policy $\pi : H_t \to A_t$ can be defined as a mapping from histories $H_t \equiv A_0 \times O_1 \times ... \times A_{t-1} \times O_t$ of past actions and observations to actions A_t, however this definition is problematic for an infinite horizon since histories may be arbitrarily long. Two approaches are often used to circumvent this issue: i) replace histories by finite length sufficient statistics such as beliefs or ii) represent policies as finite state controllers, which are mappings from *cyclic* histories to actions.

A belief $b(s)$ is a distribution over states reflecting the decision maker's belief that the process may be in each state s. We can update a belief b after executing a and observing o according to Bayes' theorem:

$$b^{ao}(s') \propto \sum_s b(s) \Pr(s'|s,a) \Pr(o|s',a) \; \forall s' \qquad (1)$$

Given the initial belief b_0 and a history $h_t = \langle a_0, o_1, ..., a_{t-1}, o_t \rangle$, we can compute the belief b_t at time step t by repeatedly applying the above equation for each action-observation pair in the history. Hence, we can equivalently define policies as mappings $\pi : B \to A$ from beliefs to actions. The value $V^\pi(b_0)$ of policy π when starting in b_0 is the discounted sum of expected rewards $V^\pi(b_0) = \sum_{t=0}^\infty \gamma^t R(b_t, \pi(b_t))$ where $R(b,a) = \sum_s b(s) R(s,a)$.

We can also consider policies represented by a finite state controller $\pi = \langle N, \phi, \psi \rangle$, which is defined by a set N of nodes n, a mapping $\phi : N \to A$ indicating which action a to execute in each node n and a mapping $\psi : N \times O \to N$ indicating that the edge rooted at n and labeled by o should point to n'. A controller is executed by alternating between executing the action $\phi(n)$ of the current node n and moving to the next node $\psi(n, o)$ by following the edge rooted at n that is labeled with the current observation o. The value α_n of the controller when starting in n is an $|S|$-dimensional vector computed as follows:

$$\alpha_n(s) = R(s, \phi(n)) + \gamma \sum_{s',o} \Pr(s'|s,a) \Pr(o|s',a) \alpha_{\psi(n,o)}(s') \; \forall n, s \qquad (2)$$

Policy optimization algorithms can be classified in two broad categories: i) *offline* techniques that pre-compute a policy before the start of the execution [14,15,16] and ii) *online* techniques that perform all their computation at run time by searching for the best action to execute after receiving each observation [3]. Online techniques can take advantage of the history so far to focus their computation only on the current belief. When computational resources are not constrained and there is sufficient time between decisions to search for the next action to execute, online techniques can perform very well and can scale to very large problems. In contrast, offline techniques do not scale as well, but permit the deployment of POMDP policies on mobile and/or embedded devices with severe resource constraints due to energy, memory or CPU limitations.

Among the offline techniques, we can further classify algorithms based on the type of policies (belief mapping or finite state controller) that they produce. Algorithms that produce belief mappings often exploit the fact that the value V^* of an optimal policy satisfies Bellman's equation:

$$V^*(b) = \max_a \sum_s b(s) \left[R(s,a) + \gamma \sum_{s',o} \Pr(s'|s,a) \Pr(o|s',a) V^*(b^{ao}) \right] \forall b \qquad (3)$$

The continuous nature of the belief space prevents us from performing value iteration at all beliefs and therefore the important class of point-based techniques performs point-based Bellman backups only at a finite set of beliefs [15]. An approximation of the value function at all beliefs is obtained by computing the gradient in addition to the value at each belief. This allows the formation of a set of linear value functions that are often represented by α-vectors, similar to the value functions of controller nodes. While the details of point-based value iteration are not important for the rest of this paper (see [17] for more information), what is important to know is that they produce a set Γ of $\langle \alpha_i, b_i, a_i \rangle$-tuples that associate each α_i with an action a_i and a witness belief b_i (i.e., belief for which α_i yields the highest value: $\alpha_i(b_i) \geq \alpha_j(b_i)\ \forall j$ where $\alpha(b) = \sum_s b(s)\alpha(s)$). The policy π induced by Γ is obtained by computing

$$\pi(b) = a_{best} \text{ where } best = \arg\max_i \alpha_i(b) \tag{4}$$

Although point-based value iteration techniques compute the set Γ offline, they still require a certain amount of computation at each decision point. The belief must be updated after each action and observation according to Eq. 1 (complexity $O(|S|^2)$) and the best α-vector must be identified according to Eq. 4 (complexity $O(|S||\Gamma|)$). This amount of computation may still be prohibitive when S and Γ are large and there isn't enough memory, time or energy.

Alternatively, the other group of offline techniques produces policies represented as finite state controllers [10]. Since the execution of a controller merely consists of a table lookup, they are the most convenient type of policies for deployment in resource constrained applications. Unfortunately, they do not scale as well as point-based techniques and they often lack robustness due to local optima issues. Instead of directly optimizing a finite state controller, in this paper we propose two techniques to compile policies into approximately equivalent controllers. This has the benefit that we can use existing scalable algorithms such as point-based value iteration to quickly obtain a good policy. In addition, the controller compilation allows those policies to be executed on devices that are much more constrained.

3   Controller Compilation

Kaelbling et al. [5] observed that an optimal controller can be extracted from an optimal value function. Unfortunately, the best value functions found by state of the algorithms are approximate/suboptimal for most problems. Hansen wrote "it is unclear how to construct suboptimal controllers from [such value functions]" [18]. Hence, for the past 15 years, research has focused on directly optimizing controllers. We propose two approaches to compile suboptimal policies into approximately equivalent controllers. The first approach is limited to policies implicitly represented by sets of α-vectors as produced by point-based value iteration techniques. The second approach works with arbitrary policies.

3.1   Compiling Controllers from Alpha Vectors

As explained in Sec. 2, point-based value iteration techniques produce a set Γ of $\langle \alpha_i, b_i, a_i \rangle$-tuples from which a belief mapping policy is extracted. Alg. 1 shows how to compile Γ into an approximately equivalent controller $\langle N, \phi, \psi \rangle$. We create a node n_i for each vector α_i (Line 4). Each node n_i is labeled with the action $\phi(n_i) = a_i$ associated with α_i (Line 5). To determine where the edge rooted at n_i and labeled with o should point to, we update the witness b_i of belief according to Eq. 1 based on action a_i and observation o. Let the resulting belief be $b_i^{a_i, o}$. We then find which α-vector has the highest value at $b_i^{a_i, o}$ (Line 9) and assign the corresponding node to $\psi(n_i, o)$ (Line 10). The complexity of this compilation technique is $O(|\Gamma|^2 |O| |S|^2)$, however in practice the dependence on $|O|$ and $|S|$ can often be reduced by exploiting sparsity. The overall running time is typically a fraction of the time taken by point-based value iteration to obtain Γ. The quality of the resulting controller varies. The compilation technique ensures that the actions selected at the first two time steps are identical to that of the policy induced by Γ. This can be observed by noting that the action associated with the best α-vector is selected at the first two time steps which correspond exactly to what would be done in a policy induced by Γ. If the set of α-vectors in Γ corresponds to the optimal value function, then the resulting controller will also be optimal. However, when the set of α-vectors is suboptimal, which is the case most of the time, then actions selected after the second time step may be different than those selected by the policy induced by Γ, leading to a controller that may be better or worse. In the next section we describe an approach that ensures that the resulting controller is at least as good as the original policy in the limit.

Algorithm 1. Compilation of α-vectors into an approximately equivalent controller $\langle N, \phi, \psi \rangle$

ALPHA2FSC(Γ)

1. Let Γ be a set of $\langle \alpha_i, b_i, a_i \rangle$-tuples
2. $N \leftarrow \emptyset$
3. **for** $i = 1$ **to** $|\Gamma|$ **do**
4. $\quad N \leftarrow N \cup \{n_i\}$
5. $\quad \phi(n_i) \leftarrow a_i$
6. **for** $i = 1$ **to** $|\Gamma|$ **do**
7. $\quad$ **for all** $o \in O$ **do**
8. $\quad\quad$ **if** $\Pr(o|b_i, a_i) > 0$ **then**
9. $\quad\quad\quad best \leftarrow \arg\max_j \alpha_j(b_i^{a_i, o})$
10. $\quad\quad\quad \psi(n_i, o) \leftarrow n_{best}$
11. $\quad\quad$ **else**
12. $\quad\quad\quad \psi(n_i, o) \leftarrow n_i$
13. **return** $\langle N, \phi, \psi \rangle$

Algorithm 2. Policy Tree Generation

POLICYTREE$(\pi, b, depth)$

1. $N \leftarrow \emptyset$
2. $j \leftarrow 1$
3. $queue \leftarrow \{\langle b, 0, j \rangle\}$
4. **while** $\neg isEmpty(queue)$ **do**
5. $\quad \langle b, d, i \rangle \leftarrow removeFirst(queue)$
6. $\quad N \leftarrow N \cup \{n_i\}$
7. $\quad \phi(n_i) \leftarrow \pi(b)$
8. $\quad$ **if** $d = depth$ **then**
9. $\quad\quad \psi(n_i, o) \leftarrow * \; \forall o \in O$
10. $\quad$ **else**
11. $\quad\quad$ **for all** $o \in O$ **do**
12. $\quad\quad\quad j \leftarrow j + 1$
13. $\quad\quad\quad addLast(queue, \langle b^{\phi(n_i)o}, d+1, j \rangle)$
14. $\quad\quad\quad \psi(n_i, o) \leftarrow n_j$
15. **return** $\langle N, \phi, \psi \rangle$

3.2   Compiling Controllers from Arbitrary Policies by Simulation

We describe an approach to compile arbitrary policies into approximately equivalent controllers. The approach simulates the policy up to a certain depth and ensures that the controller will execute the same actions up to that depth. In the limit, with an infinite depth, we obtain a controller that matches the policy exactly. Although, as we show in the experiments, we can often obtain a controller that is at least as good by simulating up to a reasonable depth.

The approach works in two steps: i) first we generate a policy tree up to a certain depth, then ii) we compress the policy tree into a controller by detecting matching subtrees. Alg. 2 shows how to generate a policy tree up to a certain depth by simulating the policy. Since simulation does not require the policy to be in any format, the approach works with arbitrary policies. We just need to generate the next action given the current observation at each time step, which is always possible since this is how all policies are executed in practice. To be concrete, Alg. 2 shows how to generate a policy tree for policies that are belief mappings, but we could easily modify the algorithm to work with policies that are represented as history mappings or any other type of mapping. The algorithm generates a policy tree in breadth first order, which will become handy in the compression step. Since leaves do not have edges, we set $\psi(n, o)$ to $*$ for all edges rooted at a leaf n (Line 9).

Fig. 1 shows the policy tree generated by Alg. 2 up to a depth of 5 for the classic tiger problem [19]. In this problem there are three actions (listen, open-right and open-left), two observations (tiger-right, tiger-left). Nodes are labeled with actions and edges are labeled with observations. Nodes are also numbered according to the breadth-first order in which they were generated.

In the second step, the policy tree is compressed into a controller by identifying matching conditional plans. Each node of the policy tree is the root of a conditional plan. Conditional plans rooted at each node are compared to conditional plans rooted at previous nodes in the breadth-first order. When two conditional plans match, we replace the node with highest breadth-first index by the node with the lowest breadth-first index. Two conditional plans are said to match when they select the same actions in each path up until a leaf is encountered. Hence, conditional plans with different depths can still match since we stop the verification as soon as a leaf is encountered in a path. Alg. 3 shows how to verify whether two conditional plans match. Alg. 4 uses this verification procedure to prune nodes whose conditional plans match the conditional plan of an earlier node in the breadth-first order. This process gives rise to a controller that is often much smaller than the original policy tree and yet ensures that the same actions are executed up to the depth of the original policy tree.

Fig. 2 shows again the policy tree for the tiger problem with additional dashed edges indicating that the parent node is replaced by the child node due to matching conditional plans. For instance, node 4 will be replaced by node 0 since their conditional plans match. Fig. 3 shows the resulting reduced controller once all node substitutions indicated by dashed edges in Fig. 2 are performed. Since leaf nodes have a trivial one-step conditional plan and they are last in the

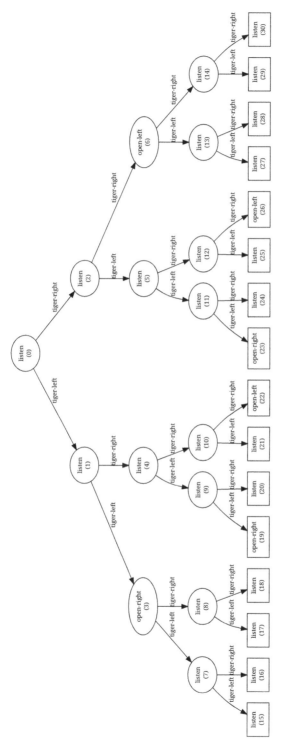

Fig. 1. Policy tree up to a depth of 5 for the classic tiger problem

Algorithm 3. Equivalent Conditional Plans	**Algorithm 4.** Compilation of arbitrary π into approx. equivalent controller $\langle N, \phi, \psi \rangle$

EQUIVALENTCP(n_1, n_2, ϕ, ψ)

1. **if** $\phi(n_1) \neq \phi(n_2)$ **then**
2.   **return** $false$
3. **for all** $o \in O$ **do**
4.   **if** $\psi(n_1, o) \neq *$ **and** ¬EQUIVA-LENTCP$(\psi(n_1, o), \psi(n_2, o), \phi, \psi)$ **then**
5.   **return** $false$
6. **return** $true$

POLICY2FSC$(\pi, b, depth)$

1. $\langle N, \phi, \psi \rangle \leftarrow$ POLICYTREE$(\pi, b, depth)$
2. **for all** $n_i \in N$ in increasing index i **do**
3.   **for all** $n_j \in N$ such that $j < i$ **do**
4.   **if** EQUIVALENTCP(n_i, n_j, ϕ, ψ) **then**
5. $N \leftarrow N \setminus \{n_i, descendents(n_i)\}$
6.   **for all** $n \in N, o \in O$ **do**
7.   **if** $\psi(n, o) = n_i$ **then**
8. $\psi(n, o) \leftarrow n_j$
9. **return** $\langle N, \phi, \psi \rangle$

breadth-first order, they will be replaced by interior nodes as long as there is an interior node with the same action. Since actions eventually repeat in a large enough tree, the compilation procedure generally produces controllers without leaves (i.e., all nodes have a full set of edges). The breadth-first order also ensures that Alg. 3 terminates since in each pair of conditional plans that we compare, the one rooted at the node with the highest index is necessarily a tree of finite depth (i.e., no loop). In addition, when we replace the node with the highest index we can delete the entire subtree below it since there is no way to reach that subtree other than through the node that is being replaced. This pruning greatly improves the running time. Finally, the breadth first order also helps to produce a small controller since nodes are always replaced by nodes with a lower index and therefore earlier in the tree.

The complexity of Alg. 4 is quadratic in the size of the policy tree. However, due to the pruning of subtrees each time a node is replaced, we can show that the complexity is really linear in the size of the policy tree times the size of the reduced controller. The experiments show that the reduced controller is often significantly smaller than the policy tree, yielding a substantial speed up. That being said, the linear dependence on the size of the policy tree is still significant since the size of policy trees is exponential in the depth (i.e., $O(|O|^{depth})$). We can often reduce the base $|O|$ of the exponential by exploiting sparsity or considering only observations with a probability greater than some threshold.

4   Controller Compression

Once a policy is compiled to a controller, it often contains redundant or dominated nodes. Redundant nodes often occur when some observations have zero probability, leading to multiple conditional plans with the same value. Dominated nodes often occur when the original policy is suboptimal and the compilation process generates suboptimal conditional plans. We describe a technique to compress a controller while ensuring that its value does not decrease and in some cases it increases. The idea is to prune all nodes with α-vectors that are dominated in value by other α-vectors. This approach was first used by Hansen

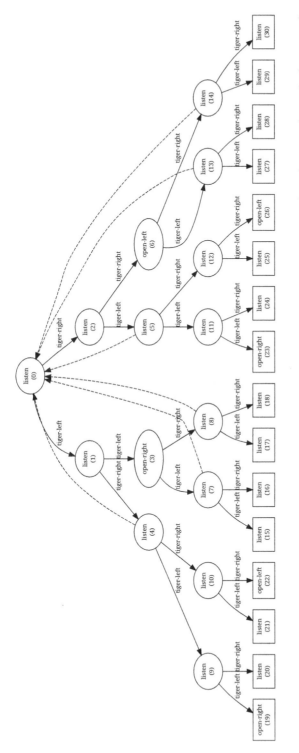

Fig. 2. Policy tree up to a depth 5 for the classic tiger problem with dashed edges indicating nodes whose conditional plans match according to Alg. 3

Algorithm 5. FSC Compression

FSCCOMPRESSION(N, ϕ, ψ)

1. **repeat**
2.   Eval controller by solving (2)
3.   **for each** $n_1 \in N$ **do**
4.   **for each** $n_2 \in N \setminus \{n_1\}$ **do**
5.   **if** $\alpha_{n_1}(s) \le \alpha_{n_2}(s)$ $\forall s$ **then**
6. $N \leftarrow N \setminus \{n_1\}$
7.   **for all** $n \in N, o \in O$ **do**
8.   **if** $\psi(n, o) = n_1$ **then**
9. $\psi(n, o) \leftarrow n_2$
10.   **break**
11.   **if** $\alpha_{n_1}(s) \ge \alpha_{n_2}(s)$ $\forall s$ **then**
12. $N \leftarrow N \setminus \{n_2\}$
13.   **for all** $n \in N, o \in O$ **do**
14.   **if** $\psi(n, o) = n_2$ **then**
15. $\psi(n, o) \leftarrow n_1$
16. **until** N doesn't change
17. **return** $\langle N, \phi, \psi \rangle$

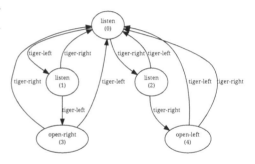

Fig. 3. Controller obtained by reducing a 5-step policy tree according to Alg. 4 for the classic tiger problem

in his policy iteration algorithm [14]. Below, Algorithm 5 describes how to repeatedly compress a controller until there are no dominated nodes. The approach alternates between policy evaluation and node substitution. The evaluation step computes the α-vector of each node by solving a system of linear equations. Then the α-vector of each node is compared to the α-vectors of the other nodes. When $\alpha_1(s) \le \alpha_2(s)$ $\forall s$ then n_1 can be replaced by n_2. Since the value of n_2 is at least as good as that of n_1 in all states, then pruning n_1 and replacing it by n_2 does not lower the value of the controller. The value will go up if there is an s such that $\alpha_2(s) > \alpha_1(s)$. The complexity of the policy evaluation step in Alg. 5 is $O(|N|^3|S|^3|O|)$, however sparsity often allows to reduce the dependence on $|S|$ and $|O|$. The complexity of the pruning step is $O(|N|^2|S|)$. Overall, compression time is a small fraction of compilation time.

5 Experiments

We evaluate our methods using policies computed by two state-of-the-art point-based POMDP algorithms: GapMin [12] and SARSOP [13]. GapMin returns $\langle \alpha_i, b_i, a_i \rangle$-tuples and therefore we can compile its policies into finite-state controllers using both of our methods. SARSOP was used to compute policies for the largest POMDP benchmarks, however it returns only α-vectors, which is sufficient to apply policy2fsc, but not alpha2fsc (witness beliefs are also needed, but SARSOP's interface does not expose them). The experiments are conducted with some benchmark problems and a real-world POMDP for smart phones. The running time of compilation algorithms—reported in the column 'time'—corresponds to the time of actual compilation. The time to compute initial policies before compilation can be found in the columns 'GapMin' or 'SARSOP'

Table 1. Compilation of GapMin policies using: (1) alpha2fsc applied to GapMin lower bound alpha vectors. (2) policy2fsc applied to GapMin lower bound policy (GM-LB) and (3) policy2fsc applied to GapMin upper bound policy (GM-UB).

POMDP	GapMin	method	depth	tree size	nodes	value	time	c
4x5x2.95 $\|S\|=39, \|A\|=4$ $\|O\|=4, \gamma=0.95$	GM-lb=2.08 GM-ub=2.08 time=4.96s $\|lb\|=58$ $\|ub\|=243$	alpha2fsc			47(58)	2.08(2.08)	0.29	2
		GM-LB	8	287	10(17)	**2.08**(1.85)	0.30	1
		GM-UB	8	287	10(17)	**2.08**(1.85)	0.29	1
		B&B			5	2.02	639.9	
		EM			10	2.01 ± 0.02	66.8	
		QCLP			10	1.74 ± 0.11	7.7	
		BPI			8	0.71 ± 0.09	0.72	
aloha.10 $\|S\|=30, \|A\|=9$ $\|O\|=3, \gamma=0.999$	GM-lb=533.4 GM-ub=544.1 time=5223s $\|lb\|=158$ $\|ub\|=406$	alpha2fsc			137(158)	533.2(533.2)	11.5	1
		GM-LB	11	29525	390(1116)	**537.6**(537.5)	83.6	2
		GM-UB	11	29525	402(1148)	537.6(537.6)	94.5	1
		B&B			10	529.0*	24h	
		EM			40	534.8 ± 0.25	2739	
		QCLP			25	534.37 ± 0.52	99.2	
		BPI			5	112.4 ± 1.59	0.69	
chainOfChains3 $\|S\|=10, \|A\|=4$ $\|O\|=1, \gamma=0.95$	GM-lb=157 GM-ub=157 time=0.86s $\|lb\|=10$ $\|ub\|=1$	**alpha2fsc**	10	10	10(10)	**157**(157)	0.26	0
		GM-LB	11	11	10(10)	**157**(157)	0.42	0
		GM-UB	11	11	10(10)	**157**(157)	0.26	0
		B&B			10	**157**	1.69	
		EM			10	0.17 ± 0.06	6.9	
		QCLP			10	0 ± 0	0.16	
		BPI			10	25.7 ± 0.77	4.25	
cheese-taxi $\|S\|=34, \|A\|=7$ $\|O\|=10, \gamma=0.95$	GM-lb=2.481 GM-ub=2.481 time=1.88s $\|lb\|=22$ $\|ub\|=13$	**alpha2fsc**			17(22)	**2.476**(2.476)	0.29	1
		GM-LB	15	167	17(24)	**2.476**(2.476)	0.56	1
		GM-UB	15	167	17(24)	**2.476**(2.476)	0.55	1
		B&B			10	-19.9*	24h	
		EM			17	-12.16 ± 2.08	337.9	
		QCLP			17	-18.22 ± 1.77	227.4	
		BPI			16	-18.1 ± 0.39	7.18	
lacasa2a $\|S\|=320, \|A\|=4$ $\|O\|=12, \gamma=0.95$	GM-lb=6714.6 GM-ub=6717.6 time=54s $\|lb\|=5$ $\|ub\|=14$	alpha2fsc			5(5)	6714.0(6714.0)	5.15	0
		GM-LB	5	22621	106(421)	**6715.0**(6715.0)	933.9	1
		GM-UB	5	22621	100(517)	6714.1(6714.1)	256.4	1
		B&B			3	6710.0	493.8	
		EM			11	6710 ± 0.11	6485	
		QCLP			2	6699.9 ± 5.5	181	
		BPI			26	6709.3 ± 0.2	121.5	
lacasa3.batt $\|S\|=1920, \|A\|=6$ $\|O\|=36, \gamma=0.95$	GM-lb=293.4 GM-ub=294.7 time=5386s $\|lb\|=26$ $\|ub\|=48$	alpha2fsc			25(26)	292.4(292.4)	399.7	1
		GM-LB	4	12601	47(60)	293.1(292.7)	1451	2
		GM-UB	4	12697	41(48)	**293.2**(293.1)	1030	2
		B&B			5	287.0*	24h	
		EM			5	**293.2 ± 0.03**	13331	
		BPI			9	**293.2 ± 0.12**	2102	
lacasa4.batt $\|S\|=2880, \|A\|=6$ $\|O\|=72, \gamma=0.95$	GM-lb=291.1 GM-ub=292.6 time=8454s $\|lb\|=10$ $\|ub\|=23$	alpha2fsc			10(10)	285.5(285.5)	302	0
		GM-LB	3	745	19(22)	287.3(287.1)	3652	1
		GM-UB	4	23209	87(94)	**290.8**(290.8)	3681	1
		B&B			10	285.0*	24h	
		EM			3	290.2 ± 0.0	19920	
		BPI			6	290.6 ± 0.2	4124	
hhepis6obs_woNoise $\|S\|=20, \|A\|=4$ $\|O\|=6, \gamma=0.99$	GM-lb=8.64 GM-ub=8.64 time=2.6s $\|lb\|=18$ $\|ub\|=7$	**alpha2fsc**			14(18)	**8.64**(8.64)	0.49	1
		GM-LB	12	21	14(18)	**8.64**(8.64)	0.89	1
		GM-UB	12	21	14(18)	**8.64**(8.64)	0.74	1
		B&B			8	**8.64**	4.48	
		EM			14	0.0 ± 0.0	49.2	
		QCLP			14	0.16 ± 0.10	26	
		BPI			13	0.0 ± 0.0	1.68	
machine $\|S\|=256, \|A\|=4$ $\|O\|=16, \gamma=0.99$	GM-lb=62.38 GM-ub=66.32 time=3784s $\|lb\|=39$ $\|ub\|=243$	alpha2fsc			5(39)	54.61(54.09)	5.53	1
		GM-LB	9	376	26(41)	62.92(62.84)	18.5	1
		GM-UB	12	2864	11(159)	**63.0**(60.29)	86.8	2
		B&B			6	62.6	52100	
		EM			11	62.93 ± 0.03	1757	
		QCLP			11	62.45 ± 0.22	4636	
		BPI			10	35.7 ± 0.52	2.14	

depending on which solver was used. High time limits were selected in order to compute policies of high quality. Thus, this time could be considerably shorter if one stops the planning algorithms as soon as a policy of sufficient quality is obtained. This could lead to a substantial reduction of the planning/initialization time since longer planning times (e.g., 10^4 seconds instead of 10^3 seconds in the case of SARSOP [13]) do not usually lead to dramatically improved policies.

5.1 LaCasa Domain

We tested our approaches on three instantiations of the LaCasa domain [20,11], which is a real-world planning task where a smart phone estimates the risk of wandering by a dementia patient and when necessary assists the patient with wayfinding or calls a caregiver. In this domain, it is particularly important to minimize energy consumption since the smart phone won't be able to assist the patient once the battery runs out. Offloading computation to a cloud service is not desired either since it requires a data connection (which may not always be available or reliable) and wireless communication uses a non-negligible amount of energy. A controller offers the best solution since computation consists of negligible table lookups and no data connection is required.

5.2 Results

Tables 1 and 2 compare the results obtained by compiling policies produced by GapMin and SARSOP respectively to four techniques that directly optimize controllers: bounded policy iteration (BPI) with escape [8], quadratically constrained linear programming (QCLP) [7], expectation maximization (EM) with forward search [21] and branch&bound (B&B) with isomorph pruning [11]. Policy2fsc was used in an iterative deepening fashion, starting from depth 2, up to a depth where the resulting controller was at least as good as the original policy or a time limit was exceeded. Hence, the time reported for policy2fsc is the cumulative time (seconds) to process all compilations from depth 2 up to the depth reported in column depth. Tree size is the size of the policy tree for that depth (note that edges with zero probability reduce the size of the policy tree considerably). Column 'nodes' displays the number of nodes in the final controller after compression (before compression in the parentheses). Column 'value' shows the value of controllers after compression (analogously, before compression in the parentheses). Column 'c' indicates the number of iterations of the compression until there is no compression possible. The absence of any result for QCLP, EM and BPI for some problems indicates that 3Gb of memory was not sufficient. A * besides the value of B&B indicates that B&B did not complete its search in 24h and that the value reported is for the best controller found in 24h.

Table 1 compares our two compilation methods for policies computed by GapMin. Method GM-LB stands for policy2fsc applied to the GapMin lower bound policy whereas GM-UB to the upper bound policy. Results confirm that our methods are successful in compiling POMDP policies into finite-state controllers of approximately equivalent quality. The highest value found for each problem

Table 2. Compilation and compression of SARSOP policies

POMDP	SARSOP	method	depth	tree size	nodes	value	time	c
baseball	time 122.7s	policy2fsc	7	175985	10(47)	**0.641**(0.641)	78.22	1
|S|=7681, |A|=6	|α|=1415	B&B			5	0.636*	24h	
|O|=9, γ = 0.999	UB=0.642	EM			2	0.636 ± 0.0	48656	
	LB=0.641	BPI			9	0.636 ± 0.0	445	
elevators_inst_pomdp_1	time 11,228s	policy2fsc	11	419	20(24)	**-44.41**(-44.41)	1357	1
|S|=8192, |A|=5	|α|=78035	B&B			10	-149.0*	24h	
|O|=32, γ = 0.99	UB=-44.31							
	LB=-44.32							
tagAvoid	time 10,073s	policy2fsc	28	7678	91(712)	**-6.04**(-6.04)	582.2	1
|S|=870, |A|=5	|α|=20326	B&B			10	-19.9*	24h	
|O|=30, γ = 0.95	UB=-3.42	EM			9	-6.81 ± 0.12	19295	
	LB=-6.09	QCLP			2	-19.99 ± 0.0	12.9	
		BPI			88	-12.42 ± 0.13	1808	
underwaterNav	time 10,222s	policy2fsc	51	1242	52(146)	745.3(745.3)	5308	1
|S|=2653, |A|=6	|α|=26331	B&B			10	747.0*	24h	
|O|=103, γ = 0.95	UB=753.8	EM			5	**749.9** ± 0.02	31611	
	LB=742.7	BPI			49	748.6 ± 0.24	14758	
rockSample-7_8	time 10,629s	policy2fsc	31	2237	204(224)	**21.58**(21.58)	1291	1
|S|=12545, |A|=13	|α|=12561	B&B			10	11.9*	24h	
|O|=2, γ = 0.95	UB=24.22	BPI			5	7.35 ± 0.0	78.8	
	LB=21.50							

is bolded. Alpha2fsc compiles $|lb|$ α-vectors into controllers with similar value, though sometimes the value is significantly worse (e.g., lacasa4.batt and machine). In contrast, policy2fsc finds better controllers by simulating the input policy to a larger depth, but this takes more time. It was stopped as soon as the value of the controller matches GapMin's lower bound or 1h was reached. In many cases, the number of nodes is still less than or equal to the size of the input policy (e.g., 4x5x2.95, cheese-taxi, lacasa2, machine). The direct optimization techniques (B&B, QCLP, EM, BPI) generally take much longer and/or do not consistently produce good controllers.

Table 2 summarizes the results for some problems that are among the largest available benchmarks for point-based value iteration techniques that do not exploit factored representations. In this case, SARSOP was used to obtain a lower bound policy that is then compiled by policy2fsc. Even though SARSOP returned value functions with thousands of α-vectors, we compiled those policies into considerably smaller controllers (up to 3 orders of magnitude reduction) of the same or better quality (e.g., underwaterNav) demonstrating that our method scales to large problems. Policy2fsc produced the best value for all problems except underwaterNav where the direct optimization techniques produced better controllers. This simply indicates that the policy compiled from SARSOP was not the best as opposed to any weakness in policy2fsc.

6   Conclusion

We have presented two novel methods for compiling policies for partially observable Markov decision processes (POMDPs) into approximately equivalent finite state controllers (FSCs). Our motivation is that these FSC representations are very useful in resource-constrained applications such as on mobile or wearable devices. Methods that can create FSC policies open up new possibilities for using POMDP controllers on these devices, where battery, computation and memory resources are at a premium. We showed how we can get very compact, yet equivalent representations for POMDP policies as those generated by two state-of-the-art offline planners.

Acknowledgments. This work was supported by the Ontario Ministry of Research and Innovation, NSERC, Toronto Rehabilitation Institute and the Alzheimer's Association grant ETAC-10-173237.

References

1. Williams, J.D., Young, S.: Scaling POMDPs for spoken dialog management. IEEE Trans. on Audio, Speech, and Language Processing 15(7), 2116–2129 (2007)
2. Hoey, J., Boutilier, C., Poupart, P., Olivier, P., Monk, A., Mihailidis, A.: People, sensors, decisions: Customizable and adaptive technologies for assistance in healthcare. ACM Transactions on Interactive Intelligent Systems 2(4) (2012)
3. Ross, S., Pineau, J., Paquet, S., Chaib-draa, B.: Online planning algorithms for POMDPs. Journal of Artificial Intelligence Research 32, 663–704 (2008)
4. Silver, D., Veness, J.: Monte-Carlo planning in large POMDPs. In: NIPS (2010)
5. Kaelbling, L., Littman, M., Cassandra, A.: Planning and acting in partially observable stochastic domains. Artificial Intelligence 101(1-2), 99–134 (1998)
6. Braziunas, D., Boutilier, C.: Stochastic local search for POMDP controllers. In: AAAI, pp. 690–696 (2004)
7. Amato, C., Bernstein, D., Zilberstein, S.: Optimizing fixed-size stochastic controllers for POMDPs and decentralized POMDPs. JAAMAS (2009)
8. Poupart, P., Boutilier, C.: Bounded finite state controllers. In: Thrun, S., Saul, L.K., Schölkopf, B. (eds.) NIPS. MIT Press (2003)
9. Toussaint, M., Harmeling, S., Storkey, A.: Probabilistic inference for solving (PO)MDPs. Technical Report EDI-INF-RR-0934, School of Informatics, University of Edinburgh (2006)
10. Meuleau, N., Kim, K.E., Kaelbling, L.P., Cassandra, A.R.: Solving POMDPs by searching the space of finite policies. In: UAI, pp. 417–426 (1999)
11. Grześ, M., Poupart, P., Hoey, J.: Isomorph-free branch and bound search for finite state controllers. In: Proc. of IJCAI (2013)
12. Poupart, P., Kim, K.E., Kim, D.: Closing the gap: Improved bounds on optimal POMDP solutions. In: ICAPS (2011)
13. Kurniawati, H., Hsu, D., Lee, W.: SARSOP: Efficient point-based POMDP planning by approximating optimally reachable belief spaces. In: RSS (2008)
14. Hansen, E.: An improved policy iteration algorithm for partially observable MDPs. In: NIPS (1998)
15. Pineau, J., Gordon, G., Thrun, S.: Point-based value iteration: An anytime algorithm for POMDPs. In: IJCAI, pp. 1025–1032 (2003)
16. Spaan, M.T.J., Vlassis, N.: Perseus: Randomized point-based value iteration for POMDPs. Journal of Artificial Intelligence Research 24, 195–220 (2005)
17. Shani, G., Pineau, J., Kaplow, R.: A survey of point-based POMDP solvers. Journal of Autonomous Agents and Multi-Agent Systems 27(1), 1–51 (2013)

18. Hansen, E.A.: Finite-memory control of partially observable systems. PhD thesis, University of Massachusetts Amherst (1998)
19. Kaelbling, L.P., Littman, M.L., Cassandra, A.R.: Planning and acting in partially observable stochastic domains. Artificial Intelligence 101, 99–134 (1998)
20. Hoey, J., Yang, X., Quintana, E., Favela, J.: LaCasa: Location and context-aware safety assistant. In: Pervasive Comp. Techn. for Healthcare, pp. 171–174 (2012)
21. Poupart, P., Lang, T., Toussaint, M.: Analyzing and escaping local optima in planning as inference for partially observable domains. In: Gunopulos, D., Hofmann, T., Malerba, D., Vazirgiannis, M. (eds.) ECML PKDD 2011, Part II. LNCS, vol. 6912, pp. 613–628. Springer, Heidelberg (2011)

Learning CP-net Preferences Online from User Queries

Joshua T. Guerin[1], Thomas E. Allen[2], and Judy Goldsmith[2]

[1] The University of Tennessee at Martin
jguerin@utm.edu
[2] University of Kentucky
{teal223,goldsmit}@cs.uky.edu

Abstract. We present an online, heuristic algorithm for learning Conditional Preference networks (CP-nets) from user queries. This is the first efficient and resolute CP-net learning algorithm: if a preference order can be represented as a CP-net, our algorithm learns a CP-net in time n^p, where p is a bound on the number of parents a node may have. The learned CP-net is guaranteed to be consistent with the original CP-net on all queries from the learning process. We tested the algorithm on randomly generated CP-nets; the learned CP-nets agree with the originals on a high percent of non-training preference comparisons.

1 Introduction

To support decision making, an intelligent agent often requires some way to *learn* what a human user prefers and *concisely represent* those preferences. CP-nets [1] offer a potentially compact qualitative representation of human preferences that operate under *ceteris paribus* ("with all else being equal") semantics. In this paper we present a novel algorithm through which an agent learns the preferences of a user. CP-nets are used to represent such preferences and are learned online through a series of queries generated by the algorithm. Our algorithm builds a CP-net for the user by creating nodes and initializing Conditional Preference Tables (CPTs), then gradually adding edges and forming more complex CPTs consistent with responses to queries until a confidence parameter is reached or no further progress can occur. The algorithm does not always converge to the original CP-net, but our experiments show that it can learn a CP-net that closely tracks with the original for a set of outcome comparison queries not used in the learning phase. While one could treat the model learning process as one of replicating the structure of the original (latent) CP-net, we assume here that this is unnecessary, as long as the *preferences modeled by* the two networks differ by very little.

The problem of learning CP-nets from example data is known to be hard in general, even for acyclic, binary CP-nets (the class we consider here), and also for *separable* CP-nets (such as we use in the first phase of our algorithm; see Def. 1) [2,3]. This has led to a diversity of proposed methods for learning CP-nets: a regression-based approach [4,5], Anguin-style query learning [6,7], learning via reduction to

P. Perny, M. Pirlot, and A. Tsoukiàs (Eds.): ADT 2013, LNAI 8176, pp. 208–220, 2013.
© Springer-Verlag Berlin Heidelberg 2013

2-SAT [8], and learning a CP-net indirectly via the exponentially larger preference graph [9]. Our work builds upon this body of previous research, particularly that of [2,3] and [8], but differs in several notable respects. First, our method is designed for an *active, online* environment in which the agent directly interacts with the user, rather than one in which the user's actions can be observed over time. Second, in our queries we allow for arbitrary outcome comparisons rather than (as in [6,7]) restricting to *swap* comparisons where outcomes differ in at most one variable. We believe our approach better reflects the rich environment in which human decisions are often made. Third, our algorithm is *robust*; unlike [8], for example, it will always output a CP-net. Finally, given a constant bound on the number of parents, our algorithm can learn a CP-net in polynomial time. This differs in particular from that of [9], which in the worst case is of double-exponential complexity in the number of variables.

The remainder of our paper is organized as follows: In Sec. 2 we review the use of CP-nets to model preferences. In Sec. 3 we present the algorithm itself, followed by analysis in Sec. 4. Section 5 consists of a series of experiments and significant results. We conclude with opportunities for future research.

2 Modeling Preferences with CP-nets

Consider a recommender system that assists customers in purchasing a guitar. The customer surely cannot consider every possible guitar, but will buy one that is satisfactory, given her preferences. Various factors differentiate the possibilities, such as BRAND, BODY, and COLOR. We assume the customer can consistently rank alternatives: presented two guitars, she can say, "I prefer the first to the second." We further assume the customer can make more general comparisons, such as, "In general, I prefer the *Fender* BRAND to *Gibson*." But offered several alternatives, that customer may ultimately prefer a *specific* Gibson guitar based on other factors; that is, the user's preferences may be *conditional*.

More formally, by *preference*, we mean a strict partial order $\succ$ over a finite set of *outcomes* $\mathcal{O}$ by a user. Such outcomes can be factored into *variables* $\mathcal{V}$ with associated (binary) *domains* Dom($\mathcal{V}$): $\mathcal{O} = v_1 \times v_2 \times \cdots \times v_n$. We call $o[i]$ the *projection* of outcome o onto variable v_i and $o[U]$ the projection onto a *set* of variables $U \subseteq \mathcal{V}$. Note that the number of outcomes and orderings is exponential in the number of variables. CP-nets can offer a more compact representation.

Definition 1. *A CP-net $\mathcal{N}$ is a directed graph. Each node v_i represents a preference over a finite domain. An edge (v_i, v_j) indicates that the preference over v_j depends on the value of v_i. If a node has no incoming edges, the preference involving its variable is not conditioned on values represented elsewhere in the graph. A separable CP-net is one in which no variable depends on any other.*

Definition 2. *A conditional preference table (CPT) is associated with each node v_i and specifies the preference over Dom(v_i) as a function of the values assigned to its parent nodes Pa(v_i). If a CPT has an entry for every combination of values from the domains of its parents, we say it is* complete.

We define the *size* of a CPT as its number of rows. The size of a CP-net is the sum of the sizes of its CPTs. One can observe that if a node v_j has incoming edges from k parents, each representing binary variables, then $size(CPT(v_j)) = 2^k$. Thus size grows exponentially in the number of parents. To guarantee tractability, we make some simplifying assumptions: 1. Cycles are disallowed. 2. We restrict to binary domains. 3. A maximum bound p is placed on the number of parents a node may have: We conjecture that most human preferences are conditioned on 3–5 nodes and thus feel justified in assuming such a bound.

Example 1. *Consider the CP-net to the right representing a conditional preference: The edge from* BRAND *to* COLOR *indicates that the customer's preference of* COLOR *depends on* BRAND. *The CPTs associated with each node provide the ordering. In general, the customer prefers* Fender *to* Gibson. *If a* Fender *is available, she prefers* red, *but if only a* Gibson *is available, she prefers* gold.

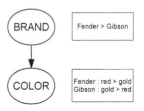

Fig. 1. Simple CP-net

3   Algorithm

Our algorithm consists of two phases. First, it constructs a separable CP-net with default CPTs. Next, it successively attempts to refine the model, adding edges and learning more complex CPTs consistent with evidence from *user queries.* (See LEARN-CP-NET and its subroutine FIND-PARENTS [Alg. 1 and 2]).

3.1   Phase 1: The Separable CP-net Basis

In Phase 1 (Alg. 1, lines 1–9), our algorithm constructs a *separable CP-net basis* by asking the user to provide a default preference for each $v_i \in \mathcal{V}$. This initial CP-net could be characterized as a first impression of the user's preferences.

Definition 3. *Let v_i be a binary variable with* $\mathrm{Dom}(v_i) = \{x_i, y_i\}$. *An attribute comparison query is one in which we ask the user whether $x_i \succ y_i$ or $y_i \succ x_i$.*

Here we assume that the user is able to reflect on possible outcomes and discern what she prefers most of the time. (In our experiments, as discussed in Sec. 5.1, we *simulate* the user's response to such queries by reporting a preference that occurs most frequently in $CPT(v_i)$.) The result is a CP-net with default CPTs and no edges. If we were *confident* that the user's preference over each attribute did not depend on the value of any other attribute, we could return $\mathcal{N}$ as *the* CP-net that consistently modeled the user's preferences. At this point, however, we are *unconfident* of each node's parentage. We maintain disjoint sets *Confident* and *Unconfident* such that $v_i \in$ *Confident* iff enough data has been collected via user queries to conclude that the preferences over v_i are conditioned only by its parent variables in the graph of $\mathcal{N}$; otherwise $v_i \in$ *Unconfident*. Initially *Unconfident* $= \mathcal{V}$ and *Confident* $= \emptyset$.

Example 2. *Nodes* BODY *and* BRAND *have been inserted into* $\mathcal{N}$ *along with their default CPTs. A node and CPT must now be created for* COLOR*. The agent asks the customer, "In general, do you prefer a **guitar** that is red or gold?" The user replies that gold is usually preferred, and a node and CPT is inserted. The resulting CP-net basis is shown.*

Fig. 2. Separable CP-net Basis

Algorithm 1. LEARN-CP-NET($\mathcal{V}$, p, q)

Input: $\mathcal{V}$ a set of binary variables
        p maximum number of parents
        q confidence threshold parameter

Global: *Comparisons*   responses to user queries
        Confident   set of learned nodes

Output: $\mathcal{N}$ the CP-net learned from the user

1: $\mathcal{N} \leftarrow \emptyset$
2: *Comparisons* $\leftarrow \emptyset$
3: *Confident* $\leftarrow \emptyset$
4: *Unconfident* $\leftarrow \mathcal{V}$
5: **for** $v_i \in \mathcal{V}$ **do**
6:    query user: do you prefer $x_i \succ y_i$ or $y_i \succ x_i$?
7:    v_i.CPT $\leftarrow$ default CPT based on user response
8:    insert v_i into $\mathcal{N}$
9: **end for**
10: **repeat**
11:    **for** $r \leftarrow 0$ **to** p **do**
12:      **for** $v_i \in$ *Unconfident* **do**
13:        $(P, C) \leftarrow$ FIND-PARENTS(v_i, r, q)
14:        **if** $C \neq$ FAIL **then**
15:          v_i.CPT $\leftarrow C$
16:          add edges from all P to v_i
17:          move v_i from *Unconfident* to *Confident*
18:        **end if**
19:      **end for**
20:    **end for**
21: **until** no parents added this iteration
22: **return** $\mathcal{N}$

3.2   Phase 2: Refining the CP-net Model

In Phase 2 we refine $\mathcal{N}$ by discovering such conditional relationships as may exist between variables by asking the user's preference over pairs of outcomes.

Algorithm 2. FIND-PARENTS(v_i, r, q)

Input: v_i node representing ith variable
        r the number of parents in the trials
        q confidence threshold parameter

Global: *Comparisons*   responses to user queries
        Confident   set of learned nodes

Output: C a CPT with 2^r rows or FAIL
        P newly parents discovered for v_i or $\emptyset$

1: **for** $P \in \{$all subsets of *Confident* of size $r\}$ **do**
2:     $(C, evidCount) \leftarrow$ CREATE-CPT(v_i, P)
3:     **while** $(C \neq$ FAIL$)$ **and** $(evidCount < q)$ **do**
4:         $(o_1, o_2) \leftarrow$ generate random query for v_i
5:         query user: do you prefer outcome o_1 or o_2?
6:         add o_1, o_2 to *Comparisons* in specified order
7:         $(C, evidCount) \leftarrow$ CREATE-CPT(v_i, P)
8:     **end while**
9:     **if** $C \neq$ FAIL **then**
10:         **return** (P, C)
11:     **end if**
12: **end for**
13: **return** $(\emptyset,$ FAIL$)$

Definition 4. *In an* outcome comparison query, *we provide the user a pair of outcomes,* $\{o_1, o_2\} \in \mathcal{O}$ *such that* $o_1 \neq o_2$. *The user responds with* $o_1 \succ o_2$, $o_2 \succ o_1$ *or* $o_1 \bowtie o_2$, *respectively indicating that she strictly prefers the first outcome to the second, the second to the first, or is unable to state a preference.*

If the user is able to answer an outcome comparison query with either $o_1 \succ o_2$ or $o_2 \succ o_1$, we treat the response as *evidence* of the user's underlying preference model. If the user cannot state a preference, we treat that as an indication that the two outcomes are *incomparable*. We expect that the user will provide consistent answers to the same outcome query and hence ensure that each unordered pair of outcomes is asked at most once. Responses are stored in a *Comparisons* database, gradually adding to the evidence used to construct the model CP-net. Outcome pairs are generated *randomly*, with the constraint that the pair must be *relevant* to the node under consideration and not already in *Comparisons*.

Definition 5. *For any given* v_i, random query *is an outcome comparison query in which* o_1 *and* o_2 *are selected uniformly randomly from their domains, with the requirement that the query must be* relevant *to node* v_i: $o_1[i] \neq o_2[i]$. *A random adaptive query* adds the additional requirement that for all $v_j \in$ Confident, $o_1[j] = o_2[j]$.

Random adaptive queries provide a heuristic that may reduce the search space for a CP-net by not continuing to analyze nodes once they are labeled *Confident*.

Using random queries, our search proceeds as follows. In the `repeat–until` loop of LEARN-CP-NET (Alg. 1, lines 10–21), we search first for nodes that do not need parents (i.e., nodes that represent features for which the user's preferences are unconditional). For each *target node* $v_i \in \mathcal{V}$, we call FIND-PARENTS. If FIND-PARENTS is confident that the preferences over v_i are unconditional, it returns $(\emptyset, C)$, where C is a CPT for v_i consistent with all queries stored in *Comparisons*; otherwise, it returns $(\emptyset, \text{FAIL})$, indicating failure. In the former case, the default CPT of v_i is replaced with C and v_i is reclassified as *Confident*; in the latter, the search continues with the next *Unconfident* node.

As long as *Unconfident* $\neq \emptyset$, we continue trying to refine our model with new conditional relationships, represented as edges and correspondingly more complex CPTs. For each remaining *Unconfident* node, we iterate over potential sets of parent nodes, starting with single-parent relationships, then two parents, three, and so on, up to the bound on parents p. If a set P of parents can be found, edges are added from each newly discovered parent to target node v_i, the default CPT is replaced with C, and v_i is reclassified as *Confident*. If at any point, however, we iterate from 0 to p and fail to add parents to *any* node, we stop refinement and are satisfied with the CP-net that we have thus generated to that point, even if some nodes remain *Unconfident*. For such nodes, the default CPT assigned in Phase 1 would be output in the finalized CP-net. Indeed, in the worst case, it is possible that *all* of the nodes could be output with the CPTs in their default state. However, in practice we find that this rarely happens. In all cases, though, LEARN-CP-NET will return a CP-net; it will never output failure.

Example 3. *Algorithm* LEARN-CP-NET *has determined from outcome comparison queries that* BODY *and* BRAND *each has 0 parents. As such, their status was moved from* Unconfident *to* Confident, *and the entries in their CPTs are thus finalized. However, the status of* COLOR *at this point is still undetermined.*

An inspection of Comparisons indicates that the preference over COLOR is not independent. Sometimes the customer prefers gold, sometimes red. Additional queries are generated; e.g., "Do you prefer a heavy gold Gibson or a heavy red Gibson?" Nonetheless, all attempts to prove that BODY or BRAND is the sole parent of COLOR fail to yield a CPT.

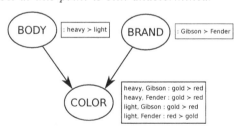

Fig. 3. The Fully Learned CP-net

The algorithm next attempts to establish whether both BODY and BRAND are parents of COLOR. This time a CPT is produced and sufficient evidence is gathered. While the user does indeed generally prefer gold, she prefers a red **guitar** only if it is light and a Fender. Figure 3 shows completed CP-net $\mathcal{N}$.

Subroutine FIND-PARENTS iterates over all subsets of *Confident* of the specified size r. For each subset P, FIND-PARENTS calls CREATE-CPT to determine whether P is consistent with all queries stored in *Comparisons*. For a given target node and set of possible parents, CREATE-CPT constructs a 2-SAT instance such that (1) a satisfying assignment tells us that the target node's values are consistent with the given set of parents and (2) the assignment to variables gives us the entries of the target node's CPT. Our method for this closely follows [8], to which the reader is referred for specifics. It should be noted that this method converges to the original network only for a restricted class of CP-nets; however, our objective is not to recover the original network, but to learn one that closely approximates the original for possible outcome comparison queries. If the CPT returned from CREATE-CPT is not complete, the default values from Phase 1 are used to provide the ordering. FIND-PARENTS continues generating random queries and calling CREATE-CPT until the *evidence* to support P as the parents of v_i reaches a specified confidence threshold q. Specifically, the *evidence count* is the number of queries in *Comparisons* relevant to node v_i, and *confidence parameter* q is the number of outcome pairs required before we are confident that P is in fact the parent set of v_i. (We discuss the use of the q threshold for minimum evidence in Section 3.2 and metrics for confidence in 5.)

4 Analysis

Theorem 1. LEARN-CP-NET *is resolute—that is, it is guaranteed to output a consistent CP-net $\mathcal{N}$.*

Proof. (Sketch.) An initial $\mathcal{N}$ is created in the algorithm's phase 1. The repeat–until loop that follows will iterate at most once per variable. Since $\mathcal{V}$ is finite, the algorithm will terminate and output a CP-net. Moreover, since edges are added only from nodes in *Confident* to those in disjoint set *Unconfident*, $\mathcal{N}$ will not contain a cycle; hence it will be consistent.   □

Theorem 2. *The learning algorithm* LEARN-CP-NET *is time polynomial in n^p and q in the worst case.*

Proof. (Sketch.) Using Tarjan's algorithm, we implement the CREATE-CPT 2-SAT algorithm in linear time [10]. Given this, the polynomial time complexity of LEARN-CP-NET and FIND-PARENTS follows directly from Alg. 1 and 2.   □

5 Experiments

5.1 Experiment Design

We evaluate the effectiveness of our algorithm through an experimental approach. First, we generate a random CP-net training model $\mathcal{N}_T$ that simulates the preferences of a hypothetical user. We then apply LEARN-CP-NET to the model, posing queries that are answered by an agent on the basis of $\mathcal{N}_T$. Finally, we evaluate the success of the learned model $\mathcal{N}_L$ in terms of the likelihood that it will correctly predict the response of $\mathcal{N}_T$ to a randomly drawn pair of outcomes from $\mathcal{O}$. We assume here that a user's preferences can be modeled by a CP-net. (Whether human preferences can actually be modeled by CP-nets is a subject the authors are presently exploring through interdisciplinary research.) We further assume that the attribute variables and their binary values are common knowledge and not additional parameters that must be learned.

A number of questions guided our experiments, such as: How many queries are required to learn a CP-net model? Does the choice of outcome comparison querying strategies (random and adaptive) affect CP-net learnability? How does quality of CP-net learning change as the size and density of its graphical structure changes? Do algorithm runtimes conform to expectations?

To evaluate the algorithm, we generated a set of *random CP-nets*. Since variables and domains are common knowledge, the nodes of $\mathcal{N}_T$ are an input to the generation algorithm, as are the number of directed edges e in the graph and the maximum number of parents p. (Note that while $\mathcal{V}$ is the same for $\mathcal{N}_T$ and $\mathcal{N}_L$, e and p may differ.) The edges of $\mathcal{N}_T$ are inserted at random with equal probability, such that no node has more than p parents and no cycles are introduced. Complete CPTs for each node are then generated by selecting, again uniformly randomly, an entry $x_i \succ y_i$ or $y_i \succ x_i$ for each row (with the provision that the CPT implies dependence on the value of each parent node).

Algorithm 3. GENERATE-RANDOM-CP-NET$(\mathcal{V}, e, p)$

Input: $\mathcal{V}$ a set of binary variables
             e number of edges in graph
             p maximum number of parents

Output: $\mathcal{N}_T$ the randomly generated CP-net
1: initialize $\mathcal{N}$: nodes $\mathcal{V}$, no edges, empty CPTs
2: **for** $k \leftarrow 1$ **to** e **do**
3:     add a randomly selected edge (v_i, v_j) to $\mathcal{N}_T$ s.t. $1 \leq i < j \leq n$ and $|\mathrm{Pa}(v_j)| \leq p$
4: **end for**
5: **for** $k = 1$ **to** $|\mathcal{V}|$ **do**
6:     **for** each of the $2^{|\,\mathrm{Pa}(v_k)|}$ entries in CPT(v_k) **do**
7:         insert $x_k \succ y_k$ or $y_k \succ x_k$ at random
8:     **end for**
9: **end for**
10: **return** $\mathcal{N}_T$

Our simulations do not employ human subjects. The learning algorithm queries an agent that answers on the basis of a given CP-net model $\mathcal{N}_T$. Recall that in Phase 1, an *attribute comparison query* is asked for each binary variable: "In general, do you prefer x_i or y_i?" We assume that the human user has the capacity to reflect on such outcomes and determine whether she usually prefers x_i or y_i. An agent that simulates such a capacity we call *attribute aware*.

Definition 6. *An agent is attribute aware if it replies to an attribute comparison query with the preference that occurs in the majority of its CPT entries. If* $x_i \succ y_i$ *and* $y_i \succ x_i$ *occur equally often, the agent outputs one of these at random.*

LEARN-CP-NET also asks the user a series of *outcome comparison queries*: "Do you prefer $o_1 \succ o_2$, $o_2 \succ o_1$, or neither?" In response, the agent consults its CP-net to determine if a preference is entailed. It has been shown that, in general, determining whether one outcome dominates another given a CP-net ($\mathcal{N} \models o_1 \succ o_2$) is hard [11]. However, [1] introduce the weaker notion of an *ordering query*, which we employ in our simulations.

Definition 7. *Given outcomes* o_1 *and* o_2, *we say* o_1 *is consistently orderable over* o_2 *with respect to* $\mathcal{N}$ *iff there exists a node* $v_i \in \mathcal{N}$ *such that* (1) o_1 *and* o_2 *assign the same value to all ancestors of* v_i *and* (2) o_1 *assigns a more preferred value to* v_i *than* o_2. *A search for such a node is called an* ordering query.

Rather than asking if $\mathcal{N} \models o_1 \succ o_2$, an ordering query asks if $\mathcal{N} \not\models o_2 \succ o_1$. Note that while $\mathcal{N} \models o_1 \succ o_2$ implies $\mathcal{N} \not\models o_2 \succ o_1$, the reverse does not hold: it may be that $\mathcal{N} \not\models o_1 \succ o_2$ and $\mathcal{N} \not\models o_2 \succ o_1$. As Boutilier, et al. [1] showed, ordering queries can be answered in time polynomial in size($\mathcal{N}$). In response to outcome queries in our simulations, $\mathcal{N}_T$'s agent answers $o_1 \succ o_2$ if o_1 is consistently orderable over o_2, $o_2 \succ o_1$ if the reverse is true, and $o_1 \bowtie o_2$ if a preference cannot be determined. Note that, according to [1]'s definition of *consistently orderability*, we could have o_1 consistently orderable over o_2 based on some variables, and vice versa based on others. For purposes of answering ordering queries, we check all variables, and say that o_1 is consistently orderable over o_2 if there is some variable for which that holds, and *no* variables that witness that o_2 is consistently orderable over o_1. This is a stronger condition than [1]'s. However, the disadvantage of using ordering queries instead of dominance testing is that an agent using the former is more likely to report that $o_1 \bowtie o_2$, especially as n grows large.

Our primary measure of how well the learned CP-net $\mathcal{N}_L$ agrees with the training model $\mathcal{N}_T$ is to compare directly the preference ordering induced by $\mathcal{N}_L$ with that of $\mathcal{N}_T$ for all unordered pairs of possible outcomes.

Definition 8. *Given an outcome comparison involving* o_1 *and* o_2, *we say that CP-net* $\mathcal{N}_L$ *agrees with* $\mathcal{N}_T$, *disagrees, or is indecisive as follows:*

$\mathcal{N}_T$	$\mathcal{N}_L : o_1 \succ o_2$	$o_2 \succ o_1$	$o_1 \bowtie o_2$
$o_1 \succ o_2$	agrees	disagrees	indecisive
$o_2 \succ o_1$	disagrees	agrees	indecisive

The agreement metric M *is then a vector representing the percentage of total outcome comparisons for which* $\mathcal{N}_L$ *agrees, disagrees, or is indecisive w.r.t.* $\mathcal{N}_T$.

Note that we do not include in our counts outcome comparisons for which *training* model $\mathcal{N}_T$ is indecisive. Because of this, the agreement metric is not symmetric; that is, in general $M(\mathcal{N}_T, \mathcal{N}_L) \neq M(\mathcal{N}_L, \mathcal{N}_T)$.

Since the number of outcomes is 2^n for n binary variables, the number of unordered outcome pairs $\{o_1, o_2\}$ is $O(2^{2n})$. Hence, while our algorithm is of polynomial complexity, our method of evaluation is exponential in the number of variables if we calculate M *exactly*. However, we can *approximate* M satisfactorily through *sampling*. For our sample size we chose $20\,000 > z_{\alpha/2}^2/(4\varepsilon^2)$, where $z_{\alpha/2}$ is obtained from the normal distribution, ε is the desired bound of $\pm 0.5\%$, and $1 - \alpha$ provides a 95%-confidence interval. We calculate M exactly for $n \leq 7$ and estimate through sampling for $n > 7$.

5.2   Experimental Results

Tables 1 and 2 show metrics of $\mathcal{N}_L$ w.r.t. $\mathcal{N}_T$ over a series of experiments for n nodes and confidence threshold q. Density $\delta = e/n$ is the desired ratio of edges to nodes ($\delta = \infty$ implies a maximally dense acyclic graph, given the bound p on parents). The results shown are for $p = 5$ and represent averages over 30 trials. The data reflect random adaptive queries and an attribute aware agent; our results for random non-adaptive queries (not shown) reflected slightly lower levels of agreement [12]. Table 2 shows agreement for a higher granularity of q, along with the choice of q in this range that maximized agreement ($q*$).

Table 1. Agreement of $\mathcal{N}_L$ with $\mathcal{N}_T$

	Agreement				Disagreement				Indecision			
n	q=5	10	15	20	q=5	10	15	20	q=5	10	15	20
						$\delta = 1$						
4	0.96	0.98	0.98	0.98	0.03	0.01	0.02	0.01	0.01	0.00	0.00	0.00
6	0.79	0.94	0.97	0.98	0.07	0.04	0.02	0.02	0.14	0.02	0.00	0.00
8	0.69	0.77	0.77	0.75	0.07	0.03	0.02	0.02	0.24	0.20	0.21	0.23
10	0.65	0.65	0.58	0.53	0.04	0.02	0.02	0.01	0.31	0.33	0.41	0.46
12	0.57	0.65	0.56	0.42	0.02	0.02	0.01	0.01	0.41	0.34	0.43	0.58
						$\delta = 2$						
4	0.92	0.98	0.98	0.98	0.06	0.02	0.02	0.02	0.02	0.00	0.00	0.00
6	0.72	0.97	0.98	0.98	0.13	0.03	0.02	0.02	0.16	0.00	0.00	0.00
8	0.60	0.76	0.76	0.76	0.11	0.04	0.03	0.02	0.29	0.20	0.21	0.22
10	0.53	0.64	0.60	0.52	0.09	0.04	0.02	0.01	0.38	0.33	0.38	0.47
12	0.52	0.64	0.41	0.36	0.07	0.03	0.02	0.01	0.41	0.33	0.57	0.63
						$\delta = 3$						
4	0.94	0.97	0.98	0.98	0.04	0.03	0.02	0.02	0.02	0.00	0.00	0.00
6	0.82	0.95	0.98	0.98	0.11	0.04	0.02	0.02	0.08	0.01	0.00	0.00
8	0.63	0.80	0.73	0.76	0.13	0.04	0.02	0.01	0.24	0.16	0.25	0.22
10	0.61	0.65	0.48	0.47	0.13	0.04	0.01	0.01	0.26	0.31	0.51	0.52
12	0.57	0.62	0.44	0.28	0.13	0.04	0.02	0.01	0.30	0.34	0.55	0.72
						$\delta = \infty$						
4	0.91	0.98	0.98	0.98	0.06	0.02	0.02	0.02	0.03	0.00	0.00	0.00
6	0.84	0.96	0.97	0.98	0.11	0.03	0.03	0.02	0.05	0.01	0.00	0.00
8	0.66	0.76	0.76	0.75	0.13	0.03	0.02	0.01	0.21	0.21	0.22	0.24
10	0.62	0.56	0.48	0.40	0.13	0.04	0.02	0.01	0.25	0.39	0.50	0.59
12	0.54	0.54	0.37	0.26	0.13	0.04	0.01	0.01	0.33	0.42	0.61	0.74

Table 2. Agreement for various choices of q

					Agreement $\delta = 1$					
n	q=6	7	8	9	10	11	12	13	14	$q*$	max
6	0.90	0.90	0.91	0.96	0.96	0.97	0.97	0.97	0.97	13	97%
7	0.77	0.80	0.85	0.87	0.83	0.87	0.84	0.89	0.89	14	89%
8	0.69	0.75	0.77	0.83	0.77	0.84	0.77	0.80	0.79	11	85%
9	0.76	0.76	0.75	0.70	0.81	0.74	0.78	0.70	0.72	10	81%
10	0.61	0.69	0.70	0.72	0.66	0.70	0.69	0.68	0.60	9	72%

Table 3. Effect of decreasing p for $\mathcal{N}_L$ below that of $\mathcal{N}_T$

	Agreement					Disagreement					Indecision				
n	p=1	2	3	4	5	p=1	2	3	4	5	p=1	2	3	4	5
						$\mathcal{N}_T$: $p = 5$, $\delta = \infty$									
6	0.33	0.42	0.67	0.86	0.96	0.05	0.03	0.03	0.03	0.03	0.63	0.55	0.30	0.11	0.00
7	0.25	0.33	0.44	0.70	0.90	0.04	0.03	0.02	0.03	0.03	0.71	0.64	0.54	0.28	0.07
8	0.19	0.24	0.34	0.55	0.76	0.03	0.02	0.02	0.02	0.03	0.78	0.74	0.64	0.43	0.21
9	0.14	0.20	0.27	0.45	0.60	0.02	0.02	0.02	0.02	0.03	0.83	0.79	0.71	0.53	0.37

Overall, the learned model exhibited a high level of agreement with the training model. For $n \leq 10$ agreement was 70–90% or higher with the proper choice of q. Significantly, we found that the learned model rarely *disagreed* with the training model. As n increases, however, the learned model is increasingly likely to be indecisive about a preference over which the training model is able to reason. For example, for $n = 20$ nodes, we found that agreement ranged from 50 to 60% for $q = 10$, depending on density δ, but disagreement was $< 1\%$. As discussed in Sec. 5.1, this increased indecision as n grows results in part from the use of ordering queries instead of dominance testing as the primary metric.

A question of particular interest was the number of queries per node required. One can observe that an exponential number of outcome comparison queries could be required in the worst case. However, we found that often just a few queries—8 to 14 (even for larger values of n)—proved optimal. The choice of q is something of an art. As the data show, increasing the number of queries required to become confident about a node sometimes has an *adverse* effect on the agreement of the learned model with the training model. We take this to be an indication that if q is too high, then overfitting can occur.

We also explored the effect of decreasing the maximum number of parents for the learned model below that of the training model (see Table 3). We found, for example, that for $\mathcal{N}_T$ a maximally dense 7-node graph with maximum 5 parents ($n = 7$, $p = 5$, $\delta = \infty$) and p is set to 5 for $\mathcal{N}_L$, agreement is 90%. If p for $\mathcal{N}_L$ is then reduced to 4, agreement decreases to 70%—still a modestly good result. Furthermore, while indecision is increased, disagreement remained static.

We were also interested in the computation time required, since it is known that learning a CP-net is NP-hard in the general case. The relative running times

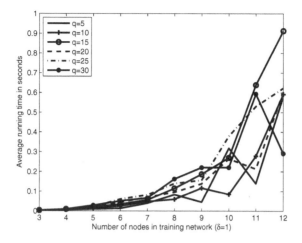

Fig. 4. Algorithm Running Times ($\delta = 1$)

for inputs of various size (Fig. 4 provides data for the case of $\delta = 1$) coincide with our expectations. While the size of the training graph is the primary determinate of running time, the number of queries required by q also plays a role. We also observed that running time can vary significantly from model to model depending on the preference relation that is being learned. Notice, also, that for some q values, time complexity does not grow monotonically. When we generate queries in order to learn the CPT for v_i, those queries may be relevant to other nodes v_j in the *Unconfident* set. It may be that, when we come to v_j, we already have q many relevant comparisons.

6 Summary and Future Research

We have presented an algorithm for learning CP-nets from queries that is efficient and is guaranteed to produce output. Our tests show that the output CP-nets are close approximations to the underlying CP-nets used to generate answers to queries, particularly if we have a close match between the parameters δ and p.

The next step in our research plan is to extend our algorithm to handle noisy and possibly inconsistent responses to queries.

Acknowledgments. We would like to thank the anonymous referees for their valuable comments. This material is based upon work supported by the National Science Foundation under Grants No. CCF-1215985 and CCF-1049360. The content reflects the views of the authors and not necessarily those of NSF.

References

1. Boutilier, C., Brafman, R.I., Hoos, H.H., Poole, D.: Reasoning with conditional ceteris paribus preference statements. In: UAI 1999, pp. 71–80 (1999)
2. Lang, J., Mengin, J.: Learning preference relations over combinatorial domains. In: NMR 2008 (2008)
3. Lang, J., Mengin, J.: The complexity of learning separable ceteris paribus preferences. In: IJCAI 2009, pp. 848–853. Morgan Kaufmann, San Francisco (2009)
4. Eckhardt, A., Vojtáš, P.: How to learn fuzzy user preferences with variable objectives. In: IFSA/EUSFLAT, pp. 938–943 (2009)
5. Eckhardt, A., Vojtáš, P.: Learning user preferences for 2CP-regression for a recommender system. In: van Leeuwen, J., Muscholl, A., Peleg, D., Pokorný, J., Rumpe, B. (eds.) SOFSEM 2010. LNCS, vol. 5901, pp. 346–357. Springer, Heidelberg (2010)
6. Koriche, F., Zanuttini, B.: Learning conditional preference networks with queries. In: IJCAI 2009, pp. 1930–1935 (2009)
7. Koriche, F., Zanuttini, B.: Learning conditional preference networks. Artificial Intelligence 174, 685–703 (2010)
8. Dimopoulos, Y., Michael, L., Athienitou, F.: Ceteris paribus preference elicitation with predictive guarantees. In: IJCAI 2009, pp. 1890–1895. Morgan Kaufmann, San Francisco (2009)
9. Liu, J., Xiong, Y., Wu, C., Yao, Z., Liu, W.: Learning conditional preference networks from inconsistent examples. TKDE PP (99), 1 (2012)
10. Knuth, D.E.: The Art of Computer Programming, vol. 4A. Addison–Wesley (1997)
11. Goldsmith, J., Lang, J., Truszczyński, M., Wilson, N.: The computational complexity of dominance and consistency in CP-nets. JAIR 33, 403–432 (2008)
12. Guerin, J.T.: Graphical Models for Decision Support in Academic Advising. PhD thesis, University of Kentucky (2012)

A Stochastic Simulation of the Decision to Retweet

Ronald Hochreiter and Christoph Waldhauser

WU Vienna University of Economics and Business
Department for Finance, Accounting and Statistics
{ronald.hochreiter,christoph.waldhauser}@wu.ac.at

Abstract. Twitter is a popular microblogging platform that sees a vast increase in use as a marketing communication tool. For any marketing campaign to be successful, word-of-mouth is an essential component. The equivalent of word-of-mouth propagation in Twitter is the retweeting of a message. So far, little focus has been put on how Twitter users arrive at deciding which tweets to retweet and which ones to ignore. This contribution offers a stochastic decision function that models a nodes decision process. This model is embedded in a simulation of an entire communication network. The contained nodes characterizations are derived from genuine Twitter data. A genetic algorithm is used to find a message that is retweeted by a maximum number of nodes. We find that the stochastic nature of the retweeting decision contributes to a large amount of uncertainty. However, the genetic algorithm is able to increase the scale on which a message is being retweeted significantly.

Keywords: Twitter, social network, message style, genetic algorithm, deterministic optimization.

1 Introduction

Twitter is a popular microblogging platform, that has been frequently at the focal point of research. Of special interest has been the complex network structure that characterizes Twitter networks and the specifics that govern the propagation of information within Twitter networks. But how can Twitter users style their messages, so that they reach furthest? The answer to this question can be put to good use in marketing and campaigning but also life-saving after disasters.

In this paper we aim at making use of that research by building a simulation framework to enable researchers to investigate more closely the propagation of information on Twitter. The centerpiece of the simulation is a stochastic decision function that each node uses to determine if a tweet shall be retweeted or not. The simulation framework is being put to the test by tasking a genetic algorithm with composing a tweet that reaches furthest in different metrics. In that, we differ from [7] seminal contribution by optimizing message contents instead of optimizing target audience. The latter approach in only of limited use in the online scenario, as Twitter authors cannot influence who will follow them.

P. Perny, M. Pirlot, and A. Tsoukiàs (Eds.): ADT 2013, LNAI 8176, pp. 221–229, 2013.

This paper is structured as follows. First relevant research regarding Twitter's networking structure and information diffusion is being reviewed. We then introduce the simulation framework and describe the algorithm that was used to obtain optimal tweets. Finally we present the results and offer some conclusions.

2   Message Diffusion in Twitter Networks

When communicating an actor seeks to get her message across [14]. A central aspect of this process is to ensure that a message is not only received by the original audience, but also that this audience spreads that message further on their own accounts [10]. This process has been researched rather thoroughly from very different aspects: medical epidemiology [5,17,13] and system dynamics [4] to name but a few approaches fielded to tackle this complex problem. While findings and insights differ, a common denominator is that message recipients will resend a message if it passes a recipient's filter, i.e. is to her liking [12]. These filters are domain specific but the common principle of message diffusion remains true for very diverse domains.

The advent of microblogging has greatly simplified access to message diffusion data. By looking at e.g. Twitter data, connection structure as well as message contents and meta data are readily available in a machine readable format. This has produced a wealth of studies relating to message diffusion on Twitter. In the following, we will survey recent contributions to the field to distill key factors that influence message diffusion on Twitter.

In Twitter, users post short messages that are publicly viewable online and that get pushed to other users following the original author. It is common practice to cite (retweet) messages of other users and thus spread them within ones own part of the network. Messages can contain up to 140 characters including free text, URL hyperlinks and marked identifiers (hashtags) that show that a tweet relates to a certain topic. Metadata associated with each tweet is the point of origin, i.e. the user that posted the tweet, the time it was posted and the user agent or interface used to post it. On top of that, the tweets relation to other tweets is available. For each user, additional meta data is available like the age of the account, the number of followers, a description and the location.

Twitter networks are typical for the networks of human communication. They are more complex (i.e. structured and scale-free) than randomly linked networks with certain users functioning as hubs with many more connections than would be expected under uniform or normal distributions. It is useful to think of Twitter networks as directed graphs with nodes being Twitter users and the following of a user being mapped to the edges [8]. A tweet then travels from the original author to all directly connected nodes. If one of the nodes chooses to retweet the message, it is propagated further down the network.

For average users, Twitter networks' degree distribution follows a power law and [8,6] report the distribution's exponent to be 2.3 and 2.4 respectively, therefore well within the range of typical human communication networks. However, there are extremely popular Twitter authors (celebrities, mass media sites) that have many more followers than would be expected even under a power-law distribution.

A distinguishing feature of Twitter is its small-world property. Most users are connected to any other user using only a small number of nodes in between. See [8] for an overview of Twitter's small world properties. Despite their findings that for popular users, the power-law distribution is being violated and average path lengths between users being shorter than expected, they underscore that homophily (i.e. similar users are more likely to be in contact) can be indeed observed and that geographically close users are more likely to be connected.

Following the notion of message filtering introduced above, it is clear that Twitter users are selecting messages for propagating them further according to specific preferences. Applying these preferences for filtering purposes, they can make use of the message contents available as listed above. Besides the number of URLs [18,16] and hashtags [16] contained, also the free text contents are of importance. According to [15,2], a key aspect in filtering free text is the polarity and the emotionality of the message. [15] also point to the length of tweet being an important predictor for its retweetability.

Beside message specific filtering criteria, also author specific filtering can occur. For instance, a Twitter user that has a past record of being retweeted often, will be more likely to be retweeted in the future [18,19]. However, when styling a single tweet for maximum retweetability, factors like past popularity or even number of followers [16] cannot be influenced and are therefore not represented in the model used.

When modeling retweet decisions, [11] focus on three elements: tweet, user and relationship features. However, they do not directly model the decision function, but rather consider it an unobservable effect hidden within a Markov network. [9] follow a different trajectory, when modeling retweeting decisions as a function of temporal, topographical or thematic proximity between nodes and message properties. Unlike our own work, they ignore message style.

Shifting the focus from the message recipient to the message sender, spreading a message as far as possible is a key goal. The success of a message can be measured using different metrics. In their seminal work, [18] list three possibilities: One is the (average) speed a tweet achieves in traversing a network. Another popularity metric is the scale, that is total number of retweets achieved. Finally, range can be considered a popularity metric as well. Here range is the number of edges it takes to travel from the original author to the furthest retweeter.

In this section we reviewed the latest research related to message diffusion on Twitter. Key factors influencing the probability of a tweet being retweeted are the polarity and emotionality of a tweet, its number of included hyperlinks and hashtags. There are other factors influencing retweet probability, however they are beyond the control of a message sender and therefore do not apply to the problem at hand. In the next section we will introduce a simulation framework that can be used to establish a Twitter-like network to analyze the diffusion principles of messages governing them.

3 Simulation Framework

This paper uses the concept of message filtering to simulate the diffusion of messages in networks and Twitter serves as an example for this. As detailed above, Twitter users are considered nodes, their following relationships edges in the network graph. Messages they send travel from node to node along the graph's edges. The topographical features of this network, i.e. the distribution of edges, follow the specifics of scale-free, small-world networks as described above. The nodes have individual preferences that govern if a message is being passed on or ignored. In the following we will describe the simulator used to simulate this kind of network.

Twitter networks exhibit a number of characteristics that we discussed above. The simulator uses these properties to generate an artificial network that very similar to Twitter networks. To this end, the number of connections a node has is drawn from a power-law distribution. In accordance with the findings reported above, the distribution's exponent is fixed 2.4. From these figures, an adjacency matrix is constructed. As Twitter's following relations are not required to be reciprocal, the resulting graph is directed. Since Twitter contains many isolated nodes, the resulting graph based on a Twitter-like power-law distribution also contains a number of isolated nodes. However, these nodes are irrelevant for the problem at hand, and are thus removed.

Every node is then initialized with a set of random message passing preferences. These distributions and their parameters setting these message preferences have been established using 2,500 harvested tweets from late 2012 and spring 2013. The dimensions, and the parameters describing the distributions are given in Table 1.

Table 1. Message and node preferences and their distributions

Parameter	Mean	SD
Polarity	9.82	14.8
Emotionality	2.23	0.48
Length	112.07	30.27
# URLs	0.54	0.52
# Hashtags	1.94	2.04

When a message is sent out from the original authoring node, it is passed on initially to all first-degree followers of that node. Each follower is then evaluated, if she will pass on the message or not. This process is repeated until all nodes that have received the message have been evaluated.

A node's decision on passing the message or not is based on the preferences of that node. However, the decision is not deterministic. Rather, the probability of a message being passed on is binomially distributed and being influenced by the mean absolute difference between the node's preferences (n) and and the message's properties (m):

$$\mathbb{P}(pass) = p = \frac{\overline{|\overrightarrow{m} - \overrightarrow{n}|}}{\Delta_{max}} \tag{1}$$

This stochastic decision model is based on the suggestions made by [3] in their seminal work. While utility of retweeting a specific message remains constant for any given node, the outcome is altered by the inclusion of random error. Normative interpretation of choice theory would conclude that errors have been made on the side of the nodes, should they chose not to retweet a message that is to their liking. We, however, go with [3]'s fourth possible interpretation: including the modeling of chance in the decision process.

The simulation framework described above was used to generate an artificial Twitter-like network for use in this simulation study. To focus on the principles of message propagation, only a small network with initially 250 nodes was generated. After removing isolated nodes, 234 nodes with at least 1 connection remained. The average path length of that network was 6.52. The maximum of first degree connections was observed to be at 46 nodes. This is much larger than median and mean observed to be at 2 and 3.1, respectively.

In this section we described how an artificial Twitter-like network was built using a power-law distribution. This network was paired with node preferences with respect to the passing on of messages. Using a stochastic decision function, each node uses its own preferences and a message's properties to decide on whether to pass it on or not. In the following we will describe a genetic algorithm that was used to craft a message that will reach a maximum number of nodes within that network.

4   Algorithm

In the simulated network, nodes pass on any message they encounter according to the message properties and their own preferences regarding these properties. If a sender now wants to maximize the effect a message has, i.e. to maximize the retweets a tweet will experience, she has to write a message that meets the expectations of the right, i.e. highly connected nodes. While topical choices are obviously important as well, also the right choices regarding message style influence the probability of a message being retweeted. In this section we present a genetic algorithm that styles messages so that a maximum number of nodes retweet it.

The algorithm's chromosome are the message properties as described in Table 1. An initial population of size 50 was initialized with random chromosomes. Using the standard genetic operators of mutations and crossover, the algorithm was tasked to maximize the number of nodes that received the message. In the terms introduced above, this relates to the scale of a message spreading.

To ensure that successful solutions are carried over from one generation to the next, the top 3 solutions were cloned into the next generation. This approach of elitism was shown by [1] to positively impact a genetic algorithm's runtime behavior. Ten percent of every generation was reseeded with random values to ensure enough fresh material in the gene pool. The remaining 85 percent of a generation was created by crossing over the chromosomes of two solutions. To identify solutions eligible for reproduction, tournament selection using a tournament size of 5 was implemented. Children's genes were mutated at random. The probability of a child being mutated was set to be at 0.05.

In this section we described a genetic algorithm that can maximize the retweetability of a tweet. Using state of the art genetic operators and selection mechanisms, a message is being styled so that it will reach a maximum number of nodes. In the following we describe the success the algorithm had in fulfilling its task using sender nodes with a high, medium and low number of first-degree connections.

5   Results

The genetic algorithm as described above was used to find optimal message composition with respect to retweetability for three different sender nodes. The sender nodes differed in the number of first degree connections they had. The genetic algorithm described above was allowed to search for an optimum for 750 generations. Each optimization run was replicated 50 times with random start values. The reported result are averages and standard errors across those 50 replications.

To evaluate the algorithm's performance, two factors are key: the number of generations it takes to arrive at markedly more successful messages and the stability of the discovered solutions. While the former is important to gauge the algorithm's runtime behavior and suitability for real-world deployment, the latter can reveal insights on how easy findings can be generalized across different networks. In the following, these the results relating to these two factors across all three node types are being described.

For highly connected nodes, the optimization quickly contributes to vastly more successful messages. For nodes with fewer connections, there is also an optimization effect, albeit one that takes much longer to develop. Given enough generations, even nodes with a low degree of connectedness will be able to produce tweets that propagate further. Figure 1 depicts the clearly visible trend. Note that the fitness development over time is not monotone as would be expected from genetic algorithms with elitism. This is rooted in the fact, that each node decides stochastically if it retweets a message or not. So a successful message in one generation may be very unsuccessful in the next generation.

Turning the attention towards stability, the last generation's best solution should be similar across all 50 replications. Table 2 gives the means and their standard errors for all three kinds of nodes.

We will discuss these results in the next section and offer some concluding remarks.

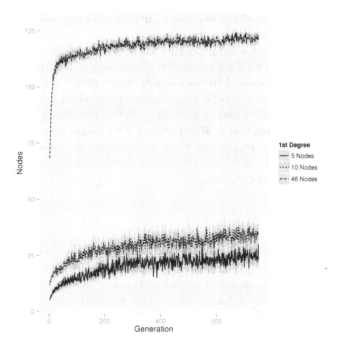

Fig. 1. Mean fitness as improving over generations for three different kinds of sender nodes. Shaded area is a 95% confidence interval derived from replicating the optimization 50 times.

Table 2. Solution stability. Mean and Standard Error (in parentheses).

Parameter	5 nodes	10 nodes	46 nodes
Polarity	0.82 (0.16)	0.64 (0.29)	0.83 (0.13)
Emotionality	0.87 (0.13)	0.87 (0.12)	0.91 (0.07)
Length	110.2 (22.3)	101.2 (13.89)	112.28 (14.96)
# URLs	0.46 (0.58)	1.12 (0.63)	0.36 (0.53)
# Hashtags	0.68 (0.68)	1.98 (1.26)	1.36 (0.69)

6   Discussion

The evaluation results provided in the previous section exhibit a number of peculiarities. Most striking is perhaps, that the genetic algorithm can much quicker improve message styles of highly connected than for lesser connected nodes. This phenomenon is rooted directly in the stochastic decision function used in the retweeting model. For a highly connected node, the probability of encountering nodes that retweet a message is higher, as there are more connections. The algorithm, therefore receives better feedback on the quality of a solution. With a lesser connected node, a promising solution might get discarded too quickly, if it is unfortunate enough to be rejected by too large a proportion of connected nodes. As a result, the algorithm requires a large number of generations to arrive at noticeable improvements.

It is also obvious, that for higher connected nodes the effect of uncertainty is much less pronounced. This becomes visible in the relative lack of fluctuations of fitness between generations and the variance of fitness within a single generation.

In many applications of genetic algorithms, the stability of identified optimal solutions across replications is a decisive factor. For the problem at hand, stability is of lesser importance. When styling a message, apparently different methods lead to nearly equal performance of the message. However, especially the factor of emotionality appears to be very stable across replications and even irrespective of node connectedness.

7   Conclusion

In this paper we introduced a genetic algorithm to optimize the retweetability of tweets. To do this, we simulated a Twitter-like network and associated each node with a set of preferences regarding message retweeting behavior. Any node's decision is stochastic, based on message properties coming close to the node's own preferences. The genetic algorithm succeeded in styling messages so that they became retweeted more widely. Dependent on the number first-degree connections of the sender node, the fitness of the algorithm's terminal solution and the speed of optimization varied.

This contribution is but a first step in an endeavor to understand the precise mechanics of message propagation on Twitter. Previous work was focused on sender node properties. By taking message properties into account when assessing retweetability, we not only ventured into uncharted territory, we also discovered new insights regarding the feasibility of message optimization.

References

1. Bhandari, D., Murthy, C.A., Pal, S.K.: Genetic algorithm with elitist model and its convergence. International Journal of Pattern Recognition and Artificial Intelligence 10(6), 731–747 (1996)
2. Cha, M., Haddadi, H., Benevenuto, F., Gummadi, P.K.: Measuring user influence in Twitter: The million follower fallacy. In: Proceedings of the Fourth International Conference on Weblogs and Social Media, ICWSM 2010. The AAAI Press (2010)
3. Davidson, D., Marschak, J.: Experimental tests of a stochastic decision theory, pp. 233–269. John Wiley and Sons (1959)
4. Goldenberg, J., Libai, B., Muller, E.: Talk of the network: A complex systems look at the underlying process of word-of-mouth. Marketing Letters 12(3), 211–223 (2001)
5. Gruhl, D., Guha, R., Liben-Nowell, D., Tomkins, A.: Information diffusion through blogspace. In: Proceedings of the 13th International Conference on World Wide Web, WWW 2004, pp. 491–501. ACM (2004)

6. Java, A., Song, X., Finin, T., Tseng, B.L.: Why we twitter: An analysis of a microblogging community. In: Zhang, H., Spiliopoulou, M., Mobasher, B., Giles, C.L., McCallum, A., Nasraoui, O., Srivastava, J., Yen, J. (eds.) WebKDD 2007. LNCS, vol. 5439, pp. 118–138. Springer, Heidelberg (2009)

7. Kempe, D., Kleinberg, J.M., Tardos, É.: Maximizing the spread of influence through a social network. In: Proceedings of the Ninth ACM SIGKDD International Conference on Knowledge Discovery and Data Mining, KDD 2003, pp. 137–146. ACM (2003)

8. Kwak, H., Lee, C., Park, H., Moon, S.B.: What is Twitter, a social network or a news media? In. In: Proceedings of the 19th International Conference on World Wide Web, WWW 2010, pp. 591–600. ACM (2010)

9. Macskassy, S.A., Michelson, M.: Why do people retweet? anti-homophily wins the day! In. In: Proceedings of the Fifth International Conference on Weblogs and Social Media, 2011. The AAAI Press (2011)

10. McNair, B.: An Introduction to Political Communication. Routledge (2011)

11. Peng, H.-K., Zhu, J., Piao, D., Yan, R., Zhang, Y.: Retweet modeling using conditional random fields. In: IEEE 11th International Conference on Data Mining Workshops (ICDMW 2011), pp. 336–343. IEEE (2011)

12. Rogers, E.M.: Diffusion of Innovations. Free Press (2010)

13. Salathé, M., Bengtsson, L., Bodnar, T.J., Brewer, D.D., Brownstein, J.S., Buckee, C., Campbell, E.M., Cattuto, C., Khandelwal, S., Mabry, P.L., et al.: Digital epidemiology. PLoS Computational Biology 8(7), e1002616 (2012)

14. Shannon, C., Weaver, W.: The Mathematical Theory of Communication. University of Illinois Press (2002)

15. Stieglitz, S., Dang-Xuan, L.: Political communication and influence through microblogging-an empirical analysis of sentiment in Twitter messages and retweet behavior. In: 45th Hawaii International International Conference on Systems Science (HICSS-45 2012), pp. 3500–3509. IEEE Computer Society (2012)

16. Suh, B., Hong, L., Pirolli, P., Chi, E.H.: Want to be retweeted? Large scale analytics on factors impacting retweet in Twitter network. In: Proceedings of the 2010 IEEE Second International Conference on Social Computing, SocialCom / IEEE International Conference on Privacy, Security, Risk and Trust, PASSAT 2010, pp. 177–184. IEEE Computer Society (2010)

17. Venkatachalam, S., Mikler, A.: Towards computational epidemiology: Using stochastic cellular automata in modeling spread of diseases. In: Proc. of the 4th Annual International Conference on Statistics (2005)

18. Yang, J., Counts, S.: Predicting the speed, scale, and range of information diffusion in Twitter. In: Proceedings of the Fourth International Conference on Weblogs and Social Media, ICWSM 2010. The AAAI Press (2010)

19. Zaman, T.R., Herbrich, R., Van Gael, J., Stern, D.: Predicting information spreading in Twitter. In: Workshop on Computational Social Science and the Wisdom of Crowds, NIPS 2010 (2010)

Judgment Aggregation Rules and Voting Rules

Jérôme Lang[1] and Marija Slavkovik[2]

[1] Université Paris-Dauphine, France
lang@lamsade.dauphine.fr
[2] University of Liverpool, UK, and University of Bergen, Norway
marija.slavkovik@infomedia.uib.no

Abstract. Several recent articles have defined and studied judgment aggregation rules based on some minimization principle. Although some of them are defined by analogy with some voting rules, the exact connection between these rules and voting rules is not always obvious. We explore these connections and show how several well-known voting rules such as the top cycle, Copeland, maximin, Slater or ranked pairs, are recovered as specific cases of judgment aggregation rules.

1 Introduction

Judgment aggregation studies the problem of finding collective judgments that represent a collection of individual judgments on several logically interrelated issues. Originating from law and studied in social choice theory, it has now become clear that judgment aggregation also relates to various fields of knowledge representation, such as belief merging or nonmonotonic reasoning.

The literature on judgment aggregation has, until recently, focused much more on impossibility or possibility theorems than on the study of specific rules, which departs from the (admittedly much older) field of voting theory. However, several recent, independent papers have started to explore the zoo of interesting, concrete judgment aggregation rules, in particular [MO09, EGP12, NPP11, LPSvdT11, DP12, Die12].

Some of these rules were obviously defined by analogy with a well-known voting rule; for instance, the so-called Young rule in [LPSvdT11], that looks for a minimum number of agents to remove so that the resulting profile becomes majority-consistent, is the obvious counterpart of the Young voting rule. For a few others, the analogy remains clear, but the formal connection is less trivial to establish; as an example of such result, [EGP12] show that the distance-based procedure proposed in [MO09] (and close to the distance-based majoritarian merging operator proposed in [KPP02]), corresponds in some sense to the Kemeny rule. For a few other rules, the analogy itself is not obvious.

The formal connection between judgment aggregation rules and voting rules makes use of the *preference agenda* [DL07]: given a set of alternatives C, this agenda is composed of propositions of the form "x is preferred to y", where x and y are alternatives in C; a profile corresponds to a set of individual judgments, whose consistency condition corresponds to the transitivity of the individual votes. A nontrivial question is whether the collective judgment set should be consistent with the transitivity constraint, or only with the constraint expressing the existence of an undominated alternative.

P. Perny, M. Pirlot, and A. Tsoukiàs (Eds.): ADT 2013, LNAI 8176, pp. 230–243, 2013.
© Springer-Verlag Berlin Heidelberg 2013

Section 2 introduces the judgment aggregation framework we are using. Section 3 gives some background on judgment aggregation rules, while Section 4 gives some background on voting rules. Section 5 addresses the question of relations between voting and judgment aggregation rules in full detail: we define a formal way of mapping a judgment aggregation rule into *two* voting rules, obtained by requiring the collective judgment to be consistent with one constraint or the other.

It is rather intriguing to see which pairs of well-known voting rules correspond to the same judgment aggregation rule. For instance, as we show, the Copeland rule comes together with the Slater rule, whereas the maximin rule comes together with the "ranked pairs" rule. Section 6 discusses some implications of our results as well as further research issues.

2 Judgment Aggregation: General Definitions

Let $\mathcal{L}$ be a set of well-formed propositional logical formulas, including $\top$ (tautology) and $\bot$ (contradiction). An *issue* is a pair of formulas φ, $\neg\varphi$ where $\varphi \in \mathcal{L}$ and φ is neither a tautology nor a contradiction. An *agenda* $\mathcal{A}$ is built up from a finite set of issues, and has the form $\mathcal{A} = \{\varphi_1, \neg\varphi_1, \ldots, \varphi_m, \neg\varphi_m\}$. The *preagenda* $[\mathcal{A}]$ associated with $\mathcal{A}$ is $[\mathcal{A}] = \{\varphi_1, \ldots, \varphi_m\}$. A *judgment* on $\varphi \in [\mathcal{A}]$ is one of φ or $\neg\varphi$. A *judgment set* J is a subset of $\mathcal{A}$. It is *complete* iff for each $\varphi \in [\mathcal{A}]$, either $\varphi \in J$ or $\neg\varphi \in J$.

Constraints can be specified to explicitly represent logical dependencies enforced on agenda issues. Since we have a finite $\mathcal{L}$, without loss of generality we can assume that the constraints consist of *one* propositional formula (typically the conjunction of several simpler constraints). The constraint associated to an agenda $\mathcal{A}$ is thus a consistent formula $\Gamma \in \mathcal{L}$. When not otherwise specified, Γ is the tautology $\top$. Involving constraints in judgment aggregation has already been considered in a few places, such as [DL08, GE13].

A judgment set J (and more generally, a set of propositional formulas) is Γ-*consistent* if and only if $J \cup \{\Gamma\} \nvDash \bot$. Let $\mathcal{D}(\mathcal{A}, \Gamma)$ be the set of all Γ-consistent judgment sets (for agenda $\mathcal{A}$) and $\mathbb{D}(\mathcal{A}, \Gamma) \subset \mathcal{D}(\mathcal{A}, \Gamma)$ be the set of all judgment sets that are also *complete*. We omit specifying $\mathcal{A}$ and Γ when they are clear from the context.

A *profile* $P = \langle J_1, \ldots, J_n \rangle \in \mathbb{D}^n(\mathcal{A}, \Gamma)$ is a collection of complete, Γ-consistent individual judgment sets. Given $I \subseteq \{1, \ldots, n\}$, the sub-profile P_I is the collection $P_I = \langle J_i \mid i \in I \rangle$. In the whole paper (except at one place), we assume we have *an odd number n of voters*.

A *sub-agenda* is a subset of issues from $\mathcal{A}$, that is, a subset of $\mathcal{A}$ of the form $\{\varphi_j, \neg\varphi_j \mid j \in J\}$. A *sub-preagenda* is a subset of $[\mathcal{A}]$. Given a sub-agenda Y, the projection of J on Y is $J^{\downarrow Y} = J \cap Y$. Given a profile $P = \langle J_1, \ldots, J_n \rangle$, the projection of P on Y is $P^{\downarrow Y} = \langle J_1^{\downarrow Y}, \ldots, J_n^{\downarrow Y} \rangle$. An example is given in Figure 1. For $\varphi \in \mathcal{A}$, the set of agents in P with judgment sets that contain φ is $N(P, \varphi) = \#\{i \mid \varphi \in J_i\}$.

An *irresolute judgment aggregation rule*, for n voters, is a function $F_\Gamma : \mathbb{D}^n \to 2^{\mathcal{D}} \setminus \{\emptyset\}$, *i.e.*, F_Γ maps a profile of complete judgment sets to a nonempty set of judgment sets. When Γ is omitted, *i.e.*, when we note F instead of F_Γ, we assume that F is defined for any possible constraint Γ (F then defines a family of judgment

$$P = \begin{pmatrix} P_I \begin{array}{|c|c c|} \{p & q & r\} \\ \{p & q & r\} \\ \{p & \neg q & \neg r\} \\ \{p & \neg q & \neg r\} \\ \hline \{\neg p & q & \neg r\} \\ \{\neg p & q & \neg r\} \\ \{\neg p & q & \neg r\} \end{array} \end{pmatrix} \begin{array}{c} 1 \\ 2 \\ 3 \\ 4 \\ 5 \\ 6 \\ 7 \end{array}$$

$$P^{\downarrow Y}$$

Fig. 1. A profile P for 7 agents for $[\mathcal{A}] = \{p, q, r\}$ and $\Gamma = (p \wedge q) \leftrightarrow r$. The grey shaded area depicts sub-profile P_I for $I = \{1, 2, 3, 4\}$, while the dotted lined area corresponds to the projection $P^{\downarrow Y}$ for $Y = \{q, \neg q, r, \neg r\}$. We have, for instance, $N(P, q) = 5$.

aggregation rules – one for each Γ – but by a slight abuse of language we will still call F a judgment aggregation rule).[1]

The majoritarian judgment set associated with profile P contains all elements of the agenda that are supported by a majority of judgment sets in P, i.e.,

$$m(P) = \{\varphi \in \mathcal{A} \mid N(P, \varphi) > \frac{n}{2}\}.$$

A profile P is *(Γ)-majority-consistent* iff $m(P)$ is Γ-consistent. A judgment aggregation rule F_Γ is *majority-preserving* iff, for every Γ-majority-consistent profile $P \in \mathbb{D}^n$, $F(P) = \{m(P)\}$.

Given a set of formulas Σ, $S \subseteq \Sigma$ is a maximal Γ-consistent subset of Σ if S is Γ-consistent and no S' such that $S \subset S' \subseteq \Sigma$ is Γ-consistent; and $S \subseteq \Sigma$ is a maxcard (for "maximal cardinality") Γ-consistent subset of Σ if S is Γ-consistent and no $S' \subseteq \Sigma$ such that $|S| < |S'|$ is Γ-consistent. $MaxCons(m(P), \Gamma)$ denotes the set of all maximal Γ-consistent subsets of $m(P)$. $MaxCardCons(m(P), \Gamma))$ denotes the maxcard set of Γ-consistent subsets of $m(P)$.

3   Judgment Aggregation Rules

We recall four minimization-based judgment aggregation rules. We reuse the names from [LPSvdT11] and indicate when a rule has appeared elsewhere with a different name. Let $P = \langle J_1, \ldots, J_n \rangle$ from $\mathbb{D}(\mathcal{A}, \Gamma)^n$.

Definition 1 (Maximal and maxcard sub-agenda rules). *The maximal sub-agenda (MSA) and the maxcard sub-agenda (MCSA) rules are defined as follows:*

$$MSA_\Gamma(P) = MaxCons(m(P), \Gamma), \tag{1}$$

$$MCSA_\Gamma(P) = MaxCardCons(m(P), \Gamma). \tag{2}$$

---

[1] We could have opted for the more complex notation $F_{n, \mathcal{A}, \Gamma}$. However, omitting n and $\mathcal{A}$ will not lead to any ambiguity.

The MSA rule is called "Condorcet admissible set", and the $MCSA$ "Slater rule", in [NPP11].

Definition 2 (Ranked agenda). *Let $\succsim_P$ be the weak order on $\mathcal{A}$ defined by: for all $\psi, \psi' \in \mathcal{A}$, $\psi \succsim_P \psi'$ iff $N(P, \psi) \geq N(P, \psi')$. For $\mathcal{A} = \{\psi_1, \ldots, \psi_{2m}\}$ and a permutation σ of $\{1, \ldots, 2m\}$, let $>_\sigma$ be the linear order on $\mathcal{A}$ defined by $\psi_{\sigma(1)} > \ldots > \psi_{\sigma(2m)}$. We say that $>_\sigma$ is compatible with $\succsim_P$ if $\psi_{\sigma(1)} \succsim_P \ldots \succsim_P \psi_{\sigma(2m)}$. The ranked agenda rule RA_Γ is defined as $J \in RA_\Gamma(P)$ iff there exists a permutation σ such that $>_\sigma$ is compatible with $\succsim_P$ and such that $J = J_\sigma$ is obtained by the following procedure:*

$$
\begin{array}{ll}
S := \emptyset; & 1 \\
\text{for } j = 1, \ldots, 2m \ \text{do} & 2 \\
\quad \text{if } S \cup \{\psi_{\sigma(j)}\} \text{ is } \Gamma\text{-consistent, then } S := S \cup \{\psi_{\sigma(j)}\} & 3 \\
\text{end for;} & 4 \\
J_\sigma := S. & 5
\end{array}
$$

The RA rule is called by the name "leximin rule" by [NPP11].

The next rule is defined as the distance-based rule in [EGP12], "maxweight sub agenda" rule in [LPSvdT11], "Prototype" in [MO09], "median rule" in [NPP11], and "simple scoring rule" in [Die12] has received much more attention that the others. Its relationship to the Kemeny rule is considered in [EGP12] (see also [EM05]).

Definition 3 (Maxweight sub-agenda rule). *Let $J \in \mathbb{D}(\mathcal{A}, \Gamma)$. The maxweight sub-agenda rule MWA is defined as* [2]

$$
MWA(P) = \arg \max_{J \in \mathbb{D}(\mathcal{A}, \Gamma)} W_P(J) \quad \text{where} \quad W_P(J) = \sum_{\varphi \in J} N(P, \varphi).
$$

Definition 4 (Young rule). *Let $MSP(P)$ be the set of all maxcard Γ-majority-consistent sub-profiles $P_I \in \mathbb{D}^{|I|}(\mathcal{A}, \Gamma)$ of $P \in \mathbb{D}^n(\mathcal{A}, \Gamma)$, namely,*

$$
MSP(P) = \{P_I \mid \text{there is no } I' \text{ such that } |I| < |I'| \text{ and } m(P_{I'}) \in \mathbb{D}(\mathcal{A}, \Gamma)\}.
$$

The Young judgment aggregation rule is defined as

$$
Y_\Gamma(P) = \{m(P_I) \mid P_I \in MSP(P)\}.
$$

4 Voting Rules

Let $C = \{x_1, \ldots, x_q\}$ be a set of alternatives. An *n-voter profile* over C (recall that n is assumed to be odd) is a collection $V = \langle \succ_1, \ldots, \succ_n \rangle$ of linear orders on C, called

---

[2] Alternatively the rule MWA can be defined as a (Hamming) distance based rule

$$
R^{d_H, \Sigma}(P) = \arg \min_{J \in \mathbb{D}(\mathcal{A}, \Gamma)} \sum_{i=1}^n d_H(J_i, J).
$$

The equivalence between these two definitions was shown in [LPSvdT11].

votes. An *irresolute voting rule* (or *voting correspondence*) is a function R mapping every profile V into a nonemptyset of alternatives $R(V) \in 2^C \setminus \{\emptyset\}$. For every pair of alternatives $(x, y) \in C$ and profile V, let $n_V(x, y)$ be the number of votes in V ranking x above y, and let $M(V)$ be the majority graph associated with V, whose vertices are C and containing edge (x, y) iff $n_V(x, y) > \frac{n}{2}$. The alternative $x \in C$ is a *Condorcet winner* for V if there is an outgoing edge in $M(V)$ from x to every $y \neq x$.

We now define several (irresolute) voting rules.

The *Top-cycle* (TC) rule maps every profile V to the set of alternatives $x \subseteq C$ such that for all $y \in C \setminus x$, there exists a path in $M(V)$ that goes from x to y. Equivalently, $TC(P)$ is the smallest set S such that for every $x \in S$ and $y \in C \setminus S$, we have $(x, y) \in M(V)$.

A *Slater order* for V is a linear order $\succ$ over C maximizing the number of (x, y) s.t. $x \succ y$ iff $(x, y) \in M(V)$. The *Slater* rule maps a profile V to the set of all alternatives that are dominating in some Slater order for $M(V)$.

The *Copeland* rule maps V to the set of alternatives maximizing the number $n_c(x)$ of outgoing edges from x in $M(V)$.

The *ranked pairs* rule [Tid87] is defined as follows. We define first its non-neutral version: given a tie-breaking priority, that is, a linear order ρ over $\{(x, y) \in C^2, x \neq y\}$, the linear order $>_\rho$ on $\{(x, y) \in C^2, x \neq y\}$ is constructed as follows: $(x, y) >_\rho (x', y')$ iff either (a) $n_V(x, y) > n_V(x', y')$ or (b) if $n_V(x, y) = n_V(x', y')$ and ρ gives priority to (x, y) over (x', y'). Then all pairs (x, y) are considered in sequence according to $>_\rho$, and we build a linear order $\succ_\rho$ over C starting with the pair on top of $>_\rho$, and iteratively adding the current pair to $\succ_\rho$ if it does not make it cyclic. The ranked pairs winner for V according to ρ is the unique undominated element in $\succ_\rho$. Now, x is a winner of the neutral ranked pairs rule for V iff it is a winner of the non-neutral ranked pairs rule for some ρ. (See [BF12] for a recent discussion on neutral and non-neutral variants of ranked pairs.)

The *maximin* rule maps V to the set of alternatives that maximize

$$mm(x, V) = \min_{y \in C \setminus \{x\}} n_V(x, y).$$

Let $S_Y(x, V)$ be the minimal number of votes whose removal from V makes x a Condorcet winner. The *Young* (voting) rule maps V to the set of alternatives that minimize $S_Y(x, V)$.

5 From Judgment Aggregation to Voting Rules

In this Section, we assume that judgment profiles contain an *odd number n* of individual judgments. The reason for this assumption is that the connections to voting rules are much easier to state, and more natural, under this assumption.

A specific type of agenda is the *preference agenda* associated with a set of alternatives C [DL07] whose propositions are of the form xPy ("x preferred to y").

Definition 5. *The preference agenda associated with* $C = \{x_1, \ldots, x_q\}$ *is* $\mathcal{A}_C = \{x_i P x_j \mid 1 \leq i < j \leq q\}$.

When $j > i$, $x_i P x_j$ is not a proposition of $\mathcal{A}_C$, but we will write $x_j P x_i$ as a shorthand for $\neg(x_j P x_i)$.

Definition 6. *Let $V = \langle \succ_1, ..., \succ_n \rangle$ be an n-voter profile over C. With every individual vote $\succ_i$ we associate the individual judgment set*

$$J(\succ_i) = J_i = \{x P y \mid x \succ_i y, \text{ for } x, y \in C\}.$$

The judgment aggregation profile associated with V is $P(V) = \langle J_1, \ldots, J_n \rangle$.
Conversely, given a judgment set J on $\mathcal{A}_C$, the binary relation $\succ_J$ over C is defined by: for all $x_i, x_j \in C$, $x_i \succ_J x_j$ if $x_i P x_j \in J$ and $x_j \succ_J x_i$ if $\neg x_i P x_j \in J$.

Now we define two preference constraints: the transitivity constraint Tr and the dominating alternative, or "winner", constraint W.

Definition 7. *We define the transitivity Tr and dominating alternative W constraints:*

- $Tr = \bigwedge_{i,j,k \in \{1,...,m\}} \left((x_i P x_j) \wedge (x_j P x_k) \rightarrow (x_i P x_k) \right)$
- $W = \bigvee_{i \leq m} \bigwedge_{j \neq i} (x_i P x_j)$

For complete judgment sets, Tr is stronger than W, therefore, any complete Tr-consistent judgment set is also W-consistent.

Lemma 1. *Let J be a judgment set on $\mathcal{A}_C$.*

- *J is Tr-consistent iff $\succ_J$ is acyclic;*
- *J is W-consistent iff $\succ_J$ has at least one undominated element.*

The proof is almost straightforward from Definition 6: J is Tr-consistent if $\succ_J$ can be completed into a transitive order, *i.e.*, iff $\succ_J$ is acyclic; J is W-consistent if some x can be made a winner by adding the missing propositions $x P y$, which is possible iff some x is undominated in $\succ_J$.

As a consequence of Lemma 1, any Tr-consistent judgment is also W-consistent. Note also that $\succ_J$ is a linear order if and only if J is complete and Tr-consistent.

For instance, let $J = \{a P b, a P c, b P c, d P b, c P e, e P b\}$; then

$$\succ_J = \{(a, b), (a, c), (b, c), (d, b), (c, e), (e, b)\}$$

J is not Tr-consistent because $b P c \wedge c P e \wedge Tr \models \neg e P b$ (or equivalently, $\succ_J$ contains the cycle $b \succ_J c \succ_J e \succ_J b$). However, it is W-consistent: a and d are both undominated in $\succ_J$.

For each $x \in C$ we define $W(x) = \bigwedge_{y \in C, y \neq x}(x P y)$. Note that W is equivalent to $\bigvee_{x \in C} W(x)$ and that J is $W(x)$-consistent iff x is undominated in $\succ_J$.

Since each vote $\succ_i$ is a linear order, the individual judgment sets J_i are complete and consistent with Tr (and *a fortiori* with W). The collective judgment will sometimes be required to be consistent with respect to Tr and sometimes only to be consistent with respect to W. Lemma 2 is straightforward from Definition 6.

Lemma 2. *Given a voting profile V, for all $x, y \in C$, $x P y$ is in $m(P(V))$ iff $(x, y) \in M(V)$.*

Proposition 1. *A voting profile V has a Condorcet winner iff $m(P(V))$ is W-consistent.*

Proof. From Lemma 2, xPy is in $m(P(V))$ iff $M(V)$ contains (x, y). Since n is odd, $m(P(V))$ contains either x_iPx_j or x_jPx_i for all $i \neq j$, therefore $m(P(V)) \cup \{W\} \nvDash \bot$ iff there exists $x \in C$ s.t. $m(P(V))$ contains $\{xPy \mid y \neq x\}$, *i.e.,*, by Lemma 2again, iff V has a Condorcet winner.

Note that for an even n, W-consistency would be equivalent to the existence of a *weak* Condorcet winner.

Definition 8. *Let*
$$Win(J) = \{x \mid J \cup W(x) \nvDash \bot\}$$
Let $\Gamma \in \{Tr, W\}$ and F be a judgment aggregation rule. The voting rule $R_{F,\Gamma}$ induced from F and Γ is defined as $x \in R_{F,\Gamma}(P(V))$ if there is a $J \in R_{F,\Gamma}(P(V))$ such that $x \in Win(J)$, or equivalently:
$$R_{F,\Gamma}(P) = \bigcup_{J \in F_\Gamma(P(V))} Win(J).$$

Note that $J \cup \{W\} \nvDash \bot$ or $J \cup \{Tr\} \nvDash \bot$, then $Win(J) \neq \emptyset$, therefore Definition 8 is well-founded.

Thus, for every judgment aggregation rule F we have two voting rules, obtained by requiring the collective judgment set to be acyclic, *i.e.*, consistent with Tr, or to have a undominated element, *i.e.*, consistent with W.

Example 1. Let $V = \langle a \succ_1 b \succ_1 c \succ_1 d, b \succ_2 c \succ_2 a \succ_1 d, d \succ_3 c \succ_3 a \succ_3 b \rangle$. We have $P(V) = \langle J_1, J_2, J_3 \rangle$ with $J_1 = \{aPb, aPc, aPd, bPc, bPd, cPd\}$, $J_2 = \{bPa, bPc, bPd, cPa, cPd, aPd\}$ and $J_3 = \{dPa, dPb, dPc, cPa, cPb, aPb\}$; and we have $m(P(V)) = \{aPb, bPc, cPa, aPd, bPd, cPd\}$.   .
Let us choose $F = MSA$ and $\Gamma = Tr$.
We have $F_{Tr}(P(V)) = \{J, J', J''\}$, where $J = \{aPb, bPc, aPd, bPd, cPd\}$, $J' = \{aPb, cPa, aPd, bPd, cPd\}$ and $J'' = \{bPc, cPa, aPd, bPd, cPd\}$.
Now, $Win(J) = \{a\}$, $Win(J') = \{c\}$ and $Win(J'') = \{b\}$.
Therefore, $R_{MSA,Tr}(P(V)) = Win(J) \cup Win(J') \cup Win(J'') = \{a, b, c\}$.

Proposition 2.

1. $R_{MSA,Tr} = TopCycle$
2. $R_{MSA,W} = \begin{cases} \{c\} & \text{if } V \text{ has a Condorcet winner } c \\ C & \text{otherwise} \end{cases}$

Proof. We prove the first correspondence. From Lemmas 1 and 2, $J \in MaxCons(m(P), Tr)$ iff $\succ_J$ is a maximal acyclic sub-graph of $M(V)$. Let $x \in TC(V)$; then there exists an acyclic subrelation G of $M(V)$ containing, for all $y \neq x$, a path from x to y. G can be completed into a maximal acyclic subrelation G' of $M(V)$, and x is undominated in G' (because adding an edge to any $y \neq x$ would create a cycle), therefore G' corresponds to a maximal Tr-consistent subset J of $m(P(V))$, consistent with $W(x)$, which means that $x \in R_{MSA,Tr}(V)$. Conversely, if there is a $J \in R_{MSA,Tr}(V)$ such that $x \in Win(J)$, then $\succ_J$ is a maximal acyclic subrelation of $M(V)$ in which x does not have any incoming edge. Assume $x \notin TC(V)$; then there is an y such that there is no path from x to y in $M(V)$. Obviously, $(x,y) \notin M(V)$, therefore, since $M(V)$ is complete, $(y,x) \in M(V)$. Adding (y,x) to $\succ_J$ results in an acyclic subrelation of $M(V)$ that contains $\succ_J$, therefore $\succ_J$ is not a maximal acyclic subset of $M(V)$, contradiction.

Now we prove the second correspondence. Assume there is no Condorcet winner. Let $x \in C$. Let $S(x)$ be the subset of $m(P(V))$ defined by $\{yPz | z \neq x, yPz \in m(P(V))\}$. $S(x)$ is W-consistent, because it is consistent with $W(x)$. Assume $S(x)$ is not maximal: then there is some element of $m(P(V)) \setminus S(x)$ that can be added to $S(x)$ without violating W-consistency; now, every element of $m(P(V)) \setminus S(x)$ is of the form yPx. Let $S' = S(x) \cup \{yPx\}$. S' is not consistent with $W(x)$. Therefore, since it is W-consistent, it must be consistent with $W(z)$ for some $z \neq x$. This implies that there is no $tPz \in S'$, therefore, no $tPz \in S(x)$. Now, by construction of $S(x)$, this means that there is no $tPz \in m(P(V))$, which implies that z is a Condorcet winner: contradiction.

Proposition 3.

1. $R_{MCSA,Tr} = Slater$
2. $R_{MCSA,W} = Copeland$

Proof. For point 1, let $J \in MCSA_{Tr}(P(V))$, hence $J \in MaxCardCons(m(P),Tr)$ and $\succ_J$ is an acyclic subrelation of $M(V)$. Let $>$ be a linear order extending $\succ_J$. The number of edge reversals needed to obtain $>$ from $\succ_J$ is $|m(P(V)) \setminus J|$. This number is minimal iff J has a maximal cardinality. Consequently, $>$ is a Slater order for V. Conversely, let $>$ be a Slater order for V and let $J = \{xPy \mid x > y \text{ and } xPy \in m(P(V))\}$. Because $>$ is a linear order, J is Tr-consistent. Moreover, $|m(P(V)) \setminus J|$ is the number of edge reversals needed to obtain $>$ from $M(V)$. Since $|m(P(V)) \setminus J|$ is minimal, $|J|$ is maximal and therefore $J \in MCSA_{Tr}(P(V))$. This one-to-one correspondence between Slater orders for V and maxcard acyclic subgraphs of $P(V)$ allows us to conclude.

For point 2, let $J \in MaxCardCons(m(P),W)$. From $J \cup \{W\} \nvdash \bot$ it follows that there exists a $x \in C$ s.t. for every $y \in C$, $yPx \notin J$. For every $y \in C$, consider $z \in C$, $z \neq x$, such that $yPz \in m(P(V))$. Adding yPz to J results in a judgment set which is still W-consistent, therefore the maximum W-consistent subsets of $m(P(V))$ are of the form $J_x = m(P(V)) \setminus \{yPx, y \neq x\}$ for some $x \in C$, and such a judgment set J_x is a maxcard W-consistent subset of $m(P(V))$ iff $|\{y \mid xPy \in m(P(V))\}|$ is maximal, *i.e.*, using Lemma 2, iff $x \in Copeland(V)$.

Example 2. Let V be such that $M(V) = \{(a,b),(a,c),(b,c),(b,d),(c,d),(d,a)\}$, *i.e.*, $m(P(V)) = \{aPb, aPc, bPc, bPd, cPd, dPa\}$. The only maxcard Tr-consistent subset of $m(P(V))$ is $J = \{aPb, aPc, bPc, bPd, cPd\}$, and $Win(J) = \{a\}$; a is also the only Slater winner for P. Now, $m(P(V))$ has two maxcard W-consistent subsets: J and $J' = \{aPc, bPc, bPd, cPd, dPa\}$; $Win(J') = \{b\}$; a and b are also the Copeland winners for V.

Proposition 4.

1. $R_{RA,Tr} = ranked\ pairs$.
2. $R_{RA,W} = maximin$.

Proof. The proof of point (1) is simple, due to the similarity of the definitions of ranked pairs and RA, and observing that adding xPy to a current Tr-consistent judgment set without violating Tr corresponds to adding (x, y) to a current acyclic graph without creating a cycle. The proof of point (2) is more interesting. The candidate x is a maximin winner if it maximizes $mm(x, V)$, or equivalently, if it minimizes $\max_y n_V(y, x)$. Let $\beta = \min_x \max_y n_V(y, x)$. (Note that we have $\beta \geq \frac{1}{2}$ when there is no Condorcet winner.) Assume that x is a Maximin winner for V. In order to show that $x \in R_{RA,W}(V)$, we have to construct a linear order $\succ = \succ_\sigma$ on $\{xPy \mid (x, y) \in C^2, x \neq y\}$, compatible with $\succsim_P$, such that the judgment set J_σ obtained by following $\succ_\sigma$ is such that $x \in Win(J_\sigma)$. Let $\succ_\sigma$ be as follows:

1. the first propositions of $\succ_\sigma$ are all uPv such that $n_V(u, v) > \beta$, with ties broken in an arbitrary manner;
2. the propositions that follow in $\succ_\sigma$ are all yPz such that $n_V(y, z) = \beta$ and $z \neq x$;
3. the following propositions are all yPx such that $n_V(y, x) = \beta$;
4. the rest of $\succ_\sigma$ does not matter.

We now follow step by step the construction of J_σ. During step (1) – corresponding to considering one by one the proposition in (1) above – we consider all the propositions uPv such that $n_V(u, v) > \beta$, and all are added to S, because the resulting judgment set is consistent with $W(x)$, and *a fortiori* with W (otherwise it would be the case that for all y, $n_V(y, x) > \beta$, contradicting $\min_x \max_y n_V(y, x) = \beta$). During step (2) all propositions yPz such that $n_V(y, z) = \beta$ and $z \neq x$ are considered one by one, and they are all added to S, because the resulting judgment set is, each time, consistent with $W(x)$ and *a fortiori* with W. After steps (1) and (2), due to the fact that $\beta = \min_x \max_y n_V(y, x)$, S contains some yPz for all $z \neq x$. Step (3) considers all yPx such that $n_V(y, x) = \beta$, and does not add them to S, because this would make it inconsistent with W. Finally, the propositions considered in Step (4) are not of the form yPx. Therefore, $x \in Win(J_\sigma)$ and $x \in R_{RA,W}(V)$.

Conversely, let $x \in R_{RA,W}(V)$. Let $>$ be the order refining $\succsim_P$ such that the judgment set obtained is J, with $x \in Win(J)$. First, all formulas uPv such that $N(P, uPv) > \beta$ are added to S without creating any inconsistency with W. Then, $>$ must consider all propositions zPy such that $N(P, zPy) = \beta$ and $y \neq x$, and add them all to S; at this point, for any $y \neq x$, a proposition zPy has been considered and added to S, otherwise there would be an y such that for no z it holds that $n_V(z, y) \geq \beta$, which would contradict $\beta = \min_x \max_y n_V(y, x)$. Therefore, no propositions zPx will be added to S (or else W would be violated). Therefore, x is such that $\min_x \max_y n_V(y, x) \leq \beta$, hence $\min_x \max_y n_V(y, x) = \beta$: x is a maximin winner.

Example 3. Let $n = 9$ and V such that n_V is as follows:

$$
\begin{array}{c|cccc}
n_V & a & b & c & d \\
\hline
a & - & 6 & 2 & 4 \\
b & 3 & - & 5 & 6 \\
c & 7 & 4 & - & 2 \\
d & 5 & 3 & 7 & - \\
\end{array}
\tag{3}
$$

The weak order $\succsim_P$ starts with cPa and dPc (tied), then aPb and bPd, then bPc and dPa, etc. Applying RA with $\Gamma = W$ starts by adding cPa and dPc, whatever the choice of the linear order $\succ_\sigma$ refining $\succsim_P$. Next, there is a choice between aPb or bPd. If aPb is considered first (that is, if $aPb \succ_\sigma bPd$), then it is added to S, bPd is not (because it would violate W-consistency), and then all other propositions except aPd, bPd and cPd are added. The other choice is similar, replacing d by b. Therefore, $RA(P(V), W)$ contains the two judgment sets

$$J_1 = \{dPa, dPB, dPc, aPb, bPa, aPc, cPa, bPc, cPb\}$$

and

$$J_2 = \{bPa, bPc, bPd, aPc, cPa, aPd, dPa, cPd, dPa\},$$

with $Win(J_1) = d$ and $Win(J_2) = b$. We check that b and d are also the maximin winners for V.

Applying RA with $\Gamma = W$ first adds cPa and dPc. Next, there is a choice between aPb or bPd. If aPb is considered first, then it is added to S, bPd is not, then *all* other propositions except aPd, bPd and cPd are added.

For MWA, it is already known that the choice of the transitivity constraint leads to the Kemeny rule, *i.e.*, $R_{MWA,Tr} = Kemeny$. The proof can be found in [EGP12].

The choice of the W constraint leads to an unknown voting rule, for which, interestingly, the winners maximizes the sum of the Borda score and a second term: if $S_B(P, x)$ is the Borda score of P for profile P, $R_{MWA,W}$ is the voting rule defined by

$$S_{MWA,W}(x) = S_B(P, x) + \sum_{y \neq z \neq x} \max(N_P(y, z), N_P(z, y))$$

and

$$R_{MWA,W}(P) = \underset{x \in X}{\operatorname{argmax}}\, S_{MWA,W}(x).$$

We give in Appendix an example of winner determination for this rule, which shows that it differs from Borda.

The connection between the Young judgment aggregation rule and the Young voting rule is less clear that is seems at first glance: because the removal of judgments (and votes) can make the number of judgments (votes) even, the voting rule obtained from Y together with W is not Young (even for n odd), but a weak version of Young: the *weak Young* voting rule is defined as the voting rule, except that we look for a minimal number of votes whose removal in V makes x a *weak* Condorcet winner (where x is a weak Condorcet winner if for any $y \neq x$, at least half of the voters prefer x to y).

Proposition 5. $R_{Young,W} = WeakYoung$.

Proof. Removing a minimal number of judgments from $P(V)$ so as to make it consistent is equivalent to removing a minimal number of votes from V so that the majority graph contains an undominated outcome, *i.e.*, so that there exists a weak Condorcet winner.

$R_{Young,Tr}$ does not appear to be a known voting rule. It consists of the dominating candidates in maximum cardinality sub-profiles of $P(V)$ whose majoritarian aggregation is acyclic.

Another judgment aggregation rule defined in [LPSvdT11] is the distance-based rule $R_{d_H,max}$. Because the voting rules we obtain from it are not known voting rules, we omit the corresponding results.

6   Discussion

We have obtained a number of correspondences between judgment aggregation rules and pairs of voting rules. It is especially interesting to see which pairs come together. We summarize the results here.

$R_{F,\Gamma}$	$F = MSA$	$F = MCSA$	$F = MWA$	$F = RA$	$F = Y$
$\Gamma = Tr$	Top Cycle	Slater	Kemeny	ranked pairs	weak Young
$\Gamma = W$		Copeland		maximin	

Note that if the assumption that profiles have an odd number of judgments sets is relaxed, then the voting rules obtained are generally be *weak* versions of the usual voting rule, strict majority being replaced by weak majority. In particular, $R_{MCSA,W}$ would be $Copeland^0$, where ties count as much as victories.

What do these results tell us? After all, these judgment aggregation rules have not been widely studied yet (although some of them have been introduced independently in several papers), and one may argue that they were defined in such a way that their specialization to the preference agenda correspond to such or such voting rule, and one may advocate that this makes these correspondence results rather pointless. This is an important point: while the premise is not entirely false (at least for some of the rules, such as RA), we would strongly disagree with the conclusion. The definition and study of judgment aggregation rules is only starting, and knowing that a judgment aggregation rule specializes to a well-known voting rules (sometimes, to *two* well-known voting rules) is a hint that the judgment aggregation rule is a natural generalization of interesting voting rules, which is a first justification for studying it. Also, it gives insights about the properties it may satisfy. In particular, a challenging question is the *axiomatization* of judgment aggregation rules, and for this, a good start could be to start with the axiomatization (when it exists) of the voting rule(s) into which the judgment aggregation rule degenerates.

In our correspondence results, a voting rule is defined from two elements: a judgment aggregation operator and a constraint. This is reminiscent of a recent research stream on the *distance rationalizability* of voting rules (see [Bai87, Kla05a, Kla05b, MN08] for early works and [EFS09, EFS12] for a systematic study). There, one seeks to define voting rules via a distance between profiles and a consensus class. In some sense, our judgment aggregation rules play the role of distances whereas the constraint plays the role of the consensus class. More precisely, the Tr and W constraint more or less correspond to, respectively, the *strong unanimity* and *Condorcet* consensus classes, with a noticeable difference: the definition of a consensus class bears on a profile, whereas a constraint bears on an (individual or collective) judgment set (this may explains why we don't have any constraint that corresponds to the unanimity and majority consensus classes). This is consistent with the fact that the two rules we obtain by letting $\Gamma = W$ are also rationalizable for the Condorcet consensus class (see [Kla05b] for Copeland and [EFS09] for maximin).

The discussion about distance rationalizability leads to a very intriguing question. One of the key questions in [EFS12] is a systematic study of which rules can be axiomatized by a given consensus class or a given distance. They not only show that some rules are indeed axiomatizable via a given consensus class or a given distance function, but also that some rules are *not*. This leads us to ask the following question: which voting rules are definable from a judgment aggregation rule by specializing to the preference agenda and imposing the Tr or the W constraint? Asked this way, this question is trivial; the judgment aggregation rule can be defined such as its application to the preference agenda behaves exactly like the voting rule we started from. However, suppose we ask the judgment aggregation rule to be *neutral with respect to propositional symbols*, which means that if σ is a permutation of the set of propositional symbols, J_σ the judgment set obtained by applying σ in every $\varphi \in J$ and $P(V)_\sigma = \{J_\sigma | J \in P(V)\}$, then $F(P(V)_\sigma) = F(P(V))_\sigma$. (Note that this is the case of all the rules we study here.) Then the question becomes highly nontrivial, and we suspect that some well-known rules will not be definable this way.

Finally, as argued three paragraphs above, the definition of judgment aggregation rules which specialize to well-known voting rules (and thereby give them a justification) is a bottom-up process. A subsequent top-down process would consists in applying these judgment aggregation rules (obtained as a generalization from voting rules) to other specific agendas and/or with other constraints. We give here two examples. A first example would consist in keeping the preference agenda and to consider constraints that are intermediate between Tr and W, such as the judgment set being transitive on the top k candidates; for instance, $k = 2$ this would be $\bigvee_{x \neq y} \left(\bigwedge_{z \neq x} xPz \wedge \bigwedge_{z \neq x, y} yPz \right)$. A second example would consist in keeping an preference agenda of the form $\{xPy, x, y \in X\}$, but with a very different meaning, where xPy means that x and y are in the same equivalence class, and choose Γ as the expression of an equivalence relation; this process will give interesting rules for aggregating equivalence relations.

Acknowledgements. We thank Gabriella Pigozzi, Leon van der Torre and Srdjan Vesic for helpful discussions and comments, as well as the anonymous ADT-13 referees for their useful reviews.

References

[Bai87]   Baigent, N.: Metric rationalization of social choice functions according to principles of social choice. Mathematical Social Science 14(1), 59–65 (1987)

[BF12]   Brill, M., Fischer, F.A.: The price of neutrality for the ranked pairs method. In: AAAI (2012)

[Die12]   Dietrich, F.: Scoring rules for judgment aggregation. MPRA paper, University Library of Munich, Germany (2012)

[DL07]   Dietrich, F., List, C.: Arrow's theorem in judgment aggregation. Social Choice and Welfare 29(1), 19–33 (2007)

[DL08]   Dietrich, F., List, C.: Judgment aggregation under constraints. In: Boylan, T., Gekker, R. (eds.) Economics, Rational Choice and Normative Philosophy, Routledge (2008)

[DP12]   Duddy, C., Piggins, A.: A measure of distance between judgment sets. Social Choice and Welfare 39, 855–867 (2012)

[EFS09]   Elkind, E., Faliszewski, P., Slinko, A.: On distance rationalizability of some voting rules. In: TARK, pp. 108–117 (2009)

[EFS12]   Elkind, E., Faliszewski, P., Slinko, A.: Rationalizations of condorcet-consistent rules via distances of hamming type. Social Choice and Welfare 39(4), 891–905 (2012)

[EGP12]   Endriss, U., Grandi, U., Porello, D.: Complexity of judgment aggregation. Journal Artificial Intelligence Research (JAIR) 45, 481–514 (2012)

[EM05]   Eckert, D., Mitlöhner, J.: Logical representation and merging of preference information. In: Proceedings of the IJCAI 2005 Multidisciplinary Workshop on Preference Handling (2005)

[GE13]   Grandi, U., Endriss, U.: Lifting integrity constraints in binary aggregation. Artificial Intelligence, 199–200, 45–66 (2013)

[Kla05a]   Klamler, C.: Borda and condorcet: some distance results. Theory and Decision 59(2), 97–109 (2005)

[Kla05b]   Klamler, C.: The copeland rule and condorcet's pirnciple. Economic Theory 25(3), 745–749 (2005)

[KPP02]   Konieczny, S., Pino-Pérez, R.: Merging information under constraints: a logical framework. Journal of Logic and Computation 12(5), 773–808 (2002)

[LPSvdT11]   Lang, J., Pigozzi, G., Slavkovik, M., van der Torre, L.: Judgment aggregation rules based on minimization. In: TARK, pp. 238–246 (2011)

[MN08]   Meskanen, T., Nurmi, H.: Closeness counts in social choice. In: Power, Freedom, and Voting, pp. 289–306. Springer (2008)

[MO09]   Miller, M.K., Osherson, D.: Methods for distance-based judgment aggregation. Social Choice and Welfare 32(4), 575–601 (2009)

[NPP11]   Nehring, K., Pivato, M., Puppe, C.: Condorcet admissibility: Indeterminacy and path-dependence under majority voting on interconnected decisions (July 2011), http://mpra.ub.uni-muenchen.de/32434/

[Tid87]   Tideman, T.N.: Independence of clones as a criterion for voting rules. Social Choice and Welfare 4, 185–206 (1987)

Appendix

$R_{MWA,W}$

Let $X = \{a, b, c, d, e\}$, $n = 25$, and let V be the profile containing: 3 votes $abcde$, 2 votes $bcade$, 2 votes $cabde$, 2 votes $edabc$, 2 votes $edbca$, 2 votes $edcab$, 2 votes $adebc$, 2 votes $deabc$, 2 votes $eadbc$, 2 votes $cbade$, 2 votes $cbdea$, 2 votes $cbead$. The weighted majority graph associated with V is:

$$
\begin{array}{c|c|c|c|c|c}
n_V & a & b & c & d & e \\
\hline
a & - & 15 & 11 & 15 & 11 \\
\hline
b & 10 & - & 15 & 13 & 13 \\
\hline
c & 14 & 10 & - & 13 & 13 \\
\hline
d & 10 & 12 & 12 & - & 14 \\
\hline
e & 14 & 12 & 12 & 11 & - \\
\end{array}
\tag{4}
$$

The Borda scores are respectively 52 for a (Borda winner), 51 for b, 50 for c, 48 for d and 49 for e.

For all $x \in X$, let

$$S_{MWA,W}(x) = S_B(P, x) + \sum_{y \neq z \neq x} \max(N_P(y, z), N_P(z, y)).$$

We get

$$S_{MWA,W}(a) = 52 + 81 = 133;$$
$$S_{MWA,W}(b) = 51 + 82 = 133;$$
$$S_{MWA,W}(c) = 50 + 84 = 134;$$
$$S_{MWA,W}(d) = 48 + 83 = 131;$$
$$S_{MWA,W}(e) = 49 + 85 = 134.$$

The co-winners are c and e.

Aggregating Conditionally Lexicographic Preferences Using Answer Set Programming Solvers

Xudong Liu and Miroslaw Truszczynski

University of Kentucky
Lexington, KY 40506, USA
liu@cs.uky.edu, mirek@cs.engr.uky.edu

Abstract. We consider voting over combinatorial domains, where alternatives are binary tuples. We assume that votes are specified as *conditionally lexicographic preference trees*, or *LP trees* for short. We study the aggregation of LP tree votes for several positional scoring rules. Our main goal is to demonstrate that answer-set programming tools can be effective in solving the *winner* and the *evaluation* problems for instances of practical sizes. To this end, we propose encodings of the two problems as answer-set programs, design methods to generate LP tree votes randomly to support experiments, and present experimental results obtained with ASP solvers *clingo* and *clingcon*.

Keywords: conditionally lexicographic preferences, social choice theory, positional scoring voting rules, answer set programming.

1 Introduction

Preferences are an essential component in areas including constraint satisfaction, decision making, and social choice theory. Modeling and reasoning about preferences are crucial to these areas. Consequently, several preference formalisms have been developed, such as penalty logic, possibilistic logic, and conditional preference networks (*CP nets*, for short) [6]. Each of these formalisms provides the user with a concise way to express her preferences.

The problem of aggregating preferences of a group of users (often referred to as *voters*) is central to decision making and has been studied extensively in social choice theory. While in the cases when the number of alternatives is small the problems of computing the winner and of deciding whether there is an outcome with the score higher than a given threshold are polynomially solvable for the majority of commonly considered voting rules, the situation changes when we consider votes over combinatorial domains of binary p-tuples. Since the number of outcomes grows exponentially with p, applying voting rules directly is often infeasible. Issue-by-issue voting approximates voting rules by considering only the first choices in votes, but experiments show that the winners selected issue by issue are often not the winners selected by common voting rules [4].

When votes are represented as *conditionally lexicographic preference trees*, or LP trees, for short [1], the problem of computing the winning alternative, or

P. Perny, M. Pirlot, and A. Tsoukiàs (Eds.): ADT 2013, LNAI 8176, pp. 244–258, 2013.

the *winner* problem, is generally NP-hard, while a related problem to decide whether there is an alternative with the score exceeding a given threshold, the so called *evaluation* problem, is NP-complete [8]. Nevertheless, the two problems arise in practice and computational tools to address them effectively are needed. In this work we study Borda, k-approval and 2-valued (k, l)-approval scoring rules. For the 2-valued (k, l)-approval, we obtain new complexity results. For all three rules we encode the problems in answer-set programming [9] and study the effectiveness of ASP solvers *clingo* [5] and *clingcon* [10] in solving the winner and the evaluation problems. We chose these two solvers as they represent substantial different approaches to computing answer sets. The former, *clingo*, is a native ASP solver developed along the lines of satisfiability solvers. The latter, *clingcon*, enhances *clingo* with specialized treatment of some common classes of numeric constraints by delegating some reasoning tasks to a CP solver *Gecode* [12]. As problems we are considering involve numeric constraints, a comparison of the two solvers is of interest. To support the experimentation we propose and implement a method to generate LP votes, of some restricted form, randomly.

The main contributions of our work are: (1) new complexity results for the winner and the evaluation problems for a class of positional scoring rules; and (2) demonstration that ASP is an effective formalism for modeling and solving problems related to aggregation of preferences given as LP trees.

2   Technical Preliminaries

A *vote* over a set of *alternatives* (or *outcomes*) $\mathcal{X}$ is a *strict total* order $\succ$ on $\mathcal{X}$. Here we consider votes over alternatives from combinatorial domains determined by a set $\mathcal{I} = \{X_1, X_2, \ldots, X_p\}$ of p binary *issues*, with each issue X_i having a binary domain $D(X_i) = \{0_i, 1_i\}$. The *combinatorial domain* in question is the set $\mathcal{X}(\mathcal{I}) = D(X_1) \times D(X_2) \times \ldots \times D(X_p)$. If $\mathcal{I}$ is implied by the context, we write $\mathcal{X}$ instead of $\mathcal{X}(\mathcal{I})$. For instance, let $\mathcal{I} = \{X_1, X_2, X_3\}$. A 3-tuple $(0_1, 1_2, 1_3)$ is an alternative from $\mathcal{X}(\mathcal{I})$, or simply, an alternative over $\mathcal{I}$. We often write it as $0_1 1_2 1_3$ or just as 011. It assigns 0 to X_1, 1 to X_2, and 1 to X_3.

Clearly, the cardinality of $\mathcal{X}(\mathcal{I})$, which we denote by m throughout the paper, is 2^p. Thus, even for moderately small values of p, eliciting precise orders over all alternatives and representing them directly may be infeasible. Instead, in cases when preferences have some structure, that structure can be exploited to give rise to concise preference expressions [2,8].

In this work we study the case when votes (preferences) are given as LP trees [1]. An *LP tree* T over a set $\mathcal{I}$ of p binary issues $X_1, \ldots, X_p$ is a *binary tree*. Each node t in T is labeled by an issue from $\mathcal{I}$, denoted by $Iss(t)$, and with *preference information* of the form $a > b$ or $b > a$ indicating which of the two values a and b comprising the domain of $Iss(t)$ is preferred (in general the preference may depend on the values of issues labeling the ancestor nodes). We require that each issue appears exactly once on each path from the root to a leaf.

Intuitively, the issue labeling the root of an LP tree is of highest importance. Alternatives with the preferred value of that issue are preferred over alternatives with the non-preferred one. The two subtrees refine that ordering. The left subtree determines the ranking of the preferred "upper half" and the right subtree determines the ranking of the non-preferred "lower half." In each case, the same principle is used, with the root issue being the most important one. We note that the issues labeling the roots of the subtrees need not be the same (the relative importance of issues may depend on values for the issues labeling the nodes on the path to the root).

The precise semantics of an LP tree T captures this intuition. Given an alternative $x_1 x_2 \ldots x_p$, we find its preference ranking in T by traversing the tree from the root to a leaf. When at node t labeled with the issue X_i, we follow down to the left subtree if x_i is preferred according to the preference information at node t. Otherwise, we follow down to the right subtree.

It is convenient to imagine the existence of yet another level of nodes in the tree, not represented explicitly, with each node in the lowest explicitly represented level "splitting" into two of these implicit nodes, each standing for an alternative. Descending the tree given an alternative in the way described above takes us to an (implicit) node at the lowest level that represents precisely that alternative. The more to the left the node representing the alternative, the more preferred it is, with the one in the leftmost (implicit) node being the most desirable one as left links always correspond to preferred values.

To illustrate these notions, let us consider an example. A group of friends in Lexington want to make vacation plans for the next year. Having brainstormed for a while, the group decided to focus on three binary issues. The *Time* (X_1) of the travel could be either *summer* (s or 1_1) or *winter* (w or 0_1), the *Destination* (X_2) could be either *Chicago* (c or 1_2) or *Miami* (m or 0_2) and the mode of *Transportation* (X_3) could be to *drive* (d or 1_3) or *fly* (f or 0_3).

Jane, a member of the group, prefers a summer trip to a winter trip, and this preference on the issue *Time* is the most important one. Then for a summer trip, the next most important issue is *Destination* and she prefers Chicago to Miami, and the least important issue is *Transportation*. Jane prefers driving to flying if they go to Chicago, and flying, otherwise. For a winter trip, the importance of the remaining two issues changes with *Transportation* being now more important than the destination – Jane does not like driving in the winter. As for the *Destination*, she prefers to go to Miami to avoid the cold weather. These preferences can be captured by the LP tree T in Figure 1. We note that the trees ordering the vacation plans for summer and for winter are determined by trees with different assignments of issues to nodes. For instance, for the summer trips, the destination is the most important factor while for the winter trips the mode of transportation. The tree shows that the preferred vacation plan for Jane is to drive to Chicago in the summer and the next in order of preference is to fly to Chicago in the summer. The least preferred plan is to drive to Chicago in the winter.

Sometimes LP trees can be represented in a more concise way. For instance, if for some node t, its two subtrees are identical (that is, the corresponding nodes are assigned the same issue), they can be collapsed to a single subtree, with the

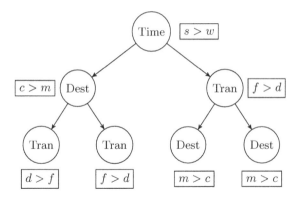

Fig. 1. An LP tree T

same assignment of issues to nodes. To retain preference information, at each node t' of the subtree we place a *conditional preference table*, and each preference in it specifies the preferred value for the issue labeling that node given the value of the issue labeling t. In the extreme case when for every node its two subtrees are identical, the tree can be collapsed to a path.

Since the preferred issue at a node depends on the values of issues above, the conditional preference table for the node t located at distance i from the root has possibly as many as 2^i rows (in general, 2^j rows, where j is the number of ancestor nodes with one child only), with each row specifying a combination of values for the ancestor issues together with the preferred value for $Iss(t)$ given that combination. Thus, collapsing subtrees alone does not lead to a smaller representation size. However, it can be achieved if there are nodes whose preferred value depends only on a limited number of issues labeling their single-child ancestor nodes as in such cases the conditional preference table can be simplified.

Formally, given an LP tree (possibly with some subtrees collapsed), for a node t, let $NonInst(t)$ be the set of ancestor nodes of t whose subtrees were collapsed into one, and let $Inst(t)$ represent the remaining ancestor nodes. A *parent* function $\mathcal{P}$ assigns to each node t in T a set $\mathcal{P}(t) \subseteq NonInst(t)$ of *parents* of t, that is, the nodes whose issues may have influence on the local preference at $Iss(t)$. Clearly, the conditional preference table at t requires only $2^{|\mathcal{P}(t)|}$ rows, possibly many fewer than in the worst case. In the extreme case, when an LP tree is a path and each node has a bounded (independent of p) number of parents, the tree can be represented in $O(p)$ space.

If for every node t in an LP tree, $\mathcal{P}(t) = \emptyset$, all (local) preferences are unconditional and conditional preference tables consist of a single entry. Such trees are called *unconditional preference* LP trees (UP trees, for short). Similarly, LP trees with all non-leaf nodes having their subtrees collapsed are called an *unconditional importance* LP trees (UI trees, for short). This leads to a a natural classification of LP trees into four classes: unconditional importance and unconditional preference LP trees (UI-IP trees), unconditional importance and conditional preference trees (UI-CP trees), etc. The class of CI-CP trees comprises all LP trees, the class of UI-UP trees is the most narrow one.

The LP tree T in Figure 1 can be represented more concisely as a (collapsed) CI-CP tree v in Figure 2. Nodes at depth one have their subtrees collapsed. In the tree in Figure 1, the subtrees of the node at depth 1 labeled *Tran* are not only identical but also have the same preference information at every node. Thus, collapsing them does not incur growth in the size of the conditional preference table.

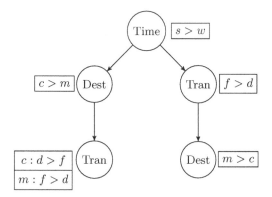

Fig. 2. An CI-CP LP tree v

A set of votes (collected from, say, n voters) over a domain $\mathcal{X}$ is called a *profile*. Among many rules proposed to aggregate a profile into a single preference ranking representing the group, positional scoring rules have received particular attention. For profiles over a domain with m alternatives, a *scoring vector* is a sequence $w = (w_0, \ldots, w_{m-1})$ of integers such that $w_0 \geq w_1 \geq \ldots \geq w_{m-1}$ and $w_0 > w_{m-1}$. Given a vote v with the alternative o in position i ($0 \leq i \leq m - 1$), the score of o in v is given by $s_w(v, o) = w_i$. Given a profile $\mathcal{V}$ of votes and an alternative o, the score of o in $\mathcal{V}$ is given by $s_w(\mathcal{V}, o) = \sum_{v \in \mathcal{V}} s_w(v, o)$. These scores determine the ranking generated from $\mathcal{V}$ by the scoring vector w (assuming, as is common, some independent tie breaking rule). In this paper we consider three positional scoring rules:

1. Borda: $(m - 1, m - 2, \ldots, 1, 0)$
2. k-approval: $(1, \ldots 1, 0, \ldots 0)$ with k the number of 1's
3. 2-valued (k, l)-approval: $(a, \ldots, a, b, \ldots, b, 0 \ldots, 0)$, where a and b are constants ($a > b$) and the numbers of a's and b's equal to k and l, respectively.

3   The Problems and Their Complexity

Here we consider only *effective implicit* positional scoring rules, that is, rules defined by an algorithm that given m (the number of alternatives) and i, $0 \leq i \leq$

$m - 1$ (an index into the scoring vector) returns the value w_i of the scoring vector and works in time polynomial in the sizes of i and m. Borda, k-approval and (k, l)-approval are examples of effective implicit positional scoring rules.

Let us fix an effective implicit positional scoring rule $\mathcal{D}$ with the scoring vector w. Given an LP profile $\mathcal{V}$, the *winner* problem for $\mathcal{D}$ consists of computing an alternative $o \in \mathcal{X}$ with the maximum score $s_w(\mathcal{V}, o)$. Similarly, given a profile $\mathcal{V}$ and a positive integer R, the *evaluation* problem for $\mathcal{D}$ asks if there exists an alternative $o \in \mathcal{X}$ such that $s_w(\mathcal{V}, o) \geq R$. In each case, w is the scoring vector of $\mathcal{D}$ for m alternatives; we recall that it is given implicitly in term of an algorithm that efficiently computes its entries.

We apply the voting rules listed above to profiles consisting of LP trees or *LP profiles*, for short. We distinguish four classes of profiles, UI-UP, UI-CP, CI-UP and CI-CP depending on the type of LP trees they consist of.

In the most restrictive case of UI-UP profiles, the evaluation problem for the Borda rule is in P and it is NP-complete for the three other classes of profiles [8]. The picture for the the k-approval rule is more complicated. If $k = 2^{p-1}$ the evaluation problem is in P for all four classes of profiles. However, if k equals 2^{p-2} or 2^{p-3}, the problem is NP-complete, again for all four LP profile types [8] (in fact, the result holds for a larger set of values k, we refer for details to Lang et al. [8]). Clearly, in each case where the evaluation problem is NP-complete, the winner problem is NP-hard.

To the best of our knowledge, the complexity of the 2-valued (k, l)-approval rule has not been studied. It is evident that (k, l)-approval is an effective implicit positional scoring rule. It turns out that, as with the k-approval rule, for some values of the parameters, the evaluation problem for (k, l)-approval is NP-complete. We describe two such cases here: (1) $k = l = 2^{p-2}$, and (2) $k = l = 2^{p-3}$. We note that if $a = 1$ and $b = 0$, case (1) reduces to 2^{p-2}-approval and case (2) to 2^{p-3}-approval. If $a = 2$ and $b = 1$, we refer to the rule in case (1) as *2K-approval*.

Theorem 1. *The following problem is NP-complete: decide for a given UI-UP profile $\mathcal{V}$ and an integer R whether there is an alternative o such that $s_w(\mathcal{V}, o) \geq R$, where w is the scoring vector of the $(2^{p-2}, 2^{p-2})$-approval rule.*

Proof. We can guess in polynomial time an alternative $o \in \mathcal{X}$ and verify in polynomial time that $S_w(\mathcal{V}, o) \geq R$ (this is possible because (k, l)-approval is an effective implicit scoring rule; the score of an alternative in a vote can be computed in polynomial time once its position is known, and the position can be computed in polynomial time be traversing the tree representing the vote). So membership in NP follows. Hardness follows from a polynomial reduction from the problem 2-*MINSAT*[1] [7], which is NP-complete. Given an instance $\langle \Phi, l \rangle$ of the 2-MINSAT problem, we construct the set of issues $\mathcal{I}$, the set of alternatives $\mathcal{X}$, the profile $\mathcal{V}$ and the threshold R.

---

[1] Let N be an integer ($N > 1$), the N-MINSAT problem is defined as follows. Given a set Φ of n N-clauses $\{c_1, \ldots, c_n\}$ over a set of propositional variables $\{X_1, \ldots, X_p\}$, and a positive integer l ($l \leq n$), decide whether there is a truth assignment that satisfies at most l clauses in Φ.

Important observations are that o is among the top first quarter of alternatives in an LP tree $\mathcal{L}$ if and only if the top two most important issues in $\mathcal{L}$ are both assigned the preferred values; and that o is among the second top quarter of alternatives if and only if the most important issue is assigned the preferred value and the second most important one is assigned the non-preferred one.

(1). We define $\mathcal{I} = \{X_1, \ldots, X_p\}$, where X_is are all propositional letters occurring in Φ. Clearly, the set $\mathcal{X}$ of all alternatives over $\mathcal{I}$ coincides with the set of truth assignments of variables in $\mathcal{I}$.

(2). Let Ψ be the set of formulas $\{\neg c_i : c_i \in \Phi\}$. For each $\neg c_i \in \Psi$, we build $a + b$ UI-UP trees. For instance, if $\neg c_i = X_2 \wedge \neg X_4$, then we proceed as follows. Firstly, we build $a - b$ duplicate trees shown in Figure 3a. Secondly, we construct b duplicate trees shown in Figure 3b. Thirdly, we build another b duplicate trees shown in Figure 3c. (In all three figures we only indicate the top two issues since the other issues can be ordered arbitrarily.) Denote by $\mathcal{V}_i$ the set of these $a + b$ UI-UP trees for formula $\neg c_i$. Then $\mathcal{V} = \bigcup_{1 \leq i \leq n} \mathcal{V}_i$ and has $n * (a + b)$ votes.

(3). Finally, we set $R = (n - l) * (a^2 - ab + b^2) + l * ab$.

Note that the construction of $\mathcal{V}$ ensures that if $o \models \neg c_i$, $S_w(\mathcal{V}_i, o) = a^2 - ab + b^2$; otherwise if $o \not\models \neg c_i$, $S_w(\mathcal{V}_i, o) = ab$. We have $a^2 - ab + b^2 > ab$ since $(a - b)^2 > 0$. Hence, there is an assignment satisfying at most l clauses in Φ if and only if there is an assignment satisfying at least $n - l$ formulas in Ψ if and only if there is an alternative with the $(2^{p-2}, 2^{p-2})$-approval score of at least R given the profile $\mathcal{V}$.

Since the first equivalence is clear, it suffices to show the second. Let o be an assignment satisfying l' formulas in Ψ. We have $S_w(\mathcal{V}, o) - R = (l' + l - n) * (a^2 - ab + b^2) + (n - l' - l) * ab = (l' + l - n) * (a^2 - 2ab + b^2) = (l' + l - n) * (a - b)^2$. It follows that $S_w(\mathcal{V}, o) \geq R$ if and only if $l' + l - n \geq 0$ if and only if $l' \geq n - l$. $\square$

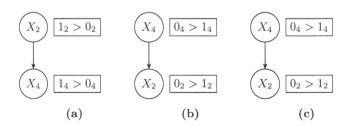

Fig. 3. UI-UP LP trees

This hardness proof applies to more general classes of LP trees, namely UI-CP, CI-UP and CI-CP, and the winner problem for those cases is NP-hard. The evaluation problem according to $(2^{p-3}, 2^{p-3})$-approval for the four classes of LP trees is also NP-complete. In this case, the hardness is proven by a reduction from an NP-complete version of the 3-*MINSAT* problem [11].

4   The Problems in Answer-Set Programming

The winner and the evaluation problems are in general intractable in the setting we consider. Yet, they arise in practice and computational tools to handle them are needed. We develop and evaluate a computational approach based on *answer-set programming* (ASP) [9]. We propose several ASP encodings for both problems for the Borda, k-approval, and (k, l)-approval rules (for the lack of space only the encodings for Borda are discussed). The encodings are adjusted to two ASP solvers for experiments: *clingo* [5], and *clingcon* [10] and demonstrate the effectiveness of ASP in modeling problems related to preference aggregation. We selected

Encoding LP Trees as Logic Programs. In the winner and evaluation problems, we use LP trees only to compute the ranking of an alternative. Therefore, we encode trees as program rules in a way that enables that computation for a given alternative. In the encoding, an alternative o is represented by a set of ground atoms $eval(i, x_i)$, $i = 1, 2, \ldots, p$ and $x_i \in \{0, 1\}$. An atom $eval(i, x_i)$ holds precisely when the alternative o has value x_i on issue X_i.

If X_i is the issue labeling a node t in vote v at depth d_i^v, $CPT(t)$ determines which of the values 0_i and 1_i is preferred there. Let us assume $\mathcal{P}(t) = \{t_1, \ldots, t_j\}$ and $Inst(t) = \{t_{j+1}, \ldots, t_\ell\}$, where each t_q is labeled by X_{i_q}. The location of t is determined by its depth d_i^v and by the set of values $x_{i_{j+1}}, \ldots, x_{i_\ell}$ of the issues labeling $Inst(t)$ (they determine whether we descend to the left or to the right child as we descend down the tree). Thus, $CPT(t)$ can be represented by program rules as follows. For each row $u : 1_i > 0_i$ in $CPT(t)$, where $u = x_{i_1}, \ldots, x_{i_j}$, we include in the program the rule

$$vote(v, d_i^v, i, 1) :\text{-} \, eval(i_1, x_{i_1}), \ldots, eval(i_j, x_{i_j}),$$
$$eval(i_{j+1}, x_{i_{j+1}}), \ldots, eval(i_\ell, x_{i_\ell}) \tag{1}$$

(and similarly, in the case when that row has the form $u : 0_i > 1_i$).

In this representation, the property $vote(v, d_i^v, i, a_i)$ will hold true for an alternative o represented by ground atoms $eval(i, x_i)$ *precisely when* (or *if*, denoted by ":-" in our encodings) that alternative takes us to a node in v at depth d_i^v labeled with the issue X_i, for which at that node the value a_i is preferred. Since, in order to compute the score of an alternative on a tree v all we need to know is whether $vote(v, d_i^v, i, a_i)$ holds (cf. our discussion below), this representation of trees is sufficient for our purpose.

For example, the LP tree v in Figure 2 is translated into the logic program in Figure 4 (*voteID(v)* identifies the id of the vote (LP tree)).

```
1  voteID(1).
2  vote(1,1,1,1).
3  vote(1,2,2,1)  :- eval(1,1).
4  vote(1,3,3,1)  :- eval(2,1), eval(1,1).
5  vote(1,3,3,0)  :- eval(2,0), eval(1,1).
6  vote(1,2,3,0)  :- eval(1,0).
7  vote(1,3,2,0)  :- eval(1,0).
```

Fig. 4. Translation of v in ASP

Encoding the Borda Evaluation Problem in *Clingo*. The evaluation and the winner problems for Borda can be encoded in terms of rules on top of those that represent an LP profile. Given a representation of an alternative and of the profile, the rules evaluate the score of the alternative and maximize it or test if it meets or exceeds the threshold.

We first show the encoding of the Borda evaluation problem in *clingo* (Figure 5). Parameters in the evaluation problem are defined as facts (lines 1-4):

```
1   issue(1). issue(2). issue(3).
2   numIss(3).
3   val(0). val(1).
4   threshold(5).
5   1{ eval(I,M) : val(M) }1 :- issue(I).
6   wform(V,I,W) :- vote(V,D,I,A), eval(I,A), numIss(P), W=#pow(2,P-D).
7   wform(V,I,0) :- vote(V,D,I,A), eval(I,M), A != M.
8   goal :- S = #sum [ wform(V,I,W) = W ], threshold(TH), S >= TH.
9   :- not goal.
```

Fig. 5. Borda evaluation problem encoding in *clingo*

predicates *issue/1*s representing three issues, *numIss/1* the number of issues, *threshold/1* the threshold value, together with *val/1*s the two values in the issues' binary domains. Line 5 generates the search space of all alternatives over three binary issues. It expresses that if X is an issue, exactly one of *eval(X,Y)* holds for all *val(Y)*, i.e., exactly one value Y is assigned to X.

Let o be an alternative represented by a set of ground atoms $eval(i, x_i)$, one atom for each issue X_i. Based on the representation of trees described above, for every tree v we get the set of ground atoms $vote(v, d_i^v, i, a_i)$. The Borda score of an alternative in that tree corresponds to the rank of the leaf the alternative leads to (in a "non-collapsed" tree), which is determined by the direction of descent (left or right) at each level. Roughly speaking, these directions give the binary representation of that rank, that is, the Borda score of the alternative. Let us define $s_B(v, o)$ as a function that computes the Borda score of alternative o given one vote v. Then one can check that

$$s_B(v, o) = \sum_{i=1}^{p} 2^{p-d_i^v} \cdot f(a_i, x_i), \tag{2}$$

where $f(a_i, x_i)$ returns 1 if $a_i = x_i$, 0 otherwise. Thus, to compute the Borda score with regard to a profile $\mathcal{V}$, we have

$$s_B(\mathcal{V}, o) = \sum_{v=1}^{n} \sum_{i=1}^{p} 2^{p-d_i^v} \cdot f(a_i, x_i). \tag{3}$$

In the program in Figure 5, lines 6 and 7 introduce predicate *wform/3* which computes $2^{p-d_i^v} \cdot f(a_i, x_i)$ used to compute Borda score. According to equation (3), if issue I appears in vote V at depth D and A is its preferred value, and if the value of I is indeed A in an alternative o, then the weight W on I in V is

2^{P-D}, where P is the number of issues; if issue I is assigned the less preferred value in o, then the weight W on I in V is 0. The Borda score of the alternative is then equal to the sum of all the weights on every issue in every vote, and this is computed using the aggregate function *#sum* built in the input language of *clingo* (rule 8). Rule 9 is an *integrity constraint* stating that contradiction is reached if predicate *goal/0* does not hold in the solution. Together with rule 8, it is ensured that the Borda evaluation problem is satisfiable if and only if there is an answer set in which *goal/0* holds.

The encoding for the Borda winner problem for *clingo* replaces rules 7 and 8 in Figure5 with the following single rule:

```
#maximize[ wform(V,I,W) = W ].
```

The *#maximize* statement is an optimization statement that maximizes the sum of all weights (W's) for which *wform(V,I,W)* holds.

Encoding the Borda Evaluation Problem in *Clingcon*. In this encoding, we exploit *clingcon*'s ability to handle some numeric constraints by specialized constraint solving techniques (by means of the CP solver *Gecode* [12]). In Figure 6 we encode the Borda evaluation problem in *clingcon*.

```
1  $domain(1..4).
2  issue(1). issue(2). issue(3).
3  numIss(3).
4  val(0). val(1).
5  threshold(5).
6  1{ eval(I,M) : val(M) }1 :- issue(I).
7  wform(V,I,W) :- vote(V,D,I,A), eval(I,A), numIss(P), W=#pow(2,P-D).
8  wform(V,I,0) :- vote(V,D,I,A), eval(X,M), A != M.
9  weight(V,I) $== W :- wform(V,I,W).
10 $sum{ weight(V,I) : voteID(V) : var(I) } $>= TH :- threshold(TH).
```

Fig. 6. Borda evaluation problem encoding using *clingcon*

Lines 2-8 are same as lines 1-7 in Figure 5. Line 9 defines the constraint variable *weight(V,I)* that assigns weight W to each pair (V,I) and line 10 defines a global constraint by use of *$sum* declares that the Borda score must be at least the threshold. Line 1 restricts the domain of all constraint variables (only *weight/2* in this case) to $[1,4]$ as weights of issues in an LP tree of 3 issues are 2^0, 2^1 and 2^2.

The encoding for the Borda winner problem for *clingcon* replaces rules 10 in Figure 6 with the following one rule:

```
$maximize{weight(V,I):voteID(V):issue(I)}.
```

The *$maximize* statement is an optimization statement that maximizes the sum over the set of constraint variables *weight(V,I)*.

5   Experiments and Results

To experiment with the programs presented above and with *clingo* and *clingcon* solvers, we generate logic programs that represent random LP trees and profiles of random LP trees. Our algorithm generates encodings of trees from the most general class CI-CP under the following restrictions: (1) Each LP tree has exactly two paths with the splitting node appearing at depth $d_s = \lfloor \frac{p}{2} \rfloor$; (2) Each non-root node at depth $\leq d_s + 1$ has exactly one parent; (3) Each node at depth $> d_s + 1$ has exactly two parents, one of which is at depth $< d_s$.[2]

The algorithm starts by randomly selecting issues to label the nodes on the path from the root to the splitting node and then, similarly, labels the nodes on each of the two paths (different labelings can be produced for each of them). Then, for each non-root node, the algorithm selects at random one or two parent nodes (as appropriate based on the location of the node). Finally, the algorithm decides local preferences (for each combination of values of the parent issues) randomly picking one over the other. In each step, all possible choices are equally likely. We call CI-CP LP trees satisfying these restrictions *simple*. Each simple LP tree has size linear in p. Figure 7 depicts a CI-CP tree of 4 issues in this class.

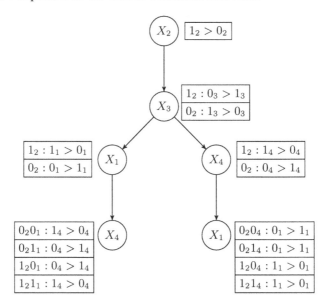

Fig. 7. A CI-CP tree of 4 issues

The goals of experimentation are to demonstrate the effectiveness of ASP tools in aggregating preferences expressed as LP trees, and to compare the performance of *clingo* and *clingcon*. We focus on three voting systems, Borda, 2^{p-2}-approval and $2K$-approval, for both the winner and the evaluation problems.

---

[2]   The restrictions are motivated by the size of the representation considerations. They ensure that the size of generated LP trees is linear in the number of issues.

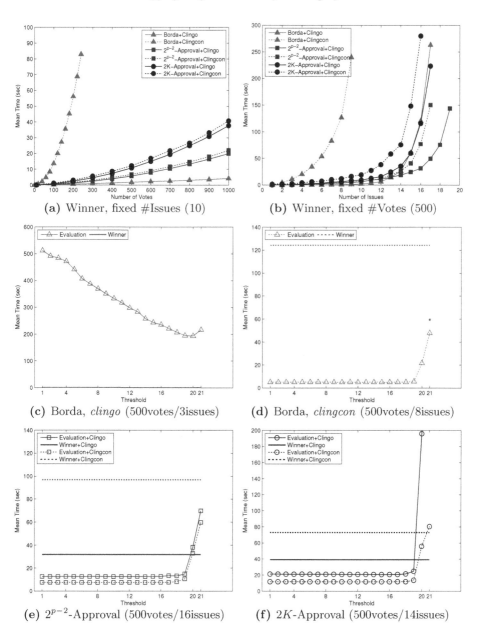

(a) Winner, fixed #Issues (10)

(b) Winner, fixed #Votes (500)

(c) Borda, *clingo* (500votes/3issues)

(d) Borda, *clingcon* (500votes/8issues)

(e) 2^{p-2}-Approval (500votes/16issues)

(f) $2K$-Approval (500votes/14issues)

Fig. 8. Aggregating *simple* LP trees

All our experiments were performed on a machine with an Intel(R) Core(TM) i7 CPU @ 2.67GHz and 8 GB RAM running Ubuntu 12.04 LTS. Each test case (using *clingo 3.0.5* or *clingcon 2.0.3*) was performed with a limit of 10 minutes.

We first consider the winner problem. In the study, we consider the computation time with a fixed number of issues (5/10/20) and for each number of issues we range the number of votes in a profile up to 1000 for $\{Borda, 2^{p-2}\text{-}approval, 2K\text{-}approval\} \times \{clingcon, clingo\}$. Then we fix the number of votes (500) and vary the number of issues up to 20, again for same set of settings. Each time result in seconds is computed as the mean of 10 tests over different randomly generated profiles of *simple* LP trees.

For the evaluation problem, we compare its experimental complexity with that of the winner problem. For each of the 10 randomly generated profiles, we compute the winning score WS and set the threshold for the evaluation problem with a percentage of WS, starting with 5% and incremented by 5% for the following tests until we reach the full value of WS. We run one more test with the threshold $WS + 1$ (there is no solution then and the overall method allows for the experimental comparison of the hardness of the winner and evaluation problems). That allows us to study the effectiveness of the maximization construct in *clingo* (the main difference between the *winner* and the *evaluation* problems is in the use of that construct in the encoding of the former). We again present and compare average time results.

Varying the Number of Issues and the Number of Votes. Our experiments on the winner problem for the three voting rules with the fixed number of issues are consistent with the property that the problem is solvable in polynomial time. Both *clingo* and *clingcon* scale up well. Figure 8a depicts the results for the cases with 10 issues. When we fix the number of votes and vary the number of issues the time grows exponentially with p (cf. Figure 8b), again consistently with the computational complexity of the problems (NP-hardness).

Generally *clingo* is better compared to *clingcon* in solving the winner problem for the three scoring rules. We attribute that first to the use of the optimization construct in *clingo*, which allows us to keep the size of the ground propositional theory low, and second to the effective way in which optimization constructs are implemented in that system. Thus, in these examples, the main benefit of *clingcon*, its ability to avoid grounding and preprocessing by "farming out" some of the solving job to a dedicated constraint solver, does not offer *clingcon* the edge. Finally, for both *clingo* and *clingcon*, Borda is the hardest rule to deal with, especially when the number of issues is large.

Comparison of the Problems: Evaluation vs Winner. The evaluation problem can be reduced to the winner problem, as an evaluation problem instance has an answer *YES* if and only if the score of the winner equals or exceeds the threshold. Thus, the evaluation problem is at most as complex as the winner problem.

We compared the two problems by first solving the winner problem, and then solving the evaluation problem on the same instances with the value of the threshold growing at the step of 5% of the winner score. That gives us 20 normalized points for each instance. In the last run (point 21 on the x-axis) we used the winner's score plus 1 as the threshold to determine the optimality of the winner's score. These experiments allow us to compare the hardness of the winner problem (more precisely, the effectiveness of solvers) with that of the evaluation problem.

First, we note that for *clingo*, the evaluation problem is harder than the winner problem in the entire range for Borda (Figure 8c), and for the two other rules, when the threshold is close to the winner's score or exceeds it (Figures 8e and 8f). We attribute that to the fact that the encodings of the evaluation problem have to model the threshold constraint with the *#sum* rule which, in *clingo*, leads to large ground theories that it finds hard to handle. In the winner problem encodings, the *#sum* rule is replaced with an optimization construct, which allows us to keep the size of the ground theory low.

For *clingcon* the situation is different. Figures 8d, 8e and 8f show that the evaluation problem is easier than the winner problem when the threshold values are smaller than the winning score and the evaluation problem becomes harder when the thresholds are close to it. It seems to suggest that the constraint solver used by *clingcon* performs well in comparison with the implementation of the optimization constructs in *clingcon*. Finally, in all cases *clingcon* outperforms *clingo* on the evaluation problems. It is especially clear for Borda, where the range of scores is much larger than in the case of approval rules. That poses a challenge for *clingo* that instantiates the *#sum* rule over that large range, which *clingcon* is able to avoid.

6   Conclusions and Future Work

Aggregating votes expressed as LP trees is a rich source of interesting theoretical and practical problems. In particular, the complexity of the winner and evaluation problems for scoring rules is far from being fully understood. First results on the topic were provided by Lang et al. [8]; our work exhibited another class of positional scoring rules for which the problems are NP-hard and NP-complete, respectively. However, a full understanding of what makes a positional scoring rule hard remains an open problem.

Importantly, our results show that ASP tools are effective in modeling and solving the winners and the evaluation problems for some positional scoring rules such as Borda, 2^{p-2}-approval and $2K$-approval. When the number of issues is fixed the ASP tools scale up consistently with the polynomial time complexity. In general, the tools are practical even if the number of issues is up to 15 and the number of votes is as high as 500. This is remarkable as 15 binary issues determine the space of over 30,000 alternatives.

Finally, the preference aggregation problems form interesting benchmarks for ASP that stimulate advances in ASP solver development. As the preference aggregation problems involve large domains, they put to the test those features of ASP tools that attempt to get around the problem of grounding programs over large domains. Our results show that the optimization statements in *clingo* in general perform well. When they cannot be used, as in the evaluation problem, it is no longer the case. The solver *clingcon*, which reduces grounding and preprocessing work by delegating some tasks to a constraint solver, performs well in comparison to *clingo* on the evaluation problem, especially for the Borda rule (and we conjecture, for all rules that result in large score ranges).

In the future work we will expand our experimentation by developing methods to generate richer classes of randomly generated LP trees. We will also consider the use of ASP tools to aggregate votes given in other preference systems such as CP-nets [2] and answer set optimization (ASO) preferences [3].

Acknowledgments. This work was supported by the NSF grant IIS-0913459.

References

1. Booth, R., Chevaleyre, Y., Lang, J., Mengin, J., Sombattheera, C.: Learning conditionally lexicographic preference relations. In: ECAI, pp. 269–274 (2010)
2. Boutilier, C., Brafman, R., Domshlak, C., Hoos, H., Poole, D.: CP-nets: A tool for representing and reasoning with conditional ceteris paribus preference statements. Journal of Artificial Intelligence Research 21, 135–191 (2004)
3. Brewka, G., Niemelä, I., Truszczynski, M.: Answer set optimization. In: IJCAI, pp. 867–872 (2003)
4. Fargier, H., Conitzer, V., Lang, J., Mengin, J., Schmidt, N.: Issue-by-issue voting: an experimental evaluation. In: MPREF (2012)
5. Gebser, M., Kaminski, R., Kaufmann, B., Ostrowski, M., Schaub, T., Schneider, M.: Potassco: The Potsdam answer set solving collection. AI Communications 24(2), 105–124 (2011)
6. Kaci, S.: Working with Preferences: Less Is More. In: Cognitive Technologies. Springer (2011)
7. Kohli, R., Krishnamurti, R., Mirchandani, P.: The minimum satisfiability problem. SIAM J. Discrete Math. 7(2), 275–283 (1994)
8. Lang, J., Mengin, J., Xia, L.: Aggregating conditionally lexicographic preferences on multi-issue domains. In: Milano, M. (ed.) CP 2012. LNCS, vol. 7514, pp. 973–987. Springer, Heidelberg (2012)
9. Marek, V.W., Truszczynski, M.: Stable models and an alternative logic programming paradigm. In: Apt, K.R., Marek, V.W., Truszczynski, M., Warren, D.S. (eds.) The Logic Programming Paradigm: a 25-Year Perspective, pp. 375–398. Springer, Berlin (1999)
10. Ostrowski, M., Schaub, T.: ASP modulo CSP: The clingcon system. TPLP 12(4-5), 485–503 (2012)
11. Papadimitriou, C., Yannakakis, M.: Optimization, approximation, and complexity classes. In: Proceedings of the Twentieth Annual ACM Symposium on Theory of Computing, STOC 1988, pp. 229–234. ACM, New York (1988)
12. Schulte, C., Tack, G., Lagerkvist, M.Z.: Modeling and programming with gecode (2010)

PREFLIB: A Library for Preferences
HTTP://WWW.PREFLIB.ORG

Nicholas Mattei and Toby Walsh

NICTA and UNSW
Sydney, Australia
{nicholas.mattei,toby.walsh}@nicta.com.au

Abstract. We introduce PREFLIB: A Library for Preferences; an online resource located at http://www.preflib.org. With the emergence of computational social choice and an increased awareness of the applicability of preference reasoning techniques to areas ranging from recommendation systems to kidney exchanges, the interest in preferences has never been higher. We hope to encourage the growth of all facets of preference reasoning by establishing a centralized repository of high quality data based around simple, delimited data formats. We detail the challenges of constructing such a repository, provide a survey of the initial release of the library, and invite the community to use and help expand PREFLIB.

Keywords: Preferences, Computational Social Choice, Empirical Analysis.

1 Introduction

To date, research in computational social choice has been largely theoretical. There is, however, a growing realization of the limitations of a purely theoretical approach to computational questions in social choice. For example, worst-case results about the hardness of manipulation may not reflect the cost in practice to compute manipulations [22–24]. On the other hand, average-case analysis (e.g. [6, 16, 25]) may need to make additional assumptions, which can be rather simplistic and unrepresentative of preferences met in practice. Many such analyses assume that all preferences are equally likely; which is not supported by studies in behavioral social choice [15, 17] or studies on real data [12, 18, 21].

While our main interest is in computational voting there are other fields which fall in the area of preference handling and computational social choice including recommender systems [19], matching [8], and fair division problems [14]. These areas have found new and exciting application areas in modern life including matching kidney donors [5] and allocating students to seats in classrooms. There is a growing movement in the computational social choice community to identify and use **real** preference data to test algorithms and assumptions about voting systems (e.g. [10, 11, 13, 20]). To encourage and facilitate more empirical studies, we discuss building a library of preferences, PREFLIB guided by our experiences building CSPLIB [7].

Initiatives such as the UCI Machine Learning Repository [1] have fostered a greater sense of sharing and collaboration in the machine learning and data mining communities. The contents of the UCI database focus on a broad range of problems. We hope to

P. Perny, M. Pirlot, and A. Tsoukiàs (Eds.): ADT 2013, LNAI 8176, pp. 259–270, 2013.
© Springer-Verlag Berlin Heidelberg 2013

provide a similar level of community, exposure, and sharing, with PREFLIB while taking to heart the lessons learned by other communities that have created and maintained shared tools and data.

2 Motivation

Whilst one of the prime motivations for building a preference library is to encourage and facilitate more empirical studies in computational social choice, there are some other related motivations.

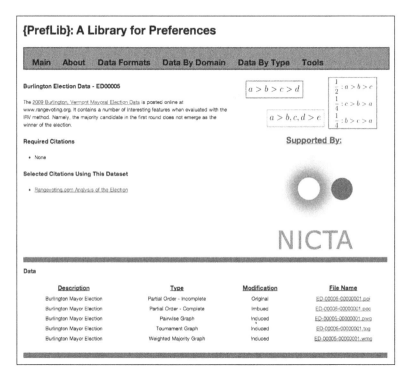

Fig. 1. An example page from http://www.preflib.org with data from the 2009 Burlington, Vermont mayoral elections

Benchmarking: A library can provide a common set of problems on which different research groups can quickly compare their algorithms. For example, we can compare the ability of different heuristics to compute solutions to NP-hard problems like finding a Borda manipulation [4].

Competitions: The Netflix Prize [2] demonstrate the benefits to research that a common set of preference data can have. While the large cash prize undoubtably had a strong impact, the Netflix data continues to be used extensively today despite the prize having been awarded.

Realism: As argued before, real world preference data could help direct the research community onto more practical computational issues in social choice. Representation and learning of complex preference models can happen more readily with a large corpus of preference data that is easily available.

Challenges: A benchmark library can be a forum for challenges that can help push the technology onto new heights. For instance, it can include open problems that may help drive research.

Insularity: The research community looking into computational social choice is rather insular: most people work on their own problems and their own data. A common problem library can encourage people to tackle a common set of problems, and help break down many barriers.

The construction of a benchmark library appears to be a common rite of passage for many research communities [1, 7]. For many reasons, it appears a good time for the computational social choice community and, generally, the preference handling community, to take this step.

3   Challenges

There are a number of challenges in building a preference library.

Variety: Preferences come in many shapes and forms. There are qualitative and quantitative preferences. There are voting preferences which might be simple plurality ballots, lists of approved candidates, lists of vetoes or complete rankings of the candidates. There are preferences for matching problems like hospital-resident problems and kidney exchanges. There are preferences over products in recommender system. There are temporal preferences for scheduling problems. When domains are large, there are combinatorial preferences which might be expressed using CP-nets or one of the many compact preference formalisms. While we wish to include all these in PREFLIB it will be hard to find and post high-quality datasets spanning the entire range of preferences formalisms and domains.

Elicitation: Preferences are difficult to elicit. Users will only answer a limited number of questions about their preferences before they expect a system to start making good recommendations. In addition, users often have difficulty in articulating their true preferences and may not reply truthfully. These problems are well known and somewhat understood by other disciplines such as psychology. While we can learn lessons from these other disciplines, elicitation remains a key challenge in the preference handling community.

Modeling: Part of the challenge in social choice is modeling users' preferences. It is unlikely that your or my preferences are actually a CP-net [3] or even a linear order. These are just formalisms to approximate the complete preference functions we actually have. The existence of a preference library may distract attention from such important modeling issues.

Over-fitting: If the library is small, we run the risk of over-fitting. On the other hand, collecting a lot of preference data to avoid over-fitting may require considerable time and effort.

Privacy and Data Silos: Sharing of many datasets may be precluded or difficult for a variety of reasons. Medical and admissions data may need to be cleaned or anonymized in a suitable way before it can be shared. Additionally, some researchers may not want to share data in order to maintain exclusivity of a research topic. While we hope that all researchers understand the benefit of posting data openly and sharing, there are and will be bumps in the road.

Fortunately, none of these problems are insurmountable. In fact, the answer to many of these problems can be found in the community rallying around a common, open standard and library. With enough contributors we can ensure a large and rich cross section of problem instances, models, and elicitation procedures. With enough contributors the individual time investment will be minimal. And with a focus on sharing researchers will, hopefully, take careful consideration of their methods in order to create datasets.

Taking a page from the UCI Machine Learning Repository [1] we are maintaining a list of research publications that use individual datasets as well as research publications that should be cited along with the use of any particular dataset. By providing credit and exposure to researchers who give back to the community and creating a common resource for research *within* the community we hope that more groups will fully and publicly share their data.

4　Structure of the Library

PREFLIB is currently divided into a four large sections according to overarching data type. The following list is not exhaustive, and is just a starting point for the library. For instance, it does not include preferences in fair division problems or preferences in facility location problems However, we expect that PREFLIB will eventually grow to include such preferences.

Currently PREFLIB holds over 2,000 datasets describing elections, ratings, and matchings. We have 100's of preferences from the Netflix Challenge, 100's of examples of matching data from kidney matching markets. We have several real elections including the 2007 Glasgow City Council elections, Mayoral Elections from the United States, and elections held in Dublin. We are still expanding PREFLIB and we hope to bring more datasets online in the coming weeks and months.

Election Data: Election data includes data from real elections and other instances where rank orders are elicited from individuals. Currently we host a variety of rank order preference information with sources as varied as NASA spacecraft path selection to real ballots from mayoral elections in the United States.

Matching Data: Matching problems include two-sided markets (specifically stable marriage, hospital/resident and hospital/resident with couples problems), and one-sided markets (specifically room-mate and kidney exchange problems). We currently host synthetic data about kidney matching problems and real data related to university course selections.

Combinatorial Data: Large, combinatorial domains introduce interesting issues regarding representation. Combinatorial preferences subdivide into CP-nets, GAI-nets, and lexicographical preferences. Additionally we host single and multi-attribute rating data such as TripAdvisor data. Quantitative and qualitative multi-attribute rating data is of interest to researchers in the recommendation systems area and we hope to bring more datasets in this area online in the near future. **Optimization Data:** Optimization problems subdivide into max-SAT, max-CSP, weighted-CSP and fuzzy-CSP problems.

Each dataset is posted in its original format as well as several easily induced (derived from) or imbued (data added) formats. For instance, for each set of rank ordered preference data we also include an imbued instance where each unranked candidate is placed tied, at the end, of each ranking. We also include an induced tournament graph for each set of rank ordered preference data. We have clearly marked each dataset as original, induced, or imbed, respectively. We encourage caution when drawing broad conclusions from studies on imbued or induced data, (see, e.g., [15,17] for a discussion of potential pitfalls). However the data is interesting for testing of algorithmic results.

5 Preference Data

An important aspect of building a preference library is ensuring the preference data is in an easily accessible and computer readable form. Here we can learn from other domains. For instance, the propositional satisfiability community has a very successful library, SATLIB which grew out of the Second DIMACS challenge in 1992 to 1993. The DIMACS format is widely accepted as the standard for Boolean formulae in CNF. Indeed, every satisfiability solver that we know about will read problems in the DIMACS format.

Why did DIMACS format become a standard? First, the format was in the right place at the right time. The format was proposed at the time that there was a lot of interest in developing new SAT solvers. There was therefore a very immediate need to compare solvers on a set of common benchmarks. Second, the format was quickly adopted by SATLIB and by the semi-annual SAT competition. Third, the format is very simple. Each clause is a line in the input, made up of a sequence of positive and negative numbers terminated by a zero. It doesn't matter what computer language you write in, it takes just a few minutes to write a parser to read such problems.

Another successful format is the TPTP dataset for first order theorem proving. This is a slightly more complex format (but that is perhaps inevitable as first order problems are more complex to specify than purely propositional problems). The TPTP library includes a very useful tool, TPTP2X that converts TPTP problems into all the different formats used by the main first order theorem provers. This helps compensate for the greater complexity of the TPTP format. It means that users can quickly read problems into whatever theorem prover they might want to try out.

Based on these experiences, we use very simple formats for expressing preference data. We have attempted, as much as possible, to preserve a basic comma separated (CSV) format as is possible. This has several advantages including human readability and interoperability with outside data handling programs such as Excel, R, and Matlab. For example, candidates in an election are represented by the numbers 1 to m, and each preference ballot is represented by a permutation of these numbers in a comma separated format.

6   Datasets, Numbering and Tools

Our data comes from a variety of sources and locations. In general, we want data that is honest and comes from real decision makers regarding things that they care about and are incentivized to answer honestly. For example, while an anonymous surveys are good, there is no guarantee that respondents will respond truthfully (or something not completely random). Datasets that are derived from real elections, or real preference data (such as Netflix), or judging on real competitions, can have far more value than random surveys.

To understand the formatting and presentation of the data we present a full element of one of our datasets. This particular dataset provides a partial order over the 20 skaters in the women's 1998 world championships B group qualifier according to their ratings by 9 individual judges.

```
20
1,Maria Butyrskaya
2,Silvia Fontana
3,Vanessa Gusmeroli
4,Yankun Du
5,Anna Wenzel
6,Anna Rechnio
7,Olga Vassiljeva
8,Elena Liashenko
9,Rocia Salas
10,Tanja Szewczenko
11,Valeria Trifancova
12,Marta Andrade
13,Tatyana Malinina
14,Lucinda Ruh
15,Diana Poth
16,Mojca Kopac
17,Zuzana Paurova
18,Roxana Luca
19,Helena Pajovic
20,Yulia Lavrenchuk
9,9,9
1,1,6,20,8,13,10,3,15,12,2,17,14,16,5,4,11,7,19,9,18
1,1,6,8,10,20,3,13,14,2,15,16,17,7,5,12,11,4,18,19,9
1,1,6,3,13,8,10,20,15,2,17,12,5,7,14,16,11,4,19,18,9
1,1,6,10,13,20,3,8,2,14,15,16,17,{11,19},5,7,12,4,18,9
1,1,8,6,3,20,13,10,15,14,12,16,2,17,5,4,18,7,11,19,9
1,1,6,8,10,20,3,13,15,2,14,17,{12,16},4,5,7,11,19,18,9
1,1,6,13,8,20,10,15,3,17,5,2,16,12,7,4,14,11,18,19,9
1,1,6,13,8,20,10,3,16,15,2,17,4,14,5,12,7,11,9,19,18
1,1,6,10,8,13,3,20,15,2,14,12,17,5,16,7,4,11,19,18,9
```

The first line contains the number of candidates or items in this instances. The next set of lines are a number for each of the candidates and the real name or label for the candidate.

The first line under the list of candidates contains information about the number of voters. the first number is the number of actual ballots cast in the instance. The second number is the sum of the preference count (the number of preferences expressed). In most cases the number of ballots is the same as the sum of the vote preference count, except where for example, we have induced a relation like generating a pairwise graph from a set of linear orders. In this case we would have some number n of voters over m alternatives but we would have $\binom{n \cdot m}{2}$ as the sum of preferences since each voter expresses a relation between each pair of elements. The final number of this line is the number of unique preferences expressed.

The remaining lines in the file are the all of the format: count, preference list. The first element is the number of voters expressing the preference list. In the preference list each element is separated by a comma, and we close indifferent alternatives in { }.

In the example, each voter has selected skater 1, Maria Butyrskaya, as the best skater in the pool. Each unique order is indicated by a single line that is comma separated. This allows our data to be easily ported between different applications as it is delimited in a very simple manner.

6.1 Numbering

In order to make navigating particular datasets in PREFLIB easier every individual datafile has a unique identifier which has a common numbering format. Below is the number for the woman's ice-skating world championship dataset shown above along with an explanation of the fields.

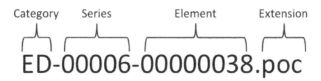

Category   Series   Element   Extension

ED-00006-00000038.poc

Category: Is a 2 letter category code; ED for election data, MD for matching data, CD for combinatorial data, and OD for optimization data.

Series: Is a 5 digit Series Code which specifies the source of the data. The Skate data shown above is number ED-00006.

Element: Is an 8 digit Element Number for each individual file of a particular extension. The example from the Skate data is 00000038, signifying that it is the 38th dataset from the Skate set with the same file extension.

Extension: Is a unique file extension to described the type of data in the file. The list of extensions is updated every time we obtain data in a new domain. For example, we use **soc** for datasets that are complete strict orders (all candidates are ranked with no ties between candidates) and **poc** for complete partial orders (all candidates are ranked but there are some ties between some candidates).

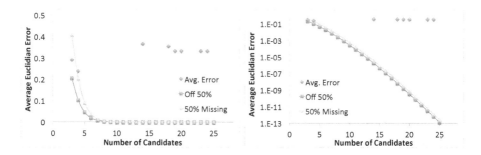

Fig. 2. Comparison of IC and the PREFLIB dataset. Both graphs show the average Euclidian Distance between the empirical distribution given some number of candidates in PREFLIB and IC. Additionally, we have plotted two hypothetical distributions: 50% MISSING assumes probability 0 for 50% of the possible strict orders and OFF 50% assumes each probability of observing a strict order is 50% different than the prediction made by IC. The left plot is linear while the right plot is a log scale.

6.2 Tools

In addition to the preference data on the site we plan to have a small set of tools available to the community. At this time we have no plans to create a monolithic tool chain like Weka [9]. All of the existing toolchain is written in Python3 and includes the ability to read, write, and process all of the data formats present on the site. Additionally, the toolchain includes functions to generate synthetic preferences according to a number of well studied preference cultures including the Impartial Culture, the Impartial Anonymous Culture, the Urn Model, and others [12, 21]. We look forward to adding support for other program languages and models in the future as we receive feedback on the requirements of the research community.

7   Distributions of Preferences

In collecting such a large and diverse set of preference data we hope researchers can begin to ask questions that were not possible before due to lack of data. A central question to the social choice community is testing the validity of generative models of preferences. In particular, we can start to look at the Impartial Culture (IC) assumption with more rigor than previously possible [12, 17, 21].

When looking at the data available in PREFLIB, one of the first observations that we can make is that research in computational social choice may need expand to generalizations of many current results for strict orders to strict orders that are not complete rankings. The current version of PREFLIB contains 220 instances of complete strict order ranking data. However, it also contains 118 incomplete strict order (SOI) ranking data. In fact, the SOI data contains many instances of actual elections from Dublin, Glasgow, and trade unions in the EU. These instances contain, on average, 80% incomplete preference relations, with many votes only expressing a top alternative. There are a number of hazards associated with dropping the incomplete votes or randomly

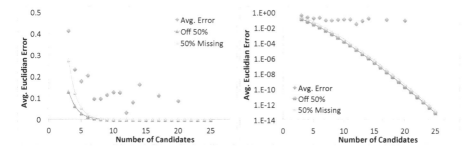

Fig. 3. Comparison of IIC and the PREFLIB dataset. Both graphs show the average Euclidian Distance between the empirical distribution given some number of candidates in PREFLIB and IIC. Additionally, we have plotted two hypothetical distributions: 50% MISSING assumes probability 0 for 50% of the possible strict orders and OFF 50% assumes each probability of observing a strict order is 50% different than the prediction made by IIC. The left plot is linear while the right plot is a log scale.

extending them as these procedures introduce many assumptions about the underlying data [15]. Without generalizing our thinking to include SOI data we may leave real-world behaviors unstudied in computational social choice.

Figures 2, 3, and 4 we compare the Impartial Culture with the SOI and SOC data currently in the PREFLIB data base. While IC is well defined for complete strict orders, we needed a suitable generalization to incomplete strict orders. For this, we make as few assumptions as possible to create the Incomplete-Impartial Culture (IIC): every ordering (including truncated orderings) has an equal probability of occurring. The probability of observing a given ranking r for n candidates is:

$$Pr(r) = \left(\sum_{i=1}^{n} i! \binom{n}{i} \right)^{-1}.$$

We follow the same procedures as Tideman and Plassmann [21] and Mattei et al. [11, 12] in our study. In order to compare an empirical distribution to a generative one we reorder the empirical distribution such that the preference order of the most frequent vote is the labeling for the candidates (we use the most frequent complete order for re-labeling in IIC). This procedure ensures distributions from different empirical scenarios are comparable by giving them a uniform shape. Once we have done this we compute the Euclidian Distance between the empirical distribution and IC or IIC, respectively. We call this number the Euclidian Error and it gives us an idea of how near or far two distributions are from each other.

Figure 2 and 3 shows the error of the empirical distribution on linear and log plots for a given number of candidates. We only plot points where PREFLIB has 2 or more unique datasets. Additionally, we have plotted two hypothetical distributions: 50% MISSING assumes probability 0 for 50% of the possible strict orders and OFF 50% assumes each

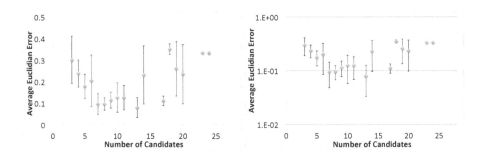

Fig. 4. Comparison of IC/IIC and the PREFLIB dataset. Both graphs show the average Euclidian Distance between the empirical distribution given some number of candidates in PREFLIB and IC/IIC respectively, including standard error bars. The left plot is linear while the right plot is a log scale.

probability of observing a strict order is 50% different than the prediction made by IC/IIC.

We have plotted these additional distributions to give some perspective on just how different IC/IIC is from our empirical distribution. Most would agree that a distribution that is *always* off by 50% to be a fairly poor estimate. When we look at the SOI and SOC data we see that this *bad* distribution is significantly closer to IC/IIC than the empirical distribution found in PREFLIB. These distributions show us just how much IC/IIC diverges with SOI: falling completely outside of the projected curve for either of the "bad" distributions once we have more than 4 or 5 candidates.

Figure 4 shows the SOC and SOI combined average difference and standard error on linear and log plots. Here we can see that, in general, for a given number of candidates, the empirical distribution in PREFLIB is (1) extremely variable and (2) a reasonable distance from IC/IIC. Even with all the data we have collected so far, we are under-sampling. Thus making it unwise to draw too many conclusions from this data. While we are still collecting data we see that what we have currently does not support the simplistic IC/IIC assumptions.

8   How to Contribute

One way to increase the usefulness of PREFLIB is to build a community around the datasets. What we have presented here is only the beginning; we hope that interested researchers will contact us with donations of data or pointers to datasets that we may have missed while constructing the first version of the site.

In order to contribute data please contact Nicholas Mattei; we host all the data so that it is available in a central location. We work collaboratively with all our data donors to convert the data into a simple, CSV-like format. We post links and citations to any donor suggested papers or external websites as well as citations that are requested to accompany the use of particular datasets. We want to make sure that donors who take the time and effort to work with us on posting datasets receive the recognition they deserve for taking the time to support the community.

We make no claims on ownership of data on the website. While we have worked hard to only include high quality, accurate datasets we make no explicit warranties or guarantees about the data and distribute the data "as is."

9   Conclusion

We have introduced the first version of PREFLIB and an associated toolchain for working with preference data. We hope to provide a ongoing and valuable service to not only the computational social choice community, but the preference handling and reasoning community writ large. To support this mission we must have the support and donations of the research community. We encourage anyone with interesting datasets to contact us; we will work with you on encoding and hosting interesting data. Please help us to grow the empirical side of of preference handling.

Acknowledgements. The authors are supported by the Australian Government's Department of Broadband, Communications and the Digital Economy, The Australian Research Council and the Asian Office of Aerospace Research and Development through grant AOARD-124056.

References

1. Bache, K., Lichman, M.: UCI machine learning repository, University of California, Irvine, School of Information and Computer Sciences (2013),
   `http://archive.ics.uci.edu/ml`
2. Bennett, J., Lanning, S.: The Netflix prize. In: Proceedings of the KDD Cup and Workshop (2007)
3. Boutilier, C., Brafman, R.I., Domshlak, C., Hoos, H.H., Poole, D.: CP-nets: A tool for representing and reasoning with conditional ceteris paribus preference statements. Journal of Artificial Intelligence Research (JAIR) 21, 135–191 (2004)
4. Davies, J., Katsirelos, G., Narodytska, N., Walsh, T.: Complexity of and algorithms for Borda manipulation. In: Proceedings of the Twenty-Fifth AAAI Conference on Artificial Intelligence (AAAI 2011), pp. 657–662. AAAI Press (2011)
5. Dickerson, J.P., Procaccia, A.D., Sandholm, T.: Optimizing kidney exchange with transplant chains: Theory and reality. In: Proceedings of the 11th International Conference on Autonomous Agents and Multiagent Systems (AAMAS 2012), pp. 711–718 (2012)
6. Friedgut, E., Kalai, G., Nisan, N.: Elections can be manipulated often. In: Proceedings of the 49th Annual IEEE Symposium on Foundations of Computer Science (FOCS 2008), pp. 243–249. IEEE Computer Society Press (2008)
7. Gent, I.P., Walsh, T.: CSPlib: A benchmark library for constraints. In: Jaffar, J. (ed.) CP 1999. LNCS, vol. 1713, pp. 480–481. Springer, Heidelberg (1999)
8. Gusfield, D., Irving, R.W.: The Stable Marriage Problem: Structure and Algorithms. MIT Press, Cambridge (1989)
9. Hall, M., Frank, E., Holmes, G., Pfahringer, B., Reutemann, P., Witten, I.H.: The WEKA data mining software: an update. ACM SIGKDD Explorations Newsletter 11(1), 10–18 (2009)
10. Lu, T., Boutilier, C.: Robust approximation and incremental elicitation in voting protocols. In: Proceedings of the Twenty-Second International Joint Conference on Artificial Intelligence, pp. 287–293. AAAI Press (2011)

11. Mattei, N.: Empirical evaluation of voting rules with strictly ordered preference data. In: Brafman, R. (ed.) ADT 2011. LNCS, vol. 6992, pp. 165–177. Springer, Heidelberg (2011)
12. Mattei, N.: Decision Making Under Uncertainty: Theoretical and Empirical Results on Social Choice, Manipulation, and Bribery. Ph.D. thesis, University of Kentucky (2012)
13. Mattei, N., Forshee, J., Goldsmith, J.: An empirical study of voting rules and manipulation with large datasets. In: Fourth International Workshop on Computational Social Choice (COMSOC 2012). Springer (2012)
14. Moulin, H.: Fair Division and Collective Welfare. The MIT Press (2004)
15. Popova, A., Regenwetter, M., Mattei, N.: A behavioral perspective on social choice. Annals of Mathematics and Artificial Intelligence (to appear, 2013)
16. Procaccia, A.D., Rosenschein, J.S.: Average-case tractability of manipulation in voting via the fraction of manipulators. In: Proceedings of 6th International Joint Conference on Autonomous Agents and Multiagent Systems (AAMAS 2007), pp. 718–720. IFAAMAS (2007)
17. Regenwetter, M., Grogman, B., Marley, A.A.J., Testlin, I.M.: Behavioral Social Choice: Probabilistic Models, Statistical Inference, and Applications. Cambridge Univ. Press (2006)
18. Regenwetter, M., Kim, A., Kantor, A., Ho, M.H.R.: The unexpected empirical consensus among consensus methods. Psychological Science 18(7), 629–635 (2007)
19. Ricci, F., Rokach, L., Shapira, B.: Introduction to recommender systems handbook. In: Recommender Systems Handbook, pp. 1–35. Springer (2011)
20. Skowron, P., Faliszewski, P., Slinko, A.: Achieving fully proportional representation is easy in practice. In: Proceedings of 12th International Joint Conference on Autonomous Agents and Multiagent Systems (AAMAS 2013), pp. 399–406 (2013)
21. Tideman, T.N., Plassmann, F.: Modeling the outcomes of vote-casting in actual elections. In: Electoral Systems, pp. 217–251. Springer (2012)
22. Walsh, T.: Where are the really hard manipulation problems? The phase transition in manipulating the veto rule. In: Proceedings of the 21st International Joint Conference on Artificial Intelligence (IJCAI 2009), pp. 324–329 (2009)
23. Walsh, T.: An empirical study of the manipulability of single transferable voting. In: Proceedings of the 19th European Conference on Artificial Intelligence (ECAI 2010). Frontiers in Artificial Intelligence and Applications, vol. 215, pp. 257–262. IOS Press (2010)
24. Walsh, T.: Where are the hard manipulation problems? Journal of Artificial Intelligence Research 42, 1–39 (2011)
25. Xia, L., Conitzer, V.: Generalized scoring rules and the frequency of coalitional manipulability. In: EC 2008: Proceedings of the 9th ACM conference on Electronic Commerce, pp. 109–118. ACM (2008)

How to Decrease the Degree of Envy in Allocations of Indivisible Goods

Trung Thanh Nguyen and Jörg Rothe

Institut für Informatik, Heinrich-Heine-Universität Düsseldorf, 40225 Düsseldorf,
Germany

Abstract. We consider the problem of fairly distributing a number of
indivisible goods among agents with additive utility functions. Among
the common criteria of fairness, we focus on envy-freeness and its weaker
notions. Instead of concentrating on envy-free allocations (which might
not always exist), we seek to find an allocation with minimum envy.
Based on a notion introduced by Chevaleyre et al. [7], we define several
problems of minimizing the degree of envy and study their approximability.

1 Introduction

We study the problem of fairly allocating a set of indivisible goods among several
agents that are assumed to have additive utility functions over bundles of goods.
This problem has received much attention in both economics and computer science during last few years, especially due to its many applications in multiagent
resource allocation (see the survey by Chevaleyre et al. [6]).

As a most prominent interpretation of fairness, we focus on envy-freeness,
which means that no agent wants to swap her bundle of goods in an allocation
with another agent. Much of the work so far has investigated envy-freeness in
the setting where one *divisible* good is to be divided among the agents (the so-
called *cake-cutting* problem, see [5,12]). Here, however, we focus on allocating
indivisible, nonshareable goods. While envy-free allocations of a divisible good
always exist (and can even be guaranteed by a finite bounded procedure [4]),
an envy-free allocation of indivisible goods may not exist in general, assuming
that all goods must be assigned to the agents (see Example 1). Therefore, we
seek to compute allocations that ensure envy to be as small as possible. There
are several ways to define a measure of envy for a given allocation. Chevaleyre
et al. [7] proposed a framework for defining the degree of envy of an allocation
based on the degree of envy among individual agents. More formally, let $\pi =
(\pi_1, \ldots, \pi_n)$ be an allocation amongst n agents, where π_i is the bundle assigned
to agent a_i, $1 \leq i \leq n$. Agent a_i's envy regarding agent a_j's bundle is determined
by $u_i(\pi_j) - u_i(\pi_i)$, where u_i is a_i's utility function, and a_i's envy with respect to
allocation π is defined by using the aggregation functions max $(\max_{j \neq i}\{u_i(\pi_j) -
u_i(\pi_i)\})$ and sum $(\sum_{j \neq i}(u_i(\pi_j) - u_i(\pi_i)))$. Finally, the envy of the allocation π is
aggregated from the envy of individual agents via the aggregation functions max

P. Perny, M. Pirlot, and A. Tsoukiàs (Eds.): ADT 2013, LNAI 8176, pp. 271–284, 2013.
© Springer-Verlag Berlin Heidelberg 2013

and sum. Considering the optimization problems based on this measure of envy, a drawback of this approach is that, unless P = NP, there are no approximation algorithms for them, since the objective function might be zero (see the work of Lipton et al. [9]). We circumvent this by defining similar notions of degree of envy based on max and product ($\prod_{j \neq i} u_i(\pi_j)/u_i(\pi_i)$), and study approximability of the corresponding optimization problems.

A number of papers in the literature has investigated the computation of envy-free allocations for indivisible goods. These papers focus on either centralized or distributed protocols, or take into account the way in which the agents' preferences are expressed: cardinal or ordinal. Chevaleyre et al. [7] studied a distributed protocol in which agents negotiate on the exchange of goods to reach an allocation that is envy-free or has minimal envy. Regarding centralized protocols, Bouveret, Endriss, and Lang [2] dealt with the problem where the agents' preferences are represented ordinally by using so-called SCI-nets, while Bouveret and Lang [3] considered the logical aspects of representation and related complexity issues. Lipton et al. [9] addressed the problem of computing allocations with minimal envy when agents have numerical additive preferences, which corresponds exactly to our problem of minimizing $er^{\max,\max}$ (which will be defined in Section 2). Among other results, they provided a *polynomial-time approximation scheme* (PTAS) for the case of agents with identical utility functions and mentioned that one can obtain even a *fully polynomial-time approximation scheme* (FPTAS[1]) for this case if the number of agents is fixed, thus extending the corresponding result of Bazgan et al. [1] for the problem SUBSET-SUMS RATIO: Given a set $S = \{b_1, \ldots, b_n\}$ of positive integers, the goal is to find a partition of S into two subsets, S_1 and S_2, with $\sum_{b_i \in S_1} b_i \geq \sum_{b_i \in S_2} b_i$ such that the ratio $\sum_{b_i \in S_1} b_i / \sum_{b_i \in S_2} b_i$ is minimized. However, the method that gives an FPTAS for such a problem cannot be applied for the case of agents with different utilities.

The rest of this paper is organized as follows. Section 2 defines the notions of degree of envy and models four optimization problems. In Section 3 we present the main result of this paper: an FPTAS for each of our problems. In addition, we provide a hardness of approximation result in Section 4 and an exact polynomial-time algorithm for a restricted case in Section 5. Finally, Section 6 provides some conclusions about the results obtained so far and lists some open questions for future work.

2   Degree of Envy

Let $\mathcal{A} = \{a_1, \ldots, a_n\}$ be a set of agents and $\mathcal{G} = \{g_1, \ldots, g_m\}$ be a set of indivisible goods. Each agent a_i has an *additive utility function* $u_i : 2^{\mathcal{G}} \to \mathbb{Q}^+$,

---

[1] A minimization problem has an FPTAS if for each ε, $0 < \varepsilon < 1$, it can be approximated to a factor of $1 + \varepsilon$ by an algorithm whose running time is polynomial in both the input size and $1/\varepsilon$. A PTAS is defined similarly, except that the running time is required to be polynomial in the input size only. For more details on approximation algorithms, we refer the reader to the textbook by Vazirani [13] and the references cited therein.

i.e., for every subset $B \subseteq \mathcal{G}$, $u_i(B) = \sum_{g_j \in B} u_i(g_j)$. An *allocation* is a partition π of $\mathcal{G}$ into n subsets $(\pi_1, \ldots, \pi_n)$, where π_i is assigned to agent a_i. We now adopt the approach of Chevaleyre et al. [7], who introduced the notion of *degree of envy* in society. Their definition proceeds in three stages by first defining envy between any two agents, then envy of any agent with respect to all other agents, and finally envy of society. With respect to a given allocation $\pi = (\pi_1, \ldots, \pi_n)$, the three stages of the process are as follows:

– **Stage 1 (envy between any two agents):** For each i and $j \neq i$, *agent a_i's envy with respect to agent a_j* is defined as the ratio between a_i's utility for the bundle assigned to a_j and a_i's utility for the bundle assigned to herself:

$$er(i,j) = \frac{u_i(\pi_j)}{u_i(\pi_i)}.$$

– **Stage 2 (degree of envy of any agent):** Here we measure how envious any agent a_i is with respect to anyone else, where we consider two aggregation functions, product and max:

$$er^{\mathrm{pro}}(i) = \prod_{j \neq i} er(i,j) \quad \text{and} \quad er^{\mathrm{max}}(i) = \max_{j \neq i} er(i,j).$$

– **Stage 3 (degree of envy of society):** Based on the degree of envy of individual agents, we define the *degree of envy of society* for allocation π, by again considering the two aggregation functions max and product. While the max function focuses on the most envious agent of society, the product measures envy of society as a whole:

$$er^{\mathrm{pro,opt}}(\pi) = \prod_{i=1}^{n} er^{\mathrm{opt}}(i) \quad \text{and} \quad er^{\mathrm{max,opt}}(\pi) = \max_{i=1}^{n} er^{\mathrm{opt}}(i),$$

where $\mathrm{opt} \in \{\mathrm{max}, \mathrm{pro}\}$.

Here we only consider these two operators, *max* and *product*, for aggregating degrees of envy of individual agents. However, there are also other potential alternatives for the aggregation of individual preferences, e.g., using the *leximin ordering* (see Moulin [11]).

Note that one of the two measures of envy of society given by Lipton et al. [9] corresponds to $er^{\mathrm{max,max}}(\pi)$.

Now, given these notions of the degree of envy of society, we define the following optimization problems, where we let $\mathrm{opt}_1, \mathrm{opt}_2 \in \{\mathrm{max}, \mathrm{pro}\}$.

MINIMUM ENVY $(\mathrm{opt}_1, \mathrm{opt}_2)$

Input: A set of m indivisible goods and a set of n agents, each having an additive utility function over the bundles of goods.

Output: An allocation π that minimizes $\max\{1, er^{\mathrm{opt}_1, \mathrm{opt}_2}(\pi)\}$.

Note that we minimize $\max\{1, er^{\mathrm{opt}_1, \mathrm{opt}_2}(\pi)\}$ rather than $er^{\mathrm{opt}_1, \mathrm{opt}_2}(\pi)$ in order to make the problem fit with the common definition of objective functions in optimization problems so that approximation algorithms can be applied to it and analyzed in a standard manner.

Example 1. Consider an instance with three agents and six single goods, where each agent has an additive utility function which is given in Table 1.

Table 1. Utilities of the agents for Example 1

Resources	Agent a_1	Agent a_2	Agent a_3
r_1	1	**5**	5
r_2	2	5	4
r_3	0	0	1
r_4	3	1	**6**
r_5	**4**	1	4
r_6	**3**	0	2

The numbers in boldface show an allocation whose envy is minimized: $\pi = (\{r_5, r_6\}, \{r_1\}, \{r_2, r_3, r_4\})$ with

$$er^{\max,\max}(\pi) = \left\{ \frac{1}{7}, \frac{5}{7}, \frac{1}{5}, \frac{6}{5}, \frac{6}{11}, \frac{5}{11} \right\} = \frac{6}{5}.$$

However, there is no envy-free allocation for this instance.

In comparison with other notions of fairness, it is worth noting that allocations with minimum envy do not always optimize either egalitarian social welfare (the utility of the agent who is worst off) or social welfare by Nash (the product of the individual agent utilities), and vice versa. For instance, the allocation π shown in Example 1 has the Nash product $7 \cdot 5 \cdot 11 = 385$ and thus does not maximize Nash social welfare. Also, it is not an optimal allocation with respect to the egalitarian social welfare. Conversely, the allocation $\pi' = (\{r_5, r_6\}, \{r_1, r_2\}, \{r_3, r_4\})$ maximizes both the egalitarian and Nash social welfare, but its envy $er^{\max,\max}(\pi') = 9/7$ is not minimal.

3   Approximation Schemes

In this section, we prove that there is a fully polynomial-time approximation scheme for the problem of minimizing envy, for any pair $(\mathrm{opt}_1, \mathrm{opt}_2)$, where $\mathrm{opt}_i \in \{\max, \mathrm{pro}\}$, for a bounded number of agents. Our proof proceeds as follows: We first design a pseudo-polynomial-time algorithm for our optimization problem by using dynamic programming and then modify it in a suitable way to obtain an FPTAS.

Let $M = (\mathcal{A}, \mathcal{G})$ be an instance of the problem in which each agent a_i, $1 \leq i \leq n$, has an additive utility function over the bundles of goods. We denote by s_{ij} the value of good g_j for agent a_i. Without loss of generality, we assume that $s_{ij} \in \mathbb{N}$ for all i and j. In Algorithm 1 below, V_j denotes the set of all possible allocations of $\mathcal{G}_j = \{g_1, \ldots, g_j\}$, which assigns the first j goods in $\mathcal{G}$ to agents. Each allocation in V_j is represented by a vector $\boldsymbol{v} = (v_{ip})$ in which v_{ip}

is the evaluation of a_p's bundle by agent a_i for all $i, p \in \{1, \ldots, n\}$. The envy of the allocation corresponding to a vector $\boldsymbol{v} \in V_j$ is denoted by $er^{\text{opt}_1, \text{opt}_2}(\boldsymbol{v})$. We denote by $\boldsymbol{\mu}_{k,i}$, $1 \leq k, i \leq n$, a vector of dimension n^2 with a 1 in the coordinate t_{ki} and a 0 everywhere else. Note that there are totally n^2 such vectors.

Lemma 1. *Algorithm 1 is a pseudo-polynomial-time algorithm.*

Proof. First, it is easy to see that the running time of Algorithm 1 is $\mathcal{O}(m \sum_{j=1}^m \|V_j\|)$, where $\|V_j\|$ denotes the size of the set V_j that is created after Step 6. Let $B = \max_{1 \leq i \leq n} \sum_{j=1}^m s_{ij}$. Then it is important to note that the coordinates of any vector of V_j are nonnegative integers not exceeding B. Therefore, the size of any set V_j is always bounded by B^{n^2}. Finally, the running time of Algorithm 1 is $\mathcal{O}(mB^{n^2})$ and thus it runs in pseudo-polynomial time. $\square$

Algorithm 1. Pseudo-polynomial-time algorithm

1: $V_0 := \{\boldsymbol{0}\}$;
2: **for** $j := 1$ **to** m **do**
3:     $V_j := \emptyset$;
4:     **for each** $\boldsymbol{v} \in V_{j-1}$ **do**
5:         **for** $i := 1$ **to** n **do**
6:             $V_j := V_j \cup \{\boldsymbol{v} + \sum_{k=1}^n s_{kj} \cdot \boldsymbol{\mu}_{k,i}\}$
7:         **end for**
8:     **end for**
9: **end for**
10: **return** vector $\boldsymbol{v} \in V_m$ that minimizes $er^{\text{opt}_1, \text{opt}_2}(\boldsymbol{v})$.

Let ε be any fixed number such that $0 < \varepsilon < 1$, and let

$$\lambda = 1 + \frac{\varepsilon}{4m\delta},$$

where δ depends on n and will be determined later. Algorithm 2 below modifies Algorithm 1 by changing the sets of vectors V_j once they have been created. This will help us to keep the number of vectors as low as possible. Of course, that may also make Algorithm 1 return inexact solutions, but we will still obtain a good approximation of the solution. In more detail, we will delete some unnecessary vectors in V_j in a way such that all remaining vectors of V_j are not *equivalent* in the sense defined below.

Let $K = \lceil \log_\lambda B \rceil$ and $L_k = [\lambda^{k-1}, \lambda^k]$, where $1 \leq k \leq K$. The equivalence relation $\equiv$ on the set V_j is defined as follows. Any two vectors $\boldsymbol{x} = (x_1, \ldots, x_{n^2})$ and $\boldsymbol{y} = (y_1, \ldots, y_{n^2})$ are equivalent, denoted by $\boldsymbol{x} \equiv \boldsymbol{y}$, if for every ℓ, $1 \leq \ell \leq n^2$, we have either $x_\ell = y_\ell = 0$ or $x_\ell, y_\ell \in L_k$ for some k. One can easily check that this relation is reflexive, symmetric, and transitive and, thus, is an equivalence relation. By this relation, each set of vectors V_j can be divided into equivalence classes, i.e., any two vectors from the same class are equivalent with respect to $\equiv$. We claim that if $x \equiv y$ then we have

$$\frac{y_\ell}{\lambda} \leq x_\ell \leq \lambda y_\ell \quad \text{and} \quad \frac{x_\ell}{\lambda} \leq y_\ell \leq \lambda x_\ell \qquad (1)$$

for all ℓ, $1 \le \ell \le n^2$. Indeed, the statement is obviously true if $x_\ell = y_\ell = 0$. Consider now the case that $x_\ell, y_\ell \in L_k$ for some k, that is, $\lambda^{k-1} \le x_\ell, y_\ell \le \lambda^k$. We have

$$\frac{x_\ell}{y_\ell} \ge \frac{\lambda^{k-1}}{\lambda^k} = \frac{1}{\lambda} \quad \text{and} \quad \frac{x_\ell}{y_\ell} \le \frac{\lambda^k}{\lambda^{k-1}} = \lambda.$$

Algorithm 2. Fully polynomial-time approximation scheme

1: $V_0 := \{\mathbf{0}\}; V_0^* := \{\mathbf{0}\};$
2: **for** $j := 1$ **to** m **do**
3:     $V_j := \emptyset;$
4:     **for each** $\mathbf{v}^* \in V_{j-1}^*$ **do**
5:         **for** $i := 1$ **to** n **do**
6:             $V_j := V_j \cup \{\mathbf{v}^* + \sum_{k=1}^n s_{kj} \cdot \boldsymbol{\mu}_{k,i}\}$
7:         **end for**
8:     **end for**
9:     $V_j^* := \text{REDUCE}(V_j)$
10: **end for**
11: **return** vector $\mathbf{v}^* \in V_m^*$ that minimizes $er^{\text{opt}_1, \text{opt}_2}(\mathbf{v}^*)$.

By the procedure $\text{REDUCE}(V_j)$, the set of vectors V_j is divided into equivalence classes due to the relation $\equiv$ and then some vectors are removed such that each class contains only one vector. It is easy to see that in Algorithm 2, we do the same steps as in the Algorithm 1, but this time the set of vectors V_j will be created from V_{j-1}^* rather than from V_{j-1} and then is modified by the procedure REDUCE applied to V_j to get the reduced set V_j^*. Note that this procedure also ensures that the number of vectors in V_j^* is always bounded by K^{n^2}. In the following, we will show the relationship between the two sets V_j and V_j^*.

Lemma 2. *Let V_j and V_j^* be the two sets of vectors that have been created by Algorithms 1 and 2, respectively. For each vector $\mathbf{v} = (v_{ip}) \in V_j$, there always exists a vector $\mathbf{v}^* = (v_{ip}^*) \in V_j^*$ such that for all $i \ne p, 1 \le i, p \le n$:*

$$v_{ii}^* \ge \frac{v_{ii}}{\lambda^j} \quad \text{and} \quad v_{ip}^* \le \lambda^j \cdot v_{ip}.$$

Proof. The proof is by induction on j. The case with $j = 1$ is true by Equation (1) and the fact that $V_1^* = V_1$. Assume that the statement is true for $j - 1$. Consider an arbitrary vector $\mathbf{v} = (v_{ip}) \in V_j$. This vector $\mathbf{v}$ must be created in line 6 of Algorithm 1 from some vector $\mathbf{w} = (w_{ip})$ of the set V_{j-1}. Without loss of generality, we assume that $\mathbf{v}$ has the form

$$(w_{11} + s_{1j}, w_{12}, \ldots, w_{1n}, \ldots, w_{n1} + s_{nj}, w_{n2}, \ldots, w_{nn}),$$

where $v_{i1} = w_{i1} + s_{ij}$ for $i \in \{1, \ldots, n\}$ and $v_{ip} = w_{ip}$ for $p \ne 1$. Using the inductive assumption above, there exists some vector $\mathbf{w}^* = (w_{ip}^*) \in V_{j-1}^*$ such that

$$w_{ii}^* \ge \frac{w_{ii}}{\lambda^{j-1}} \quad \text{and} \quad w_{ip}^* \le \lambda^{j-1} \cdot w_{ip} \quad (i \ne p) \tag{2}$$

for all $i, p \in \{1, \ldots, n\}$. On the other hand, note that the vector

$$(w_{11}^* + s_{1j}, w_{12}^*, \ldots, w_{1n}^*, \ldots, w_{n1}^* + s_{nj}, w_{n2}^*, \ldots, w_{nn}^*)$$

will also be created for V_j but may be removed by the procedure REDUCE(V_j), which outputs V_j^*. However, there must be another vector $\boldsymbol{v}^* = (v_{ip}^*) \in V_j^*$ such that $\boldsymbol{v}^* \equiv \boldsymbol{w}^*$. This yields

$$v_{11}^* \geq \frac{1}{\lambda} \cdot (w_{11}^* + s_{1j}) \geq \frac{1}{\lambda^j} \cdot w_{11} + \frac{1}{\lambda} \cdot s_{1j} \geq \frac{1}{\lambda^j} \cdot (w_{11} + s_{1j}) = \frac{1}{\lambda^j} \cdot v_{11}.$$

For $i \neq 1$, we have

$$v_{ii}^* \geq (1/\lambda) \cdot w_{ii}^* \geq (1/\lambda^j) \cdot w_{ii} = (1/\lambda^j) \cdot v_{ii}$$

and

$$v_{i1}^* \leq \lambda \cdot (w_{i1}^* + s_{ij}) \leq \lambda^j \cdot w_{i1} + \lambda \cdot s_{ij} \leq \lambda^j \cdot (w_{i1} + s_{ij}) = \lambda^j \cdot v_{i1}.$$

For any $p \neq 1$ and $i \neq p$, we have

$$v_{ip}^* \leq \lambda \cdot w_{ip}^* \leq \lambda^j \cdot w_{ip} = \lambda^j \cdot v_{ip}.$$

This completes the proof.   ❑

Lemma 3. *The running time of Algorithm 2 is bounded polynomially in both the input size and $1/\varepsilon$.*

Proof. We prove that Algorithm 2 has a running time that is polynomial in m, $|M|$, and $1/\varepsilon$. First, since the set V_m^* has at most K^{n^2} vectors, the running time of the algorithm is in $\mathcal{O}(mK^{2n^2})$. On the other hand, we have

$$K = \lceil \log_\lambda B \rceil = \left\lceil \frac{\ln B}{\ln \lambda} \right\rceil = \left\lceil \frac{\ln B}{\ln (1 + \varepsilon/4m\delta)} \right\rceil < \left\lceil \left(1 + \frac{4m\delta}{\varepsilon}\right) \ln B \right\rceil.$$

The above inequality follows, since $f(a) = \ln a - 1 + 1/a$ is a continuous, increasing function on the interval $(1, \infty)$. This function is increasing on this interval, as $f'(a) = 1/a - 1/a^2 > 0$ for all $a > 1$. Hence, we have $f(a) > f(1) = 0$ for all $a > 1$. By choosing $a = \alpha$, the inequality follows.

Furthermore, we have $|M| \geq \log B = (\log e)(\ln B)$ and thus:

$$K \leq \left(1 + \frac{4m\delta}{\varepsilon}\right) \frac{|M|}{\log e}.$$

Note that δ depends only on n and thus is constant. This implies that K is bounded by the input size and $1/\varepsilon$, completing the proof.   ❑

We are now ready to prove the main result of this section.

Theorem 1. *For any fixed $\varepsilon > 0$ and any pair $(\mathrm{opt}_1, \mathrm{opt}_2)$ with $\mathrm{opt}_i \in \{\max,$ pro$\}$, Algorithm 2 always produces an allocation whose envy is within a factor of $1 + \varepsilon$ of the optimum.*

Proof. Let $\boldsymbol{v} = (v_{ip}) \in V_m$ be a vector returned by Algorithm 1. By Lemma 2, there exists a vector $\boldsymbol{v}^* = (v_{ip}^*) \in V_m^*$ such that

$$v_{ii}^* \geq \frac{1}{\lambda^m} \cdot v_{ii} \quad \text{and} \quad v_{ip}^* \leq \lambda^m \cdot v_{ip} \quad (i \neq p)$$

for all $i, p \in \{1, \ldots, n\}$. Assume Algorithm 2 outputs a vector $\boldsymbol{x}$. By Lemma 2 and the fact that $er^{\mathrm{opt}_1, \mathrm{opt}_2}(\boldsymbol{v}^*) \geq er^{\mathrm{opt}_1, \mathrm{opt}_2}(\boldsymbol{x}) \geq er^{\mathrm{opt}_1, \mathrm{opt}_2}(\boldsymbol{v})$, it is easy to see that $er^{\mathrm{opt}_1, \mathrm{opt}_2}(\boldsymbol{x}) = \infty$ if and only if $er^{\mathrm{opt}_1, \mathrm{opt}_2}(\boldsymbol{v}) = \infty$. In case $er^{\mathrm{opt}_1, \mathrm{opt}_2}(\boldsymbol{v}) < \infty$, we have $v_{ii}, v_{ii}^* \neq 0$ for all i, and so

$$\frac{v_{ip}^*}{v_{ii}^*} \leq \lambda^{2m} \cdot \frac{v_{ip}}{v_{ii}} \quad (i \neq p)$$

for all $i, p \in \{1, \ldots, n\}$. By choosing δ appropriately, we can show that for each pair $(\mathrm{opt}_1, \mathrm{opt}_2)$:

$$\max\{1, er^{\mathrm{opt}_1, \mathrm{opt}_2}(\boldsymbol{x})\} \leq (1 + \varepsilon) \max\left\{1, er^{\mathrm{opt}_1, \mathrm{opt}_2}(\boldsymbol{v})\right\}. \tag{3}$$

Indeed, we prove the claim for the four possible cases below. Note that we have $er^{\mathrm{opt}_1, \mathrm{opt}_2}(\boldsymbol{x}) \leq er^{\mathrm{opt}_1, \mathrm{opt}_2}(\boldsymbol{v}^*)$. The inequality below is very helpful and not difficult to prove (but we omit the proof due to the space limit):

$$\lambda^{2m\delta} = \left(1 + \frac{\varepsilon}{4m\delta}\right)^{2m\delta} \leq e^{\varepsilon/2} \leq 1 + \varepsilon. \tag{4}$$

The first inequality follows from the known inequality $(1 + x/z)^z \leq e^x$ for all $z \geq 1$. The second inequality can be proven easily as follows. Consider the function $f(x) = e^x - 1 - 2x$ on the interval $[0, 1]$. The derivative $f'(x) = 0$ if and only if $x = \ln 2$. Therefore,

$$\max_{x \in [0,1]} f(x) = \max\{f(0), f(1), f(\ln 2)\} = f(0) = 0.$$

It follows that $f(x) = e^x - 1 - 2x \leq 0$, or equivalently, $e^x \leq 1 + 2x$ for all $x \geq 0$. Finally, choosing $x = \varepsilon/2$ the proof of inequality (4) is completed.

Case 1: $(\mathrm{opt}_1, \mathrm{opt}_2) = (\mathrm{pro}, \mathrm{pro})$. In this case, we have

$$\max\{1, er^{\mathrm{pro}, \mathrm{pro}}(\boldsymbol{x})\} \leq \max\left\{1, \prod_{i=1}^n \prod_{p \neq i} \frac{v_{ip}^*}{v_{ii}^*}\right\}$$

$$\leq \max\left\{1, \lambda^{2mn(n-1)} \prod_{i=1}^n \prod_{p \neq i} \frac{v_{ip}}{v_{ii}}\right\}$$

$$\leq \lambda^{2mn(n-1)} \max\{1, er^{\mathrm{pro}, \mathrm{pro}}(\boldsymbol{v})\}.$$

Choosing $\delta = n(n - 1)$ and applying inequality (4), we obtain (3).

Case 2: $(\mathrm{opt}_1, \mathrm{opt}_2) = (\max, \mathrm{pro})$. Here we obtain

$$
\max\{1, er^{\max,\mathrm{pro}}(\boldsymbol{x})\} \leq \max_{i=1}^{n}\left\{1, \prod_{p\neq i}\frac{v_{ip}^*}{v_{ii}^*}\right\}
$$

$$
\leq \max_{i=1}^{n}\left\{1, \lambda^{2m(n-1)}\prod_{p\neq i}\frac{v_{ip}}{v_{ii}}\right\}
$$

$$
\leq \lambda^{2m(n-1)}\max\left\{1, er^{\max,\mathrm{pro}}(\boldsymbol{v})\right\}.
$$

Choosing $\delta = n - 1$ and applying inequality (4), we again obtain (3).

Case 3: $(\mathrm{opt}_1, \mathrm{opt}_2) = (\mathrm{pro}, \max)$. Now we have

$$
\max\{1, er^{\mathrm{pro},\max}(\boldsymbol{x})\} \leq \max\left\{1, \prod_{i=1}^{n}\max_{p\neq i}\frac{v_{ip}^*}{v_{ii}^*}\right\}
$$

$$
\leq \max\left\{1, \lambda^{2m(n-1)}\prod_{i=1}^{n}\max_{p\neq i}\frac{v_{ip}}{v_{ii}}\right\}
$$

$$
\leq \lambda^{2m(n-1)}\max\left\{1, er^{\mathrm{pro},\max}(\boldsymbol{v})\right\}.
$$

Choosing $\delta = n - 1$ and applying inequality (4), we obtain (3).

Case 4: $(\mathrm{opt}_1, \mathrm{opt}_2) = (\max, \max)$. Finally, we have

$$
\max\{1, er^{\max,\max}(\boldsymbol{x})\} \leq \max\left\{1, \max_{i=1}^{n}\max_{p\neq i}\frac{v_{ip}^*}{v_{ii}^*}\right\}
$$

$$
\leq \max\left\{1, \lambda^{2m}\max_{i=1}^{n}\max_{p\neq i}\frac{v_{ip}}{v_{ii}}\right\}
$$

$$
\leq \lambda^{2m}\max\left\{1, er^{\max,\max}(\boldsymbol{v})\right\}.
$$

Choosing $\delta = 1$ and applying inequality (4), we again obtain (3).

This completes the proof. □

4 An Inapproximability Result

In this section we prove that if the number of agents is part of the input, the problem MINIMUM ENVY $(\max, \max)$ does not have a polynomial-time approximation scheme. This result can be extended without difficulty to the other cases of $(\mathrm{opt}_1, \mathrm{opt}_2)$ with $\mathrm{opt}_i \in \{\max, \mathrm{pro}\}$.

Theorem 2. *There exists a constant α such that unless* P $=$ NP*, there is no polynomial-time approximation algorithm within a factor of α for the problem* MINIMUM ENVY $(\max, \max)$.

Proof. The proof is by a reduction from the well-known NP-complete problem EXACT COVER BY THREE SETS (X3C, for short): Given a finite set B with $\|B\| = 3q$ and a collection $C = \{S_1, \ldots, S_n\}$ of 3-element subsets of B, does there exist a subcollection $C' \subseteq C$ such that every element of B occurs in exactly one of the sets in C'? Let $(B, \{S_1, \ldots, S_n\})$ with $\|B\| = 3q$ be an instance of X3C, where we may assume that $n \geq q$. We construct an instance M as follows: There are n agents, each corresponding to one set S_i, $1 \leq i \leq n$; the set of goods contains $2q + n$ single goods, where $3q$ "real" goods correspond to the $3q$ elements of B and there are $n - q$ "dummy" goods. For each i, $1 \leq i \leq n$, agent a_i has utility 1 for each good in S_i and utility 0 for each good in $B \setminus S_i$. Every dummy good has utility 3 for all agents. We also denote by u_i the additive utility function of agent a_i.

Suppose that $(B, \{S_1, \ldots, S_n\})$ is a yes-instance of X3C. Then there exists a set $I \subseteq \{1, \ldots, n\}$, $\|I\| = q$, such that $S_i \cap S_j = \emptyset$ for all $i, j \in I$, $i \neq j$, and $\bigcup_{i \in I} S_i = B$. Hence, we assign the bundle S_i to agent a_i for each $i \in I$, and each dummy good to one of the $n - q$ remaining agents. This allocation has an envy of 1 and thus is optimal.

Conversely, if $(B, \{S_1, \ldots, S_n\})$ is a no-instance of X3C, we show that any optimal allocation for M has always envy of at least $3/2$. Indeed, let $\pi = (\pi_1, \ldots \pi_n)$ be an optimal allocation for M, and consider the following two cases for π. First, if there is some agent a_i whose bundle π_i contains at least two dummy goods, then there must be another agent a_k owning a bundle π_k of value at most 3. This implies that

$$er^{\max,\max}(\pi) \geq \frac{u_k(\pi_i)}{u_k(\pi_k)} \geq 2$$

and thus the envy of π is at least 2 in this case. Second, consider the case that the $n - q$ dummy goods are assigned to $n - q$ distinct agents and let a_i be one of these. Since there are at most $q - 1$ disjoint sets from $S_1, \ldots, S_n$, there must be at least one agent a_k who is assigned a bundle π_k of value at most 2. Furthermore, the bundle π_i of a_i has utility at least 3 for agent a_k. Hence,

$$er^{\max,\max}(\pi) \geq \frac{u_k(\pi_i)}{u_k(\pi_k)} \geq \frac{3}{2}$$

To sum up, an approximation algorithm with a factor better than $3/2$ will distinguish the yes- from the no-instances of X3C in polynomial time, contradicting NP-hardness of X3C unless P = NP. The proof is completed. ❏

5   A Restricted Case

We next consider the problem of minimizing the envy in the special case when there are as many agents as goods. By applying a matching technique, we will show that one can solve this restricted case in polynomial time. The crucial point here is that each agent will get exactly one good, for otherwise there would be at least one agent owning nothing and thus the envy of the allocation would be infinite. Hence, one can transfer our optimization problem into a problem of finding a suitable maximum matching in a weighted bipartite graph. We consider the following two variants of the matching problem: MIN-MAX-MATCHING and MAX-PRO-MATCHING.

MIN-MAX-MATCHING
Input: A bipartite graph $G = (X \cup Y, E)$ and a weight function $w : E \to \mathbb{R}^+$.
Output: A *min-max-matching*, i.e., a maximum matching $\mathcal{M} \subseteq E$ such that $\max_{e \in \mathcal{M}} w(e) \leq \max_{e \in \mathcal{M}'} w(e)$ for any other maximum matching $\mathcal{M}'$ of G.

MAX-PRO-MATCHING
Input: A bipartite graph $G = (X \cup Y, E)$ and a weight function $w : E \to \mathbb{R}^+$.
Output: A *max-pro-matching*, i.e., a maximum matching $\mathcal{M} \subseteq E$ that maximizes $\prod_{e \in \mathcal{M}} w(e)$.

Lemma 4. *Given a weighted bipartite graph G, a min-max-matching of G can be found in polynomial time.*

Proof. Given a weighted bipartite graph G, a min-max-matching of value k can be found in polynomial time as follows. Let G_k be a subgraph of G that contains only the edges of weight less than or equal to k. Obviously, G has a min-max-matching of value k if and only if G_k has a maximum matching. The smallest value of k can be found by binary search. □

Lemma 5. *Given a weighted bipartite graph G, a max-pro-matching of G can be found in polynomial time.*

Proof. We may assume that G is a complete bipartite graph, since if there exist two vertices $x, y \in V$ with $(x, y) \notin E$, we can add the edge (x, y) of weight zero to E. Moreover, by multiplying every weight $w(e)$ by a large enough number it suffices to consider the case when $w(e) \in \mathbb{R}_{\geq 1} \cup \{0\}$. The basic idea is to transform the given MAX-PRO-MATCHING instance into an instance of the problem of finding a maximum weighted maximum matching (i.e., a maximum matching of maximum weight), which can be solved in polynomial time (for more details, see [10]). The weight function $w' : E \to \mathbb{R}$ is defined as follows:

$$w'(e_i) = \begin{cases} \log w(e_i) & \text{if } w(e_i) \neq 0 \\ -k \log \Delta & \text{otherwise,} \end{cases}$$

where $\Delta = \max_{e_i \in E} w(e_i)$ and $k = \min\{\|X\|, \|Y\|\}$.

Without loss of generality, we can assume that $\mathcal{M} = \{e_1, \ldots, e_k\}$ is a maximum matching of G with weight function w' so that $\sum_{i=1}^{k} w'(e_i)$ is maximal. We now prove that $\mathcal{M}$ is exactly a matching of G that maximizes $\prod_{i=1}^{k} w(e_i)$. It is easy to see that $\sum_{i=1}^{k} w'(e_i) \leq 0$ if and only if there exists an edge $e_i \in \mathcal{M}$ with negative weight $w'(e_i) = -k \log \Delta$, i.e., $w(e_i) = 0$. In this case, for every maximum matching $\mathcal{M}$ of G, the product of the edge weights in $\mathcal{M}$ is zero with respect to the weight function w. Now suppose that $w(e_i) > 0$ for all $i \in \{1, \ldots, k\}$. Assume that there exists another matching $\mathcal{M}' = \{e_1', \ldots, e_k'\}$ of G such that $\prod_{i=1}^{k} w(e_i') > \prod_{i=1}^{k} w(e_i)$. This implies that

$$\log\left(\prod_{i=1}^{k} w(e_i')\right) > \log\left(\prod_{i=1}^{k} w(e_i)\right)$$

or, equivalently,

$$\sum_{i=1}^{k} \log w(e_i') > \sum_{i=1}^{k} \log w(e_i)$$

which in turn is equivalent to

$$\sum_{i=1}^{k} w'(e_i') > \sum_{i=1}^{k} w'(e_i)$$

This is a contradiction.   ❑

Theorem 3. *For any pair* $(\mathrm{opt}_1, \mathrm{opt}_2)$ *with* $\mathrm{opt}_i \in \{\max, \mathrm{pro}\}$, *an allocation of minimum envy can be found in polynomial time if the number of agents and the number of goods are the same.*

Proof. Let $\mathcal{A} = \{a_1, \ldots, a_n\}$ be a set of agents and $\mathcal{G} = \{g_1, \ldots, g_n\}$ be a set of goods, where we assume that each agent a_i has an additive utility function u_i over the set of goods. Consider the following two cases:

Case 1: $\mathrm{opt}_1 = \max$. We construct a weighted bipartite graph $G = (X \cup Y, E)$ in which X and Y correspond to the set of agents $\mathcal{A}$ and the set of goods $\mathcal{G}$, respectively. The weight w function is defined as

$$w((a_i, g_j)) = \begin{cases} \max_{k \neq j}\left\{\dfrac{u_i(g_k)}{u_i(g_j)}\right\} & \text{if } \mathrm{opt}_2 = \max \\[2ex] \prod_{k \neq j} \dfrac{u_i(g_k)}{u_i(g_j)} & \text{if } \mathrm{opt}_2 = \mathrm{pro} \end{cases}$$

It is not difficult to see that the optimal allocation for instance M corresponds exactly to a min-max-matching $\mathcal{M}$ of G, which can be solved exactly by Lemma 4.

Case 2: $\mathrm{opt}_1 = \mathrm{pro}$. We construct a weighted bipartite graph $G = (X \cup Y, E)$ similarly as in the case 1, but this time the optimal allocation for instance M corresponds exactly to a max-pro-matching $\mathcal{M}$ of G, which can be solved exactly by Lemma 5.

This completes the proof.   ❑

6 Conclusions and Future Work

We have studied the problem of computing minimum envy allocations of indivisible goods when agents have additive utilities over bundles of goods. Building on the work of Lipton et al. [9] and Chevaleyre et al. [7], we have analyzed an alternative metric to measure the envy between two agents and the envy of society as well. Based on these measures, we model the optimization problems of minimizing envy and study their approximability. Our main result shows that these problems admit an FPTAS for the case when the number of agents is not part of the input. We have also provided a hardness of constant-factor approximation result. For the restricted case when there are as many agents as goods, we have presented a nontrivial polynomial-time algorithm that computes exact allocations of minimum envy.

As future work, it would be interesting to find a (constant) approximation algorithm for our optimization problems in the general case. Note that our problem MINIMUM ENVY (max, max) is closely related to the problem MINIMUM MAKESPAN on unrelated machines, which is known to be approximable to a factor of 2, by using integer linear programming (see [8]). Therefore, we conjecture that MINIMUM ENVY (max, max) can be also approximated to within certain constant factor. Regarding the negative results, there is still room for improving the $(3/2)$-hardness factor obtained in Section 4.

Acknowledgements. This work was supported in part by DFG grant RO-1202/14-1, a DAAD grant for a PPP project in the PROCOPE programme, and a fellowship from the Vietnamese government.

References

1. Bazgan, C., Sántha, M., Tuza, Z.: Efficient approximation algorithms for the SUBSET-SUMS EQUALITY problem. In: Larsen, K.G., Skyum, S., Winskel, G. (eds.) ICALP 1998. LNCS, vol. 1443, pp. 387–396. Springer, Heidelberg (1998)
2. Bouveret, S., Endriss, U., Lang, J.: Fair division under ordinal preferences: Computing envy-free allocations of indivisible goods. In: Proceedings of the 19th European Conference on Artificial Intelligence, pp. 387–392. IOS Press (August 2010)
3. Bouveret, S., Lang, J.: Efficiency and envy-freeness in fair division of indivisible goods: Logical representation and complexity. Journal of Artificial Intelligence Research 32, 525–564 (2008)
4. Brams, S., Taylor, A.: An envy-free cake division protocol. The American Mathematical Monthly 102(1), 9–18 (1995)
5. Brams, S., Taylor, A.: Fair Division: From Cake-Cutting to Dispute Resolution. Cambridge University Press (1996)
6. Chevaleyre, Y., Dunne, P., Endriss, U., Lang, J., Lemaître, M., Maudet, N., Padget, J., Phelps, S., Rodríguez-Aguilar, J., Sousa, P.: Issues in multiagent resource allocation. Informatica 30, 3–31 (2006)
7. Chevaleyre, Y., Endriss, U., Estivie, S., Maudet, N.: Reaching envy-free states in distributed negotiation settings. In: Proceedings of the 20th International Joint Conference on Artificial Intelligence, IJCAI 2007, pp. 1239–1244 (January 2007)

8. Lenstra, J., Shmoys, D., Tardos, E.: Approximation algorithms for scheduling un-related parallel machines. Mathematical Programming 46(1), 259–271 (1990)
9. Lipton, R., Markakis, E., Mossel, E., Saberi, A.: On approximately fair allocations of indivisible goods. In: Proceedings of the 5th ACM Conference on Electronic Commerce, pp. 125–131. ACM Press (2004)
10. Mehlhorn, K., Näher, S.: LEDA: A Platform for Combinatorial and Geometric Computing. Cambridge University Press (1999)
11. Moulin, H.: Axioms of Cooperative Decision Making. Cambridge University Press (1988)
12. Robertson, J., Webb, W.: Cake-cutting algorithms: Be fair if you can. AK Peters (1998)
13. Vazirani, V.: Approximation Algorithms. 2edn. Springer (2003)

Descriptive Profiles for Sets of Alternatives in Multiple Criteria Decision Aid

Alexandru-Liviu Olteanu[1,2,3], Patrick Meyer[2,3], and Raymond Bisdorff[1]

[1] CSC/ILIAS, FSTC, University of Luxembourg
[2] Institut Télécom, Télécom Bretagne, UMR CNRS 6285 Lab-STICC, Technopôle
Brest Iroise, CS 83818, 29238 Brest Cedex 3, France
[3] Université Européenne de Bretagne

Abstract. In the context of Multiple Criteria Decision Aid, a decision-maker may be faced at any time with the task of analysing one or several sets of alternatives, irrespective of the decision he is about to make. As in this case the alternatives may express contrasting gains and losses on the criteria on which they are evaluated, and while the sets that are presented to the decision-maker may potentially be large, the task of analysing them becomes a difficult one. Therefore the need to reduce these sets to a more concise representation is very important.

Classically, profiles that describe sets of alternatives may be found in the context of the sorting problem, however they are either given beforehand by the decision-maker or determined from a set of assignment examples. We would therefore like to extend such profiles, as well as propose new ones, in order to characterise any set of alternatives. For each of them, we present several approaches for extracting them, which we then compare with respect to their performance.

Keywords: multiple criteria decision aid, descriptive profiles, central profiles, bounding profiles, separating profiles, outranking relations, meta-heuristics.

1 Introduction

The field of Multiple Criteria Decision Aid (MCDA) focuses on modelling the preferences of decision-makers and aims at helping them in reaching certain decisions. This is not an easy task, as the entities that make the object of these decisions are defined on multiple dimensions and in many cases express contrasting gains and losses on them. Many different models have been proposed in order to reflect the subjective perspective of the decision-maker (DM) over these entities, or alternatives. Hence we are able to distinguish at least three types of relations between them [7]: indifference, strict preference and incomparability.

Classically, three well known types of decision problems have been defined in MCDA [9]: choice, ranking and sorting. The first consists in finding the best alternative, or the set containing the best ones, the second looks to build an order, partial or weak, over the set of decision alternatives, while the last problem outputs an assignment of the alternatives to a predefined set of classes, which may

P. Perny, M. Pirlot, and A. Tsoukiàs (Eds.): ADT 2013, LNAI 8176, pp. 285–296, 2013.
© Springer-Verlag Berlin Heidelberg 2013

be ordered or not. Furthermore, another type of problem, that of clustering, has begun to receive increasing interest recently, having been specifically redefined in the context of MCDA [5].

In all of the mentioned MCDA problem types we may be faced at some point with one or several sets of alternatives, either as a final recommendation or during the decision aiding process. Due to the multidimensional nature of these alternatives and the possibility of having large sets, being able to describe them using concise information becomes very important. In the case of the sorting problem, the profiles that are used to describe the classes already serve this purpose, however this is not the case for the other types of problems. We mention the central profiles [6,1] that are used for sorting into nominal classes, as well as the delimiting profiles [11,9] for sorting into ordered classes.

In this paper we extend the profiles that have been used in conjunction with the problem of sorting, as well as we propose new ones, in order to be able to reduce the information given by one or several sets of alternatives to a more concise representation. These profiles may then be used in order for the DM to better understand the sets of alternatives that he is confronted with, while considering his preferences over them.

We begin by first defining the proposed profiles in Section 2. We consider only the case where a preference model is based on an outranking relation [8]. In Section 3 we first present several exact approaches for extracting these profiles. In order to deal with complexity issues that would be faced when the sets of alternatives are of large cardinality, we especially focus on the use of meta-heuristic approaches for constructing these descriptive profiles. The proposed approaches are validated and compared in Section 4 over a large set of benchmarks that contain increasingly contrasting alternatives. Finally, we conclude with a series of remarks and perspectives for the presented work.

2   Defining the Profiles

Before defining the profiles we first present the working context. Let X be the set of all decision alternatives that can be constructed on a set of criteria $F = \{1, ..., p\}$. We denote with A and B two subsets of X and their cardinalities with n and m. The evaluation of any alternative $x \in X$ on any criterion $i \in F$ is denoted by x_i.

We consider that the DM's preferences are modelled using an outranking relation, which we denote with S [8]. For the purpose of formally defining the profiles, any outranking relation may be considered. While an outranking relation is used to reflect whether the DM considers an alternative to be at least good as another, we may additionally use it to express judgements with respect to the notions of indifference, strict preference or even incomparability [7]. Two alternatives x and y are thus indifferent if simultaneously x outranks y and y outranks x. An alternative x is said to be strictly preferred to another alternative y when x outranks y and y does not outrank x. These alternatives are incomparable when neither of them outranks the other.

We will consider three profiles for characterising one or several sets of alternatives. These are the central, bounding and separating profiles.

We define a **central profile** for a set A as an alternative, real or fictive, which is indifferent to as many of the alternatives in A as possible. Based on this definition, a central profile may be used to substitute all the alternatives in A, due to that fact that the DM considers it indifferent to them, therefore he cannot distinguish between them.

We model the fitness of a central profile c_A with respect to the set A through:

$$f^c(c_A, A) = \frac{1}{n} \cdot |\{x \in A : x\,\mathrm{I}\,c_A\}| \tag{1}$$

The function above is straightforward, giving the proportion of alternatives in A that are indifferent to the central profile, inside a $[0, 1]$ value range.

When the alternatives in A are mostly indifferent to each other, a central profile may be able to represent them with a high degree of confidence. However, if we consider that the alternatives in A have rather contrasting evaluations, a central profile may not be able to represent them faithfully. For this purpose we will consider the second type of descriptive profile, the bounding profiles.

The bounding profiles may be seen as the best and worst alternatives in A, bounding all the rest between them with respect to the preferences of the DM. We define the **upper bounding profile** of a set A as an alternative which is either strictly preferred or indifferent to any alternative in A, but not strictly preferred by them. In this way we may state that no alternative in A is better than the upper bounding profile. Similarly, the **lower bounding profile** of A is either strictly preferred by any alternative in A, or indifferent to them, therefore no alternative in A may be said to be worse than the lower bounding profile. We denote these profiles with b_A^+ and b_A^- respectively.

When the upper or lower bounding profiles cannot be selected from A we may proceed to construct them using the following functions to model their quality:

$$f^{b+}(b_A^+, A) = \frac{1}{n \cdot (n+1)} \cdot \left(n \cdot |\{x \in A : b_A^+\,\mathrm{S}\,x\}| + |\{x \in A : x\,\mathrm{S}\,b_A^+\}| \right) \tag{2}$$

$$f^{b-}(b_A^-, A) = \frac{1}{n \cdot (n+1)} \cdot \left(|\{x \in A : b_A^-\,\mathrm{S}\,x\}| + n \cdot |\{x \in A : x\,\mathrm{S}\,b_A^-\}| \right) \tag{3}$$

Each of the two fitness measures counts the number of alternatives from A that the considered bounding profile outranks in the first term of the sum, but also the number of alternatives in A that are outranked by it in the second term.

Since an upper bounding profile mainly has to outrank all the alternatives in A (hence it will be either strictly preferred or indifferent to them), the first term has been weighted so that it dominates the second. We would also like to have an upper bounding profile that is indifferent to as many alternatives in A as possible, therefore the second term is also necessary.

Similarly, the lower bounding profile reverses the importance of the two terms, as it mainly needs to be outranked by the alternatives in A (hence it will be either strictly preferred by them or indifferent). Nevertheless, if this first condition is met, then the lower bounding profile should also outrank the alternatives in A in order to be indifferent to them.

Finally, we consider two sets of alternatives, A and B, and a relation of strict preference of the first over the second. In such a case we may consider defining a profile that separates the alternatives between the two sets as well as possible.

We define a **separating profile** between a set A that is strictly preferred to a set B, as an alternative, real or fictive, which is strictly preferred by the alternatives in A, or at least indifferent to them, while in turn it is strictly preferred to the alternatives in B, or at least indifferent to them.

The fitness measure for such a profile is:

$$f^s(s_B^A, A, B) = \frac{(n+m)\left(\left|\{x \in A : x S s_B^A\}\right| + \left|\{x \in B : s_B^A S x\}\right|\right) + \left|\{x \in A : s_B^A \not S x\}\right| + \left|\{x \in B : x \not S s_B^A\}\right|}{(n+m)(n+m+1)}$$

(4)

The first term, which is multiplied with $(n + m)$, counts the number of alternatives in A that outrank the separating profile and the number of alternatives in B that are outranked by it. If all the alternatives in A and B are counted, then the separating profile is not strictly preferred to any of the alternatives in A, while none of the alternatives in B are strictly preferred to it. In this case, the separating profile may be said to have been placed between the two sets of alternatives. However, we may have certain alternatives from both sets that are indifferent to the separating profile. In this case the separating profile may not be considered to properly separate A and B. For this reason we have added the second term from Equation (4), which counts the number of alternatives from A that are not outranked by s_B^A, and the number of alternatives from B that do not outrank s_B^A. If this term is also maximized then all the alternatives from A will be strictly preferred to the separating profile, while all the alternatives from B will be strictly preferred by it.

3 Algorithmic Approaches to Determine the Profiles

Several approaches to constructing the presented profiles may be considered, from very simple ones to others that are more complex. Some of them are independent of the preference model that is used in order to reflect the perspective of the DM over the set of alternatives, while others are tailored for a particular type of outranking relation. We will consider in the case of the latter, the outranking relation from [2], although the approaches that we will present may easily be adapted to other outranking relations.

For the selected relations, the "at least as good as" comparisons are characterized for all pairs of alternatives x and y and for all criteria $i \in F$ by:

$$C_i(x, y) = \begin{cases} 1 \text{ if } & y_i < x_i + q_i\,; \\ -1 \text{ if } & y_i \geq x_i + p_i\,; \\ 0 \text{ otherwise}\,, \end{cases}$$

(5)

where $0 \leq q_i$ (resp. $p_i \geq q_i$) is a constant indifference (resp. preference) threshold associated with the ith criterion. A weight $w_i > 0$ is associated with each criterion i, and the overall concordance index $C(x, y)$ is defined as the weighted sum of the marginal concordances. A veto threshold v_i for each criterion i is also introduced in order to invalidate the outranking in case a very large difference of evaluations on at least one criterion is detected in favour of the overall less preferred alternative. Consequently, an alternative x outranks an alternative y ($x\,S\,y$) iff $C(x,y) > 0$ and $y_i - x_i < v_i\ \forall i \in F$.

3.1   Exact Approaches

One of the simplest approaches is to **select** an existing alternative based on how well it performs with respect to the considered fitness measures. For instance a central profile is selected as follows:

$$c_A = \arg\max_{x \in A} f^c(x, A). \tag{6}$$

Not only will such an approach be very fast, but it will also give the DM a result with which he is familiar, as the profiles are real alternatives.

Nevertheless, profiles that are selected from the existing alternatives may not always be of good quality, considering the fitness measures we have proposed. This may easily be imagined for sets containing very contrasting alternatives.

Another simple and quick approach for building these profiles is to **construct** them directly from the evaluations of the alternatives. In the case of a central profile we may consider a simple mean operator as follows:

$$c_{Ai} = \frac{1}{n} \sum_{x \in A} x_i, \forall i \in F. \tag{7}$$

This approach is only suited when the criteria are defined on quantitative scales, however we may use a median operator when confronted with ordinal scales.

In the case of bounding profiles, since the upper bounding profile should mainly outrank the alternatives in A, while the lower bounding profile should mainly be outranked by them, we may use the max and min operators:

$$b^+_{Ai} = \max_{x \in A} x_i, \forall i \in F, \qquad b^-_{Ai} = \min_{x \in A} x_i, \forall i \in F. \tag{8}$$

A separating profile may be given as the mean between the central profiles of the two sets:

$$s^A_{Bi} = \frac{1}{2}\Big(\frac{1}{n}\sum_{x \in A} x_i + \frac{1}{m}\sum_{x \in B} x_i\Big), \forall i \in F. \tag{9}$$

While these approaches for building central, bounding or separating profiles are simple and fast, nothing guarantees that they will find a good result with respect to the fitness measures defined in Section 2.

A third approach is to use **mathematical programs** that model the outranking relations between alternatives to extract the central, bounding or separating profiles which are optimal with respect to the fitness measures defined in Section 2. We have considered an extension of the work of [4], which may be used to model the outranking relation presented earlier in this section, in order to determine the optimal profiles in an exact way. However, due to complexity issues, such an approach quickly becomes impractical when considering larger sets of alternatives. As we will consider such cases for the empirical validation of the presented algorithmic approaches in Section 4, we do not elaborate further on the topic of using a mathematical program.

3.2   Meta-heuristic Approach

An alternative to finding an optimal central, bounding or separating profile is to perform a trade-off between the quality of the profile and the time required by the approach in order to find it. Hence, we may use meta-heuristic approaches [10], which find results that are close to the optimal one in a fraction of the time required by exact approaches.

In our case, any single-solution meta-heuristic may be used. We present the outline of these approaches below [10]:

Algorithm 1. Single-solution meta-heuristic

Input: Initial solution s_0.
1: $t = 0$;
2: **while not** Stopping criterion satisfied **do**
3:    $N(s_t) = $ GENERATE(s_t); /* Generate candidate solutions from s_t */
4:    $s_{t+1} = $ SELECT$(N(s_t))$; /* Select a solution to replace the current one */
5:    $t = t + 1$;
Output: Best solution found.

The initial solution may either be constructed randomly, or may be guided towards a good solution. In our case we will be using the first approach of selecting an existing alternative that maximizes the considered fitness measure.

The neighbours of the current solution will be those that contain an evaluation change on only one criterion. This change will be either the smallest increase or the smallest decrease of the evaluation, which would change the way in which the profile compares on a particular criterion to the alternatives that it tries to describe. We motivate this by the desire to be able to explore the search space from one neighbouring solution to the next, without performing large changes to a profile, which may lead us to miss potentially better intermediate solutions.

The selection of the new solution generally depends on the actual type of meta-heuristic used. Nevertheless, in many cases the neighbouring solutions are evaluated based on the fitness measure and then a selection procedure is applied. However, it may be the case that assessing the fitness of all neighbouring solutions, or even constructing them, will increase the execution time of the approach. In such cases, certain heuristics may be used to assess the quality of each change on the current solution. We will propose in what follows different heuristic measures for each of the three types of profiles that have been defined in this paper. The outranking relation which we use here is the one defined in the beginning of this section. Note that similar heuristics can be given for other definitions of the outranking relation.

We begin with the heuristic for increasing the evaluation of a central profile on a particular criterion $i \in F$, considering the alternatives in set A:

$$h^c(c_A, i) = \left| \{x \in A : x_i - c_{Ai} > q_i \wedge c_A \not\succeq x\} \right| - \left| \{x \in A : x_i - c_{Ai} < -q_i \wedge c_A \not\succeq x\} \right| \quad (10)$$

Since the central profile should be indifferent to the alternatives in A, the heuristic in Equation (10) may be seen as a voting procedure where each alternative in A votes in favour of increasing the evaluation of c_A on criterion i, in disfavour, or refrains from voting. This is reflected by the two terms in this equation. The first term counts the number of alternatives which have an evaluation higher than that of c_A by more than the q_i threshold. This means that those alternatives are not considered indifferent to c_A on criterion i. Moreover, those alternatives are preferred to it, therefore, from their perspective, the evaluation of c_A should be increased in order for them to become indifferent. The second term counts in a similar way the alternatives from whose perspective the evaluation of c_A on criterion i should be decreased. The alternatives which are already indifferent with c_A on criterion i do not require an increase or decrease in the evaluation of c_A Furthermore, the alternatives in A that are already overall indifferent to c_A do not take part in this process, even if their evaluations on criterion i are not indifferent to that of c_A, as this would not increase the fitness of the central profile. The heuristic is valued in the $[-n, n]$ interval.

The heuristic of decreasing the evaluation of c_A on criterion i is $-h^c(c_A, i)$.

In Figure 1 we illustrate the way in which the heuristic works, considering a set containing only four alternatives.

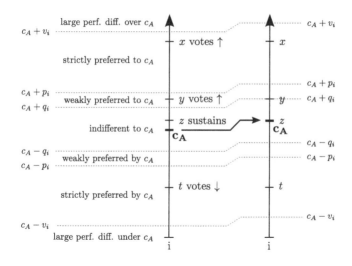

Fig. 1. Detailing the heuristic for changing c_A for a set A of 4 alternatives

In this example we consider a set $A = \{x, y, z, t\}$ of four alternatives and their central profile c_A. We consider that none of these alternatives are at this point overall indifferent to c_A, therefore they all take part in the voting process. It is evident that the evaluation of c_A should be increased, as two alternatives from A are in favour of this change, one is against and another refrains from voting, therefore giving a positive value to the heuristic measure. However, we would only add to the evaluation of c_A the smallest amount which changes at least one of the comparisons between it and the alternatives in A. The first alternative, x, would require c_A to be increased by an amount that brings the first dotted line below the evaluation of x on i just above it. This amount is $x_i - c_{Ai} - p_i + \epsilon$, where $\epsilon > 0$ and $\epsilon \ll 1$, as in this case x would become only weakly preferred to c_{A_i}. However, this amount can be seen to be larger than the amounts that would be required in order for the other alternatives compare differently to c_A, therefore we will not increase c_{A_i} by this amount. The use of ϵ is necessary following the definition of the outranking relation. The smallest amount that would impact the way in which at least one alternative compares to c_A on i is equal to $y_i - c_{A_i} - q_i$, which would make y become indifferent to c_A on criterion i, while all the other alternatives will remain in the same state as before. Therefore the increase of c_{A_i} would be this amount.

Having a positive fitness value for the heuristic in Equation (10) does not imply that we would increase its evaluation. All the operations of both increasing and decreasing the evaluations of c_A on all criteria, characterised through the described heuristic measure, are used in the meta-heuristic approach.

We continue with the heuristic functions for increasing the evaluations of the bounding profiles on a particular criterion in Equations (11) and (12).

$$h^{b+}(b_A^+, i) = n \cdot \left| \{x \in A : x_i - b_{A_i}^+ > q_i \wedge b_A^+ \not\!S x\} \right| - \left| \{x \in A : x_i - b_{A_i}^+ < -q_i \wedge x \not\!S b_A^+\} \right| \quad (11)$$

$$h^{b-}(b_A^-, i) = \left| \{x \in A : x_i - b_{A_i}^- > q_i \wedge b_A^- \not\!S x\} \right| - n \cdot \left| \{x \in A : x_i - b_{A_i}^- < -q_i \wedge x \not\!S b_A^-\} \right| \quad (12)$$

We find that these heuristics are defined similarly to the one for a central profile. The first term from both counts the number of alternatives that do not outrank each profile but have an evaluation that is above that of the profile by more than the indifference threshold. In this case the evaluation of the profile should be increased so that it would outrank the considered alternatives on criterion i. Similarly, in the second term the alternatives that are not outranked by the bounding profiles and that have their evaluations lower by more than the indifference threshold require the evaluations of the profiles to be decreased. The two terms are weighted so that one of them dominates the other, as is the case with the fitness measures for these profiles.

Finally, we present the heuristic for increasing the evaluation of a separating profile, considering the two sets A and B:

$$h^s(s_B^A, i) = (n+m) \cdot \left(\left| \{x \in B : x_i - s_{B_i}^A > q_i \wedge s_B^A \not\!S x\} \right| - \left| \{x \in A : x_i - s_{B_i}^A < -q_i \wedge x \not\!S s_B^A\} \right| \right)$$
$$+ \left| \{x \in B : x_i - s_{B_i}^A > -q_i \wedge x \, S \, s_B^A\} \right| - \left| \{x \in A : x_i - s_{B_i}^A < q_i \wedge s_B^A \, S \, x\} \right| \quad (13)$$

The first term counts the alternatives from B that require the evaluation of s_B^A on i to be increased in order for it to outrank them, while the second term counts the alternatives from A that require that this evaluation is lowered in order for them to outrank the separating profile. These terms are weighted, as they account for the most important part of the definition of a separating profile. The following two terms account for the alternatives in B that require an increase in the evaluation of s_B^A, and those from A that require a decrease.

4   Empirical Validation

In order to be able to compare the performance of the proposed approaches for extracting each type of profile, we have generated a large number of problem instances containing one or two sets of alternatives. We have fixed the size of these sets of alternatives to 50, making them very difficult for a DM to analyse directly.

4.1 Constructing the Benchmarks

The alternatives are defined on a number of 11 criteria which are valued on ratio scales in the interval $[0, 1]$. This number has been chosen in order for the alternatives to resemble those from real problems that are considered to be difficult, but also allowing us to construct very diverse ones. In order to model a wide range of potential problems, we also generate the evaluations of the alternatives in each set so that they are increasingly contrasting. A total of ten generators are used, which we denote alphabetically from $\mathcal{A}$ to $\mathcal{J}$. While the first builds each alternative using a normal distribution centred at the median level on every criterion, the following four randomly pick for each alternative normal distributions that are increasingly spaced apart. The following generators are the same as the first five except that very good and very bad performance evaluations are additionally inserted. For each alternative, two distinct criteria are randomly picked and with a 50% probability the evaluation on the first criterion is maximized, while with the same probability the evaluation on the second criterion is minimized. Using each generator we have built 5 problem instances.

The perspective of a fictive DM on these sets of alternatives is modelled using the outranking relation from [2]. The criteria have been given equal importance weights as we are not dealing with real decision problems, but also due to the fact that by reducing the significance of certain criteria in favour of others we reduce the impact that they would have on the way in which the alternatives compare to each other. By maintaining the criteria importance weight the same for all of the criteria, we are assuring that the benchmarks have the highest diversity in their structures as possible. The discrimination thresholds are selected so that evaluations that are generated using the same normal distribution are in a high percentage indifferent. Only one veto threshold is used, which is set to three quarters of the value range, making veto situations appear very rarely inside the instances constructed using the first five generators.

4.2 Results

For each of the 50 problem instances that we have generated, we have constructed the central, bounding and separating profiles using the three approaches proposed in this paper. The approaches of selecting existing alternatives and that of constructing them from the evaluations of the alternatives in the sets have been executed only once on each benchmark, as they are deterministic.

In the case of the meta-heuristic approaches, we have selected a simulated annealing implementation [3]. The initialisation step is given the solution of the first of the previous approaches, while the cooling rate is fixed so that the algorithm will run at most for one minute. This limit has been set in order to simulate real-life conditions where the approaches of constructing these profile need to quickly output good results. Furthermore, a strategy using restarts had been additionally applied. This approach has been executed 50 times over each benchmark in order for the results with respect to the average fitness of the profiles to be significant.

The average fitness of the three types of profiles, as well as the standard deviations, where relevant, are presented for each of the ten types of benchmarks in Figure 2.

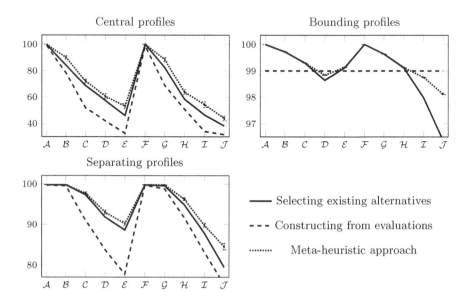

Fig. 2. Average fitness of the central, bounding and delimiting profiles

Certain conclusions may be drawn for the results of finding any type of profile. First of all, we notice that the approaches find profiles that are less fit with respect to the considered fitness measures as we tackle problems instances that contain increasingly contrasting alternatives. This is seen through the decrease in fitness from the first type of benchmarks up to the fifth, as well as from the sixth and up to the last. The two sets of benchmarks resemble strongly each other, except that in the case of the second set we have added large performance gains or losses for certain alternatives in the sets.

Secondly, we may notice that the approaches of building the profiles using simple operation on the evaluations of the alternatives in each set generally perform poorer than all the rest. A few exceptions occur when constructing bounding profiles, where the proposed approach is always able to build an upper bounding profile that outranks all the alternatives in the set and a lower bounding profile that is outranked by them. Any fitness results that are above these highlight the fact that we have constructed bounding profiles that come closer to the alternatives in the set, i.e. indifferent to them.

The meta-heuristic approaches improve on the results given by the approach of selecting an existing alternative from the dataset. The largest improvements may be seen to occur for the approach of constructing a central profile, however this is due to the nature of the fitness measure. Nevertheless, in this case we are able to improve the results of the approach of selecting an existing alternative by as much as 10%.

For the other types of profiles the fitness measures model two objectives, the first dominating the other, and thus improvements over the less important objective are less visible. Nevertheless, when the first objective is maximized the second one becomes also very important. This is especially the case for the bounding profiles, which we additionally want to become indifferent to as many of the alternatives in the set as possible. We find that in this case the first approach is already performing quite well for the first types of benchmarks, and the meta-heuristic is not able to improve on its results.

Finally, in the case of the separating profiles, the meta-heuristic approach performs quite well, for certain benchmark type being able to find separating profiles that are strictly preferred by the alternatives in the first set and strictly preferred to the alternatives in the second set.

5   Conclusions and Perspectives

In this paper we have proposed three types of profiles that may be used in order to describe one or several sets of alternatives. Two of these profiles, the central and separating profiles, have been extended from the context of the problem of sorting, while the bounding profiles are new. Through them we are able to reduce one or two sets of alternatives to a condensed representation, which would aid a DM in understanding and dealing with these sets as a whole, especially when we are dealing with a large number of alternatives inside them. The definitions of these profiles make their use very intuitive.

For each of the three types of profiles we have presented three approaches for constructing them, which we have tested over a large number of benchmarks holding various difficulties. The results show that in most cases the approach of selecting an existing alternative performs quite well, however using a meta-heuristic we are able to find even better results. Furthermore, the approaches of constructing the profiles using simple operations over the evaluations of the alternatives in the sets in general perform worse than the others.

Although we have considered the use of a mathematical program in order to find the optimal central, bounding or separating profiles, due to the size of the sets of alternatives such an approach became highly impractical.

We would like to consider in the future extending these profiles for other definitions of outranking relations as well as additionally considering the credibility degrees that are normally associated with them.

We envision the use of these profiles mainly in the problem of clustering in MCDA [5]. As clustering may be seen as an exploratory data analysis technique, being able describe clustering results over large sets of alternatives, using a considerably smaller set of central, bounding or separating profiles, would greatly enhance the exploration and understanding of the original dataset. Furthermore, these profiles could also be used in conjunction with the problems of choice and ranking, provided that large groups alternatives are generated as results to these problems. Finally, being able to summarize the information given by a set of alternatives may additionally aid in a process of eliciting the preferences of a DM over large sets of alternatives. We will explore these topics in the future.

References

1. Belacel, N.: Multicriteria assignment method PROAFTN: Methodology and medical application. European Journal of Operational Research 125(1), 175–183 (2000)
2. Bisdorff, R., Meyer, P., Roubens, M.: Rubis: a bipolar-valued outranking method for the choice problem. 4OR, A Quarterly Journal of Operations Research 6(2), 143–165 (2008)
3. Kirkpatrick, S., Gelatt, C., Vecchi, M.: Optimization by simulated annealing. Science 220(4598), 671–680 (1983)
4. Meyer, P., Marichal, J.-L., Bisdorff, R.: Disaggregation of bipolar-valued outranking relations. In: An, L.T.H., Bouvry, P., Dinh, T.P. (eds.) MCO 2008. CCIS, vol. 14, pp. 204–213. Springer, Heidelberg (2008)
5. Meyer, P., Olteanu, A.-L.: Formalizing and solving the problem of clustering in MCDA. European Journal of Operational Research 227(3), 494–502 (2013)
6. Perny, P.: Multicriteria filtering methods based on concordance/non-discordance principles. Annals of Operations Research 80, 137–167 (1998)
7. Roubens, M., Vincke, P.: Preference Modeling. In: LNEMS, vol. 250, Springer, Berlin (1985)
8. Roy, B.: Decision science or decision-aid science? European Journal of Operational Research 66, 184–203 (1993)
9. Roy, B., Bouyssou, D.: Aide Multicritère à la Décision: Méthodes et Cas. Economica, Paris (1993)
10. Talbi, E.: Metaheuristics: From Design to Implementation. Wiley Series on Parallel and Distributed Computing. Wiley (2009)
11. Yu, W.: ELECTRE TRI: Aspects méthodologiques et manuel d'utilisation. Document du LAMSADE no 74, Université Paris-Dauphine (1992)

Estimating Violation Risk for Fisheries Regulations

Hans Chalupsky[1], Robert DeMarco[2], Eduard H. Hovy[3], Paul B. Kantor[2],
Alisa Matlin[2], Priyam Mitra[2], Birnur Ozbas[2], Fred S. Roberts[2], James Wojtowicz[2],
and Minge Xie[2]

[1] USC Information Sciences Institute, CA, USA
[2] Rutgers, The State University of New Jersey, NJ, USA
[3] Carnegie Mellon University, PA, USA
froberts@dimacs.rutgers.edu

Dedication. This paper is dedicated in memoriam to Dr. Tayfur Altiok. Without his efforts and motivation this project would not have been possible.

Abstract. The United States sets fishing regulations to sustain healthy fish populations. The overall goal of the research reported on here is to increase the efficiency of the United States Coast Guard (USCG) when boarding commercial fishing vessels to ensure compliance with those regulations. We discuss scoring rules that indicate whether a given vessel might be in violation of the regulations, depend on knowledge learned from historical data, and support the decision to board and inspect. We present a case study from work done in collaboration with USCG District 1 (HQ in Boston).

Keywords: Regulatory compliance, Coast Guard, Fisheries, Machine learning, Statistical models.

1 Introduction

This paper describes a targeted risk-based approach to enforcing fisheries laws in the United States Coast Guard First District 1 (USCG D1), based in Boston, Massachusetts. The work is a joint project of the Laboratory for Port Security (based at Rutgers University) and the Command, Control and Interoperability Center for Advanced Data Analysis (CCICADA, a US nationwide consortium headed by Rutgers).

Fisheries rules and regulations have been established through a complex process whose key aims include preservation of the fisheries biomass. The primary mission of the fisheries law enforcement program is to maintain a balanced playing field among industry participants (professional fishing companies) through effective enforcement of the regulations. Over the years USCG D1 has developed an approach to fisheries law enforcement, which among other things includes scheduling fishing vessel inspections using a scoring matrix. In this paper we describe a project aimed at validating and extending the scoring matrix by further refining the ability to determine the risk target profile of active vessels within the population of the First District.

Our research seeks a model that determines which vessels pose a higher safety risk through non-compliance with safety codes and which vessels are most likely to be

P. Perny, M. Pirlot, and A. Tsoukiàs (Eds.): ADT 2013, LNAI 8176, pp. 297–308, 2013.
© Springer-Verlag Berlin Heidelberg 2013

contravening fishing laws and regulations. The main measure of effectiveness explored here, "boarding efficiency" (BE), is defined as the fraction of recommended boardings that yield either a fishery or a safety violation. We also formulate other measures of effectiveness and study approaches to improving them.

Currently the USCG determines whether to board a fishing vessel using a rule called OPTIDE (created by LCDR Ryan Hamel and LT Ryan Kowalske of USCG D1), which constructs a score by assigning points to known factors describing a vessel, such as the time since last boarding and the vessel's history of fisheries violations. The OPTIDE system recommends boarding if the sum of points exceeds a threshold. The developers of the method used expert opinion to select the factors in the rule, and to set their relative weights. The scoring matrix was developed using expert knowledge. This paper addresses the question: Can naïve researchers using methods of data analysis approach the effectiveness of such expert rules?

The USCG made available 11 years of data on USCG boarding activities and violations incurred by commercial fishing vessels. Our project studied introducing other features, such as weather, seasonality, fish price, fish migration, key fish species, home port, and detailed vessel history. The project team worked with economic data such as fish market prices and considered socio-economic factors such as family fishing boats in comparison to large commercial fishing vessels and fishers' attitudes toward law enforcement. We looked at the seasonal variation in boardings and outcomes. In the analysis, fisheries violations were separated from safety violations.

Machine learning methods were used to seek other features, or combinations of present and added features, that might lead to decision rules increasing the BE. In addition, alternative models for the boarding decision were considered. One model poses a choice of which boat to board, within a set of K alternatives. Section 2 describes this approach. Another approach sought regression models that derive alternative weights for the same features used in OPTIDE. This method is discussed in detail in Section 3. Section 4 discusses alternative goals, including balanced deterrence, balanced policing, and balanced maintenance of safe operations. Here we discuss alternative measures of effectiveness, e.g., violations found per hour rather than per boarding. We also discuss alternative decision strategies: random strategies; varying the number of boats used based on weather, season, or economics; alternative searching protocols to find the candidate vessels for boarding.

2   RIPTIDE: A Machine Learning Approach

In this section we describe a scoring rule, RIPTIDE, which loosely stands for Rule Induction OPTIDE. RIPTIDE extends OPTIDE by learning a more fine-grained and data-driven prediction and ranking model from past activity data, using a machine learning approach. Using the best model found so far, RIPTIDE outperforms OPTIDE by up to 75% with regard to a specific scoring rule, described in more detail below. A software package implementing RIPTIDE can be used to experiment with the learned models, and can be applied to rank operational data.

The OPTIDE rule was built based on expert judgment and intuition. It is an abstraction of a set of features that a commanding officer will routinely consider when deciding whether to board a vessel. However, to our knowledge, there had been little or no optimization of the rule based on historical data.

To extend OPTIDE, we used a data-driven machine learning approach to learn a classification model from historic boarding activity data. RIPTIDE uses machine learning to automatically find regularities in past boarding activity data and encodes them in a model (or classifier) that can then be used to rank new, previously unseen candidate boarding opportunities. The classifier takes a single (new) data instance and applies the previously learned model to assign the new instance to one of two classes (e.g., "violation" or "no violation"). In doing so, the classifier estimates a probability that may be interpreted as the "confidence" of the prediction. This estimate is based on how well the model performed for similar cases on the training data. These probabilities can then be used to rank instances, as does the OPTIDE risk score.

Machine learning is built upon two core principles, data representation and generalization. First, every data instance is represented in a computer-understandable form. This is generally done by engineering a set of features or attribute/value pairs that carry relevant information and that can be either directly observed or computed from the data. In the generalization phase, the classifier uses many data instances for which the class is known as training data, and seeks regularities in that data that allow it to predict the class of a new data instance. There are many different data representation schemes and learning algorithms that can be used (see, e.g., [2, 5, 9] for an overview).

For RIPTIDE, we chose a learning algorithm called a boosted decision tree that is a good general-purpose tool for problems with a small to medium number of features. One advantage of decision trees is that the learned models are (large) 'if-then-else' statements that can be inspected by humans, and that are therefore to some extent understandable. This is useful for comparison to a rule-based approach such as OPTIDE, as the experts want to be able to decide whether they should trust such a model. Other learning methods such as support vector machines or neural nets produce largely if not completely opaque models, which can be judged only by their input/output behavior.

Classification performance can be improved by combining multiple classifiers that were trained using different algorithms, features, sections of the data, etc. One such strategy is called boosting. In boosting, instead of learning a single decision tree, we learn multiple trees on different subsets of the training data. An algorithm such as AdaBoost [4] (for Adaptive Boosting) then learns the "best" weights for combining the results of those individual decision trees into an overall boosted decision tree. For our currently best-performing classifier (Model 58), boosting improves performance on a boarding tradeoff task (described below) by about 25%.

Some 10,000 boarding activities from 2002 to the end of 2011 were used as training data and a set of about 1000 boardings in 2012 was used as a held-out test set to evaluate the models. To use a classifier such as RIPTIDE, one must set a threshold, which we can estimate from the training data. If the estimated probability of finding a violation is above the threshold, we recommend boarding a vessel; otherwise, not. Let TP be the number of true positives, that is, cases where the score is above threshold, and the boarding in fact found a violation; the remaining cases where the classifier says "board" are the false positives FP. Standard measures of effectiveness (MoEs) for classifiers are recall R (the percentage of vessels having some violation that are flagged for boarding), precision P = TP/(TP+FP) measuring the fraction of true decisions, and their harmonic mean, known as the F1 value: F1 = 2*P*R/(P+R). Picking a low probability threshold will give high recall but low precision; conversely, a high threshold will give high precision but low recall. Every choice represents a tradeoff between TP and FP, and what is acceptable depends on external factors such as task objectives and resources. Using a generic rule such as maximizing R or F1 value will generally not give the best compromise in practical applications.

The best way to compare classifiers without setting a threshold is to plot ROC (Receiver Operating Characteristic) curves. An ROC curve shows the true positive rate (or recall) plotted against the false-positive rate, that is the ratio of false predictions to the number of non-violating vessels, for each possible threshold point. The curve shows a tradeoff space showing how many more false positives one must accept to get additional true positives.

We can use the area under the ROC curve to compare different classifiers; a higher area under the curve generally means a better classifier. Figure 1 shows a comparison of ROC curves for OPTIDE and Model 58 for the held-out test data covering the year 2012. Both models have more or less identical area under the curve (AUC) of about 0.65, This shows that they are doing better than random choice (the dotted line with an AUC of 0.5), but not very much so, indicating that there is not a very strong signal in the data to begin with. Model 58 is doing significantly better at picking up the higher yield boardings (the bump at the beginning of the curve), but it loses that advantage towards lower-risk boardings. It also is much more fine-grained than OPTIDE, a feature we will explore in more detail below.

In the current formulation of OPTIDE, for values of the score, the yield distribution is very flat, which can be seen in the long straight sections of the OPTIDE ROC curve. About 84% of all boardings fall in a very narrow band of yield close to the threshold level. This means a large number of ships are apparently indistinguishable. Our analysis of the data suggests that there are no standout "red flags" that positively indicate that a ship might be in violation of some regulation. Even among vessels having the highest risk score, only one third of boardings yield a violation. This means we cannot assign a strong meaning to any of the OPTIDE risk categories.

Instead of focusing on absolute risk scores with a global interpretation, we explore an alternative MOE: How well can a model select among a small set of alternative vessels? For example, a set of ships may be encountered more or less simultaneously, calling for an informed decision as to which ships to board, given available time and resources. Technically, this calls for ranking the boats in the small candidate set.

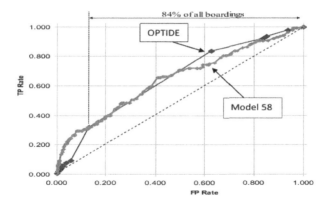

Fig. 1. ROC curves for OPTIDE and Model 58 for the held-out test data in 2012. Model 58 is a weighted combination of 20 different tree models, found using AdaBoost.

To evaluate ranking performance we consider the following MOE. Given a test set of boarding activities such as the 2012 held-out set, we randomly pick a set (or bucket) of size k and rank the elements in the bucket according to our model. We then pick the top-ranked boarding activity in the bucket (choosing randomly in case of ties) and test whether it actually had a violation or not. We repeat this experiment many times and compute the fraction of trials in which we picked a winner (i.e., a boarding with a violation). The probability of picking a winner is strongly dependent on the bucket size, since smaller buckets have a smaller chance of containing a vessel with a very high score. For example, for the held-out set of 1002 boardings of which 14% yielded a violation, the probability that a random set of two boardings contains at least one with a violation is about 26%, for 5 it is 53%, for 10 it is 78% and for 20 it is 97% (almost certain). Note that this high probability doesn't mean that it is easier to find one with a violation; that aspect still requires a good ranking function to find the best item in the bucket. Since all of our analysis is based on data collected under historical boarding policies, and, more recently, OPTIDE, the practical implications of the findings in this section remain to be explicated in future work, which our USCG partners are currently undertaking in exploration of our new ideas.

Table 1 shows the results of these experiments. It compares our currently best model, Model 58, to OPTIDE and two other models. Model 58 includes features not used in OPTIDE, such as distance to coast and vessel subtype. An alternative model (Model 57) omits a feature (distance to coast) and still a third model (Model 48) adds something called observed activity as a feature. The top of Table 1 shows standard AUC and Max-F1 metrics, and all models perform fairly similarly. In the lower portion, we show results on ranking experiments with bucket sizes ranging from 2 to 50. We find that our best model improves up to 76% over OPTIDE for a bucket size of 20, where we have an almost 45% chance to pick a winner, and even for a more realistic bucket size of 10, the improvement is still a good 38%. This shows that the apparently small advantage of RIPTIDE at higher levels of yield can become a substantial improvement if it is possible to batch the candidate vessels and choose the most likely one to board.

Table 1. Evaluation results for OPTIDE and several alternate models

	Random	OPTIDE	Model 57	Model 48	Model 58	58 vs. OPTIDE
N-Thresh		15	135	191	206	
Max-F1		0.301	0.300	0.310	0.328	+9.0%
AUC		0.648	0.626	0.656	0.646	-0.3%
Bucket Size			Choose 1 of k			
5	0.135	0.210	0.217	0.236	0.243	+15.9%
10	0.135	0.237	0.279	0.311	0.328	+38.5%
15	0.135	0.244	0.328	0.364	0.393	+60.9%
20	0.135	0.251	0.363	0.403	0.443	+76.4%
25	0.134	0.261	0.399	0.440	0.484	+85.1%
30	0.135	0.276	0.422	0.466	0.516	+86.8%
35	0.135	0.290	0.447	0.488	0.542	+86.6%
40	0.134	0.307	0.464	0.505	0.567	+84.7%
50	0.137	0.336	0.492	0.542	0.601	+78.9%

We have developed a small RIPTIDE software suite that can be used to classify and rank potential boardings based on the best models found so far, and to retrain models if necessary. RIPTIDE builds upon the Weka toolkit [5] and adds a number of methods for data translation and various other tasks. RIPTIDE is purely Java based and can be run on Linux, MacOS and Windows platforms

Using the RIPTIDE approach in practice will require the users to retrain the machine learning models at regular intervals, perhaps on a yearly basis, to ensure that significant changes in behavior are incorporated. This would be an uncomplicated task, as long as the basic set of features to consider remains the same or similar. The actual implementation of RIPTIDE is experimentally underway at the USCG.

3   DE-OPTIDE

In this section, we describe an alternative approach that utilizes regression methods in statistics and the historical data to derive alternative weights for the same features used in OPTIDE. Based on this approach, a new decision rule was developed, called Data-Enhanced OPTIDE (DE-OPTIDE). We compare its performance with the original OPTIDE rule.

An underlying assumption of OPTIDE is that probability of a violation is related to an underlying score that is a weighted sum of some predictor variables $X_1, X_2, ..., X_n$ (i.e., features used in the OPTIDE rule). The decision is made to board if the score exceeds a threshold. This assumption, plus potential random errors, leads us directly to a statistical model called a logistic regression model (see [6]). Logistic regression is an instance of a generalized linear model [1, 8]. It allows one to analyze and predict a discrete outcome (known as a response variable), such as group membership, from a set of variables (known as predictor variables) that may be continuous, discrete, dichotomous, or a mix of any of these. Generally, the response variable is dichotomous, such as presence/absence or success/failure. In our case the response variable is the violation indicator (presence/absence) of a vessel.

When sample data from such a model are available, we can perform a statistical analysis to estimate the unknown coefficients and thus estimate the relationship between the response and predictor variables. We can then use the logistic regression model to predict the category to which new individual cases are likely to belong.

We assume a violation is related to an underlying latent score S which is a weighted sum of some predictor variables plus potential errors, i.e., $S = W_1X_1 + W_2X_2 + \ldots + W_nX_n + error$, where the Ws are weights describing the contributions of the feature and the random "error" follows a normal distribution with mean 0 and variance σ^2. As with the tree-based rules, if the score of a vessel exceeds a certain threshold value, the vessel should be boarded. Mathematically, these assumptions lead to the aforementioned logistic regression [3,10]. We used logistic regression and the data set available to us to estimate the coefficients W_1, W_2, ..., W_n and we then used these weights to create a new decision rule. Since the new decision rule uses the same features as in the original OPTIDE rule but their weights are determined by the historical data, we call the new rule a Data-Enhanced OPTIDE (DE-OPTIDE) rule.

We note that in the original OPTIDE matrix, all of the features are categorical. Although some of them are naturally continuous, they are categorized or binned for the analysis, which may cause some loss of information. We therefore performed an additional analysis using the same set of features, but retaining continuous values for some of the features. Using the continuous versions does somewhat improve the performance of the DE-OPTIDE rule. In treating the features as continuous, we employed standard imputation techniques for missing data.

In our analysis, we randomly split the entire boarding data set available to us into two subsets: 50% used for training and 50% used for validating. We fit the logistic regression model to the training data and used the estimated probabilities to determine a new decision rule. Then we applied the new rule to the remaining 50% of data to assess its effectiveness. In the new decision rule, the threshold for boarding was chosen by either setting a required percentage of vessels to be boarded, or setting a target boarding efficiency. To control variation caused by the random 50-50 splitting, the calculations were repeated 10 times. Therefore, the results we describe do not correspond to a single unique boarding rule.

Starting with just categorical data, we explored the relationship between the Boarding Efficiency BE and the percentage of recorded boardings (that is, the fraction of all records in the data set for which boarding is recommended, at a given threshold). Results are shown in Figure 2. When applied to the data that was not used to train the model, DE-OPTIDE yields a somewhat higher or similar BE compared to OPTIDE for almost the entire range of recorded boarding percentages. For DE-OPTIDE, efficiency ranges from 20% to 35%, and setting the threshold to reduce the number of boardings yields higher efficiency. This is because the rule ranks vessels by their probability of yielding violations. Therefore, when fewer are boarded, the average chance of finding a violation is higher. In choosing the threshold for the decision rule one may need to take into account not just efficiency but also the fraction of recorded boardings.

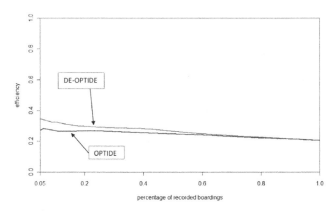

Fig. 2. Boarding Efficiency vs. percentage of recorded boardings using both OPTIDE and DE-OPTIDE for different thresholds (test data 50%). The results are based on 10 repetitions of the random selection of training data.

We also compared the efficiency of DE-OPTIDE with that of OPTIDE using another MOE. The threshold for DE-OPTIDE was chosen based on examining the efficiency of the procedure over different percentages of recorded boardings. We found that efficiency for DE-OPTIDE with a decreasing percentage of recorded boardings starts to increase when the percentage of recorded boardings is less than 10%. Thus, we chose the threshold corresponding to 10% of recorded boardings for DE-OPTIDE.

We also explored an alternative way of selecting the threshold for OPTIDE, i.e., letting threshold correspond to 10% of the recorded boardings (RBs), as we did with DE-OPTIDE. We found that the efficiency of the DE-OPTIDE procedure reaches 32%, compared to 24% efficiency of OPTIDE when using an adjusted threshold (due to our data omitting values for some of the OPTIDE features) and 27% if we use OPTIDE with threshold corresponding to 10% of RBs. We recognize that the USCG would not cut boardings to one tenth of the current level. However, some combination of this rule in a randomized or mixed strategy for boarding might be effective. Note that selecting vessels for boarding purely at random yields only 16% efficiency.

Figure 3 presents the ROC curves for both the OPTIDE and DE-OPTIDE rules. This plot helps to illustrate the performance of these two decision rules as the threshold is varied over the entire range of possible values. The ROC curve for OPTIDE has an area of 0.576 under the curve, while that for DE-OPTIDE has AUC = 0.605. Again, this indicates that the DE-OPTIDE rule is somewhat better than the OPTIDE rule. These plots are based on a single random selection of the training data. Plots from nine other repetitions are similar.

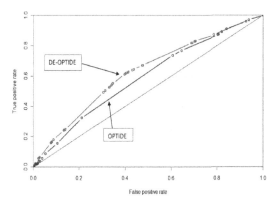

Fig. 3. The ROC curves for both the OPTIDE and DE-OPTIDE rule for various choices of thresholds (test data =50%). The plots are each based on a single run. Plots for 9 other runs show the points for DE-OPTIDE lying almost always above those for OPTIDE itself.

Next we used logistic regression treating certain features as continuous. We computed the relationship of the BE to the percentage of recorded boardings under the modified DE-OPTIDE rule using some continuous features, a rule we call DE-OPTIDE-C. DE-OPTIDE-C achieves better efficiency than OPTIDE. For OPTIDE, efficiency ranges from 20% to 30%. For DE-OPTIDE-C efficiency rises to almost 35% at levels below 10% of recorded boardings. As with the discussion of batching in Section 2, it is not known whether the set of candidates could be expanded enough for such a lower fraction of sightings to yield an acceptable number of boardings. We also compared the efficiency of DE-OPTIDE-C to that of OPTIDE using alternative ways of setting the threshold. The efficiency of the DE-OPTIDE-C procedure reaches 34%, compared to 32% for DE-OPTIDE.

4   Other Approaches

In this section we consider other MOEs, e.g., violations per hour of enforcement activity rather than violations per boarding. We also mention alternative decision strategies: random strategies; changing the number of patrol boats based on factors such as weather, season, or economics; and varying the protocols for finding candidates for boarding.

4.1   Other Ways of Measuring Effectiveness

The models discussed so far consider all violations to be equally important. From the perspective of deterrence, this is plausible. But in terms of economic impact on fisheries and lives saved it may be more appropriate to group violations into classes $i=1,2,\ldots,I$ and seek to maximize the sum $\Sigma w_i x_i$ where x_i is the number of violations in class i. For this to be meaningful the weights must be defined on an interval or ratio scale, and not be simply ordinal [12,13].

The "denominator" in the MOE has been "boardings." Alternatively, we may want to measure effectiveness against time. Time is spent both in boarding and in seeking the next candidate. The choice of which to use will lead to different decisions. Suppose (based on the scoring rule) Vessel A has estimated 12% yield (probability a violation will be found) and the predicted time for the boarding is 4 hours. Vessel B has 15% yield and predicted boarding time 6 hours. If efficiency is violations per boarding (VPB), Vessel A has 0.12 VPB, and Vessel B has 0.15 VPB. We prefer to board Vessel B. If efficiency is violations per hour (VPH), then Vessel A has 0.12/4=0.03 VPH, and Vessel B has 0.15/6=.025 VPH. So we prefer to board Vessel A. In fact, boarding time varies randomly, according to some rule that could be estimated from data. One might also include in the denominator time spent seeking the next candidate.

4.2   Other Kinds of Enforcement Strategies

The OPTIDE-class rules discussed here are deterministic. Randomized strategies make it harder for intentional violators. The variation in goals discussed in Section 4.1 might be incorporated into a randomized mixture: e.g. 30% of time use OPTIDE, 40% of time use VPB, and 30% of time use VPH.

We can model the boarding decision as a choice between boarding and seeking further targets. For simplicity we suppose that a patrol boat meets a fishing vessel every T minutes, and must immediately decide whether to board it. That the decision to board must be made immediately is based on observations from [7] that fishermen can and do modify their behavior when they observe Coast Guard boats, seeking to limit the violations found if boarded. One boat every T minutes is a simplifying model of the random rate at which a patrol will encounter fishing vessels.

Suppose the yield p varies uniformly from 0 to 1. Suppose boarding takes time tT. What value of p should be the threshold for boarding? It can be shown that under certain assumptions, the optimal choice is

$$p = \frac{(2t+2) - \sqrt{(2t+2)^2 - 4t^2}}{2t}$$

As boarding time tT increases, the threshold yield p increases. This confirms the intuition that the longer boarding takes, the pickier one must be in boarding. More realistic models for $T, t,$ and the distribution of p can be developed from log data.

Finally, we considered patrol strategies, using analogies to ecology where the limiting resource is the energy available to predators [11]. In particular, we have compared pure pursuers and pure searchers. The former expend little or no energy in seeking food; they wait until sufficiently valuable prey (sufficiently risky vessel) is in sight and then act (e.g. anolis lizards). Pure searchers (e.g., warblers) spend time and energy prowling to seek food; when they sight it they decide whether to try to catch it and in that case spend little time on pursuit. We studied when a pure searcher should adopt the patient strategy of waiting for the "best" type of food (vessel with highest risk score) or the impatient strategy of waiting for a while for the "best" type of food and then choosing what is available.

4.3   Bringing in Other Goals of Fisheries Law Enforcement

In addition to efficiency of boardings, fisheries law enforcement seeks other goals: balanced deterrence, balanced policing, and balanced maintenance of safe operations. To balance deterrence, the USCG might seek to board all vessels at least once a year. This would require, at times, boarding a low yield vessel. When should this be done? Should the rule depend on recent prior boardings? Suppose Vessel A has an estimated yield of 13% and has been boarded twice in the past year while Vessel B has a 15% yield and has been boarded six times in the past year. In some cases we might prefer to board A rather than B. We might want to board neither, and wait for some boat that has not been examined in two years.

We have developed a simple model representing a tradeoff between balance and yield. The score is based on three parameters, $y(v)$ = the yield assigned to Vessel v, $D(v)$ = days since Vessel v was last boarded, and α, a model parameter. The modified score is $S(v) = y(v) + \alpha D(v)$. The probability $y(v)$ depends on an initial class probability for that boat and on its boarding history. The class probability reflects differences that affect the probability of violation. Explicitly, we take y for a vessel with b past boardings and u "successful" past boardings to be $y = f(b,u) + .05Z$ where Z is uniformly distributed between -1 and 1, and $f(b,u)$ is presumed to come from observed data.

We ran simulations of this model, with five candidates per day, selected uniformly at random from the 100 vessels having the highest score at the start of the day. We do not simply take the five with highest scores because they might not all be accessible: the patrol might stay in a particular area and not all boats are fishing each day. Running the model 20 times for 1095 simulated days (3 years), and for each α between .0001 and .001 (incrementing α by .0001), we found the average output. A scatter plot comparing average number of observed violations over the entire 3-year period to average number of vessels boarded in the last year of the simulation can offer predictions on what the outcome might be under different scoring rubrics. Future work will consider more general scoring metrics.

5   Conclusions

Our analysis supports several conclusions. First, the existing OPTIDE approach extracts a nearly optimal rule based on the data that are used in it. The ROC curves produced by state of the art techniques for learning rules are somewhat above the curve for the existing OPTIDE rule. If the number of vessels considered could be increased, operation at a higher threshold for boarding would likely result in discovering a larger absolute number of violations per year, contributing to both fishery management and safety goals. Second, automated methods, as described in this paper, can be used to extract optimal rules by analysts who have no subject area expertise in this domain. Indeed, such methods can find decision rules that perform as well as, or somewhat better than, models that require substantial knowledge of the data and domain expertise to develop. This means that as the USCG considers adding additional variables to the rules that trigger boardings, the automated methods used here can assess, in advance, the effectiveness of using that additional data. All that is required is to develop a data set in which the values of those new variables are reported along with the existing key variables and the results of the boarding. Finally, we have identified ways in which the objectives of the scoring rule work can be made more complex and closer to the operational realities of the USCG. Preliminary theoretical work has produced simple models showing how to include those realities in the computation of the more sophisticated yield representing complex goals of fisheries law enforcement.

We presented the results described here to USCG D1 in a briefing to the highest-level Coast Guard leadership. The results were very well received and are in the process of being implemented in USCG D1. In addition, the USCG Research and Development Center is working with D1 to explore modifications in the methods that would make them applicable to other Coast Guard districts around the country.

Acknowledgements. This report was made possible by a grant from the U.S. Coast Guard District 1 Fisheries Law Enforcement Division to Rutgers University. The statements made herein are solely the responsibility of the authors.

We extend a special thanks to LCDR Ryan Hamel and LT Ryan Kowalske for working with us on this project, for their support and patience throughout this process. Thanks also to CCICADA researchers Andrew Philpot and William Strawderman.

References

1. Agresti, A.: Categorical Data Analysis. Wiley Interscience, New York (2002)
2. Bishop, C.M.: Pattern Recognition and Machine Learning. Springer, New York (2007)
3. Finney, D.J.: Probit Analysis, 3rd edn. Cambridge University Press, Cambridge (1971)
4. Freund, Y., Schapire, R.E.: Experiments with a new boosting algorithm. In: Thirteenth International Conference on Machine Learning, San Francisco, pp. 148–156 (1996)
5. Hall, M., Frank, E., Holmes, G., Pfahringer, B., Reutemann, P., Witten, I.H.: The WEKA data mining software: An update. SIGKDD Explorations 11(1) (2009)
6. Hilbe, J.M.: Logistic Regression Models. Chapman & Hall/CRC Press, London (2009)
7. King, D.M., Porter, R.D., Price, E.W.: Reassessing the value of U.S. Coast Guard at-sea fishery enforcement. In: Ocean Development & International Law, vol. 40, pp. 350–372. Taylor and Francis, London (2009)
8. McCullagh, P., Nelder, J.A.: Generalized Linear Models, 2nd edn. Chapman and Hall, London (1989)
9. Mitchell, T.: Machine Learning. McGraw Hill, New York (1997)
10. Morgan, B.J.T.: Analysis of Quantal Response Data. Chapman and Hall, London (1992)
11. Roberts, F.S., Marcus-Roberts, H.: Efficiency of energy use in obtaining food II: Animals. In: Marcus-Roberts, H., Thompson, M. (eds.) Life Science Models, pp. 286–348. Springer, New York (1983)
12. Roberts, F.S.: Limitations on conclusions using scales of measurement. In: Barnett, A., Pollock, S.M., Rothkopf, M.H. (eds.) Operations Research and the Public Sector, pp. 621–671. Elsevier, Amsterdam (1994)
13. Roberts, F.S.: Measurement Theory, with Applications to Decisionmaking, Utility, and the Social Sciences. Cambridge University Press, Cambridge (2009)

Computing Convex Coverage Sets
for Multi-objective Coordination Graphs

Diederik M. Roijers[1], Shimon Whiteson[1], and Frans A. Oliehoek[2]

[1] Informatics Institute, University of Amsterdam, The Netherlands
{d.m.roijers,s.a.whiteson}@uva.nl
[2] Dept. of Knowledge Engineering, Maastricht University, The Netherlands
frans.oliehoek@maastrichtuniversity.nl

Abstract. Many real-world decision problems require making trade-offs between multiple objectives. However, in some cases, the relative importance of the objectives is not known when the problem is solved, precluding the use of single-objective methods. Instead, multi-objective methods, which compute the set of all potentially useful solutions, are required. This paper proposes new multi-objective algorithms for cooperative multi-agent settings. Following previous approaches, we exploit loose couplings, as expressed in graphical models, to coordinate efficiently. Existing methods, however, calculate only the *Pareto coverage set* (PCS), which we argue is inappropriate for stochastic strategies and unnecessarily large when the objectives are weighted in a linear fashion. In these cases, the typically much smaller *convex coverage set* (CCS) should be computed instead. A key insight of this paper is that, while computing the CCS is more expensive in unstructured problems, in many loosely coupled settings it is in fact cheaper to compute because the local solutions are more compact. We propose *convex multi-objective variable elimination*, which exploits this insight. We analyze its correctness and complexity and demonstrate empirically that it scales much better in the number of agents and objectives than alternatives that compute the PCS.

Keywords: Multi-agent systems, Multi-objective optimization, Game theory, Coordination graphs.

1 Introduction

In cooperative multi-agent systems, agents must coordinate their behavior in order to maximize their common utility. Key to making coordination efficient is exploiting the *loose couplings* common to such tasks: each agent's actions directly affect only a subset of the other agents. Such independence can be captured in a graphical model called a *coordination graph*, and exploited using methods such as *variable elimination* [8,9]. This paper considers how to address cooperative multi-agent systems in which the agents have multiple objectives, i.e., the utility is vector-valued. Many real-world problems have multiple objectives, e.g., maximizing performance of a computer network while minimizing power consumption [16].

The presence of multiple objectives does not in itself necessitate special solution methods. In many cases, the vector-valued utility function can be *scalarized*, i.e., converted to a scalar function. Subsequently, the original problem may be solvable with existing single-objective methods. However, this approach is not applicable when the

P. Perny, M. Pirlot, and A. Tsoukiàs (Eds.): ADT 2013, LNAI 8176, pp. 309–323, 2013.
© Springer-Verlag Berlin Heidelberg 2013

parameters of the scalarization are not known in advance. For example, consider a company that produces different resources whose market prices vary. If there is not enough time to re-solve the decision problem for each price change, then we need multi-objective methods that compute a set of solutions optimal for all possible scalarizations.

This paper focuses on one-shot decision-making problems, for which several methods [5,6,12] have been developed. For instance Rollón [14] introduces an algorithm that we refer to as *multi-objective variable elimination* (MOVE), which solves multi-objective coordination graphs by iteratively solving local problems to eliminate agents from the graph. However, these methods all compute the *Pareto coverage set* (PCS), i.e., the Pareto front, of deterministic strategies.

In this paper, we argue that the PCS is often not the most appropriate solution concept. In the common case where the scalarization function is linear, the PCS is typically much larger than necessary. In addition, when joint strategies can be stochastic, the PCS is inadequate, even if the scalarization function is nonlinear.

To address these issues, we propose new methods that compute an alternative solution concept, the *convex coverage set* (CCS). The CCS is *the exact solution set* when the scalarization function is linear, and often much smaller than the PCS. In addition, it is a sufficient set of deterministic strategies from which to construct all optimal stochastic strategies. A key insight of this paper is that, while the CCS is more costly to compute than the PCS for nongraphical problems, it is often less costly to compute for loosely coupled problems because the *local* CCSs are much smaller than the local PCSs.

Thus, the main contribution of this paper is that it shows—both theoretically and empirically—that large speedups can be obtained when solving multi-objective coordination graphs by using the CCS as the solution concept. In particular, we 1) analytically show that the local CCSs can be much smaller than local PCSs, 2) present *convex MOVE* (CMOVE), an extension to MOVE that efficiently computes the CCS, 3) analyze the correctness and complexity of CMOVE in terms of the size of the coverage sets, and 4) demonstrate empirically that CMOVE scales much better than previous algorithms.[1]

2   Multi-objective Coordination Graphs

We formalize our problem setting as a multi-objective extension to coordination graphs [8]. In particular, a *multi-objective coordination graph* (MO-CoG) is a tuple $\langle \mathcal{D}, \mathcal{A}, \mathcal{U} \rangle$: $\mathcal{D} = \{1, ..., n\}$ is the set of n agents; $\mathcal{A} = \mathcal{A}_i \times ... \times \mathcal{A}_n$ is the joint action space (the Cartesian product of the finite action spaces of all agents) and $\mathcal{U} = \{\mathbf{u}^1, ..., \mathbf{u}^\rho\}$ is the set of ρ, d-dimensional *local payoff functions*. The total team payoff is the (vector) sum of local payoffs, with a limited scope e, i.e., the subset of agents that participate in it: $\mathbf{u}(\mathbf{a}) = \sum_{e=1}^{\rho} \mathbf{u}^e(\mathbf{a}_e)$. We use u_i to indicate the value of the i-th objective.

A team strategy π is a probability distribution over joint actions $\mathcal{A} \to [0,1]$. In general strategies are stochastic. Every *joint* action gets assigned a probability $0 \leq \pi(\mathbf{a}) \leq 1$, and the probabilities for all joint actions together sum to 1, $\sum_{\mathbf{a} \in \mathcal{A}} \pi(\mathbf{a}) = 1$. The value of a strategy $\mathbf{u}^\pi$ is the expected (vector-valued) utility of the strategy $\mathbf{u}^\pi = \sum_{\mathbf{a} \in \mathcal{A}} \pi(\mathbf{a}) \mathbf{u}(\mathbf{a})$. A deterministic strategy is a special case of a strategy in which one

---

[1] A preliminary version of this work was presented in [13].

joint action **a** has probability 1 and the rest probability 0. We refer to the set of all vectors for all possible strategies as $\mathcal{V}$.[2]

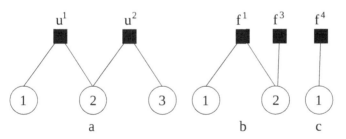

Fig. 1. (a) A MO-CoG factor graph, (b) after eliminating agent 3 by adding f^3, and (c) after eliminating agent 2 by adding f^4

The decomposition of $\mathbf{u}(\mathbf{a})$ into local payoff functions can be represented as a *factor graph* containing agents (variables) and local payoff functions (factors), with edges connecting local payoff functions to the agents in their scope. Figure 1a shows a factor graph for the payoff function $\mathbf{u}(\mathbf{a}) = \sum_{e=1}^{\rho} \mathbf{u}^e(\mathbf{a}_e) = \mathbf{u}^1(a_1, a_2) + \mathbf{u}^2(a_2, a_3)$.

We assume there exists a *scalarization function* f that converts $\mathbf{u}^\pi$ to a scalar payoff $u_{\mathbf{w}}^\pi = f(\mathbf{u}^\pi, \mathbf{w})$. This function is parameterized by a weight vector $\mathbf{w}$, which is unknown *when the MO-CoG is solved* but known when the agents must select a strategy. The solution to a MO-CoG is the *coverage set* (CS) [2], i.e., all strategies π and associated values $\mathbf{u}^\pi$ that are optimal for some $\mathbf{w}$:

$$CS(\mathcal{V}) = \left\{ \mathbf{u}^\pi \; : \; \mathbf{u}^\pi \in \mathcal{V} \; \wedge \; \exists \mathbf{w} \forall \pi' \; u_{\mathbf{w}}^\pi \geq u_{\mathbf{w}}^{\pi'} \right\}.$$

For convenience, we assume that the coverage set contains both the values and associated strategies. What the CS looks like depends on what strategies are allowed, and what we know about the scalarization function.

A minimal assumption about the scalarization function is that it is monotonically increasing in all objectives (i.e., if the value for one objective increases while the values for the other objectives stay constant, the scalarized value cannot go down). This assumption ensures that objectives are actually objectives, i.e., having more of them is better. In this case, the CS is called the *Pareto coverage set* (PCS) or *Pareto front*:

$$PCS(\mathcal{V}) = \left\{ \mathbf{u}^\pi \; : \; \mathbf{u}^\pi \in \mathcal{V} \; \wedge \; \neg \exists \pi' \; \mathbf{u}^{\pi'} \succ_P \mathbf{u}^\pi \right\},$$

where $\succ_P$ indicates *Pareto dominance* (P-dominance): greater or equal in all objectives and strictly greater in at least one objective. Note that computing P-dominance[3] requires only comparing pairs of vectors [7].

A highly prevalent scenario is that, in addition to knowing that the scalarization function is monotonically increasing, we also know that it is linear, $f = \mathbf{w} \cdot \mathbf{u}^\pi$. This is

---

[2] MO-CoGs are similar to the *multi-objective weighted constraint satisfaction problems* (MO-WCSPs) considered in [15]. However, MO-WCSPs consider only deterministic strategies and bounded, integer-valued payoffs. In addition, they consider *constraints*, the absence of which in Mo-CoGs has important implications for our complexity analysis (see Section 6).

[3] P-dominance is often called *pairwise dominance* in the POMDP literature.

the case in, e.g., clinical trials [11] or resource gathering [1]. In this case, all we need is the convex coverage set (CCS):[4]

$$CCS(\mathcal{V}) = \left\{ \mathbf{u}^\pi : \mathbf{u}^\pi \in \mathcal{V} \ \wedge \ \exists \mathbf{w} \forall \pi' \ \mathbf{w} \cdot \mathbf{u}^\pi \geq \mathbf{w} \cdot \mathbf{u}^{\pi'} \right\}.$$

Vectors not in the CCS are *C-dominated*. In contrast to P-domination, C-domination cannot be tested for with pairwise comparisons because it can (in the setting of deterministic strategies) take two or more vectors to C-dominate a vector: a vector can be dominated over the entire weight-space, but not necessarily always by the same vector, as indicated in Figure 2 (right). The CCS contains all strategies that could be optimal for some weight in a linear scalarization, i.e., all strategies that are not C-dominated. Anything in the PCS but not in the CCS is C-dominated and cannot be useful given the assumption of a linear scalarization function. Because we assume the linear scalarization is monotonically increasing, we can represent it without loss of generality as a convex combination of the objectives: i.e., the weights are positive and sum to 1. Since such linear functions are a subset of monotonically increasing functions, the CCS is a subset of the PCS.

Many multi-objective methods, e.g., [5,6,12,14] simply assume that the PCS is the appropriate solution set. However, which CS one should use depends what one can assume about how utility is defined with respect to the multiple objectives, i.e., which scalarization function is used to scalarize the vector-valued payoffs. We argue that in many situations, one can assume that the scalarization function will be linear. For example, when the different objectives are products and/or resources that need to be bought and sold on a market, every objective will be associated with a current unit price on the market, leading to linear trade-offs. In such cases one should use the CCS.

In addition, the choice of solution concept also depends on whether only deterministic strategies are considered or whether stochastic ones are also permitted. We consider this issue in the next section.

3   Deterministic versus Stochastic Strategies

When we allow only deterministic strategies, i.e., one joint action is chosen with probability 1, the PCS and CCS can be quite different. In Figure 2 (left) the values of deterministic strategies are represented as points in value-space, for a two-objective MO-CoG. The strategy A is in both the CCS and the PCS. B, however, is in the PCS, but not the CCS, because there is no weight for which a linear scalarization of B's value would be optimal, as shown in Figure 2 (right), where the scalarized value of the strategies are plotted as a function of the weight on the first objective ($w_2 = 1 - w_1$). C is in neither the CCS nor the PCS: it is Pareto-dominated by A. We refer to the deterministic PCS as the PCS of deterministic strategies, i.e., the PCS when only deterministic strategies are allowed. We refer similarly to the deterministic CCS.

As discussed in Section 2, stochastic strategies are linear combinations of deterministic strategies. The value of a stochastic strategy is thus also a linear combination of the

---

[4] The convex coverage set is often called the *convex hull*. We avoid this term because it is imprecise: the convex hull (a term from graphics) is a superset of the convex coverage set.

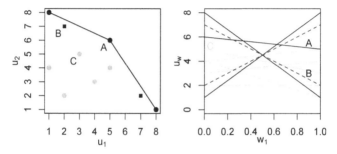

Fig. 2. The CCS (filled circles at left, and solid black lines at right) versus the PCS (filled circles and squares at left, and both dashed and solid black lines at right) for twelve random 2-dimensional payoff vectors

value vectors of the deterministic strategies it is a mixture of: $\mathbf{u}^\pi = \sum_{\mathbf{a} \in \mathcal{A}} \pi(\mathbf{a}) \mathbf{u}(\mathbf{a})$. Therefore, the optimal values (for both linear and nonlinear monotonically increasing scalarization functions [17]) lay on the convex upper surface spanned by the strategies in the deterministic CCS, as indicated by the black lines in Figure 2 (left). In the stochastic case, the PCS and CCS are thus identical. Furthermore, the values for the stochastic PCS/CCS can be constructed from the values in the deterministic CCS. The stochastic PCS/CCS is thus very different from the deterministic PCS and the deterministic CCS. While the deterministic PCS and deterministic CCS contain finite numbers of strategies, the stochastic PCS/CCS contains inifinitely many strategies.

However, when we know that the scalarization function is linear, we do not actually need the entire stochastic CCS: for each weight, there exists a deterministic strategy that is optimal. For every optimal strategy in the stochastic CCS there exists a deterministic strategy that is just as good, because a linear combination of the values of two or more deterministic strategies never yields a larger scalarized utility for any $\mathbf{w}$, than one of the constituent deterministic strategies: $\mathbf{w} \cdot \mathbf{u}^\pi = \sum_{\mathbf{a}} \pi(\mathbf{a})(\mathbf{w} \cdot \mathbf{u}(\mathbf{a}))$. By contrast, when the scalarization function is monotonically increasing (but not necessarily linear), the full stochastic PCS is required. This is a problem, because it contains infinitely many strategies. However, all values on the stochastic PCS can be attained by making a stochastic mixture from the strategies on the deterministic CCS [17]. Note that these mixtures (all points on the black lines in Figure 2 (left)) dominate all points, like B, that are in the deterministic PCS but not the deterministic CCS. Therefore the CCS can be used to create all possible values on the PCS of stochastic strategies, and is more compact than the deterministic PCS.

It might of course be the case that the problem setting is restricted to deterministic solutions. For example in the medical domain [11], it can be unethical to treat patients based on a stochastic strategy. However, in most settings, stochasticity is permissable and the aim is to optimize the expected return.

Therefore, in this paper we present methods for computing the strategies in the deterministic CCS because it is an appropriate solution concept, not only when the scalarization function is linear, but also any time stochastic strategies are considered, even if the scalarization function is nonlinear, as shown in Table 1. For brevity, in the rest of the paper, we refer to the deterministic CCS as simply the CCS, to deterministic strategies as *joint actions*, and to the set of values of all deterministic strategies as $\mathcal{V}$.

Table 1. Motivating scenarios

	Linear scalarization functions	Monotonically increasing scalarization functions
Deterministic strategies	Deterministic CCS	Deterministic PCS
Stochastic strategies	Deterministic CCS	Deterministic CCS

4   Nongraphical Convex Approach

One way to compute the CCS, is to *ignore the graphical structure*, calculate the set of all possible payoffs for all joint actions $\mathcal{V}$, and prune away the C-dominated joint actions. To determine the set $\mathcal{V}$, we first translate the problem to a set of *value set factors* (VSFs), $\mathcal{F}$. Each VSF f is a function mapping local joint actions to sets of payoff vectors. Initially, the VSFs are constructed from the local payoff functions such that $f^e(\mathbf{a}_e) = \{\mathbf{u}^e(\mathbf{a}_e)\}$, i.e., each VSF maps a local joint action to the singleton set containing only that action's local payoff. We can now define $\mathcal{V}$ in terms of $\mathcal{F}$ using the *cross-sum* operator over all VSFs in $\mathcal{F}$ for each joint action $\mathbf{a}$: $\mathcal{V}(\mathcal{F}) = \bigcup_{\mathbf{a}} \bigoplus_{f^e \in \mathcal{F}} f^e(\mathbf{a}_e)$.[5] The CCS can now be calculated by applying a pruning operator CPrune (described below) that removes all C-dominated vectors from a set of value vectors, to $\mathcal{V}$:

$$CCS(\mathcal{V}(\mathcal{F})) = \mathtt{CPrune}(\mathcal{V}(\mathcal{F})) = \mathtt{CPrune}(\bigcup_{\mathbf{a}} \bigoplus_{f^e \in \mathcal{F}} f^e(\mathbf{a}_e))$$

The CCS contains the all the vectors that are maximizing for some $\mathbf{w}$:

$$\forall \mathbf{a} \ \left(\exists \mathbf{w} \ s.t. \ \mathbf{a} = \arg\max_{\mathbf{a} \in \mathcal{A}} \mathbf{w} \cdot \mathbf{u}(\mathbf{a}) \right) \implies \mathbf{u}(\mathbf{a}) \in CCS(\mathcal{V}(\mathcal{F})) \qquad (1)$$

This is exactly the same problem as in *partially observable Markov decision processes* (POMDPs) [7], where the optimal α-vectors (corresponding to the value vectors $\mathbf{u}^\pi$) for all beliefs (corresponding to the weight vectors $\mathbf{w}$) must be found. Therefore, we can use pruning operators from the POMDP literature. Algorithm 1 describes our implementation of CPrune, which is based on [7] with the modification that, in order to improve runtime guarantees, we first pre-prune to the PCS using the PPrune operator shown in Algorithm 2, which computes the (deterministic) PCS in $O(d|\mathcal{V}_{det}||PCS|)$ by running pairwise comparisons.

Next, we maintain a partial CCS ($\mathcal{U}^*$), which is constructed as follows: we select a random vector $\mathbf{u}$ from the set of candidate vectors $\mathcal{U}'$ and test whether there is a weight vector $\mathbf{w}$ for which it is better than the vectors in $\mathcal{U}^*$ by solving the linear program shown in Algorithm 3. If so, we find the best vector $\mathbf{v}$ for $\mathbf{w}$ in $\mathcal{U}'$ and move $\mathbf{v}$ to $\mathcal{U}^*$. If there is no weight for which $\mathbf{u}$ is better, we remove $\mathbf{u}$ from $\mathcal{U}'$ (because it is C-dominated).

The runtime of the CPrune operator we use is $O(d|\mathcal{V}_{det}||PCS| + |PCS|P(d|CCS|))$, where $P(d|CCS|)$ is a polynomial in the size of the CCS and the number

---

[5] The cross-sum of two sets A and B contains all possible vectors that can be made by summing one payoff vector from each set: $A \oplus B = \{\mathbf{a} + \mathbf{b} : \mathbf{a} \in A \wedge \mathbf{b} \in B\}$.

of objectives d, which is the runtime of the linear program that tests for C-domination (Algorithm 3).

Algorithm 1. CPrune($\mathcal{U}$)	**Algorithm 2.** PPrune($\mathcal{U}$)

$\mathcal{U}' = $ PPrune($\mathcal{U}$)
 $\mathcal{U}^* = \emptyset$
 while notEmpty($\mathcal{U}'$) **do**
   select random $\mathbf{u}$ from $\mathcal{U}'$
   $\mathbf{w} \leftarrow $ findWeight($\mathbf{u}, \mathcal{U}^*$)
   if $\mathbf{w}$=*null* **then**
      remove $\mathbf{u}$ from $\mathcal{U}'$
   else
      move best $\mathbf{v}$ for weight $\mathbf{w}$ from $\mathcal{U}'$ to
      $\mathcal{U}^*$
 return $\mathcal{U}^*$
	$\mathcal{U}^* \leftarrow \emptyset$
	 while $\mathcal{U} \neq \emptyset$ **do**
	   $\mathbf{u} \leftarrow$ the first element of $\mathcal{U}$
	   foreach $\mathbf{v} \in \mathcal{U}$ **do**
	     if $\mathbf{v} \succ_P \mathbf{u}$ **then**
	       $\mathbf{u} \leftarrow \mathbf{v}$ // *Continue with $\mathbf{v}$ in-*
	       stead of $\mathbf{u}$
	   Remove $\mathbf{u}$, and all vectors Pareto-dominated by it, from $\mathcal{U}$
	   Add $\mathbf{u}$ to $\mathcal{U}^*$
	 return $\mathcal{U}^*$

5 Exploiting Loose Couplings

In the previous section, we showed that, for the nongraphical approach, computing the CCS is more expensive than computing the PCS. In this section, we show that, by exploiting the MO-CoG's graphical structure, we can often compute the CCS much more efficiently. In particular, we solve the MO-CoG as a series of local subproblems, by iteratively *eliminating* agents, and thereby manipulating $\mathcal{F}$. The key idea is, for each agent elimination, to compute a *local CCS* (LCCS), pruning away as many vectors as possible at the lowest possible level. This minimizes the number of payoff vectors that are calculated at the global level, which can greatly speed computation. Here we describe the elim operator for eliminating agents used by CMOVE in Section 6.

To eliminate agent i, we define $\mathcal{F}_i$, the set of relevant VSFs with i in scope. Then, for each possible local joint action of n_i, agent i's neighbors, we define an LCCS that contains the payoffs of the C-undominated responses of agent i to the given local joint action of n_i. In other words, it is the CCS of the subproblem that arises when considering only $\mathcal{F}_i$ and fixing a specific local joint action of n_i. To compute the LCCS, we must consider all payoff vectors and prune the dominated ones. If we fix all actions in $\mathbf{a}_{n_i}$ except a_i, the set of all payoff vectors for this subproblem is: $\mathcal{V}_i(\mathcal{F}_i, \mathbf{a}_{n_i}) = \bigcup_{a_i} \bigoplus_{f^e \in \mathcal{F}_i} f^e(\mathbf{a}_e)$, where $\mathbf{a}_e$ is formed from a_i and the appropriate part of $\mathbf{a}_{n_i}$. The corresponding LCCS is thus the undominated subset of $\mathcal{V}_i(\mathcal{F}_i, \mathbf{a}_{n_i})$:

$$LCCS_i(\mathcal{F}_i, \mathbf{a}_{n_i}) = CCS(\mathcal{V}_i(\mathcal{F}_i, \mathbf{a}_{n_i})).$$

Using these LCCSs we can define a new VSF, f^{new} conditioned on the actions of the agents in n_i: $\forall \mathbf{a}_{n_i} f^{new}(\mathbf{a}_{n_i}) \triangleq LCCS_i(\mathcal{F}_i, \mathbf{a}_{n_i})$. The elim operator replaces the VSFs in $\mathcal{F}_i$ in $\mathcal{F}$ by this new factor:

$$\text{elim}(\mathcal{F},i) = (\mathcal{F} \setminus \mathcal{F}_i) \cup \{f^{new}(\mathbf{a}_{n_i})\}.$$

Theorem 1. elim *preserves the CCS:* $\forall i\ \forall \mathcal{F}\ CCS(\mathcal{V}(\mathcal{F})) = CCS(\mathcal{V}(\text{elim}(\mathcal{F},i)))$.

Proof sketch. The linear scalarization function distributes over the local payoff functions: $\mathbf{w} \cdot \mathbf{u}(\mathbf{a}) = \mathbf{w} \cdot \sum_e \mathbf{u}^e(\mathbf{a}_e) = \sum_e \mathbf{w} \cdot \mathbf{u}^e(\mathbf{a}_e)$. Thus, when eliminating agent

Algorithm 3. findWeight($\mathbf{u}$,$\mathcal{U}$)	Algorithm 4. elim($\mathcal{F}$,i,prune1,prune2)
$\max\limits_{x,\mathbf{w}} \quad x$ subject to $\mathbf{w} \cdot (\mathbf{u} - \mathbf{u}') - x \geq 0, \forall \mathbf{u}' \in \mathcal{U}$ $\sum\limits_{i=1}^{d} w_i = 1$ if $x > 0$ return $\mathbf{w}$ else return null	$\mathcal{U}^*, n_i \leftarrow \emptyset$, set of neighboring agents of i $\mathcal{F}_i \leftarrow$ the subset of f functions involving i $f^{new}(\mathbf{a}_{n_i}) \leftarrow$ a new factor **foreach** $\mathbf{a}_{n_i} \in \mathcal{A}_{n_i}$ **do** $\quad \lfloor \; f^{new}(\mathbf{a}_{n_i}) \leftarrow LCCS_i(\mathcal{F}_i, \mathbf{a}_{n_i}, \text{prune1},$ $\qquad\qquad \text{prune2})$. $\mathcal{F} \leftarrow \mathcal{F} \setminus \mathcal{F}_i \cup \{f^{new}\}$ **return** $\mathcal{V}^*$

i, we divide the set of VSFs into non-neighbors (nn), in which agent i does not participate, and neighbors (n_i) such that: $\mathbf{w} \cdot \mathbf{u}(\mathbf{a}) = \sum_{e \in nn} \mathbf{w} \cdot \mathbf{u}^e(\mathbf{a}_e) + \sum_{e \in n_i} \mathbf{w} \cdot \mathbf{u}^e(\mathbf{a}_e)$. Now, following (1), the CCS contains $\max_{\mathbf{a} \in \mathcal{A}} \mathbf{w} \cdot \mathbf{u}(\mathbf{a})$ for all $\mathbf{w}$. elim pushes this maximization in: $\max_{\mathbf{a} \in \mathcal{A}} \mathbf{w} \cdot \mathbf{u}(\mathbf{a}) = \max_{\mathbf{a}_{-i} \in \mathcal{A}_{-i}} \sum_{e \in nn} \mathbf{w} \cdot \mathbf{u}^e(\mathbf{a}_e) + \max_{\mathbf{a}_i \in \mathcal{A}_i} \sum_{e \in n_i} \mathbf{w} \cdot \mathbf{u}^e(\mathbf{a}_e)$. elim replaces the agent-i factors by a term $f^{new}(\mathbf{a}_{n_i})$ that satisfies $\mathbf{w} \cdot f^{new}(\mathbf{a}_{n_i}) = \max_{\mathbf{a}_i} \sum_{e \in n_i} \mathbf{w} \cdot \mathbf{u}^e(\mathbf{a}_e)$ per definition, thus preserving the maximum for all $\mathbf{w}$ and thereby preserving the CCS.

Since LCCS $\subseteq$ LPCS $\subseteq \mathcal{V}_i$, where LPCS is the local PCS, elim not only reduces the problem size, it can do so more than is possible when considering only P-dominance. Consequently, focusing on the CCS can lead to considerable speedups.

6 Convex MOVE

We now present *Convex Multi-Objective Variable Elimination* (CMOVE), which implements elim using pruning operators, iteratively applies it to compute the CCS, and outputs the correct joint actions for each payoff vector in the CCS. It is an extension to Rollón's Pareto-based MOVE (which we denote PMOVE) [14].

Like PMOVE, CMOVE eliminates agents in sequence, solving local subproblems along the way. The most important difference is that CMOVE computes the CCS, which can lead to smaller subproblems and thus much better computational efficiency. In addition, we identify three places where pruning can take place, yielding a more flexible algorithm with different trade-offs. Finally, we use a *tagging scheme* instead of the *backwards pass* employed by Rollón, which greatly simplifies the algorithm without effecting its runtime.

CMOVE is also related to multi-objective methods for GAI networks [6] and influence diagrams [12]. However, like PMOVE, these methods compute only the PCS.

6.1 Algorithm

We first present an abstract version of CMOVE, which leaves the pruning operators unspecified. The choice of these operators leads to specific variants with different trade-offs between pruning effort and local problem sizes. As before, CMOVE first translates the problem into a set of *vector-set factors* (VSFs), $\mathcal{F}$. Next, it iteratively eliminates agents using elim. The elimination order can be determined using techniques devised for regular VE [10]. Algorithm 4 shows our implementation of elim, parameterized with two pruning operators, prune1 and prune2, corresponding to two different pruning locations inside $LCCS_i(\mathcal{F}_i, \mathbf{a}_{n_i}, \text{prune1}, \text{prune2})$, which is implemented as follows. First we define a new cross-sum-and-prune operator $A \hat{\oplus} B = \text{prune1}(A \oplus B)$, which we can apply sequentially in the definition of the LCCS operator:

$$LCCS_i(\mathcal{F}_i, \mathbf{a}_{n_i}, \text{prune1}, \text{prune2}) = \text{prune2}(\bigcup_{a_i} \hat{\bigoplus}_{f^e \in \mathcal{F}_i} f^e(\mathbf{a}_e)).$$

Applying prune1 to each cross-sum of two sets, via the $\hat{\oplus}$ operator, leads to *incremental pruning* [4]; prune2 prunes at a coarser level, after the union.

CMOVE applies elim iteratively until no agents remain, resulting in the CCS. An example of how this works is presented in Section 6.3.

Pruning can also be applied at the very end, after all agents have been eliminated, which we call prune3. In increasing level of coarseness, we thus have three pruning operators: incremental pruning (prune1), pruning after the union over actions of the eliminated agent (prune2), and pruning after all agents have been eliminated (prune3).

There are several ways to implement the pruning operators that lead to correct instantiations of CMOVE. One can use both PPrune (Algorithm 2) as well as CPrune (Algorithm 1) as long as either prune2 or prune3 is CPrune. (Note that if prune2 computes the CCS, prune3 is not necessary.) In this paper, we consider *Basic CMOVE*, which does not use prune1 and prune3 and only prunes at prune2 using CPrune, as well as *Incremental CMOVE*, which uses CPrune at both prune1 and prune2.

6.2 Tagging Scheme

Once CMOVE computes the CCS, we need to retrieve the joint actions that generate these values. In single-objective VE, this is typically done with a *backwards pass* that constructs a joint action by iterating through the eliminated agents in reverse order. However, doing so in the multi-objective setting is more complex, because the partial joint actions in the LCCSs need to be matched with the different values in the CCS instead of just backtracking a single optimal solution that automatically belongs to the optimal value. Consequently, the backwards pass used in Rollón's implementation of PMOVE [14] is fairly complex. However, we can obviate the need for a backwards pass by using a *tagging* scheme: when eliminating an agent i, CMOVE tags all the vectors in the LCCSs with the appropriate action of this agent. The payoff vectors are stored as a tuple containing both the payoff vector and a partial joint action. CMOVE combines the tags of agent i with the tags already present in $\mathcal{F}_i$. For example, in Figure 1, factor f^3 contains payoff vectors tagged with an action of agent 3 and factor f^4 contains tags with actions of both agents 2 and 3. Doing this for every agent in the elimination sequence builds the complete joint action for each payoff vector in the CCS. Replacing the backwards pass with this tagging scheme reduces by about half the number of lines of pseudocode needed to describe the algorithm.

6.3 Example

Consider the example in Figure 1a, using the payoffs defined by Table 2. First, CMOVE creates the VSFs f^1 and f^2 from u^1 and u^2 (not shown). To eliminate agent 3, it creates a new factor $f^3(a_2)$ by computing the LCCSs for every a_2 and tagging each element of each set with the action of agent 3 that generates it. For $\dot{a}_2$, CMOVE first generates the set $\{(3,1)_{\dot{a}_3}, (1,3)_{\bar{a}_3}\}$. Since both of these vectors are optimal for some $\mathbf{w}$, neither is removed by pruning and thus $f^3(\dot{a}_2) = \{(3,1)_{\dot{a}_3}, (1,3)_{\bar{a}_3}\}$. For $\bar{a}_2$, CMOVE first generates $\{(0,0)_{\dot{a}_3}, (1,1)_{\bar{a}_3}\}$. CPrune determines that $(0,0)_{\dot{a}_3}$ is dominated and consequently removes it, yielding $f^3(\dot{a}_2) = \{(1,1)_{\bar{a}_3}\}$. CMOVE then adds f^3 to the graph and removes f^2 and agent 3, yielding the factor graph shown in Figure 1b.

CMOVE then eliminates agent 2 by combining f^1 and f^3 to create f^4. For $f^4(\dot{a}_1)$, CMOVE must calculate the LCCS of:

$$(f^1(\dot{a}_1,\dot{a}_2) \oplus f^3(\dot{a}_2)) \cup (f^1(\dot{a}_1,\bar{a}_2) \oplus f^3(\bar{a}_2)).$$

The first cross sum is $\{(7,2)_{\dot{a}_2\dot{a}_3}, (5,4)_{\dot{a}_2\bar{a}_3}\}$ and the second is $\{(1,1)_{\bar{a}_2\bar{a}_3}\}$. Pruning their union yields $f^4(\dot{a}_1) = \{(7,2)_{\dot{a}_2\dot{a}_3}, (5,4)_{\dot{a}_2\bar{a}_3}\}$. Similarly, for $\bar{a}_1$ taking the union yields $\{(4,3)_{\dot{a}_2\dot{a}_3}, (2,5)_{\dot{a}_2\bar{a}_3}, (4,7)_{\bar{a}_2\bar{a}_3}\}$, of which the LCCS is $f^4(\bar{a}_1) = \{(4,7)_{\bar{a}_2\bar{a}_3}\}$. Adding f^4 results in the factor graph in Figure 1c.

Table 2. The two-dimensional payoff matrices for $\mathbf{u}^1(a_1, a_2)$ (left) and $\mathbf{u}^2(a_2, a_3)$ (right)

	$\dot{a}_2$	$\bar{a}_2$		$\dot{a}_3$	$\bar{a}_3$
$\dot{a}_1$	(4,1)	(0,0)	$\dot{a}_2$	(3,1)	(1,3)
$\bar{a}_1$	(1,2)	(3,6)	$\bar{a}_2$	(0,0)	(1,1)

Finally, CMOVE eliminates agent 1. Since there are no neighboring agents left, $\mathcal{A}_i$ contains only the empty action. CMOVE takes the union of $f^4(\dot{a}_1)$ and $f^4(\bar{a}_1)$. Since $(7,2)_{\{\dot{a}_1\dot{a}_2\dot{a}_3\}}$ and $(4,7)_{\{\bar{a}_1\bar{a}_2\bar{a}_3\}}$ dominate $(5,4)_{\{\dot{a}_1\dot{a}_2\dot{a}_3\}}$, the latter is pruned, leaving $CCS = \{(7,2)_{\{\dot{a}_1\dot{a}_2\dot{a}_3\}}, (4,7)_{\{\bar{a}_1\bar{a}_2\bar{a}_3\}}\}$.

6.4 Analysis

We now analyse the correctness and complexity of CMOVE.

Theorem 2. *MOVE correctly computes the CCS.*

Proof. The proof works by induction on the number of agents. The base case is the original MO-CoG, where each $f^e(\mathbf{a}_e)$ from $\mathcal{F}$ is a singleton set. Then, since elim preserves the CCS (see Theorem 1), no necessary vectors are lost. When the last agent is eliminated, only one factor remains; since it is not conditioned on any agent actions and is the result of an $LCCS$ computation, it must contain one set: the CCS.

Theorem 3. *The computational complexity of CMOVE is*

$$O(\, n \, |\mathcal{A}_{max}|^{w_a} \, (w_f \, R_1 + R_2) + R_3 \,), \tag{2}$$

where w_a is the induced agent width, i.e., the maximum number of neighboring agents (connected via factors) of an agent when eliminated, w_f is the induced factor width, i.e., the maximum number of neighboring factors of an agent when eliminated, and R_1, R_2 and R_3 are the cost of applying the prune1*,* prune2 *and* prune3 *operators.*

Proof. CMOVE eliminates n agents and for each one computes a value (set) in a new payoff function for each joint action of the eliminated agent's neighbors. CMOVE computes $O(|\mathcal{A}_{max}|^w)$ fields per iteration, calling prune1 for each adjacent factor, and prune2 once after taking the union over actions of the eliminated agent. prune3 is called only once, after eliminating all agents.

Thus, unlike nongraphical approaches, CMOVE is exponential only in the induced width, not the number of agents. In this respect, our results are similar to those for PMOVE [14]. However, those earlier complexity results do not make the effect of pruning explicit. Instead, the complexity bound makes use of problem constraints, which limit the total number of possible different value vectors. However, in practice such bounds are very loose or even impossible to define. Therefore, we instead give a description of the computational complexity that makes explicit the dependence on the effectiveness of pruning. Even though such complexity bounds are not better in the worst case (i.e., when no pruning is possible), they allow greater insight into the runtimes of the algorithms we evaluate, as is apparent in our analysis of the experimental results in Section 7.

Theorem 3 demonstrates that the complexity of CMOVE heavily depends on the runtime of its pruning operators, which in turn depends on the sizes of the input sets. The input set of prune2 is the union of what is returned by a series of applications of prune1, while prune3 uses the output of the last application of prune2. Therefore, we need to balance the effort of the lower-level pruning operators with that of the higher-level ones, which occur less often but are dependent on the output of the lower-level pruning operators. The bigger the LCCSs, the more can be gained from lower-level pruning. We compare different variants of CMOVE in the experimental section.

7 Experiments

In this section, we present an empirical analysis of CMOVE. The first goal of these experiments is to show that CMOVE, by exploiting the graphical structure to compute the CCS, can solve MO-CoGs substantially faster than both nongraphical methods and those that compute the PCS. To this end, we compare Basic CMOVE and Incremental CMOVE to the *nongraphical method* described in Section 4 and PMOVE.

We first present results on randomly generated MO-CoGs, in order to examine performance on MO-CoGs with widely varying properties. We then present results on *Mining Day*, a problem we propose as a MO-CoG benchmark, in order to establish that CMOVE performs well on a MO-CoG derived from a realistic scenario. The experiments use a C++ implementation that employs the lp_solve library (v5.5) to solve linear programs.

7.1 Random MO-CoGs

We employ a generation procedure for random MO-CoGs that is based on the following inputs: n, the number of agents; d, the number of payoff dimensions; ρ the number of local payoff functions; and $|\mathcal{A}_i|$, the action space size of the agents, which is the same for all agents. First, a fully connected graph with local payoff functions connected to two agents is created. Then, local payoff functions are randomly removed, while checking that the graph remains connected, until only ρ remain. The values in each local payoff function are real numbers drawn independently and uniformly from the interval $[0,10]$. All algorithms are tested on the same set of randomly generated MO-CoGs for each value of n, d, ρ, and $|\mathcal{A}_i|$ that is considered.

To compare CMOVE, PMOVE, and the nongraphical method, we tested them on random MO-CoGs with an increasing number of agents, with the average number of

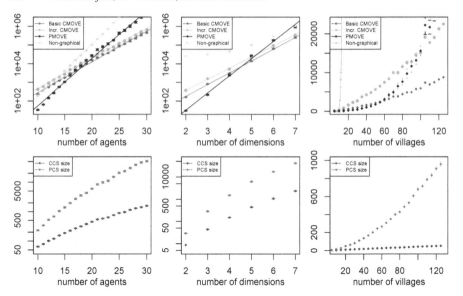

Fig. 3. Runtimes (ms) for the nongraphical method, PMOVE and CMOVE with standard errors (error bars) (top) and the corresponding number of vectors in the PCS and CCS (bottom)

factors per agent held at $\rho = 1.5n$ and the number of dimensions $d = 5$. Figure 3 (top left) shows the results, averaged over 85 MO-CoGs for each number of agents. These results demonstrate that, as the number of agents grows, using MOVE becomes key to containing the computational cost of solving the MO-CoG. CMOVE outperforms the nongraphical method from 12 agents onwards. At 25 agents, Basic CMOVE is 38 times faster. CMOVE also does significantly better than PMOVE. Though it is one order of magnitude slower with 10 agents ($238ms$ (Basic) and $416ms$ (Incremental) versus $33ms$ on average), its runtime grows much more slowly than that of PMOVE. At 20 agents, both CMOVE variants are faster than PMOVE and at 28 agents, Basic CMOVE is almost one order of magnitude faster ($228s$ versus $1,650s$ on average), and the difference increases with every agent.

While CMOVE's runtime grows much more slowly than that of the nongraphical method, it is still exponential in the number of agents, a counterintuitive result since the worst-case complexity is linear in the number of agents. There are two reasons for this. First, CMOVE is exponential in the induced width, which increases with the number of agents, from 3.1 at $n = 10$ to 6.0 at $n = 30$ on average, as a result of the MO-CoG generation procedure. Second, CMOVE's runtime is polynomial in the size of the CCS, and this size grows exponentially (Figure 3 (bottom left)). The fact that CMOVE is much faster than PMOVE can be explained by the sizes of the PCS and CCS, as the former grows much faster than the latter. At 10 agents, the average PCS size is 230 and the average CCS size is 65. At 30 agents, the average PCS size has risen to 51,745 while the average CCS size is only 1,575.

Figure 3 (top middle) compares the scalability of the algorithms in the number of objectives, on random MO-CoGs with $n = 20$ and $\rho = 30$, averaged over 100 MO-CoGs. CMOVE always outperforms the nongraphical method. Interestingly, the nongraphical method is several orders of magnitude slower at $d = 2$, grows slowly until $d = 5$, and then starts to grow with about the same exponent as Pareto MOVE. The reason is that enumeration of all the joint actions and payoff vectors takes approximately constant time while the time it takes to prune increases exponentially. When $d = 2$, CMOVE is an order of magnitude slower than PMOVE ($163ms$ (Basic) and 377 (Incremental) versus $30ms$). However, when $d = 5$, both CMOVE variants are already faster than PMOVE and at 7 dimensions they are respectively 3.7 and 2.7 times faster. This happens because the CCS grows much more slowly than the PCS (Figure 3 (bottom middle)). The difference between Incremental and Basic CMOVE decreases as the number of dimensions increases, from a factor 2.3 at $d = 2$ to 1.3 at $d = 7$.

Overall, these results indicate that CMOVE shows large speedups over PMOVE for more than a minimal number of agents. The runtime of Incremental CMOVE grows more slowly than that of Basic CMOVE and seems favorable for large numbers of agents and high dimensions.

7.2   Mining Day

In Mining Day, a mining company mines gold and silver (objectives) from a set of mines (local payoff functions) spread throughout a geographical region (Figure 4). The mine workers live in villages also spread throughout this region. The company has one van in each village (agents) for transporting workers and must determine every morning to which mine each van should go (actions). However, vans can only travel to nearby mines (graph connectivity). Workers are more efficient if there are more workers at the mine: there is a 3% efficiency bonus per worker such that the amount of each resource mined per worker is $x \cdot 1.03^w$, where x is the base rate per worker and w is the number of workers at the mine. The base rate of gold and silver are properties of a mine. Since the company aims to maximize revenue, the best strategy depends on the prices of gold and silver, which fluctuate and are not known when the plan must be computed.

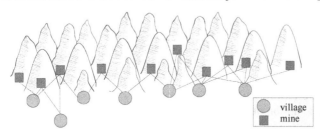

Fig. 4. The Mining Day problem

To generate a Mining Day instance with v villages (agents), we randomly assign 2-5 workers to each village and connect it to 2-4 mines. Each village is only connected to mines with a greater or equal index, i.e., if village i is connected to m mines, it is connected to mines i to $i + m - 1$. The last village is connected to 4 mines and thus the number of mines is $v + 3$. The base rates per worker for each resource at each mine are drawn uniformly and independently from $[0,10]$.

The results for the mining day problem are shown in Figure 3 (top right). The runtime of the nongraphical method grows exponentially with the number of agents. At only 13 agents, the runtime is already more than $30s$. By contrast, both CMOVE and PMOVE are able to tackle problems with over 100 agents within that timeframe. In addition, the runtime of PMOVE grows much more quickly than that of CMOVE. In this two-dimensional setting, Basic CMOVE is better than Incremental CMOVE. Basic CMOVE and PMOVE both have runtimes of around $2.8s$ at 60 agents, but at 100 agents, Basic CMOVE runs in about $5.9s$ and PMOVE in $21s$. Even though Incremental CMOVE is worse than Basic CMOVE, its runtime still grows a lot slower than PMOVE, and beats PMOVE when there are many agents.

The difference between PMOVE and CMOVE results from the relationship between the number of agents and the sizes of the CCS, which grows linearly, and the PCS, which grows polynomially (Figure 3 (bottom right)). The induced width remains around 4 regardless of v. These results demonstrate that, when the CS grows linearly (or polynomially) in the number of agents, MOVE can solve MO-CoGs with many more agents than the nongraphical approach. In problems where the CCS grows more slowly than the PCS, CMOVE can solve MO-CoGs with many more agents than PMOVE.

8   Conclusions and Future Work

In this paper, we proposed the CMOVE algorithm for multi-objective coordination graphs. Unlike previous methods, it computes the convex coverage set (CCS) rather than the Pareto coverage set (PCS). Not only does this provide the optimal solution when the scalarization function is linear or stochastic strategies are allowed, it also greatly reduces computational costs.

Using two variants of CMOVE – based on the trade-off between pruning effort and and smaller intermediate results – we analyzed CMOVE's complexity in terms of the different pruning operators that can be used to compute the local CCSs. Our empirical study showed that CMOVE can tackle multi-objective problems much faster than methods that compute the PCS. The runtime of CMOVE grows much more slowly than that of PMOVE because the CCS grows much more slowly than the PCS. Therefore, we conclude that computing the CCS is key to keeping large MO-CoGs tractable.

In future work, we hope to develop approximate techniques for MO-CoGs. The work of [5], which converts graphs to trees and applies *max-plus* [9] to approximate the PCS, could be extended to approximate the CCS. Alternatively, an efficient multi-objective version of max-plus for graphs with loops could also approximate the CCS. In addition, loosening the definition of the CCS, in the spirit of the ϵ-approximate Pareto front [3], could also yield efficient approximations. Finally, we hope to develop a multi-objective version of *sparse cooperative Q-learning* [9] that would use CMOVE as a subroutine to tackle sequential multi-objective multi-agent tasks.

Acknowledgements.   This research is supported by the NWO DTC-NCAP (#612.001.109) and NWO CATCH (#640.005.003) projects.

References

1. Barrett, L., Narayanan, S.: Learning all optimal policies with multiple criteria. In: ICML, pp. 41–47. ACM, New York (2008)

2. Becker, R., Zilberstein, S., Lesser, V., Goldman, C.V.: Transition-Independent Decentralized Markov Decision Processes. In: AAMAS (2003)

3. Brázdil, T., Brozek, V., Chatterjee, K., Forejt, V., Kucera, A.: Two views on multiple mean-payoff objectives in Markov decision processes. CoRR, abs/1104.3489 (2011)

4. Cassandra, A.R., Littman, M.L., Zhang, N.L.: Incremental pruning: A simple, fast, exact method for partially observable markov decision processes. In: UAI, pp. 54–61 (1997)

5. Delle Fave, F.M., Stranders, R., Rogers, A., Jennings, N.R.: Bounded decentralised coordination over multiple objectives. In: AAMAS, pp. 371–378 (2011)

6. Dubus, J.-P., Gonzales, C., Perny, P.: Choquet optimization using gai networks for multi-agent/multicriteria decision-making. In: Rossi, F., Tsoukias, A. (eds.) ADT 2009. LNCS, vol. 5783, pp. 377–389. Springer, Heidelberg (2009)

7. Feng, Z., Zilberstein, S.: Region-based incremental pruning for POMDPs. CoRR, abs/1207.4116 (2012)

8. Guestrin, C.E., Koller, D., Parr, R.: Multiagent planning with factored MDPs. In: NIPS (2002)

9. Kok, J.R., Vlassis, N.: Collaborative multiagent reinforcement learning by payoff propagation. J. Mach. Learn. Res. 7, 1789–1828 (2006)

10. Koller, D., Friedman, N.: Probabilistic Graphical Models: Principles and Techniques. MIT Press (2009)

11. Lizotte, D.J., Bowling, M., Murphy, S.A.: Efficient reinforcement learning with multiple reward functions for randomized clinical trial analysis. In: ICML, pp. 695–702 (2010)

12. Marinescu, R., Razak, A., Wilson, N.: Multi-objective influence diagrams. In: UAI (2012)

13. Roijers, D.M., Whiteson, S., Oliehoek, F.A.: Multi-objective variable elimination for collaborative graphical games. In: AAMAS (2013) (Extended Abstract)

14. Rollón, E.: Multi-Objective Optimization for Graphical Models. PhD thesis, Universitat Politècnica de Catalunya (2008)

15. Rollón, E., Larrosa, J.: Bucket elimination for multiobjective optimization problems. Journal of Heuristics 12, 307–328 (2006)

16. Tesauro, G., Das, R., Chan, H., Kephart, J.O., Lefurgy, C., Levine, D.W., Rawson, F.: Managing power consumption and performance of computing systems using reinforcement learning. In: NIPS (2007)

17. Vamplew, P., Dazeley, R., Barker, E., Kelarev, A.: Constructing stochastic mixture policies for episodic multiobjective reinforcement learning tasks. In: Nicholson, A., Li, X. (eds.) AI 2009. LNCS, vol. 5866, pp. 340–349. Springer, Heidelberg (2009)

Verifying Preferential Equivalence and Subsumption via Model Checking

Ganesh Ram Santhanam, Samik Basu, and Vasant Honavar

Department of Computer Science, Iowa State University, Ames, IA 50011, USA
{gsanthan,sbasu,honavar}@cs.iastate.edu

Abstract. We present a practical, model checking based, approach to verifying whether a set of ceteris paribus preference statements is equivalent to or is subsumed by another. We translate a given pair of sets of preference statements into a labeled transition system (LTS) and reduce the problem of determining whether one set of preference statements is equivalent to or is subsumed by the other to verifying the appropriate computation-tree temporal logic (CTL) formulas in the resulting LTS. Whenever the two sets of preference statements are not equivalent, our method outputs a dominance relationship that is induced by one and not by the other. Our approach is applicable to all preference languages based on ceteris paribus semantics including CP-nets, TCP-nets, CI-nets and CP-Theories.

1 Introduction

Many practical applications of preference reasoning call for effective approaches to determining whether one set preference statements is *equivalent* to another, i.e., induces the same set of preferences over a set of alternatives as another, or whether one set of preferences *subsumes* another i.e., induces a set of preferences that includes those induced by the other. For example, preference equivalence and preference subsumption testing are of central importance in determining the substitutability of preference profiles

The ceteris paribus semantics [10] interpret the preference statements in terms of an induced preference graph, where the nodes correspond to alternatives and edges or flips correspond to dominance of one alternative over another with respect to the given preferences. It is known [11] that dominance testing between alternatives with respect to a set of preference statements can be reduced to reachability in the corresponding induced preference graph. On the other hand, verifying the equivalence of two sets of preferences amounts to checking the equality of the transitive closure of the corresponding induced preference graphs, or verifying the one-to-one correspondence of dominance between all pairs of alternatives in the respective induced preference graphs. Because dominance testing is PSPACE-complete [12], the complexity of equivalence testing is arguably PSPACE-complete (because equivalence testing can be reduced to testing dominance between all-pairs of alternatives). However, the recent success of model checking [13–15] approach to dominance testing [11] and the closely related problem of preference-based ordering a set of alternatives [16] (which is NP-hard [8]) offers the hope that a similar approach could be effective for preference equivalence

P. Perny, M. Pirlot, and A. Tsoukiàs (Eds.): ADT 2013, LNAI 8176, pp. 324–335, 2013.
© Springer-Verlag Berlin Heidelberg 2013

testing and preference subsumption testing. Model checking based approach to preference reasoning benefits from (i) a succinct encoding of the dominance relations induced
by a set of preference statements without having to explicitly enumerate them; (ii) the
specialized data structures (e.g., BDDs) and algorithms used by modern model checking software [15] for efficient exploration of large state spaces.

Against this background, we present the first approach to preference equivalence
and preference subsumption testing for preference languages based on ceteris paribus
semantics which include CP-nets [17], TCP-nets [5], CP-Theories [6] and CI-nets [7].
We model the preferences induced by two sets of ceteris paribus preference statements
within a single labeled transition system [18], and generate a single computation-tree
temporal logic (CTL) [18] formula that verifies the equivalence or subsumption of the
preferences they induce over the alternatives. Whenever the induced preferences are not
equivalent, our method automatically produces a dominance relationship that is induced
by one and not by the other. Our approach is generic and can be used to verify equivalence and subsumption for any two sets of ceteris paribus statements, not necessarily
expressed in the same preference language. Preliminary experiments indicate feasibility of our approach for preferences expressed over up to 30 variables (which exceeds
the requirements of many applications in practice) in a few seconds using the NuSMV
model checker.

2   Ceteris Paribus Preference Languages

Let $V = \{X_i \mid 0 < i \leq n\}$ be a set of preference variables or attributes, each with
a domain D_i such that $X_i = v_i \in D_i$ is a valid assignment to the variable X_i. Let
$O = \Pi_i D_i$ be the set of all alternatives, where $\gamma \in O$ is a complete assignment to all
the variables, denoted by the tuple $\gamma := \langle \gamma(X_1), \gamma(X_2), \ldots, \gamma(X_n) \rangle$ s.t. $\forall X_i \in V :$
$\gamma(X_i) \in D_i$. The *partial* assignment of γ to a subset $V' \subseteq V$ of variables is denoted
$\gamma(V')$.

A direct specification of a binary preference relation $\succ$ over O requires the user to
compare up to $O(D^{2n})$ pairs of outcomes, is prohibitively expensive to be useful in
practical applications. Qualitative preference languages offer succinct expression for
the specification of preferences over the alternatives O in terms of (a) *intra-variable
preferences* for each variable over its respective domain, denoted by $\succ_i \subseteq D_i \times D_i$;
and (b) the *relative importance* of the variables, denoted by $\rhd \subseteq V \times V$. Such preferences can also be conditioned on the valuation of one or more variables. For example,
$X_j = v_j : v_i \succ_i v_i'$ represents a conditional intra-variable preference for X_i, and X_i
is said to *depend* on X_j, the *parent* of X_i. CP-nets[17] allow the specification of a set
of conditional intra-variable preferences, while TCP-nets [5] further allow conditional
relative importance preferences over pairs of variables. Other related languages include
CP-Theories [6] that allow relative importance of one variable over a set of other variables; and more recently, CI-nets [7] that allow unconditional, monotonic intra-variable
preferences, and conditional relative importance preferences over sets of attributes. For
each preference statement in a language, the ceteris paribus semantics induces a set of
flips or improvements (or analogously, worsenings) for each alternative. The semantics for the various languages differ in terms of the set of flips they induce over the
alternatives [19, 12].

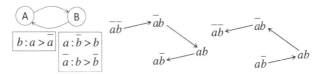

Fig. 1. P_1 and Induced Preference Graphs $\delta(P_1),\delta^-(P_1)$

To keep the discussion simple, we focus on CP-nets to introduce our approach to preference equivalence testing and preference subsumption testing. However our approach is sufficiently general to be applicable to any language based on ceteris paribus (or equivalently, flipping sequence-based) semantics including TCP-nets, CI-nets, and CP-Theories.

Definition 1 (Ceteris Paribus Semantics for CP-nets [17]). *Let P be a CP-net consisting of a set of conditional intra-variable preference statements of the form $c : v_i \succ_i v'_i$ for some $X_i \in V$, where c is an assignment to a set $C \subseteq V$ of (parent) variables. An alternative α is said to be preferred to another β with respect to a statement $c : v_i \succ_i v'_i$ in P if and only if*

    i. $(\alpha(C) = \beta(C) = c) \wedge (\alpha(X_i) = v_i) \wedge (\beta(X_i) = v'_i)$;
    ii. $\forall X_j \in V \setminus \{X_i\} : \alpha(X_j) = \beta(X_j)$.

The preference of α over β induced by a preference statement as above is called an *improving flip*[1] from β to α. Given a set P of preference statements in a, α is said to *dominate* (i.e., preferred to) β with respect to P, denoted $\alpha \succ^P \beta$, if and only if there is a sequence of alternatives $\beta = \gamma_1, \gamma_2 \ldots \gamma_n = \alpha$ such that for all $1 \leq i \leq n$, there is an improving flip from γ_i to γ_{i+1} with respect to some preference statement in P. Such a sequence is called an *improving flipping sequence* from β to α. The set of all improving flipping sequences (and hence the dominance relationships) between alternatives can be represented as a directed graph, namely the *induced preference graph*.

Definition 2 (Induced Preference Graph [17]). *Given a set P of preference statements, its* induced preference graph $\delta(P) = G(N, E)$ *is constructed as follows. The nodes N correspond to the set of all alternatives O. Each directed edge $(\alpha, \beta) \in E$ from alternative α to alternative β corresponds to an* improving flip *from α to β with respect to some preference statement in P.*

Definition 3 (Inverse Induced Preference Graph). *Given a set P of preference statements, the* inverse induced preference graph $\delta^-(P)$ *is constructed by generating $\delta(P)$ and reversing the direction of edges.*

Figure 1 shows a CP-net P_1, its induced preference graph $\delta(P_1)$ where the edges are directed toward the preferred alternatives, and its inverse induced preference graph $\delta^-(P_1)$ where the edges are directed away from the preferred alternatives. Because A's preference depends on the valuation of B and vice versa, P_1 is said to have a *dependency cycle*. When the preferences are *consistent*, i.e., no alternative dominates itself,

---

[1] Note that the improving flip is defined differently for more expressive languages such as TCP-nets, CP-Theories and CI-nets.

the induced preference graph is a DAG representing the *strict partial order* induced over the alternatives. Note that even though a P_1 has a dependency cycle (A is a parent of B and vice versa), $\delta(P_1)$ is a DAG, thus illustrating the fact that the preferences induced by a set of preference statements can be consistent, even if there are cyclic dependencies in the CP-net.

3 Preference Reasoning Using Model Checking

Because our approach to preference equivalence testing and preference subsumption testing builds on the model checking approach to dominance testing [11] and for preference-based ordering of alternatives [16], we briefly summarize the basic framework for casting problems of reasoning about preferences as problems in model checking.

The Model. Given a set P of preference statements, the induced preference graph $\delta(P)$ is first encoded as an input labeled transition system (LTS), more specifically a Kripke structure [18] $K(P)$ to a model checker such as NuSMV [15].

Definition 4 (Kripke Structure). *A Kripke structure is a tuple $\langle S, S_0, T, L \rangle$ where S is a set of states described by the valuations of a set of propositional variables P, $S_0 \subseteq S$ is a set of initial states, $T \subseteq S \times S$ is a transition relation such that $\forall s \in S : \exists s' \in S : (s, s') \in T$, and $L : S \to 2^P$ is a labeling function such that $\forall s \in S : L(s)$ is the set of propositions that are true in s.*

The central idea is to map each node in $\delta(P)$ (i.e., an alternative) to a set of states in $K(P)$ and each flip in $\delta(P)$ to a set of transitions between the corresponding states in $K(P)$. For each preference variable X_i used in P, there is a state variable x_i in $K(P)$. Thus the sets of states in $K(P)$ generated by the different valuations of x_i correspond to the set of alternatives. Another set of auxiliary variables h_is (corresponding to the x_is) are used to label the transitions between states: $h_i = 0$ in a transition implies X_i cannot change; otherwise X_i *may* change. Each of the h_is in the transition from a state s in $K(P)$ (corresponding to alternative γ in $\delta(P)$) to a state s' in $K(P)$ (corresponding to alternative γ' in $\delta(P)$) holds true if the valuation of the corresponding preference variable x_i changes in an improving flip from γ to γ' according to the ceteris paribus semantics (see Definition 1). Thus, the semantics of each preference statement in P is directly encoded as a set of guarded transitions in $K(P)$. If there is a valid flip from γ to γ' in $\delta(P)$ where a set V' of variables change values, then there is a transition from a corresponding state s to another state s' in $K(P)$. Thus, the guard conditions enable only the transitions in $K(P)$ that correspond to valid flips in $\delta(P)$. Note that a state in $K(P)$ may also contain transitions to itself (self-loops) because $h_i = 1$ does not necessarily imply that x_i will change; rather, it allows non-deterministic choice for the valuation of x_i to change. Therefore, we use another state variable g to indicate *global change*, which is set to 1 whenever *any* of the x_i's have changed in a transition; and 0 otherwise. Thus, g is used to indicate whether a transition in $K(P)$ corresponds to a flip in $\delta(P)$.

Figure 2 shows the Kripke structure $K(P_1)$ for the CP-net P_1 in Figure 1. Note that each node in $\delta(P_1)$ (i.e., an alternative) corresponds to two states in $K(P_1)$ that differ only in their valuation for g. The state $\bar{a}b\bar{g}$ is unreachable (grayed out) because the

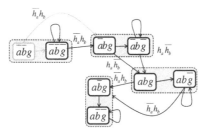

Fig. 2. $K(P_1)$: Kripke encoding for the semantics of P_1

model is initialized with $g = 0$ and there is no improving flip to $\bar{a}\bar{b}$ in $\delta(P_1)$. Each edge in $\delta(P_1)$ corresponds to a set of transitions in $K(P_1)$, e.g., the edge from ab to $a\bar{b}$ corresponds to two transitions, namely from abg and $ab\bar{g}$ to $a\bar{b}g$. The transitions are labeled to indicate the valuations of h_a and h_b (guards) that enable them. The valuations of h_is are non-deterministically set by the model checker. Whenever their valuations allow a improving flip, there is a transition from one state to another in the Kripke structure $K(P_1)$; and otherwise, there is a self-loop or a transition to a state with $\overline{(g)}$ in $K(P_1)$. In Figure 2, we only label the transitions that correspond to improving flips (showing the values of h_is that enable them); the labels for transitions that do not correspond to improving flips are not shown for easy readability.

Dominance Testing. For any path in $\delta(P)$ there exists a corresponding path in the Kripke structure $K(P)$. Leveraging this fact, Santhanam et al. test whether $\alpha \succ^P \beta$ [11] by generating a temporal logic (CTL) formula φ corresponding to a test of non-reachability of α from β in $\delta(P)$, i.e., φ has a model in $K(P)$ if and only if $\alpha \not\succ^P \beta$. Hence, whenever the dominance holds, the model checker automatically returns a path from a state in the Kripke structure $K(P)$ corresponding to β, to another state corresponding to α, which corresponds to an improving flipping sequence (a path in $\delta(P)$) from β to α. Recently, [16] used a similar idea to order over alternatives based on a given set of preferences.

4   Verifying Preference Equivalence and Preference Subsumption

We now turn to the problem of verifying the preference equivalence and preference subsumption relationships between two sets of preferences. We first formally define preference equivalence and preference subsumption.

Definition 5 (Preference Equivalence & Subsumption). *Let P_1 and P_2 be two consistent CP-nets over a set of variables V. Let $\succ^{1\star}$ and $\succ^{2\star}$ represent the transitive closures of the preference relations $\succ^1$ and $\succ^2$ induced by P_1 and P_2 respectively over the set of alternatives.*

  i. *P_1 is said to **preference subsume** P_2, denoted $P_1 \sqsupseteq P_2$ or $P_2 \sqsubseteq P_1$, iff $\forall \gamma, \gamma'$: $\gamma \succ^2 \gamma' \Rightarrow \gamma \succ^1 \gamma'$, or equivalently $\succ^{1\star} \sqsupseteq \succ^{2\star}$.*
  ii. *P_1 is said to be **preference equivalent** to P_2, denoted $P_1 \equiv P_2$ iff $P_1 \sqsupseteq P_2 \wedge P_2 \sqsupseteq P_1$.*

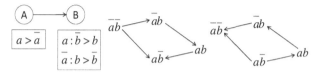

Fig. 3. P_2 and Induced Preference Graphs $\delta(P_2), \delta^-(P_2)$

In the above, equivalence and subsumption of P_1 and P_2 are defined in terms of the transitive closures of the respective induced preferences, namely $\succ^{1*}$ and $\succ^{2*}$ that represent the set of all improving flipping sequences for P_1 and P_2, and not simply in terms of $\succ^1$ and $\succ^2$ that represent only the set of improving flips induced by the respective preference statements in P_1 and P_2. This is necessary because the dominance relation is transitive.

Figure 3 shows a CP-net P_2 with its induced preference graph $\delta(P_2)$. From the preference statements of P_1 and P_2, it may appear that P_2 subsumes P_1, i.e., $P_1 \sqsubseteq P_2$, because the only difference between them is the preference over variable A: in P_1 it is conditioned on $b = 1$, whereas in P_2 it is unconditional. On the other hand, it may not be as intuitive to conclude that $P_2 \sqsubseteq P_1$. However, this is indeed the case: $P_1 \sqsubseteq P_2$ and $P_2 \sqsubseteq P_1$, i.e., $P_1 \equiv P_2$, because the induced preference graphs $\delta(P_1)$ and $\delta(P_2)$ are equivalent in terms of the reachability between any pair of alternatives. The unconditional preference over A in P_2 gives rise to an additional edge in $\delta(P_2)$ from $\bar{a}\bar{b}$ to $a\bar{b}$ that has no corresponding edge in $\delta(P_1)$, but the same has a corresponding path in $\delta(P_1)$: $\bar{a}\bar{b} \to \bar{a}b \to ab \to a\bar{b}$, which makes $\delta(P_1)$ and $\delta(P_2)$ (and hence P_1 and P_2) equivalent.

In other words, verifying the semantic equivalence of two sets P_1 and P_2 of preference statements is tantamount to checking that for each *edge* from γ to γ' in $\delta(P_1)$, there exists a corresponding *path* from γ to γ' in $\delta(P_2)$ and vice-versa. It is worth noting here that the above holds for any preference language that has a flipping-sequence based (ceteris paribus) semantics. Because dominance testing is PSPACE-complete even for CP-nets, the simplest and the least expressive among the ceteris paribus preference languages, equivalence testing is arguably PSPACE-complete for any of the more expressive preference languages such as TCP-nets, CI-nets, etc.

5   Verifying Equivalence and Subsumption via Model Checking

We now proceed to describe a novel model checking based approach to verifying the equivalence and subsumption of two sets of preferences. Given two sets of preference statements P_1 and P_2, we first consider preference subsumption testing, i.e., verifying whether $P_1 \sqsubseteq P_2$, since preference equivalence testing, i.e., checking whether $P_1 \equiv P_2$ amounts to verifying whether $P_1 \sqsubseteq P_2 \wedge P_2 \sqsubseteq P_1$. Verifying $P_1 \sqsubseteq P_2$ amounts to verifying that for each edge (γ, γ') in $\delta(P_1)$ there is a corresponding path from (γ, γ') in $\delta(P_2)$. This can be reduced to verifying a reachability property in a graph, namely the *combined induced preference graph* of P_1 and P_2, denoted $\delta(P_1, P_2)$, that embeds the semantics of P_1 and P_2.

Definition 6 (Combined Induced Preference Graph). *Given two sets P_1 and P_2 of preference statements over a set of variables V, the* combined induced preference graph

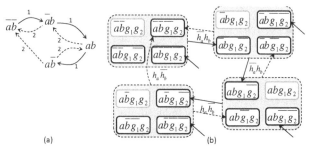

Fig. 4. Combined Induced Preference Graph $\delta(P_1, P_2)$ and its Kripke structure encoding $K(P_1, P_2)$

$\delta(P_1, P_2)$ *is a directed graph* $G(N, E)$ *with a labeling function* L *that is constructed as follows. The nodes* N *correspond to the set of alternatives generated by* V. *There is an edge* $e_{\gamma,\gamma'} = (\gamma, \gamma') \in E$ *if and only if there is an edge from* γ *to* γ' *in* $\delta(P_1)$ *or* $\delta^-(P_2)$, *and it is associated with a label*

$$
\mathscr{L}(e_{\gamma,\gamma'}) = \begin{cases} \{1\} & \text{if } \gamma' \succ^1 \gamma \text{ and } \gamma \not\succ^2 \gamma' \\ \{2\} & \text{if } \gamma \succ^2 \gamma' \text{ and } \gamma' \not\succ^1 \gamma \text{ (1)} \\ \{1,2\} & \text{if } \gamma' \succ^1 \gamma \text{ and } \gamma \succ^2 \gamma' \end{cases}
$$

Note that in the graph $\delta(P_1, P_2)$, each edge $(\gamma, \gamma') \in E$ corresponds to an improving flip from γ to γ' induced by P_1, a worsening flip from γ to γ' induced by P_2, or both. Figure 4(a) shows the combined induced preference graph $\delta(P_1, P_2)$ with respect to the CP-nets P_1 and P_2 shown in Figures 1 and 3 respectively, consisting of the edges in $\delta(P_1)$ (solid arrows, labeled 1) and those in $\delta^-(P_2)$ (dotted arrows, labeled 2).

Recall that verifying $P_1 \sqsubseteq P_2$ is equivalent to verifying that for each edge (γ, γ') in $\delta(P_1)$ there exists a corresponding path from (γ, γ') in $\delta(P_2)$, or in other words, there exists a corresponding path from (γ', γ) in $\delta^-(P_2)$. Therefore, in terms of the combined induced preference graph $\delta(P_1, P_2)$, $P_1 \sqsubseteq P_2$ holds whenever the following holds. For each edge from γ to γ' in $\delta(P_1, P_2)$ that includes label $\{1\}$, there exists a path from γ' to γ such that each edge in the path includes the label $\{2\}$. This condition forms the basis of our model checking based approach to preference subsumption testing.

Lemma 1. $P_1 \sqsubseteq P_2$ *if and only if for each edge* $e_{\gamma,\gamma'} \in \delta(P_1, P_2)$ *such that* $\mathscr{L}(e_{\gamma,\gamma'}) \supseteq \{1\}$, *there is a path* $\gamma_1 \to \gamma_2 \to \ldots \to \gamma_n$ *in* $\delta(P_1, P_2)$ *such that* $\gamma_1 = \gamma'$, $\gamma_n = \gamma$, *and* $\forall 1 \le i < n : \mathscr{L}(e_{\gamma_i,\gamma_{i+1}}) \supseteq \{2\}$.

5.1 Kripke Structure Encoding of $\delta(P_1, P_2)$

In order to verify $P_1 \sqsubseteq P_2$, we first construct a Kripke structure $K(P_1, P_2)$ that encodes the combined induced preference graph $\delta(P_1, P_2)$. The state space of $K(P_1, P_2)$ is constructed using (a) a set of *preference* variables, namely x_i's; and (b) a set of *auxiliary change* variables, namely h_i's.

Preference and Auxiliary Change Variables. The valuation of x_i in a state $s \in S$, denoted $s(x_i)$ corresponds to the valuation of the preference variable $X_i \in V$, i.e., a state s corresponds to an alternative $\gamma = \langle s(x_1), s(x_2), \ldots, s(x_{|V|}) \rangle$. The x_i's are modeled as *state* variables, while the h_i's are modeled as *input* variables, i.e., used to label the transitions in $K(P_1, P_2)$.

The transition relation $T \subseteq S \times S$ of $K(P_1, P_2)$ is implicitly specified in terms of whether and how each of the state variables (x_i's) can *individually* change values. The preference statements of P_1 and P_2 specify exactly how and under what conditions the value of each x_i can change; and the ceteris paribus interpretation requires that all other variables remain unchanged. Recall from Section 3 that the valuations of h_i's are used as *guards* to precisely enable only those transitions in $K(P_1, P_2)$ that satisfy these conditions, i.e., correspond to valid flips in $\delta(P_1, P_2)$ (see Section 3). Formally, for any two states $s, s' \in S$, we define $t = (s, s') \in T$ (denoted $s \to s'$) if one of the following conditions hold.

1. **Improving flip:**

$$\exists X_i \in V : t(h_i) = 1 \wedge \left(s'(x_i) \succ_i^1 s(x_i) \vee s(x_i) \succ_i^2 s'(x_i) \right)$$
$$\wedge \ \forall X_j \in V \setminus \{X_i\} : t(h_j) = 0 \wedge s(x_j) = s'(x_j)$$

where $s(.)$ and $t(.)$ denote the labels on the state s and the transition t respectively.

2. **Non flip:** $\qquad \forall X_i \in V : s(x_i) = s'(x_i)$

In the above, rule (1) enables a transition corresponding to a flip of a variable $X_i \in V$ with respect to the ceteris paribus interpretation of a preference statement. Rule 2 enables transitions between states corresponding to the same alternative.

Remark 1. Note that the above encoding of $\delta(P_1, P_2)$ in terms of $K(P_1, P_2)$ applies to CP-nets (Definition 1). The ceteris paribus semantics of other preference languages such as TCP-net, CI-net, CP-Theories, etc. can be similarly modeled by including additional guard conditions that enable transitions corresponding to the allowed flips in the respective languages. In particular, previous works have modeled the semantics of TCP-net [11] and CI-nets [16].

Global Change Variables. Recall that the purpose of encoding $\delta(P_1, P_2)$ in terms of $K(P_1, P_2)$ is to compute subsumption by verifying the condition of Lemma 1. To achieve this, we must be able to distinguish the transitions in the model (corresponding to flips) induced by P_1 and P_2. Hence, we introduce two additional state variables g_1 and g_2 to label the destination states of any transition as follows. For any edge from γ to γ' such that $\mathscr{L}(e_{\gamma, \gamma'}) \supseteq \{i\}$ in $\delta(P_1, P_2)$, there exists a corresponding transition from a state s to a state s' in $K(P_1, P_2)$ such that the valuations of x_i's in s and s' correspond to the alternatives γ and γ'; and $s'(g_i) = 1$. As a result, $g_1 = 1$ in the destination state of a transition in $K(P_1, P_2)$ that corresponds to an improving flip in $\delta(P_1)$, and $g_2 = 1$ in the destination state of a transition that corresponds to a worsening flip in $\delta^-(P_2)$; otherwise they are assigned to 0 in the destination state. The following rule encodes this semantics.

$$\left(s'(g_1) = 1 \Leftrightarrow (\exists X_i \in V : s'(x_i) \succ_i^1 s(x_i)) \right) \wedge$$
$$\left(s'(g_2) = 1 \Leftrightarrow (\exists X_i \in V : s(x_i) \succ_i^2 s'(x_i)) \right)$$

The variables g_1 and g_2 are modeled as *state* variables, and hence the states S of $K(P_1, P_2)$ are defined by the valuations of propositions $P_V = \{x_i | X_i \in V\} \cup \{g_1, g_2\}$. Hence, each alternative γ in $\delta(P_1, P_2)$ corresponds to a set $S^\gamma = \{s | s_{\downarrow V} = \gamma\}$ of states, where $s_{\downarrow V}$ denotes the *projection* of a state s described by P_V onto the set of variables $\{x_i | X_i \in V\} \subseteq P_V$. The variables g_1 and g_2 are initialized to 0 in $K(P_1, P_2)$, whereas x_i's are uninitialized, i.e., the model checker non-deterministically chooses and explores all possible combinations of assignments to x_i's. Hence the set of start states S_0 corresponds to the set of alternatives.

The above encoding is succinct in the sense that we do not explicitly specify each node and edge in $\delta(P_1, P_2)$ to construct the state space of $K(P_1, P_2)$. Because the h_i's are *input* variables in the Kripke model, all possible valuations of h_i's are automatically considered and non-deterministically explored by the model checker. Thus, the transitions in $K(P_1, P_2)$ correspond to valid flips in $\delta(P_1, P_2)$. The following holds by construction of $K(P_1, P_2)$ from $\delta(P_1, P_2)$.

Lemma 2. *Given CP-nets P_1 and P_2, and the Kripke structure $K(P_1, P_2) = \langle S, S_0, T, L \rangle$ (constructed from $\delta(P_1, P_2) = G(N, E)$ associated with labeling function $\mathscr{L}$),*

1. $\forall \gamma, \gamma', i \in \{1, 2\} : (\gamma, \gamma') \in E \wedge \mathscr{L}(e_{\gamma, \gamma'}) \supseteq \{i\}$
   $\Rightarrow \exists s \to s' : s_{\downarrow V} = \gamma \wedge s'_{\downarrow V} = \gamma' \wedge s'(g_i) = 1$
2. $\forall s, s' \in S : s \to s' \wedge s_{\downarrow V} \neq s'_{\downarrow V} \Rightarrow \exists i \in \{1, 2\}, \gamma, \gamma' : s_{\downarrow V} = \gamma \wedge s'_{\downarrow V} = \gamma' \wedge (\gamma, \gamma') \in E \wedge \mathscr{L}(e_{\gamma, \gamma'}) \supseteq \{i\}$

Figure 4(b) shows the Kripke structure $K(P_1, P_2)$ corresponding to $\delta(P_1, P_2)$ for our running example. The start states are marked with transitions without any source state. For clarity, the transitions between states corresponding to the same alternative are not marked, and the valuations of input variables a^0 and b^0 are not shown in the transitions. States corresponding to the same valuation of a and b are placed within dotted boxes. Further, transitions from the set of all states in a dotted box to the same destination state with a different valuation of a and b (in a different dotted box) are combinedly represented by a single arrow from dotted box to the destination state, e.g., the arrow from the dotted box containing the states corresponding to $\bar{a}\bar{b}$ indicates the presence of transitions from all states in the box to $\bar{a}bg_1\bar{g}_2$.

6   Querying $K(P_1, P_2)$ for Subsumption

We have already seen that verifying $P_1 \sqsubseteq P_2$ is equivalent to verifying that for each edge from γ to γ' in $\delta(P_1, P_2)$ that includes the label $\{1\}$, there exists a path from γ' to γ such that each edge in the path includes the label $\{2\}$ (Lemma 1). Because the set of start states S_0 in $K(P_1, P_2)$ corresponds to the set of alternatives in $\delta(P_1, P_2)$, the above reduces to verifying the following property in $K(P_1, P_2)$ by Lemma 2.

> For each state $s \in S_0$ in $K(P_1, P_2)$, if there exists a transition to a state s' with $s'(g_1) = 1$, then there exists a path $s' = s_1 \to \dots \to s_n \to s''$ in $K(P_1, P_2)$ such that $\forall 1 < i \leq n : s_i(g_2) = 1$ and $s_{\downarrow V} = s''_{\downarrow V}$.

Our objective is to express the above property in the language of Computation Tree Temporal Logic, CTL (see [18]), and automatically verify the temporal property with respect to $K(P_1, P_2)$ using a model checker such as NuSMV. One interesting and subtle challenge in realizing our objective stems from the fact that the condition requires checking the existence of paths starting from a state s and ending at a state s'' such that $s_{\downarrow V} = s''_{\downarrow V}$. However, CTL allows the specification and verification of temporal properties only with respect to states explored in the future, and therefore it is not possible to reference the start state s in the temporal property (which is necessary to ensure $s_{\downarrow V} = s''_{\downarrow V}$).

We address this challenge by introducing a set of *copy* variables, namely x_i^0's that are modeled in SMV model checker as *input* variables in $K(P_1, P_2)$ and hence not stored as part of the state. They are initialized with the valuations of the respective x_i's at the start of model exploration, and are constrained to remain invariant in the model; i.e., if the valuation of x_i at a start state $s \in S_0$ is v_i, then x_i^0 remains equal to v_i in all states along all paths starting from s. In other words, if the model checker begins exploration at state s, then the propositional formula $\psi = \bigwedge_i (x_i = x_i^0)$ can be used to refer to $s_{\downarrow V}$. Proceeding further, the following CTL formula encodes the condition for $P_1 \sqsubseteq P_2$.

$$\varphi : \mathbf{AX} \left(g_1 \Rightarrow \mathbf{EX} \ \mathbf{E} \left[g_2 \ \mathbf{U} \ (\psi \wedge g_2) \right] \right)$$

We use $\bar{p}$ to denote the negation of the proposition p. According to the semantics of CTL [18], a state s in a Kripke structure is said to satisfy (a) $\mathbf{EX} \ \psi$ if there exists a path $s = s_1 \to s_2 \dots$ such that s_2 satisfies ψ; (b) $\mathbf{AX} \ \psi$ if for all paths such that $s = s_1 \to s_2 \dots, s_2$ satisfies ψ; and (c) $\mathbf{E} \ [\psi_1 \mathbf{U} \psi_2]$ if there exists a path $s = s_1 \to s_2 \dots$ such that $\exists i \geq 1 : s_i$ satisfies ψ_2, and $\forall j < i : s_j$ satisfies ψ_1.

Therefore, φ holds in $K(P_1, P_2)$ whenever the following holds. For each transition $s \to s' \in T$ such that $s'(g_1) = 1$ (i.e., whenever $\mathbf{AX} \ g_1$ holds), there exists a transition $s' \to s''$ (i.e., $\mathbf{EX}$) such that s'' satisfies $\mathbf{E}\left[g_2 \ \mathbf{U} \ (\psi \wedge g_2) \right]$. That is, there is a path $s'' = s_1'' \to s_2'' \to s_k'' \dots \to s_n''$ such that g_2 holds in all states till a state s_n'', where ψ also holds, is reached. Recall that, propositional formula ψ is satisfied in states where the valuations of the preference variables are same as those in the start state (denoted by s in this case; see above). Note that if there are no transitions $s \to s'$ such that $s'(g_1) = 1$, then φ trivially holds in s.

Theorem 1. $K(P_1, P_2)$ *satisfies* φ *if and only if* $P_1 \sqsubseteq P_2$.

Proof. By Lemma 2, there exists a sequence of transitions in $K(P_1, P_2)$ if and only if there exists a path in $\delta(P_1, P_2)$, namely $s_{\downarrow V} = \gamma_1 \to \gamma_2 \to \dots \to \gamma_n = s_{\downarrow V}$ such that $\mathscr{L}(e_{\gamma_1, \gamma_2}) \supseteq \{1\}$ and $\forall 2 \leq i < n : \mathscr{L}(e_{\gamma_i, \gamma_{i+1}}) \supseteq \{2\}$. Because we do not initialize the x_i's in the model, the model checker verifies the satisfaction of φ with respect to each start state in S_0 (corresponding to an alternative in $\delta(P_1, P_2)$) according to the semantics of CTL. Hence the result follows from Lemma 1.

Extracting a Proof of Non-subsumption. The model checker returns true whenever φ is satisfied, i.e., $P_1 \sqsubseteq P_2$. Suppose that $P_1 \not\sqsubseteq P_2$. The model checker will then return false, and provide the justification/proof of unsatisfiability, essentially presenting a sequence that satisfies the negation of the φ (see above), which is:

$$\neg\varphi : \mathbf{EX} \left(g_1 \wedge \mathbf{AX} \ \neg\mathbf{E}\left[g_2 \ \mathbf{U} \ (\psi \wedge g_2) \right] \right)$$

The proof is presented in the form of a transition $s \to s'$ that corresponds to a flip from $\gamma = s_{\downarrow V}$ to $\gamma' = s'_{\downarrow V}$ such that (a) s' satisfies g_1 (implying that there is a path from s

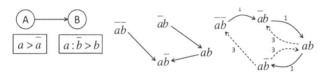

Fig. 5. P_3 and the graphs $\delta(P_3)$ and $\delta(P_1, P_3)$

to s' as per P_1) and (b) s' satisfies **AX** $\neg$**E**$\big[\, g_2$ **U** $(\psi \wedge g_2)\,\big]$ (implying that there is no path from s' back to any state s'' with $s'' \downarrow_V = s \downarrow_V$ as per P_2). In other words, the transition $s \to s'$ corresponds to a flip from γ to γ' induced by P_1 but not by P_2.

Verifying Equivalence. $P_1 \equiv P_2$ can be computed by verifying both φ and the following formula in $K(P_1, P_2)$.

$$\varphi' : \textbf{AX} \,\big(\, g_2 \Rightarrow \textbf{EX } \textbf{E}\big[\, g_1 \textbf{ U } (\psi \wedge g_1)\,\big]\,\big)$$

Note that φ' verifies $P_2 \sqsubseteq P_1$. Hence by Definition 5, $P_1 \equiv P_2$ iff $\varphi \wedge \varphi'$ is verified in $K(P_1, P_2)$.

In our running example (Figure 4), the formula $\varphi \wedge \varphi'$ is verified in $K(P_1, P_2)$, proving that $P_1 \equiv P_2$. Now consider another CP-net P_3 and its relationship with P_1 shown in Figure 5. Note that φ' is verified in $K(P_1, P_3)$, i.e., $P_3 \sqsubseteq P_1$. However φ is not, and the model checker returns false, with a path $s \to s'$ such that $s_{\downarrow V} = \bar{a}\bar{b}$ and $s'_{\downarrow V} = \bar{a}b$, which corresponds to a flip induced by P_1 but not by P_3. This provides the proof for $P_1 \not\sqsubseteq P_3$ and hence for $P_1 \not\equiv P_3$.

In summary, our approach is generic and can be used to verify equivalence and subsumption for any two sets of ceteris paribus statements, not necessarily expressed in the same preference language. Given two sets of preference statements P_1 and P_2, if there are $|V|$ nodes and $|E|$ edges in $\delta(P_1, P_2)$, then there are $O(|V|)$ states and $O(|E|)$ transitions in $K(P_1, P_2)$ by construction. Hence, the complexity of computing preference subsumption (and equivalence) is $(|V| + |E|) \times |\varphi|)$ as per the CTL model checking complexity [18]. Preliminary experiments indicate the feasibility of our approach for preferences expressed over up to 30 variables in a few seconds.

7   Summary and Discussion

We have described a novel practical approach to verifying whether a set of ceteris paribus preference statements is equivalent to or is subsumed by another. This offers, to the best of our knowledge, the first practical method to determining if the preferences held by a pair of agents agree with each other; and determining the sources of disagreement if they don't. Given two sets of preference statements, our model checking approach involves: (i) Constructing a Kripke structure that encodes the preferences induced by both sets of preference statements, and (ii) Verifying a computation-tree temporal logic (CTL) formula in the model to verify preference subsumption (and preference equivalence). Whenever the two sets of preference statements are not equivalent, our method outputs a dominance relationship that is induced by one and not by the other.

Our method can be extended to identify the set of all dominance relationships between alternatives in which the two sets of preferences differ. This can be achieved by iteratively relaxing the CTL formulas and verifying them against the model which successively produces newer proofs of non-equivalence. Preliminary experiments indicate feasibility of our approach for preferences expressed over up to 30 variables in less than a minute. In this paper, we have illustrated our model checking approach to testing preference subsumption and preference equivalence using the language of CP-nets [17]; however, it is applicable to all languages based on ceteris paribus semantics, i.e., those for which the semantics of the dominance relation can be given in terms of reachability within a graph of alternatives, including TCP-nets [5], CP-Theories [6] and CI-nets [7].

References

1. Hatfield, J.W., Immorlica, N., Kominers, S.D.: Testing substitutability. Games and Economic Behavior 75(2), 639–645 (2012)
2. Gusfield, D., Irving, R.W.: The Stable marriage problem - structure and algorithms. Foundations of computing series. MIT Press (1989)
3. Trabelsi, W., Wilson, N., Bridge, D.G., Ricci, F.: Preference dominance reasoning for conversational recommender systems: a comparison between a comparative preferences and a sum of weights approach. Int'l Journal on Art. Int. Tools 20(4), 591–616 (2011)
4. Rossi, F., Venable, K.B., Walsh, T.: mcp nets: Representing and reasoning with preferences of multiple agents. In: Proc. of Nat'l Conf. on Art. Int., pp. 729–734. AAAI Press (2004)
5. Brafman, R.I., Domshlak, C., Shimony, S.E.: On graphical modeling of preference and importance. J. Art. Intel. Res. 25, 389–424 (2006)
6. Wilson, N.: Extending cp-nets with stronger conditional preference statements. In: AAAI, pp. 735–741 (2004)
7. Bouveret, S., Endriss, U., Lang, J.: Conditional importance networks: A graphical language for representing ordinal, monotonic preferences over sets of goods. In: IJCAI, pp. 67–72 (2009)
8. Brafman, R.I., Pilotto, E., Rossi, F., Salvagnin, D., Venable, K.B., Walsh, T.: The next best solution. In: AAAI (2011)
9. Martínez, R., Massó, J., Neme, A., Oviedo, J.: On the invariance of the set of stable matchings with respect to substitutable preference profiles. Int. J. Game Theory 36(3-4), 497–518 (2008)
10. Boutilier, C., Brafman, R.I., Hoos, H.H., Poole, D.: Reasoning with conditional ceteris paribus preference statements. In: UAI, pp. 71–80 (1999)
11. Santhanam, G.R., Basu, S., Honavar, V.: Dominance testing via model checking. In: AAAI. AAAI Press (2010)
12. Goldsmith, J., Lang, J., Truszczynski, M., Wilson, N.: The computational complexity of dominance and consistency in cp-nets. J. Art. Intel. Res. 33, 403–432 (2008)
13. Queille, J.P., Sifakis, J.: Specification and verification of concurrent systems in CESAR. In: Dezani-Ciancaglini, M., Montanari, U. (eds.) International Symposium on Programming. LNCS, vol. 137, pp. 337–351. Springer, Heidelberg (1982)
14. Clarke, E.M., Emerson, E.A., Sistla, A.P.: Automatic verification of finite-state concurrent systems using temporal logic specifications. ACM TOPLAS 8(2), 244–263 (1986)
15. Cimatti, A., Clarke, E., Giunchiglia, E., Giunchiglia, F., Pistore, M., Roveri, M., Sebastiani, R., Tacchella, A.: NuSMV Version 2: An OpenSource Tool for Symbolic Model Checking. In: Brinksma, E., Larsen, K.G. (eds.) CAV 2002. LNCS, vol. 2404, pp. 359–364. Springer, Heidelberg (2002)
16. Oster, Z.J., Santhanam, G.R., Basu, S., Honavar, V.: Model checking of qualitative sensitivity preferences to minimize credential disclosure. In: Păsăreanu, C.S., Salaün, G. (eds.) FACS 2012. LNCS, vol. 7684, pp. 205–223. Springer, Heidelberg (2013)
17. Boutilier, C., Brafman, R.I., Domshlak, C., Hoos, H.H., Poole, D.: Cp-nets: A tool for representing and reasoning with conditional ceteris paribus preference statements. J. Art. Intel. Res. 21, 135–191 (2004)
18. Clarke, E., Grumberg, O., Peled, D.: Model Checking. MIT Press (January 2000)
19. Santhanam, G.R., Basu, S., Honavar, V.: Efficient dominance testing for unconditional preferences. In: KR (2010)

Learning a Majority Rule Model from Large Sets of Assignment Examples

Olivier Sobrie[1,2], Vincent Mousseau[1], and Marc Pirlot[2]

[1] École Centrale Paris
Grande Voie des Vignes
92295 Châtenay Malabry, France
olivier.sobrie@gmail.com, vincent.mousseau@ecp.fr
[2] Université de Mons, Faculté Polytechnique
9, rue de Houdain
7000 Mons, Belgique
marc.pirlot@umons.ac.be

Abstract. Learning the parameters of a Majority Rule Sorting model (MR-Sort) through linear programming requires to use binary variables. In the context of preference learning where large sets of alternatives and numerous attributes are involved, such an approach is not an option in view of the large computing times implied. Therefore, we propose a new metaheuristic designed to learn the parameters of an MR-Sort model. This algorithm works in two phases that are iterated. The first one consists in solving a linear program determining the weights and the majority threshold, assuming a given set of profiles. The second phase runs a metaheuristic which determines profiles for a fixed set of weights and a majority threshold. The presentation focuses on the metaheuristic and reports the results of numerical tests, providing insights on the algorithm behavior.

1 Introduction

Multiple criteria sorting procedures aim at assigning alternatives evaluated on multiple criteria to a category selected in a set of pre-defined and ordered categories. In this article we investigate the Majority Rule Sorting procedure (MR-Sort), a simplified version of the ELECTRE TRI sorting model [1, 2]. MR-Sort is directly inspired by the work of Bouyssou and Marchant who provide an axiomatic characterization [3, 4] of non-compensatory sorting methods. The general principle of MR-Sort is to assign alternatives by comparing their performances to those of profiles delimiting the categories. An alternative is assigned to a category "above" a profile if and only if it is at least as good as the profile on a (weighted) majority of criteria.

For using MR-Sort, several parameters need to be determined: the performance vector associated to each profile, the criteria weights and a majority threshold. It is not easy for a decision maker (DM) to assess such parameters. He often prefers to provide typical examples of assignments of alternatives to

P. Perny, M. Pirlot, and A. Tsoukiàs (Eds.): ADT 2013, LNAI 8176, pp. 336–350, 2013.
© Springer-Verlag Berlin Heidelberg 2013

categories. Several papers have been devoted to learning the parameters of such models on the basis of assignment examples. Mathematical programming techniques for learning part or all the parameters of an ELECTRE TRI model are described in [5–9] while [10] proposes a genetic algorithm designed for the same purpose. Learning the parameters of a MR-Sort model is dealt with in [11, 12].

None of these proposals can be considered suitable to our case, since we want to deal with *large* sets of assignment examples, having in mind the kind of sorting problems encountered in the field of preference learning [13], more precisely in the monotone learning subfield. [11] needs computing times as long as 25 seconds to learn the parameters of a MR-Sort model involving 5 criteria and 3 categories from a learning set containing 100 examples. This is no wonder since their algorithm is based on the resolution of a mixed integer program (MIP) in which the number of binary variables grows linearly with the number of examples. The experimental results in [11] show that the computing time increases very quickly with the number of examples.

From the previous work related to parameters learning for ELECTRE TRI models, we retain two main lessons. Firstly, learning only the weights and the majority threshold of a MR-Sort model can be done by solving a linear program without binary variables as done in [6]. On the other hand, as demonstrated in [7], learning only the profiles of such models by means of linear programming does require binary variables.

Based on these observations, we have designed an algorithm that computes the parameters of a MR-Sort model in order to assign as many as possible alternatives in the learning set to their category. This algorithm has two main components that are used repeatedly and alternatively. The first component learns optimal weights and majority threshold, in case the profiles limiting the categories are fixed. The second component adjusts the profiles for given weights and majority threshold.

To assess the new algorithm, we have set up a series of numerical experiments much in the spirit of [11]. The simulation experiments were designed in order to address the following questions:

Algorithm Performance. Given a MR-Sort model involving n criteria and p categories and a set of assignment examples compatible with this model, how fast does the algorithm find parameters of an MR-Sort model which restores the original assignments ?

Model Retrieval. Given a MR-Sort model involving n criteria and p categories and a set of assignment examples compatible with this model, how many examples are required to obtain a model that is close to the original one?

Tolerance for Errors. Given a set of assignments, obtained through a MR-Sort model, in which errors have been added, to what extent do the errors perturb the algorithm?

Idiosyncrasy. Each alternative of a set is assigned to a category by a rule that is not a MR-Sort rule (actually, they are assigned by an additive sorting rule). What's the ability of the algorithm to find a MR-Sort model restoring as many examples as possible ? In other terms, is MR-Sort flexible enough to reproduce assignments by another type of sorting rule?

In the next section of this paper, we briefly recall the precise definition of the MR-Sort procedure. In section 3, we describe the algorithm that we have developed. Numerical experiments designed for testing the algorithm are described, their results summarized and commented on in section 4. We finally conclude this paper with some perspectives for further research in view of improving the current version of the algorithm.

2   MR-Sort Procedure

The MR-Sort procedure is a simplified version of the ELECTRE TRI procedure [1, 2], based on the work of Bouyssou and Marchant developed in [3, 4].

Let X be a set of alternatives evaluated on n criteria, $F = \{1, 2, ..., n\}$. We denote by a_j the performance of alternative $a \in X$ on criterion j. The categories of the MR-Sort model, delimited by the profiles b_{h-1} and b_h, are denoted by C_h, where h denotes the category index. We convene that the best category C_p is delimited by a fictive upper profile, b_p, and the worst one by a fictive lower profile, b_0. The performances of the profiles are denoted by $b_{h,j}$, with $j = 1, ..., n$. It is assumed that the profiles dominate one another, i.e.:

$$b_{h-1,j} \leq b_{h,j} \qquad\qquad h = 1, \ldots, p; j = 1, \ldots, n.$$

Using the MR-Sort procedure (without veto), an alternative is assigned to a category if its performances are at least as good as the performances of the category's lower profile and worse than the performances of the category's upper profile on a weighted majority of criteria. In the former case, we say that the alternative is *preferred* to the profile, while, in the latter, it is not. Formally, an alternative $a \in X$ is *preferred* to profile b_h, and we denote it by aSb_h, if the following condition is met:

$$aSb_h \Leftrightarrow \sum_{j:a_j \geq b_{h,j}} w_j \geq \lambda,$$

where w_j for $j \in F$ are nonnegative weights attached to the criteria and satisfying the normalization condition $\sum_{j \in F} w_j = 1$; λ is the *majority threshold*; it satisfies $\lambda \in [1/2, 1]$. The preference relation S can be seen as an *outranking* relation without veto [2, 14, 15].

The condition for an alternative $a \in X$ to be assigned to category C_h is expressed as follows:

$$\sum_{j:a_j \geq b_{h-1,j}} w_j \geq \lambda \qquad \text{and} \qquad \sum_{j:a_j \geq b_{h,j}} w_j < \lambda \qquad (1)$$

The MR-Sort assignment rule described above involves $pn + 1$ parameters, i.e. n weights, $(p - 1)n$ profiles evaluations and one majority threshold. Note that the profiles b_0 and b_p are conventionally defined as follows: $b_{0,j}$ is a value such that $a_j \geq b_{0,j}$ for all $a \in X$ and $j \in F$; $b_{p,j}$ is a value such that $a_j < b_{p,j}$ for all $a \in X$ and $j \in F$.

A *learning set* is a subset of alternatives $A \subseteq X$ for which an assignment for each alternative is known. For $h = 1, \ldots, p$, A^h denotes the subset of alternatives $a \in A$ which are assigned to category C_h. The subsets A^h are disjoint; some of them may be empty.

3   The Algorithm

3.1   Learning of All the Parameters

As demonstrated in [11], the problem of learning the parameters of a MR-Sort model on the basis of assignment examples can be formulated as a mixed integer program (MIP) but only instances of modest size can be solved in reasonable computing times. The MIP proposed in [11] contains $m \cdot (2n + 1)$ binary variables, with n, the number of criteria, and m, the number of alternatives. A problem involving 1000 alternatives, 10 criteria and 5 categories requires 21000 binary variables. For a similar program in [12], it is mentioned that problems with less than 400 binary variables can be solved within 90 minutes. Following these observations, we understand that MIP is not suitable for the applications we want to deal with. In [10], a genetic algorithm was proposed to learn the parameters of an ELECTRE TRI model. This algorithm could be transposed for learning the parameters of a MR-Sort model. However, it is well known [16] that genetic algorithms which take the structure of the problem into account to perform crossovers and mutations give better results. It is not the case of the genetic algorithm proposed in [10] since the authors' definitions of crossover and mutation operators are standard.

Learning only the weights and the majority threshold of an MR-Sort model on the basis of assignment examples can be done using an ordinary linear program (without binary or integer variables). On the contrary, learning profiles evaluations is not possible by linear programming without binary variables. Taking these observations into account, we propose an algorithm that takes advantage of the ease of learning the weights and the majority threshold by a linear program and adjusts the profiles by means of a dedicated heuristic.

The algorithm uses the following components (see Algorithm 1):

1. a heuristic for initializing the profiles;
2. a linear program learning the weights and the majority threshold, given the profiles;
3. a dedicated heuristic adjusting the profiles, given weights and a majority threshold.

In the next subsections we describe in more detail these three elements. The algorithm uses the latter two components iteratively: starting from initial profiles, we find the optimal weights and threshold for these profiles by solving a linear program; then we adjust the profiles, using the heuristic, keeping the weights and threshold fixed; the profile adjustment operation is repeated N_{it} times. We call *main loop* the optimization of the weights and threshold followed by N_{it} iterations of the profiles adjustment operation. The main loop is executed until a stopping criterion is met.

Since the process of alternating the optimization of the weights and threshold, on the one hand, and several iterations of the (heuristic) optimization of the profiles, on another hand, is not guaranteed to converge to a good set of parameters, we implement the algorithm as an *evolutionary metaheuristic*, evolving, not a single MR-Sort model, but a population of them. The number of models in the population is denoted by N_{model}. After each application of the main loop to all models in the population, we assess the resulting population of models by using them to assign the alternatives in the learning set. The quality of a MR-Sort model is assessed by its *classification accuracy*:

$$CA = \frac{\text{Number of assignment examples restored}}{\text{Total number of assignment examples}}.$$

At this stage, the algorithm reinitializes the $\left\lfloor \frac{N_{model}}{2} \right\rfloor$ models giving the worst CA. The stopping criterion of the algorithm is met either once the classification accuracy of some model in the population is equal to 1 or after a maximum number of iterations N_o (fixed a priori).

Algorithm 1. Metaheuristic to learn all the parameters of an MR-Sort model

---

Generate a population of N_{model} models with profiles initialized with a heuristic
repeat
    for all model M of the set **do**
        Learn the weights and majority threshold with a linear program, using the current profiles
        Adjust the profiles with a metaheuristic N_{it} times, using the current weights and threshold.
    end for
    Reinitialize the $\left\lfloor \frac{N_{model}}{2} \right\rfloor$ models giving the worst CA
until Stopping criterion is met

3.2 Profiles Initialization

The first step of the algorithm consists in initializing a set of profiles so that it can be used to learn a set of weights and a majority threshold. The general idea of the heuristic we designed to set the value $b_{h,j}$ of the profile b_h on criterion j is the following. We choose this value in order to maximize the discriminating power of each criterion, relatively to the alternatives in the learning set A. More precisely, we set $b_{h,j}$ in such a way that alternatives ranked in the category above b_h (i.e. C_{h+1}) typically have an evaluation greater than $b_{h,j}$ on criterion j and those ranked in the category below b_h (i.e. C_h), typically have an evaluation smaller than $b_{h,j}$. In setting the profile values, the proportion of examples assigned to a category is taken into account so that the alternatives assigned to categories which are not often represented in the learning set have more importance. Note the initial value of a profile b_h is determined by only considering the examples assigned to the category just below and just above the profile i.e. the examples belonging respectively to the subsets A_h and A_{h+1} in the learning set A. The reason for this option is to balance the number of categories above and below the profile that are taken into account for determining this profile. For profiles b_1 and b_{p-1}, the only way of satisfying this requirement is to consider only one category above and one category below the profile. For guaranteeing an equal treatment of all profiles, we chose to consider only C_h and C_{h+1} for determining b_h.

The heuristic works as follows:

1. For each category C_h, compute the frequency π_h with which alternatives a in the learning set are assigned to category C_h : $\pi_h = \frac{|A_h|}{|A|}$.
2. For each criterion, the value of the profile $b_{h,j}$ is chosen s.t.:

$$\max_{b_{h,j}} \left[|a \in A_{h+1} : a_j \geq b_{h,j}| - |a \in A_{h+1} : a_j < b_{h,j}|\right] (1 - \pi_{h+1})$$
$$+ \left[|a \in A_h : a_j < b_{h,j}| - |a \in A_h : a_j \geq b_{h,j}|\right] (1 - \pi_h).$$

The profiles are computed in descending order.

3.3   Learning the Weights and the Majority Threshold

Assuming that the profiles are given, learning the weights and the majority threshold of a MR-Sort model from assignment examples is done by means of solving a linear program. The MR-Sort model postulates that the profiles dominate each other, i.e. $b_{h+1,j} \geq b_{h,j}$ for all h and j, and the inequality is strict for at least one j. The constraints derived from the alternatives assignments are expressed as follows:

$$\sum_{j:a_j \geq b_{h-1,j}} w_j - x_a + x'_a = \lambda \qquad \forall a \in A_h, h = 2, ..., p-1$$
$$\sum_{j:a_j \geq b_{h,j}} w_j + y_a - y'_a = \lambda - \delta \qquad \forall a \in A_h, h = 1, ..., p-2$$
$$\sum_{j=1}^{n} w_j = 1; \quad \lambda \in [0.5; 1] \quad w_j \in [0; 1] \ \forall j \in F$$
$$x_a, y_a, x'_a, y'_a \in \mathbb{R}_0^+$$

The value of $x_a - x'_a$ (resp. $y_a - y'_a$) represents the difference between the sum of the weights of the criteria belonging to the coalition in favor of $a \in A_h$ w.r.t. b_{h-1} (resp. b_h) and the majority threshold. If both $x_a - x'_a$ and $y_a - y'_a$ are positive, then the alternative a is assigned to the right category. In order to try to maximize the number of examples correctly assigned by the model, the objective function of the linear program minimizes the sum of x'_a and y'_a, i.e. the objective function is $\min \sum_{a \in A}(x'_a + y'_a)$. Note however that such an objective function does not guarantee that the maximal number of examples are correctly assigned. Failing to do so may be due to possible compensatory effects between constraints, i.e. the program may favor a solution involving many small positive values of x'_a and y'_a over a solution involving large positive values of a few of these variables. Such a compensatory behavior could be avoided, but at the cost of introducing binary variables indicating each violation of the assignment constraints. We do not consider such formulations in order to limit computing times.

3.4 Learning the Profiles

Learning the profiles by using a mathematical programming formulation requires binary variables, leading to a mixed integer program [7]. As we want to deal with problems involving large learning sets, 10 criteria and 3 to 5 categories, MIP is not an option. Therefore we opt for a randomized heuristic algorithm which is described below.

Consider a model having 2 categories and 5 criteria and assume that two alternatives are misclassified by this model. The one, a', is assigned in category C_1 by the DM and in C_2 by the model, while the other one, a'', is assigned in category C_2 by the DM and in C_1 by the model. Assuming fixed weights and majority threshold, it means that the profile delimiting the two categories, is either too high or too low on one or several criteria. In Figure 1, the arrows show the direction in which moving the profile in order to favor the correct classification of a' or a''. $\delta_j^{a'}$ (resp. $\delta_j^{a''}$) denotes the difference between the profile value $b_{1,j}$ and the alternative evaluations a'_j (resp. a''_j) on criterion j.

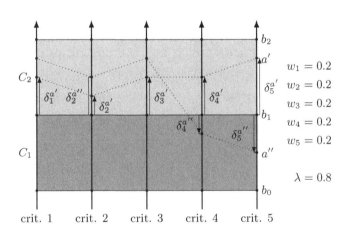

Fig. 1. Alternative wrongly assigned because of profile too low or too high

Based on these observations, we define several subsets of alternatives for each criterion j and each profile h and any positive value δ:

$V_{h,j}^{+\delta}$ (**resp.** $V_{h,j}^{-\delta}$) : the sets of alternatives misclassified in C_{h+1} instead of C_h
(resp. C_h instead of C_{h+1}), for which moving the profile b_h by $+\delta$ (resp. $-\delta$)
on j results in a correct assignment. For instance, a'' belongs to the set $V_{1,4}^{-\delta}$
on criterion 4 for $\delta \geq \delta_4^{a''}$.

$W_{h,j}^{+\delta}$ (**resp.** $W_{h,j}^{-\delta}$) : the sets of alternatives misclassified in C_{h+1} instead of C_h
(resp. C_h instead of C_{h+1}), for which moving the profile b_h of $+\delta$ (resp. $-\delta$)
on j strengthens the criteria coalition in favor of the correct classification
but will not by itself result in a correct assignment. For instance, a' belongs
to the set $W_{1,1}^{+\delta}$ on criterion 1 for $\delta > \delta_1^{a'}$.

$Q_{h,j}^{+\delta}$ (**resp.** $Q_{h,j}^{-\delta}$) : the sets of alternatives correctly classified in C_{h+1} (resp.
C_{h+1}) for which moving the profile b_h of $+\delta$ (resp. $-\delta$) on j results in a
misclassification.

$R_{h,j}^{+\delta}$ (**resp.** $R_{h,j}^{-\delta}$) : the sets of alternatives misclassified in C_{h+1} instead of C_h
(resp. C_h instead of C_{h+1}), for which moving the profile b_h of $+\delta$ (resp. $-\delta$)
on j weakens the criteria coalition in favor of the correct classification but
does not induce misclassification by itself. For instance, a'' belongs to the
set $R_{1,2}^{+\delta}$ on criterion 2 for $\delta > \delta_2^{a''}$.

$T_{h,j}^{+\delta}$ (**resp.** $T_{h,j}^{-\delta}$) : the sets of alternatives misclassified in a category higher
than C_{h+1} (resp. in a category lower than C_h) for which the current profile
evaluation weakens the criteria coalition in favor of the correct classification.

In order to formally define these sets we introduce the following notation. A_h^l
denotes the subset of misclassified alternatives that are assigned in category C_l
by the model while the DM assigns them in category C_h. $A_{<h}^{\geq l}$ denotes the subset
of misclassified alternatives that are assigned in category higher than C_l by the
model while the DM assigns them in a category below C_h. And similarly for
$A_{>h}^{\leq l}$. Finally, $\sigma(a, b_h) = \sum_{j:a_j \geq b_{h,j}} w_j$. We have, for any h, j and positive δ:

$$V_{h,j}^{+\delta} = \left\{ a \in A_h^{h+1} : b_{h,j} + \delta > a_j \geq b_{h,j} \text{ and } \sigma(a, b_h) - w_j < \lambda \right\}$$

$$V_{h,j}^{-\delta} = \left\{ a \in A_{h+1}^h : b_{h,j} - \delta < a_j < b_{h,j} \text{ and } \sigma(a, b_h) + w_j \geq \lambda \right\}$$

$$W_{h,j}^{+\delta} = \left\{ a \in A_h^{h+1} : b_{h,j} + \delta > a_j \geq b_{h,j} \text{ and } \sigma(a, b_h) - w_j \geq \lambda \right\}$$

$$W_{h,j}^{-\delta} = \left\{ a \in A_{h+1}^h : b_{h,j} - \delta < a_j < b_{h,j} \text{ and } \sigma(a, b_h) + w_j < \lambda \right\}$$

$$Q_{h,j}^{+\delta} = \left\{ a \in A_{h+1}^{h+1} : b_{h,j} + \delta > a_j \geq b_{h,j} \text{ and } \sigma(a, b_h) - w_j < \lambda \right\}$$

$$Q_{h,j}^{-\delta} = \left\{ a \in A_h^h : b_{h,j} - \delta < a_j < b_{h,j} \text{ and } \sigma(a, b_h) + w_j \geq \lambda \right\}$$

$$R_{h,j}^{+\delta} = \left\{ a \in A_{h+1}^h : b_{h,j} + \delta > a_j \geq b_{h,j} \right\}$$

$$R_{h,j}^{-\delta} = \left\{ a \in A_h^{h+1} : b_{h,j} - \delta < a_j < b_{h,j} \right\}$$

$$T_{h,j}^{+\delta} = \left\{ a \in A_{<h+1}^{>h+1} : b_{h,j} + \delta > a_j \geq b_{h,j} \right\}$$

$$T_{h,j}^{-\delta} = \left\{ a \in A_{>h}^{<h} : b_{h,j} - \delta < a_j \leq b_{h,j} \right\}$$

To avoid violations of the dominance rule between the profiles, on each criterion
j, $+\delta$ or $-\delta$ is chosen in the interval $[b_{h-1,j}, b_{h+1,j}]$. We define the value $P(b_{h,j}^{+\delta})$

which aggregates the number of alternatives contained in the sets described above as follows:

$$P(b_{h,j}^{+\delta}) = \frac{k_V |V_{h,j}^{+\delta}| + k_W |W_{h,j}^{+\delta}| + k_T |T_{h,j}^{+\delta}| + k_Q |Q_{h,j}^{+\delta}| + k_R |R_{h,j}^{+\delta}|}{d_V |V_{h,j}^{+\delta}| + d_W |W_{h,j}^{+\delta}| + d_T |T_{h,j}^{+\delta}| + d_Q |Q_{h,j}^{+\delta}| + d_R |R_{h,j}^{+\delta}|}$$

with k_V, k_W, k_T, k_Q, k_R, d_V, d_W, d_T, d_Q and d_R fixed constants. We define similarly $P(b_{h,j}^{-\delta})$. In the definition of $P(b_{h,j}^{+\delta})$ (resp. $P(b_{h,j}^{-\delta})$), the coefficients weighting the number of elements in the sets in the numerator are chosen so as to emphasize the arguments in favor of moving the value $b_{h,j}$ of profile b_h to $b_{h,j} + \delta$ (resp. $-\delta$), while the coefficients in the denominator emphasize the arguments against such a move. The values of the coefficients are empirically set as follows: $k_V = 2, k_W = 1, k_T = 0.1, k_Q = k_R = 0, d_V = d_W = d_T = 1, d_Q = 5, d_R = 1$.

The value $b_{h,j}$ of profile b_h on criterion j will possibly be moved to the value a_j of one of the alternatives a contained in $V_{h,j}^{+\delta}$, $V_{h,j}^{-\delta}$, $W_{h,j}^{+\delta}$ or $W_{h,j}^{-\delta}$. More precisely, it will be set to a_j or a value slightly below or slightly above a_j. The exact new position of the profile is chosen so as to favor a correct assignment for a.

All such values a_j are located in the interval $[b_{h-1,j}, b_{h+1,j}]$. A subset of such values is chosen in a randomized way. The candidate move corresponds to the value a_j in the selected subset for which $P(b_{h,j}^{\Delta})$ is maximal, Δ being equal to $a_j - b_{h,j}$ (i.e. a positive or negative quantity). To decide whether to make the candidate move, a random number r is drawn uniformly in the interval $[0, 1]$ and the value $b_{h,j}$ of profile b_h is changed if $P(b_{h,j}^{\Delta}) \leq r$.

This procedure is executed for all criteria and all profiles. Criteria are treated in random order and profiles in ascending order.

Algorithm 2 summarizes how this randomized heuristic operates.

Algorithm 2. Randomized heuristic used for improving the profiles

    for all profile b_h **do**
        for all criterion j chosen randomly **do**
            Choose, in a randomized manner, a set of positions in the interval $[b_{h-1,j}, b_{h+1,j}]$
            Select the one such that $P(b_{h,j}^{\Delta})$ is maximal
            Draw uniformly a random number r from the interval $[0, 1]$.
            if $r \leq P(b_{h,j}^{\Delta})$ **then**
                Move $b_{h,j}$ to the position corresponding to $b_{h,j} + \Delta$
                Update the alternatives assignment
            end if
        end for
    end for

---

4   Numerical Experiments

4.1   Performance of the Algorithm

Our first concern is to measure the performance of the algorithm and its convergence, i.e. how many iterations are needed to find a model restoring a majority of assignment examples and how much time is required to learn this model? To measure this, an experimental framework is set up:

1. An MR-Sort model M is generated randomly. The weights are uniformly generated as described in [17], i.e. $n-1$ random numbers are uniformly drawn from the interval $[0,1]$ and ranked s.t. $r_n = 1 > r_{n-1} \geq ... \geq r_1 > 0 = r_0$. Then weights are defined as follows: $w_j = r_j - r_{j-1}$, with $j = 1, ..., n$. The majority threshold is uniformly drawn from the interval $[1/2, 1]$. For the profiles evaluations, on each criterion $p-1$ random numbers are uniformly drawn from the interval $[0,1]$ and ordered s.t. $r'_{p-1} \geq ... \geq r'_1$. Profiles evaluations are determined by $b_{h,j} = r'_h$, $h = 1, ..., p-1$. Using model M as described by (1), each alternative can be assigned to a category. The resulting assignment rule is referred to as s_M.
2. A set of m alternatives with random performances on the n criteria is generated. The performances are uniformly and independently drawn from the $[0,1]$ interval. The set of generated alternatives is denoted by A. The alternatives in A are assigned using the rule s_M. The resulting assignments and the performances of the alternatives in the set A are given as input to the algorithm. They constitute the learning set.
3. On basis of the assignments and the performances of the alternatives in A, the algorithm learns a MR-Sort model which maximizes the classification accuracy. The model learned by the metaheuristic is denoted by M' and the corresponding assignment rule, $s_{M'}$.
4. The alternatives of the learning set A are assigned using the rule $s_{M'}$. The assignment resulting from this step are compared to the one obtained at step 2 and the classification accuracy $CA(s_M, s_{M'})$ is computed [1]

We test the algorithm with models having 10 criteria and 3 to 5 categories and learning sets containing 1000 assignment examples. These experiments are done on an Intel Dual Core P8700 running GNU/Linux with CPLEX Studio 12.5 and Python 2.7.3.

In Figure 2, the average value of $CA(s_M, s_{M'})$ obtained after repeating 10 times the experiment is shown. When the number of categories increases, we observe that the algorithm needs more iterations to converge to a model restoring correctly all assignment examples. This experiment shows that it is possible to find a model restoring 99% of the assignment examples in a reasonable computing time. On average two minutes are required to find the parameters of a model having 10 criteria and 5 categories with $N_{model} = 10$, $N_o = 30$ and $N_{it} = 20$.

4.2   Model Retrieval

This experiment aims at answering the following question: How many assignment examples are required to obtain the parameters of a model which restores

---

[1] To assess the quality of a sorting model, other indices are also used, like the area under curve (AUC). In this paper, we use only the classification accuracy (CA) as quality indicator.

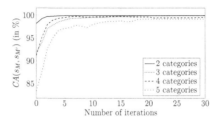

Fig. 2. Evolution of the classification accuracy of the learning set of a model composed of 10 criteria and a variable number of categories ($N_{model} = 10; N_o = 30; N_{it} = 20$)

correctly a majority of assignment examples? This question has been already covered in [11] for models having no more than 5 criteria and 3 categories. To answer this question for models with more parameters we add a step to the test procedure described above:

5. A set of 10000 random alternatives, B, is generated analogously to 2. We call this set, the generalization set. Alternatives of the set B are assigned by models M and M'. Finally the assignment obtained by models M and M' are compared and the classification accuracy $CA(s_M, s_{M'})$ is computed.

Figure 3(a) and 3(b) show the average, min and max $CA(s_M, s_{M'})$ of the generalization set after learning the parameters of models having 10 criteria and 3 or 5 categories on the basis of 100 to 1000 assignment examples. Figure 3(a) shows that 400 examples are sufficient to restore on average 95 % of the assignments for models having 3 categories, 10 criteria while 800 examples are needed for ones having 5 categories, 10 criteria (see Figure 3(b)). As expected, the higher the cardinality of the learning set, the more $CA(s_M, s_{M'})$ is high in generalization.

4.3   Tolerance for Errors

To test the behavior of the algorithm when the learning set is not fully compatible with a MR-Sort model, a step is added in the test procedure after generating the assignment examples:

2' A proportion of errors is added in the assignments obtained using the model M. For each alternative of the learning set, its assignment is altered with probability P, the altered assignment example is uniformly drawn among the other categories. We denote by $\tilde{s}_M$ the rule producing the assignments with errors.

Tolerance for errors is tested by learning the parameters of a MR-Sort model having 5 categories and 10 criteria on the basis of 1000 assignment examples generated using $\tilde{s}_M$. In Figure 4(a), the average classification accuracy of the learning set is shown for 10 test instances with 10 to 40 % of errors in the learning set. We observe that $CA(\tilde{s}_M, s_{M'})$ converges to $1 - P$ when there are

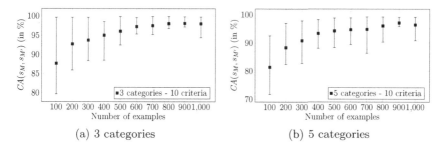

(a) 3 categories   (b) 5 categories

Fig. 3. Evolution of the classification accuracy on the generalization set. A 10 criteria with 3 or 5 categories model has been learned on the basis of learning sets containing from 100 up to 1000 assignment examples ($N_{model} = 10; N_o = 30; N_{it} = 20$).

errors in the learning set. Among the assignment examples badly assigned by the model, a majority corresponds to altered examples. To see to what extent the errors affect the algorithm, we generate a generalization set that is assigned both by the rule s_M and $s_{M'}$. The resulting sets are compared and $CA(s_M, s_{M'})$ is computed. In Figure 4(b), average, minimal and maximal $CA(s_M, s_{M'})$ are shown for 10 test instances. We observe that for small numbers of errors, i.e. less than 20 %, the algorithm tends to modify the model s.t. $CA(s_M, s_{M'})$ is altered on average by the same percentage of error in generalization. When there are more than 20% of errors in the learning set, the algorithm is able to find a model giving a smaller proportion of assignment errors in generalization.

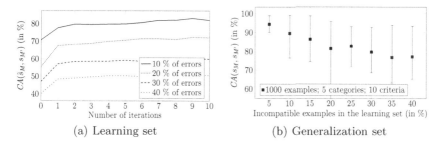

(a) Learning set   (b) Generalization set

Fig. 4. Evolution of the classification accuracy ($CA(\tilde{s}_M, s_{M'})$) for the alternatives in the learning set (a) and the generalization set (b). A 5 categories and 10 criteria model is inferred on the basis of 1000 assignment examples containing 10 to 40 % of errors ($N_{model} = 10; N_o = 30; N_{it} = 20$)

4.4   Idiosyncratic Behavior

This experiment aims at checking if an MR-Sort model is able to represent assignments that have been obtained by another sorting rule based on an additive value function (AVF-Sort model). In such a model, a marginal utility function u_j is associated to each criterion. In the chosen model, utility functions are piecewise linear and monotone. They are split in k parts in the criterion range $[g_{j*}, g_j^*]$, with g_{j*} the less preferred value and g_j^* the most preferred value on j, s.t. $u(g_{j*}) = 0$ and $u(g_j^*) = 1$. The end points of the piecewise linear functions are given by $g_j^l = g_{j*} + \frac{l}{k}(g_j^* - g_{j*})$, with $l = 0, ..., k$. Marginal utility of an alternative a on criterion j is denoted by $u_j(a_j)$. The score of an alternative is given by the global utility function which is equal to $U(a) = \sum_{j=1}^n w_j u_j(a_j)$, with $U(a) \in [0, 1]$ and $\sum_{j=1}^n w_j = 1$. The higher the value of $U(a)$, the more a is preferred. Categories are delimited by ascending global utility values β^h, s.t. an alternative a is assigned in category h iff $\beta^{h-1} \leq U(a) < \beta^h$ with $\beta^0 = 0$ and $\beta^p = 1 + \epsilon$, ϵ being a small positive value. Such an additive model is used in the UTADIS method [18, 19]. We study the ability of our metaheuristic to learn an MR-Sort model from a learning set generated by an AVF-Sort model. To do so, we replace step 1 by:

1. A sorting model M based on an additive value function is randomly generated. To generate the weights, the same rule as for the MR-Sort model is used. For each value function, $k - 1$ random are uniformly drawn from the interval $[0, 1]$ and ordered s.t. $r_k = 1 \geq r_{k-1} \geq ... \geq r_1 \geq 0 = r_0$, then end points are assigned as follows $u(g_j^l) = r_l$, with $l = 0, ..., k$. For the category limits β_h, $p - 1$ random numbers are uniformly drawn from the interval $[0, 1]$ and then ordered s.t. $r_{p-1} \geq ... \geq r_1$. Category limits are given by $\beta_h = r_h$, $h = 1, ..., p - 1$. The assignment rule is denoted by s_M^*.

Once the model generated, the alternatives are assigned by the model M and the metaheuristic tries to learn a MR-Sort model from the assignments obtained by M. To assess the ability of the heuristic to find a MR-Sort model restoring the maximum number of examples, we test it with 1000 assignment examples, on models composed of 10 criteria and 2 to 10 categories. We choose to use an AVF-Sort model in which each additive value function is composed of 3 segments.

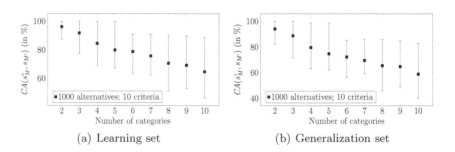

(a) Learning set   (b) Generalization set

Fig. 5. Evolution of the classification accuracy $(CA(s_M^*, s_{M'}))$ for alternatives in the learning set (a) and the generalization set (b). A 5 categories and 10 criteria model is inferred on the basis of altered assignment examples. An MR-Sort model is learned on the basis of assignment examples obtained with an AVF-Sort model having 10 criteria and 2 to 10 categories.

This experiment is repeated 10 times. Figure 5(a) presents the average, minimum and maximum $CA(s_M^*, s_{M'})$ of the learning set. The plot shows that the MR-Sort model is able to represent on average 80% of the assignment examples obtained with an AVF-Sort model when there are no more than 5 categories. We perform a generalization by assigning 10000 alternatives through the AVF-Sort model, M and through the learned MR-Sort model, M'. Figure 5(b) shows the average, minimum and maximum classification accuracy of the generalization set. These results confirm the behavior observed with the learning set. The ability to represent assignments obtained by an AVF-Sort model with an MR-Sort model is limited, even more when the number of categories increases.

5 Conclusions and Further Research

In this paper we presented an algorithm that is suitable to learn a MR-Sort model from large sets of assignment examples. Unlike the MIP proposed in [11], it is possible to learn a model composed of 10 criteria and 5 categories that restore 99% of the examples of assignments in less than two minutes for a learning set composed of 1000 alternatives with no error.

In the case of a learning set containing a proportion of assignment errors, the experimentations showed that the algorithm finds a model giving on average a smaller or equal proportion of errors in generalization. We also found that assignment examples obtained by an AVF-Sort model are quite difficult to represent with an MR-Sort model. Further researches have to be done with the AVF-Sort model to see if it is able to learn a model that restore correctly a set of assignment examples obtained by an MR-Sort model.

The metaheuristic described in this paper does not cover MR-Sort models with vetoes. Learning the parameters of a MR-Sort model with vetoes deserves to be studied and will improve a bit the ability of the model to represent assignments.

Acknowledgment. The authors thank two anonymous referees for their helpfull comments which contributed to improve the content of this paper. The usual caveat applies.

References

1. Yu, W.: Aide multicritère la décision dans le cadre de la problématique du tri: méthodes et applications. PhD thesis, LAMSADE, Université Paris Dauphine, Paris (1992)
2. Roy, B., Bouyssou, D.: Aide multicritère la décision: méthodes et cas. Economica Paris (1993)
3. Bouyssou, D., Marchant, T.: An axiomatic approach to noncompensatory sorting methods in MCDM, I: The case of two categories. European Journal of Operational Research 178(1), 217–245 (2007)
4. Bouyssou, D., Marchant, T.: An axiomatic approach to noncompensatory sorting methods in MCDM, II: More than two categories. European Journal of Operational Research 178(1), 246–276 (2007)

5. Mousseau, V., Slowinski, R.: Inferring an ELECTRE TRI model from assignment examples. Journal of Global Optimization 12(1), 157–174 (1998)
6. Mousseau, V., Figueira, J., Naux, J.P.: Using assignment examples to infer weights for ELECTRE TRI method: Some experimental results. European Journal of Operational Research 130(1), 263–275 (2001)
7. Ngo The, A., Mousseau, V.: Using assignment examples to infer category limits for the ELECTRE TRI method. Journal of Multi-criteria Decision Analysis 11(1), 29–43 (2002)
8. Dias, L., Mousseau, V., Figueira, J., Clímaco, J.: An aggregation/disaggregation approach to obtain robust conclusions with ELECTRE TRI. European Journal of Operational Research 138(1), 332–348 (2002)
9. Dias, L., Mousseau, V.: Inferring Electre's veto-related parameters from outranking examples. European Journal of Operational Research 170(1), 172–191 (2006)
10. Doumpos, M., Marinakis, Y., Marinaki, M., Zopounidis, C.: An evolutionary approach to construction of outranking models for multicriteria classification: The case of the ELECTRE TRI method. European Journal of Operational Research 199(2), 496–505 (2009)
11. Leroy, A., Mousseau, V., Pirlot, M.: Learning the Parameters of a Multiple Criteria Sorting Method Based on a Majority Rule. In: Brafman, R. (ed.) ADT 2011. LNCS, vol. 6992, pp. 219–233. Springer, Heidelberg (2011)
12. Cailloux, O., Meyer, P., Mousseau, V.: Eliciting ELECTRE TRI category limits for a group of decision makers. European Journal of Operational Research 223(1), 133–140 (2012)
13. Fürnkranz, J., Hüllermeier, E.: Preference learning: An introduction. In: Fürnkranz, J., Hüllermeier, E. (eds.) Preference Learning, pp. 1–17. Springer (2010)
14. Bouyssou, D., Pirlot, M.: A characterization of concordance relations. European Journal of Operational Research 167/2, 427–443 (2005)
15. Bouyssou, D., Pirlot, M.: Further results on concordance relations. European Journal of Operational Research 181, 505–514 (2007)
16. Pirlot, M.: General local search methods. European Journal of Operational Research 92(3), 493–511 (1996)
17. Butler, J., Jia, J., Dyer, J.: Simulation techniques for the sensitivity analysis of multi-criteria decision models. European Journal of Operational Research 103(3), 531–546 (1997)
18. Devaud, J., Groussaud, G., Jacquet-Lagrèze, E.: UTADIS: Une méthode de construction de fonctions d'utilité additives rendant compte de jugements globaux. In: European Working Group on MCDA, Bochum, Germany (1980)
19. Doumpos, M., Zopounidis, C.: Multicriteria Decision Aid Classification Methods. Kluwer Academic Publishers, Dordrecht (2002)

Roles and Teams Hedonic Game

Matthew Spradling, Judy Goldsmith, Xudong Liu,
Chandrima Dadi, and Zhiyu Li

University of Kentucky, USA
mjspra2@uky.edu, goldsmit@cs.uky.edu, liu@cs.uky.edu,
cda232@g.uky.edu, zhiyu.li@uky.edu

Abstract. We introduce a new variant of hedonic coalition formation games in which agents have two levels of preference on their own coalitions: preference on the set of "roles" that makes up the coalition, and preference on their own role within the coalition. We define several stability notions and optimization problems for this model. We prove the hardness of the decision problems related to our optimization criteria and show easiness of finding individually stable partitions. We introduce a heuristic optimizer for coalition formation in this setting. We evaluate results of the heuristic optimizer and the results of local search for individually stable partitions with respect to brute-force MaxSum and MaxMin solvers.

Keywords: coalition formation, computational complexity, hedonic games, optimization.

1 Introduction

Consider the online game, League of Legends, developed by Riot Games, Inc. According to a recent market research study, League of Legends is the most played PC video game in North America and Europe by number of hours played per month [10], with 70 million registered users and an average of 12 million daily active players [15]. Players sign on, and are matched with other players with similar Elo ratings. Once matched in a team of 3 or 5, they each choose an avatar (called a "champion") from a finite set. Each team then plays against another team, competing for Elo improvement.

The game experience could be enhanced if teams were matched on the basis of strategic combinations of champions. This is not only our hypothesis but also the observation of Riot Games. A senior user research employee for Riot Games, user name *davin*, recently commented that *"we don't have a single way of playing the game. ... So when you match people together, you'd need some way of pairing together players who have agreed on a particular strategy or want to play in a certain way."*[16] There are two criteria upon which players might express preferences: the combination of champions on which they would like to play, and the individual champion they would prefer to play on a given composition. These could be expressed separately or conditionally.

P. Perny, M. Pirlot, and A. Tsoukiàs (Eds.): ADT 2013, LNAI 8176, pp. 351–362, 2013.

Matching players by their preferences on their own teams is a hedonic coalition formation game [11]. Hedonic coalition formation games are characterized by agents' utilities depending only on the coalition they are assigned to, not on others. A game consists of a set of agents and their preferences for their possible roles and team compositions.

One of the aspects of the partitioning problem for League of Legends is the two-stage team formation: Players may be matched based on their shared interest in a team consisting of roles A, B, and C, but it may transpire that all three wish to play role A. A better partition algorithm would also use players' preferences on individual roles. We refer to this notion of a hedonic game as a *Roles and Teams Hedonic Game* (RTHG).

Recent work on hedonic coalition games has touched on notions comparable to stability in the stable marriage problem [11,4,6,14,1], etc. It is known that finding certain stable coalitions for hedonic games is NP-hard (see, for instance, [8,2]). Some papers considered restrictions on preferences that allow stable partitions, others presented heuristic algorithms for finding stable partitions.

Due to the two-stage team formation procedure in RTHG, we observe that the notions of Nash stable (NS) and individually stable (IS) partitions are quite different in this model compared to other hedonic games. We propose definitions for NS and IS partitions which address both the stability of role assignments within coalitions and permutations of agents within coalition assignments.

A different problem of optimizing social utility has also been investigated. In graphical games with unbounded treewidth, very recent work has been done to address the bi-criteria problem of maximizing both stability and social utility [13]. We provide hardness results for the decision problems related to Perfect, MaxSum and MaxMin partitions in RTHG. We define Nash stability and individual stability in this setting and show that individually stable partitions can always be found in time polynomial in the size of the input. We introduce a quadratic time greedy heuristic optimizer for coalition formation and compare to brute-force MaxSum and MaxMin solvers and the results of local search for individually stable partitions.

2   Roles and Teams Hedonic Games

Definition 1. *An RTHG instance consists of:*

- *P: a population of agents;*
- *m: a team size (we assume that $|P|/m$ is an integer);*
- *R: a set of available team member roles;*
- *C: a set of available team compositions, where a team composition is a set of m not necessarily unique roles in R;*
- *U: a utility function vector $\langle u_0, \ldots, u_{|P|-1} \rangle$, where for each agent $p \in P$, composition $t \in C$, and role $r \in R$ there is a utility function $u_p(t, r)$ with $u_p(t, r) = -\infty$ if $r \notin t$.*

A solution to an RTHG instance is a partition π of agents into teams of size m.

Table 1. Example RTHG instance with $|P| = 4, m = 2, |R| = 2$

$\langle r,t \rangle$	$u_{p_0}(r,t)$	$u_{p_1}(r,t)$	$u_{p_2}(r,t)$	$u_{p_3}(r,t)$
$\langle A, AA \rangle$	2	2	0	0
$\langle A, AB \rangle$	0	3	2	2
$\langle B, AB \rangle$	3	0	3	3
$\langle B, BB \rangle$	1	1	1	1

3 Related Work: Hedonic Partition Games

The original motivation for studying hedonic games was economic [11], but there are also many computational applications. Saad et al. have proposed hedonic coalition formation game models for a variety of multi-agent settings, including distributed task allocation in wireless agents [17], communications networks [18], and vehicular networks [19], among others.

In anonymous hedonic games [5], agents have preferences over group size and are matched to teams for a single type of activity. The group activity selection problem (GASP) includes preferences over a variety of activities given the number of agents engaged in the activity [9]. Agents in these games are homogeneous—every member of a coalition is equivalent. In RTHG, agents are heterogeneous while team size and group activity are fixed for a given instance. An RTHG agent holds preferences over its own role and the roles of its teammates. Furthermore, while GASP preferences are binary, RTHG agent preferences are not guaranteed to be.

Desirable partitioning in additively separable hedonic games (ASHG) [3] has been investigated. ASHGs allow for agents to place values on each other, making the agent population heterogeneous. The value an agent places on its coalition in such a game is the sum total value it gives other agents in its coalition. This model considers agent-to-agent valuation, but these values are fixed for any given agent-to-agent relation. ASHGs do not consider the context of the composition an agent is in. In RTHG, values are placed on team compositions and roles rather than individual agents.

Each agent has a variable role in RTHG and has preferences over which role to select for itself given a team composition.

For instances where $|C|m$ is smaller than $|P|$, the required input data for RTHG instances will be smaller than the required input for ASHG. Input for an ASHG instance requires each agent to hold a specific utility for each other agent within the population. This could be represented as a $|P| \times |P|$ matrix of utility values, U, where $U[i,j]$ is the utility that p_i holds for p_j. In RTHG, the input can be represented as a $|C|m \times |P|$ matrix. While there are millions of players in League of Legends [15], there are only around 10 basic roles to potentially fill (Healer, Mage, Assassin, etc.) and a maximum team size of 5. The input required for team formation in this setting will be orders of magnitude smaller in RTHG than if this game were treated as an ASHG.

Consider the following setting. In capstone computer science courses, students are sometimes grouped into equally-sized project teams. For a team of five students, one student may prefer a team of 2 skilled programmers, 1 designer, and 2 writers. Her second choice might be 1 programmer, 2 designers, 2 writers. In the first case, the student wants to be a programmer. In the second, she wants to be a designer, and definitely *not* a programmer.

This problem can be modeled as an RTHG. The GASP model does not apply. The ASHG model allows students to express utility values for each other, but ASHG preferences are *context-free* agent-to-agent assessments. Huxley may wish to join Clover's coalition when she needs a programmer, but not when she needs a writer. In RTHG, an agent need only express preferences on which roles and compositions she prefers. This self-evaluation may be easier to accurately poll.

Matching students to groups in a manner that optimizes utility for the class would be a useful endeavor. In a perfect world, each student would be matched to his or her most-preferred team. We show that such a *perfect* partition is not always possible in RTHG.

A *MaxSum* partition would, in a utilitarian fashion, optimize the sum total utility of the resulting coalitions. A *MaxMin* partition would take an egalitarian approach. It is unclear which metric (MaxSum or MaxMin) would best raise teaching evaluations in capstone computer science courses.

4   Evaluation of Solutions

Perfect partitions for general hedonic games have been defined such that each agent is in one of her most preferred coalitions [1].

For RTHG, we define a perfect partition to be one in which each agent gets a most-preferred coalition composition and role within that composition. Note that, in the general RTHG model, there may be multiple equivalently-valued compositions and roles. Therefore these preferences are not necessarily strict.

Definition 2. *A perfect partition is a partition of agents to coalitions so that, for each $p \in P$, $u_p(r,t) = min\{u_p(r,t) : r \in R \wedge t \in C\}$.*

A perfect partition is impossible for some RTHG instances. Consider an RTHG instance where $m = 2$ and $P = \{Alice, Bob\}$. Both Alice and Bob strictly prefer the team composition of ⟨Mage, Assassin⟩ with the role Assassin to all other ⟨r, t⟩ pairs. No perfect partition is possible. We consider the following notions of utility optimization.

Definition 3. *Given an instance I of RTHG, a MaxSum partition is one that achieves the maximum value of $\Sigma_{i < |P|} u_{p_i}$.*

MaxSum is a utilitarian optimality criterion.

Definition 4. *Given an instance I of RTHG, a MaxMin partition is one that achieves the maximum value of $\min_{p \in P} u_p$.*

MaxMin is an egalitarian optimality criterion.

In most hedonic game variants, a partition is considered *Nash stable (NS)* iff no agent p_i can benefit by moving from her coalition to another (possibly empty) coalition T. A partition is considered *individually stable (IS)* iff no agent can benefit by moving to another coalition T while not making the members of T worse off [1]. These definitions of stability do not fit well with RTHG.

Because team sizes in RTHG are fixed at m, an agent cannot simply choose to leave her coalition and join another. Rather, if an agent p_i is to move from coalition S to T, she must take the *position* (role in a particular coalition) of another agent p_j in T. This could be done as a swap, or it could be a more complex set of moves made among several agents. Note that should some $X \subseteq P$ collaboratively change positions, this permutation would not change the utilities of the compositions for the agents in $\overline{X}$. All existing compositions remain intact.

Definition 5. *A partition π is* Nash team stable (NTS) *iff no set $X \subseteq P$ of agents can improve the sum of their utilities by a new permutation of their positions in their coalitions.*

A partition π is individually team stable (ITS) *iff no set $X \subseteq P$ of agents can improve the sum of their utilities by a new permutation of their positions in their coalitions without reducing the utility of the partition for any single agent in X.*

There will always be a NTS partition π_{NTS} where all agents select the same role. In this case, no agent can improve her utility by changing positions since the new position would be identical to her previous position. Some RTHG instances may lack a *non-uniform* NTS partition $\pi_{NTS_{NU}}$, where $r_{p_i} \neq r_{p_j}$ for at least one pair of agents p_i, p_j. Consider the following RTHG instance:

Table 2. RTHG instance with $|P| = 2, m = 2, |R| = 2$ where no $\pi_{NTS_{NU}}$ exists

$\langle r, t \rangle$	$u_{p_0}(r, t)$	$u_{p_1}(r, t)$
$\langle A, AA \rangle$	1	1
$\langle A, AB \rangle$	1	1
$\langle B, AB \rangle$	0	0
$\langle B, BB \rangle$	1	1

No $\pi_{NTS_{NU}}$ exists in this instance. Consider each of the two possible non-uniform partitions:

- π_0, where $r_{p_0} = B$ and $r_{p_1} = A$. p_0 prefers to swap positions. Not NTS.
- π_1, where $r_{p_0} = A$ and $r_{p_1} = B$. p_1 prefers to swap positions. Not NTS.

To construct an individually team stable partition π_{ITS}, start with any partition π of I and iteratively improve it until no improvements are possible. At that point, the resulting partition will be ITS. To find an improvement, if one exists, construct a graph $G = \langle V, E_b \cup E_r \rangle$, where the vertices correspond to players, and there is an edge in E_b from p_i to p_j iff $u_{p_i}(r_i, t_i) = u_{p_i}(r_j, t_j)$, and there is an edge in E_r from p_i to p_j iff $u_{p_i}(r_i, t_i) > u_{p_i}(r_j, t_j)$. If there exists a cycle in the graph containing as least one edge $e_r \in E_r$, then the partition is not ITS.

Another movement option in RTHG is for an agent to remain within her coalition but change roles. This converts the existing composition to another the agent may prefer. Note that this would change the utility of the composition for her coalition, but otherwise does not affect the utility of the partition for any agent outside of her coalition.

Definition 6. *A partition π is* Nash role stable (NRS) *iff no agent p_i can improve her utility by changing from her current role r to a new role r'.*

A partition π is individually role stable (IRS) *iff no agent p_i can improve her utility by changing from her current role r to a new role r' without reducing the utility of any other agent in her coalition.*

Some RTHG instances may lack a NRS partition π_{NRS}. Consider the following RTHG instance:

Table 3. RTHG instance with $|P| = 2, m = 2, |R| = 2$ where no π_{NRS} exists

$\langle r, t \rangle$	$u_{p_0}(r, t)$	$u_{p_1}(r, t)$
$\langle A, AA \rangle$	0	1
$\langle A, AB \rangle$	1	0
$\langle B, AB \rangle$	1	0
$\langle B, BB \rangle$	0	1

No π_{NRS} exists in this instance. Consider each of the four possible partitions:

- π_0, where $r_{p_0} = A$ and $r_{p_1} = B$. p_1 prefers to switch to role A. Not NRS.
- π_1, where $r_{p_0} = B$ and $r_{p_1} = A$. p_1 prefers to switch to role B. Not NRS.
- π_2, where $r_{p_0} = A$ and $r_{p_1} = A$. p_0 prefers to switch to role B. Not NRS.
- π_3, where $r_{p_0} = B$ and $r_{p_1} = B$. p_0 prefers to switch to role A. Not NRS.

An IRS partition π_{IRS} of an RTHG instance I can be found in time polynomial in $|I|$. Given any partition π of I, perform a local search where the neighborhood is one individual in one coalition changing her role and improvement is evaluated in terms of changes to that coalition's utility. Since this improves the overall utility of the partition, there are a limited number of possible improvements; when no improvements are possible, the partition is IRS.

Definition 7. *A partition π is* Nash stable (NS) *iff it is both NTS and NRS.*
A partition π is individually stable (IS) *iff it is both ITS and IRS.*

A NS partition π_{NS} may not always exist for some RTHG instances, given that a partition π_{NRS} may not exist.

An IS partition π_{IS} of an RTHG instance I can be found in time polynomial in $|I|$. Given any partition π of I, alternatively perform IRS local search and ITS local search until neither finds an improvement. The resulting partition π' will be IS.

Theorem 1. *Every instance of RTHG has an IS partition. Not every instance of RTHG has a NS partition.*

5   Complexity

Definition 8. *An instance of Special RTHG is an instance of RTHG such that for each agent $p \in P$, each $t \in C$, and each $r \in t$; $u_p(t, r) \rightarrow \{0, 1\}$ and $u_p(t, r) = 1$ only if t is* uniform, *namely it consists of m copies of a single role r.*

In other words, each agent finds some non-empty set of single-role team compositions acceptable (utility 1), and no other types of team compositions acceptable.

Definition 9. *The language* PERFECT *RTHG consists of those instances of RTHG for which a perfect partition exists, and* PERFECT SPECIAL *RTHG consists of those instances of Special RTHG for which a perfect partition exists.*

In Special RTHG instances, the question of a perfect partition reduces to the problem of finding a MaxMin partition, or the decision problem of whether there's a partition with MaxMin value m.

Consider the EXACT COVER problem:
GIVEN a set $S \subseteq \mathcal{P}(\{1, ..., r\})$ where all elements of S have size 3,
IS THERE a subset $T \subseteq S$ such that T partitions $\{1, ..., r\}$?
EXACT COVER is NP-complete [12].

Theorem 2. PERFECT SPECIAL RTHG *is NP-complete.*

Proof. To show that PERFECT SPECIAL RTHG is in NP, consider the following NP algorithm. Given an instance of PERFECT SPECIAL RTHG, guess a partition and evaluate its MaxMin value. To compute the MaxMin value, compute the utility of each of the $|P|/m$ coalitions (time $\mathcal{O}(mt)$ for each coalition, where t is the complexity of table lookup for an individual's utility for a particular team and role), stopping and rejecting if any coalition has utility 0, else accepting. This checking is in time polynomial in the size of the input.

To show NP-hardness, we show that EXACT COVER $\leq_m^P$ SPECIAL PERFECT RTHG. In other words, given an instance $E = \langle r, S \rangle$ of EXACT COVER, we construct an instance R_E of SPECIAL PERFECT RTHG such that $E \in$ EXACT COVER iff $R_E \in$ SPECIAL PERFECT RTHG.

R_E will have the property that, for each agent, the only acceptable teams are uniform, i.e., consist of m copies of a single role. Thus, the question is whether they can be assigned to an acceptable team; the role for that team will be acceptable.

Consider $E = \langle r, S \rangle$. For each set in S, R_E will have a role and a corresponding team composition. $P = \{1, ..., r\}$. The desired team size is $m = 3$. Each agent i desires those team compositions s such that $i \in s$.

There is an exact cover of $\{1, ..., r\}$ iff there is an assignment of agents to teams of size 3 such that each team corresponds to an element of S.

Therefore, the PERFECT SPECIAL RTHG problem is NP-hard.

Corollary 1. *The general case of* PERFECT RTHG *is NP-hard.*

Proof. We observe that if there were a fast algorithm to decide the general case of PERFECT RTHG then this same algorithm would also decide PERFECT SPECIAL RTHG.

Therefore the general case of PERFECT RTHG is NP-hard.

Definition 10. *The language* MAXSUM RTHG *consists of pairs* $\langle G, k \rangle$, *where* G *is an instance of RTHG, k is an integer, and the MaxSum value of G is* $\leq k$; MAXSUM SPECIAL RTHG *consists of those instances of Special RTHG for which the MaxSum value is* $|P|$.

Definition 11. *The language* MAXMIN RTHG *consists of pairs* $\langle G, k \rangle$, *where* G *is an instance of RTHG, k is an integer, and the MaxMin value is* $\leq k$; MAXMIN SPECIAL RTHG *consists of those instances of Special RTHG for which the MaxMin value is* m.

Theorem 3. MAXMIN RTHG *and* MAXSUM RTHG *are both NP-hard.*

Proof. A Special RTHG partition π for G is perfect iff $\sum_{p \in P} u_p(\pi) = |P|$ iff MaxMin$(\pi) = m$ iff $\langle G, |P| \rangle \in$ MAXSUM RTHG iff MaxSum$(\pi) = |P|$ iff $\langle G, m \rangle \in$ MAXMIN RTHG. Therefore MAXMIN RTHG and MAXSUM RTHG are both NP-hard.

6 Greedy Heuristic Partitioning

By modeling agents as voters in an election and their preferences over team compositions and roles as votes, the *scoring* voting rule can be applied to hold a series of elections and democratically (but not necessarily optimally) assign agents to teams. A *voting rule* is a function mapping a vector a of voters' votes to one of the b candidates in a candidate set c.

Definition 12. *[7] We define scoring rules for elections as follows. Let* $a = \langle a_1, \cdots, a_m \rangle$ *be a vector of integers such that* $a_1 < a_2 < ... < a_m$. *For each voter, a candidate receives a_1 points if it is ranked first by the voter, a_2 points if it is ranked second, etc. The score s_c of candidate c is the total number of points the candidate receives .*

For our procedure, a $|C|m \times |P|$ matrix of agent utility values becomes the candidate set c. An "election" is run upon the candidate set to select the most-preferred coalition. A set of m voters with the highest utility for that coalition is selected to form a team and removed from the population. Their votes are removed, and a new election is held on the reduced candidate set. This procedure continues until all $|P|$ agents have been matched to $|P|/m$ teams. We assume that m evenly divides $|P|$. The following pseudocode describes this greedy algorithm:

Algorithm 1. GreedyRTHGPartiton(RTHG instance G, empty partition π)

---

for $|C|$ compositions $c_0 \to c_{|C|-1}$ **do**
   for m positions $r_0 \to r_{m-1} \in c_i$ **do**
      calculate the sum of agent votes on $\langle c_i, r_j \rangle$. $O(|P|)$
   end for
end for
for $|P|/m$ coalitions $t_0 \to t_{|P|/m-1}$ to assign to π **do**
   find the set of compositions C_{max} for which the sum of total votes is maximized. $O(|C| \cdot m)$
   select one composition c_i uniformly at random from within the set.
   for m positions $r_0 \to r_{m-1} \in c_i$ **do**
      find the set of agents $P_{max}(c_i, r_j)$ for whom the individual agent's vote for $\langle c_i, r_j \rangle$ is maximized. This takes time $O(|P|/m)$, given that the population shrinks by m agents as each team is formed and removed.
      select one agent p_j uniformly at random from within the set.
      add agent p_j to the coalition t_k.
      for $|C|$ compositions $c_0 \to c_{|C|-1}$ **do**
         for m positions $r_0 \to r_{m-1} \in c_i$ **do**
            remove agent p_j's vote from the population, decrementing the sum total vote on $\langle c_i, r_j \rangle$.
         end for
      end for
   end for
   append team t_k to the partition π.
end for

Observation 4. *The time complexity of GreedyRTHGPartiton is $O(|P|^2/m)$, or $O(|P| \cdot |C| \cdot m)$ if $|P| < |C| \cdot m^2$.*

7 Testing and Results

For our experiments we chose *Strictly Ordered RTHG* instances. In a Strictly Ordered RTHG instance, each agent's first choice of composition or role is weighted equivalently to other agents' first choices, as is her second choice, etc. The system does not value one agent's preferences over another.

Two hundred and forty instances of Strictly Ordered RTHG were generated by a uniformly random procedure we developed. This number of cases allowed us to test $|P|$ ranging from 6 to 15 agents, $|R|$ ranging from 3 to 6, and m ranging from 3 to 5.

We began with $|P| = 6$, $|R| = 3$, and $m = 3$ in the minimal case. Ten random preference matrices were generated with these arguments. We then incremented $|R|$ by 1 and generated ten new random preference matrices, up to $|R| = 6$. This process was repeated for $\langle m, |P| \rangle = \langle 4, 8 \rangle, \langle 5, 10 \rangle, \langle 3, 12 \rangle, \langle 4, 12 \rangle, \langle 5, 15 \rangle$. These upper bounds were chosen because larger inputs dramatically increased the time required for the brute force solvers to process the data.

Optimal results were calculated for each of these instances by MaxSum and MaxMin brute force implementations we developed. There are

$$O(|P|! \cdot (|C| + |P|/m)^{|P|/m})$$

possible partitions in an instance of RTHG. We generate all of them and find the MaxSum and MaxMin values for each instance considered. Our implementation of *GreedyRTHGPartiton* ran each instance 500 times, in order to limit random error. For the same instances, IS solutions were constructed by our implementation of *ISLocalSearch*. Fifty initial partitions were selected uniformly at random for each instance as starting points for *ISLocalSearch*. We compared the mean utilities of partitions generated by *GreedyRTHGPartiton* and *ISLocalSearch* to the optimal results as $|P|$ increased.

Computations were run on a machine using 8 GB of RAM and a 2.50 GHz Intel(R) Core(TM) i5-3210M CPU. MaxSum and MaxMin brute force algorithms were implemented in C++, while *GreedyRTHGPartiton* and *ISLocalSearch* were implemented in Python 3.3.

Results are presented in Figures 1 and 2. We show the percentages by which *GreedyRTHGPartiton* and *ISLocalSearch* overestimate optimal MaxSum and

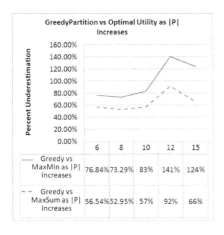

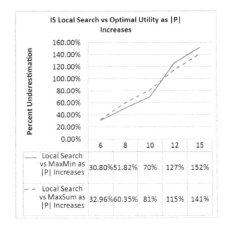

Fig. 1. Percent underestimate of optimal MaxSum and MaxMin using *GreedyRTHGPartiton* as $|P|$ increases

Fig. 2. Percent underestimate of optimal MaxSum and MaxMin using *ISLocalSearch* as $|P|$ increases

MaxMin for each test case. The lower the percent overestimation the better. Each figure shows the mean overestimation as $|P|$ increases.

$GreedyRTHGPartiton$ produces consistently better results for estimating Max-Sum compared to MaxMin. The greedy heuristic may leave a very poor coalition at the end, lowering MaxMin performance. Suppose there are 6 agents A, B, C, D, E, and F being matched to 3 teams each of size 2. The best coalition is AB, while the four worst coalitions are CD CE, CF, and EF. If A and B form a coalition together in the first iteration, then the remaining two coalitions selected will be among the worst possible. It may transpire that CD is the next team to be formed, *even if EF happens to be the worst coalition of all.*

Total utility is balanced out by strong selections made at the beginning, raising the performance against MaxSum. In our experiments, $GreedyRTHGPartiton$ underestimates MaxSum utility by 68.38% and MaxMin by 105.76% on average.

$ISLocalSearch$ performance against MaxMin and MaxSum optimal solutions is close. There is an approximately linear increase in overestimation as $|P|$ increases, because there are increasingly many local optima as the size of the input increases. In our experiments, $ISLocalSearch$ underestimates the MaxSum optimal utility by 93.62% and the MaxMin optimal by 90.88% on average.

To test the stability of $GreedyRTHGPartition$ solutions, we ran each of its outputs as input to the $ISLocalSearch$ algorithm. We included an additional 80 inputs with $|R| = 5$ and $m = 5$, with $|P|$ increased for every 10 inputs. The results are shown in Figures 3 and 4. Defining Q as the number of iterations required for $ISLocalSearch$ to form an IS partition from $GreedyRTHGPartition$, Q increases as $|P|$ increases. The speed with which Q grows relative to $|P|$ is the ratio $Q/|P|$. This is the number of local search iterations required per unit population. This ratio decreases at $|P|$ increases, suggesting that fewer local searches per unit population are required as $|P|$ grows.

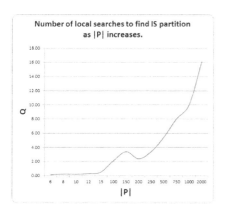

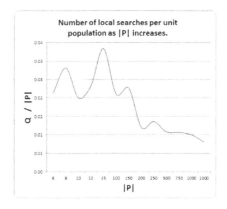

Fig. 3. Number of local searches to find IS partition as $|P|$ increases

Fig. 4. Number of local searches *per unit population* as $|P|$ increases

References

1. Aziz, H., Brandl, F.: Existence of stability in hedonic coalition formation games. In: Proc. Conference on Autonomous Agents and Multiagent Systems (AAMAS 2012) (2012)
2. Aziz, H., Brandt, F., Harrenstein, P.: Pareto optimality in coalition formation. In: Persiano, G. (ed.) SAGT 2011. LNCS, vol. 6982, pp. 93–104. Springer, Heidelberg (2011)
3. Aziz, H., Brandt, F., Seedig, H.G.: Computing desirable partitions in additively separable hedonic games. Artificial Intelligence (2012)
4. Banerjee, S., Konishi, H., Sönmez, T.: Core in a simple coalition formation game. Social Choice and Welfare 18, 135–153 (2001)
5. Banerjee, S., Konishi, H., Sönmez, T.: Core in a simple coalition formation game. Social Choice and Welfare 18(1), 135–153 (2001)
6. Bogomolnaia, A., Jackson, M.O.: The stability of hedonic coalition structures. Games and Economic Behavior 38(2), 201–230 (2002)
7. Conitzer, V., Sandholm, T.: Communication complexity of common voting rules. In: Proceedings of the 6th ACM Conference on Electronic Commerce, EC 2005, pp. 78–87. ACM, New York (2005)
8. Conitzer, V., Sandholm, T.: Complexity of constructing solutions in the core based on synergies among coalitions. Artificial Intelligence 170, 607–619 (2006)
9. Darmann, A., Elkind, E., Kurz, S., Lang, J., Schauer, J., Woeginger, G.: Group activity selection problem. In: Workshop Notes of COMSOC 2012 (2012)
10. DFC. League of Legends Most Played PC Game. DFC Intelligence (2012) http://www.dfcint.com/wp/?p=343.
11. Drèze, J.H., Greenberg, J.: Hedonic coalitions: Optimality and stability. Econometrica 48(4), 987 (1980)
12. Goldreich, O.: Computational Complexity, A Conceptual Perspective. Cambridge University Press (2008)
13. Ismaili, A., Bampis, E., Maudet, N., Perny, P.: A study on the stability and efficiency of graphical games with unbounded treewidth. In: Proceedings of the 12th International Conference on Autonomous Agents and Multiagent Systems (AAMAS 2013) (to appear, 2013)
14. Pini, M.S., Rossi, F., Venable, K.B., Walsh, T.: Stability in matching problems with weighted preferences. In: Filipe, J., Fred, A. (eds.) ICAART 2011. CCIS, vol. 271, pp. 319–333. Springer, Heidelberg (2013)
15. Riot. League of Legends' Growth Spells Bad News for Teemo—Riot Games. Riot Games Inc. (2012), http://www.riotgames.com/articles/20121015/138/ league-legends-growth-spells-bad-news-teemo
16. Riot. Let's talk about Champ Select - Page 5 - League of Legends. Riot Games Inc. (2013), http://na.leagueoflegends.com/board/showthread.php?p=35559688#35559688
17. Saad, W., Han, Z., Basar, T., Debbah, M., Hjorungnes, A.: Hedonic coalition formation for distributed task allocation among wireless agents. In: Proc. IEEE Transactions on Mobile Computing (2011)
18. Saad, W., Han, Z., Basar, T., Hjorungnes, A., Song, J.B.: Hedonic coalition formation games for secondary base station cooperation in cognitive radio networks. In: Wireless Communications and Networking Conference (WCNC), pp. 1–6 (2010)
19. Saad, W., Han, Z., Hjørungnes, A., Niyato, D., Hossain, E.: Coalition formation games for distributed cooperation among roadside units in vehicular networks. IEEE Journal on Selected Areas in Communications (JSAC), Special Issue on Vehicular Communications and Networks (2011)

Comparative Preferences Induction Methods for Conversational Recommenders

Walid Trabelsi[1,2], Nic Wilson[1], and Derek Bridge[2]

[1] Cork Constraint Computation Centre
{w.trabelsi,n.wilson}@4c.ucc.ie
[2] Department of Computer Science,
University College Cork, Ireland
d.bridge@cs.ucc.ie

Abstract. In an era of overwhelming choices, recommender systems aim at recommending the most suitable items to the user. Preference handling is one of the core issues in the design of recommender systems and so it is important for them to catch and model the user's preferences as accurately as possible. In previous work, comparative preferences-based patterns were developed to handle preferences deduced by the system. These patterns assume there are only two values for each feature. However, real-world features can be multi-valued. In this paper, we develop preference induction methods which aim at capturing several preference nuances from the user feedback when features have more than two values. We prove the efficiency of the proposed methods through an experimental study.

1 Introduction

Choosing the right or the best option is often a demanding and challenging task for the user when there are many available alternatives (e.g., a customer in an online retailer). Recommender systems aim at recommending the most suitable items to the user. However, the recommended items proposed by the system may not match the users' needs as recommender systems might miss on the users' preferences (see, e.g., [1]). One approach which ensures that the system is kept aware of the user needs is to establish a conversation between the user and the system by means of conversational recommender systems.

Preference handling is one of the core issues in the design of recommender systems and so it is important for them to catch and model the user's preferences as accurately as possible. In fact, preferences aim at offering the user the ability to express her relative or absolute satisfaction when faced with a choice between different options. One of the major approaches in today's recommender systems is utility functions which assign a numerical score to each data item [2]. A second major approach is the relational preference structures [3] as the user may wish to state simple comparisons. She may want to make no explicit quantification of preference or utility, leaving the preference purely qualitative. This could be the case for example in a travel problem, where there is a large number

P. Perny, M. Pirlot, and A. Tsoukiàs (Eds.): ADT 2013, LNAI 8176, pp. 363–374, 2013.
© Springer-Verlag Berlin Heidelberg 2013

of possible attributes involving times, transportation means and locations that vary from one user to another. In such a case, the user may want to say that she likes to travel to a country during the summer in that country, with all other attributes being equal. The user will then avoid having to communicate an accurate numerical model. It has also been claimed that the qualitative specification of preferences is more general than the quantitative one, as not all preference relations can be expressed by scoring functions [4].

In this paper, we look for elaborate and generic comparative preferences-based induction methods that we can prove to be efficient in practice with conversational recommender systems that suggest multi-valued feature products to the user.

The rest of the paper is organized as follows. In Section 2, we give an overview of conversational recommender systems and how preferences are handled in these systems. Then, Section 3 describes the framework of preference dominance that will be used in this paper. The conversational recommender system we are using is detailed in Section 4. A fundamental step in the process of recommending for conversational recommender system is preferences induction. We introduce a number of preference induction methods in Section 5. We performed experimentations which allowed us to assess the efficiency of these methods with regards to the obtained results which we discussed in Section 6. In Section 7, we conclude the paper with possible extensions of the proposed approaches.

2   Related Work

2.1   Conversational Recommender Systems

Generally speaking, people do not state their preferences up-front because initially they only have a vague idea of the product they would like to have [5]. Usually, criteria about the product the customer would like to purchase are specified during the dialogue with the seller. This is still the case even for knowledgable customers in the domains where expert users need to be assisted because available products dynamically change. A distinctive example is the list of special offers (e.g., flight tickets) which change frequently.

Conversational recommender systems [5] recognise that their users may be willing and able to provide more information on their constraints and preferences, over a dialogue. The main difference with the single-shot recommendation scenario is that in the case where the user is not satisfied she can revise her request.

2.2   Preference Handling in Conversational Recommender Systems

The acquisition of preferences is a central challenge in interactive systems like recommender systems [6]. There are two major approaches in today's recommender systems: utility functions [2] and relational preference structures [3]. A utility function assigns a numerical score to each item. Relational preference structures link pairs of items through the notions of "is preferred to" and "is equally preferable as" thus leading to qualitative preference orderings. Typically, the task of the recent online conversational recommenders is to elicit the customer requirements, while interacting with her, in a personalized way.

Critiquing [7] is an interaction model that allows users to build their preferences by examining or reviewing examples shown to her by the system. The user feedback employed in conversational recommender systems was also studied in [8] through two comparison-based recommendation approaches: *More Like This* (MLT) and *Partial More Like This* (PMLT). Their role is to induce preferences when the user reacts to the recommended items. They both generate preference statements stating the preference of features that mark the selected item over those that characterize the rejected items during an interaction stage. Information Recommendation [9] is a recommendation approach that aims at suggesting to the user how to reformulate her queries to a product catalogue in order to find the products that maximize her utility. In [9], the authors showed that, by observing the queries selected by the user among those suggested, the system can make inferences on the true user utility function and eliminate from the set of suggested queries those with an inferior utility. Authors in [10] proposed a novel use of the formalism of preference elicitation in [9]. They invoked comparative preferences-based patterns to handle the preferences deduced by the system. These patterns assume there are only two values for each feature. However, real-world features can be multiple-valued. In this paper, we investigate preference induction methods which can handle the user preferences in a conversational recommender for products with multiple-valued features.

3   CP-Tree-Based Dominance

Products in online databases need to be compared by pairs, through dominance testing, to find out which options are dominated and to eliminate them consequently . In this paper, dominance testing is based on some structure called cp-trees which were introduced in [11].

3.1   Description of a CP-Tree

A cp-tree is a directed rooted tree. Associated with each node N in the tree is a set of variables Y_N. Let γ be the maximum number of variables in Y_N. The cp-tree represents a form of lexicographic order where the importance ordering on nodes and their assignments depends on more important nodes and their assignments.

Example 1. Let $V = \{X, Y, Z\}$ be a set of variables whose domains are as follows. $\underline{X} = \{x1, x2\}$, $\underline{Y} = \{y1, y2\}$ and $\underline{Z} = \{z1, z2\}$ respectively. Figure 1 represents an example of a cp-tree with $\gamma = 1$. Each node in the cp-tree depicted in Figure 1 is labeled with a variable. The root is labeled by X as the most important variable. Each node is also associated with a preference ordering of the values of the variable. We can see the total pre-order of the outcomes below the cp-tree.

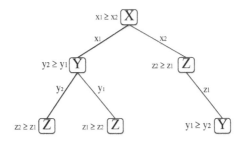

$x_1y_2z_2 \succeq x_1y_2z_1 \succeq x_1y_1z_1 \succeq x_1y_1z_2 \succeq x_2y_1z_2 \equiv x_2y_2z_2 \succeq x_2y_1z_1 \succeq x_2y_2z_1$

Fig. 1. A cp-tree σ, along with its associated ordering $\succeq_\sigma$ on outcomes, with $\gamma = 1$ (i.e., with at most one variable associated with a node)

3.2   CP-Tree-Based Dominance

Let Γ be a set of comparative preference statements. Let $\succeq_\Gamma$ be the associated preference relation of Γ on outcomes. Let α and β be two outcomes. The following definition is based on [11].

Definition 1. α *dominates* β *if and only if all possible cp-trees (every cp-tree represents a total pre-order) that satisfy* Γ, *prefer* α *over* β. *In other words,* $\alpha \succeq_\Gamma \beta$ *holds if every cp-tree* σ *that extends all preferences in* Γ *has* α *come before* β.

In this paper, we are using a dominance testing as stated in Definition 1 and which can check, in polynomial time, whether $\alpha \succeq_\Gamma \beta$.

4   The Case Study: A "Select and Get More Products" Conversational Recommender System

The product search, which needs a filter-based retrieval, can take place in tandem with preference elicitation. This motivates our present work which suggests a conversational recommender system that guides the user towards her target through a simple conversation during which the system can deduce different forms of comparative preferences from the user feedback.

4.1   The Advisor

The advisor helps the user identify a suitable product to purchase among the relatively large set of available products. The proposed system in this paper can be regarded as an instance of a kind of system in which the user is repeatedly shown products and criticises one or more of the shown products, until she finds a product that she is keen to have. During the interaction with the user, the advisor infers preference relations from the user's selections at each step. The selected potential products that are shown to the user are meant to best match these preference relations.

The user's preferences are stored in a kind of user model which is progressively updated by the system as the dialogue continues. Having that user model, the system starts to be able to determine whether certain products dominate others and to suggest to the user those that are not dominated. Dominance is computed as described in Section 3.2. The advisor asks little from the user who will only select one of the products shown to her. This will steer the user smoothly towards her target.

4.2   Products

We assume that the products are modeled with a collection of n multi-valued features $V = \{F_1, \ldots, F_n\}$. The features are intended to relate to a set of products that the user is interested in choosing between. For example, if the product is a hotel, one feature might be the size of a swimming pool in the hotel (e.g., the swimming pool can be small, medium or large).

4.3   Dialogue

Let Ω be the global set of products. Let $\mathcal{K}$ (i.e., 9) be the maximum number of products shown to the user in each step of the dialogue.

Initially, the user is shown the first $\mathcal{K}$ non-dominated products retrieved from Ω and selects a product $\mathcal{P}$. The interaction between the user and the recommender system proceeds as follows:

– The recommender system analyzes the current product $\mathcal{P}$ and induces some constraints on the user's preferences with particular regard to differences between $\mathcal{P}$ and the products the user might have selected. The nature of the induced preferences depends on the induction method used.
– From the second step of the dialogue, the system computes $\mathcal{K}$ non-dominated products among the remaining products. In fact, the system keeps retrieving products from those remaining in the database and not yet checked by the system then pruning the dominated ones until finding a set of $\mathcal{K}$ non-dominated products or there are no more products remaining in Ω. The system adds the product that the user selected in the previous step of the dialogue to the set of non-dominated products already computed. Since $\mathcal{K}$ is set to 9 in these experiments, the user is shown 10 products. If the number of products remaining in the database is less than $\mathcal{K}$ then we select the non-dominated among the remaining products and we show them to the user with the product chosen in the previous step.
– The user selects a product which becomes the new current product $\mathcal{P}$.

The sequence of steps stated above is repeated until the user is satisfied with $\mathcal{P}$ (by either choosing the same product a number of times (set to 3) or she gets the most preferred product with regards to her true preferences) or the set of remaining products in Ω is empty.

5   Induction of Constraints on Preferences within the System

Each time the user reviews a product, the recommender system, described in Section 4 induces some of the user's preferences. There are several induction methods that specify patterns of preference statements that can be induced from the user's selection. In this section, we discuss a number of these methods.

5.1   Constraint Language

Let V be a set of variables. Wilson [11] presented comparative preference theories which involve preference statements φ of the form $p > q||T$ where p and q are the respective assignments to sets of variables P and Q, and T is a set of variables ($P \subseteq V$, $Q \subseteq V$ and $T \subseteq V$). Such a statement expresses a preference for an assignment p over another assignment q with variables T held constant. We are specifically using Wilson's preference language to allow the system to handle the preferences induced from the user selection.

5.2   Preferences Deduction within the System

This section explains what the system induces when it observes the product that the user selects and the remaining products that were shown but not selected by the user. We adopt approaches which are based on comparative preference and partially inspired from *MLT* and *PMLT* approaches briefly described in Section 2.2. We have derived three patterns of preference statements that the system can induce when the user makes her selection. Let $V = \{F_1, \ldots, F_n\}$ be a set of n variables that represent features. Let C and R be two products. The combinations of feature values in the two products are denoted as follows. $C=\{f_1^C \ldots f_n^C\}$ and $R=\{f_1^R \ldots f_n^R\}$ where f_i^C and f_i^R are the two respective values that C and R have for the feature F_i (i=$\{1, \ldots, n\}$).

When the user chooses a product C and rejects another product R among a set of $\mathcal{K}$ non-dominated products that are shown to her, the system induces preference statements whose form depends on the following inference methods.

- **Basic:** A straightforward kind of preference to be induced is to express the preference of the features values combination included in C over the combination of values included in R. Thus, we model the following preference statement $f_1^C \ldots f_n^C \geq f_1^R \ldots f_n^R \; ||\emptyset$.

- **Lex-Basic:** Lexicographic preference models are regarded as simple and reasonably intuitive preference representations, and so lexicographic ordering can be well-understood by humans that use it to make preference decisions [12]. This is why we adopt a lexicographic model of the **Basic** format described above. This pattern allows the system to induce unconditional preference statements with regards to **Basic**.

Let $U \subseteq V$ be the set of features for which C and R have the same values (i.e., C and R agree on U). Let $S \subseteq V$ be the set of features for which C and R have different values. The combinations of features values of C and R can be represented by assignments us and us' respectively, with u is the assignment to U (i.e., $u \in \underline{U}$), and s and s' are the respective assignments to S of C and R which differ on each feature: $s(F) \neq s'(F)$ for all $F \in S$. Instead of stating the preference of us over us' as in **Basic** (i.e., $us > us'||\emptyset$), the idea is to induce a preference statement saying partial assignment s is preferred over partial assignment s' with all remaining features in V (i.e., features not in S) being equal. Then, the induced preference statement can be written as $s > s'||U$, the corresponding unconditional statement.

- **Every-Selected-Value:** We induce the preference statements $f_i^C \geq f_i^R||V \setminus \{F_i\}$, for every value f_i^C assigned to feature F_i in the chosen product C and for any value f_i^R assigned to feature F_i in a rejected product R.
  This states the superiority of every feature value taken by the chosen product C (i.e., f_i^C) over any other value (of the same feature) that appears at least in one rejected product R (i.e., f_i^R).

- **Cond-Selected-Value:** We induce the preference statements $\hat{f_i^C} \geq f_i^R||V \setminus \{F_i\}$. $\hat{f_i^C}$ represents every value assigned to feature F_i in the chosen product C without being present in any rejected product R. f_i^R represents every value assigned to feature F_i in a rejected product R.
  This states the superiority of every feature value $\hat{f_i^C}$ taken by the chosen product C, and which does not appear in any rejected product R, over any other possible value f_i^R (of the same feature) that appears at least in one rejected product R. The difference with *Every-Selected-Value* form is that the preferred feature value $\hat{f_i^C}$ needs to be present in the chosen product C and it should not appear in any rejected product R while *Every-Selected-Value* involves all feature values in C.

Let Φ_B, Φ_{LB}, Φ_{ESV} and Φ_{CSV} be the sets of preference statements induced by the system with **Basic**, **Lex-Basic**, **Every-Selected-Value** and **Cond-Selected-Value** respectively. We shall notice that any cp-tree that satisfies Φ_{LB} will also satisfy Φ_B as preference statements in Φ_{LB} imply statements in Φ_B. We also notice that the set of preference statements in Φ_{CSV} is included in Φ_{ESV}. Thus, all cp-trees that agree with statements in Φ_{ESV} also agree with statements in Φ_{CSV}. The set of cp-trees S_{Φ_B} that satisfy Φ_B is likely to be larger than the set of models $S_{\Phi_{LB}}$. This can explain a weaker inference for **Basic** with regards to **Lex-Basic**. The set of models that agree with statements in Φ_{ESV} will necessarily be smaller than the set of models satisfying Φ_{CSV} as $\Phi_{ESV} \subseteq \Phi_{CSV}$. Thus, this will probably make the dominance relation based on Φ_{ESV} stronger than the dominance relation based on Φ_{CSV}.

Example 2. Let $V = \{F_1, F_2, F_3\}$ be a set of features whose domains are as follows. $\underline{F_1} = \{f_1^1, f_1^2, f_1^3\}$, $\underline{F_2} = \{f_2^1, f_2^2\}$, $\underline{F_3} = \{f_3^1, f_3^2, f_3^3, f_3^4\}$. Let us assume that the user is initially shown the three following products: $f_1^1 f_2^1 f_3^2$, $f_1^2 f_2^1 f_3^3$ and $f_1^2 f_2^1 f_3^1$. Then, the user chooses $f_1^2 f_2^1 f_3^3$. The system will induce preferences in a format that depends on which among the methods introduced above is used. For **Basic**, the system induces the set of statements $\Phi_1 = f_1^2 f_2^1 f_3^3 \geq f_1^1 f_2^1 f_3^2 \,||\emptyset$, $f_1^2 f_2^1 f_3^3 \geq f_1^2 f_2^1 f_3^1 \,||\emptyset$. For **Lex-Basic**, the system induces the set of statements $\Phi_2 = \{f_1^2 f_3^3 \geq f_1^1 f_3^2 || \{F_2\}, f_3^3 \geq f_3^1 || \{F_1, F_2\}$. For **Every-Selected-Value**, the system induces the set of statements $\Phi_3 = \{f_1^2 \geq f_1^1 || \{F_2, F_3\}, f_3^3 \geq f_3^2 || \{F_1, F_2\}, f_3^3 \geq f_3^1 || \{F_1, F_2\}$. For **Cond-Selected-Value**, the system induces the set of statements $\Phi_4 = f_3^3 \geq f_3^2 || \{F_1, F_2\}, f_3^3 \geq f_3^1 || \{F_1, F_2\}$.

Let us suppose now that the system wants to show other undominated products to the user who is not yet satisfied with her selection. The system has three other products in the database and it will check whether they are dominated. Let $\alpha = f_1^2 f_2^2 f_3^3$ and $\beta = f_1^1 f_2^2 f_3^1$ be two of them.

For **Basic**, $\alpha \not\succ_{\Phi_1} \beta$ as we can identify cp-trees which satisfy preference statements in Φ_1 and prefer β over α. An example illustrating this is a cp-tree σ with root node associated with variable F_2 (and value ordering e.g., such that $f_2^1 \succ f_2^2$), and associated with value f_2^2 is a child node with variable F_1 and local ordering such that $f_1^1 \succ f_1^2$.

For **Lex-Basic**, $\alpha \succ_{\Phi_2} \beta$ as all cp-trees that satisfy preference statements in Φ_2 have nodes F_1 with a local ordering such that $f_1^2 \succeq_{F_1} f_1^1$, and nodes F_3 with a local ordering such that $f_3^3 \succeq_{F_3} f_3^1$ and $f_3^3 \succeq_{F_3} f_3^2$. All these cp-trees prefer α over β as any product that has f_1^2 and f_3^3 as values for F_1 and F_3 respectively will be preferred over any product that has f_1^1 and f_3^1 for F_1 and F_3 respectively. For **Every-Selected-Value**, with a similar justification as $\alpha \succ_{\Phi_2} \beta$, $\alpha \succ_{\Phi_3} \beta$.

For **Cond-Selected-Value**, $\alpha \not\succ_{\Phi_4} \beta$ as there exist cp-trees which satisfy Φ_4 but prefer β over α. The cp-tree σ described above is an illustrative example.

6   Experimentation and Results

This section describes experiments to assess the inference methods presented in Section 5.2. By these experiments we aim at showing the applicability and efficiency of these methods since this is the first time it is applied in the context of recommender system with multi-valued features. These experiments illustrate how a recommender system can exploit the expressiveness of comparative preferences and their relatively fast preference dominance engine.

6.1   Experiment Design

We report experiments with simulated users. The ultimate evaluation and validation of the preference dominance approaches for conversational recommender systems should be performed online. However, experiments with real users cannot be used to extensively test alternative newly-deployed interaction control algorithms. Indeed, a number of researchers pointed out the limitations of off-line experiments and their evaluation mechanisms, whereas others argued that off-line experiments are attractive because they allow comparing a wide range of approaches at an affordable cost [13].

We make assumptions concerning the behaviour of users. For a simulated user to make choices about which among the recommended products is the best one for her to have, she must be assigned a set of *true preferences*. The user's true preferences are represented either in the weights vector model by randomly generating weights vectors over product features or in the cp-tree model by randomly generating cp-trees over product features. The weights are related to product features; they are randomly selected real numbers in the interval $[0,1]$. The cp-trees representing the user's true preferences have the same structure as the cp-tree described in Section 3.1.

We have generated random products with n (e.g., 10) variables having three values each. Four recommenders use the four induction methods while other four recommenders consider four combinations of these approaches. For each pairing of a user with a recommender system, we ran 1,000 simulated dialogues. In total then, we are reporting results for 8 ways of inducing the user's preferences × 2 ways of representing the user's true preferences × 1,000 dialogues, which is 16,000 runs of the system. Experiments were run as a single thread on Dual Quad Core Xeon CPU, running Linux 2.6.25 x64, with overall 11.76 GB of RAM, and processor speed 2.66 GHz.

6.2   Pruning

The recommender system considered in this work will keep only those products which are not dominated regarding the user's preferences collected so far during the dialogue between the user and the system. In the experiments, we compare the pruning rates achieved by the eight recommender systems. As mentioned in Section 4.3, in each step of the dialogue, the goal of the system is to show a (predefined) number of non-dominated products to the user. Thus, the system selects $\mathcal{K}$ non-dominated products from a subset of $\mathcal{L}$ products among those remaining in the global set of products and not yet retrieved by the system. The pruning rate is defined as the proportion of $\mathcal{K}$ in $\mathcal{L}$.

6.3   Discussion of Results

The capability of pruning dominated combinations of features is an important success key of a conversational recommender system. But, the pruning capacity is not sufficient to make a conversational recommender system prevail over another. For instance, when the system prunes a large number of products, the user-system dialogue could be longer and the user might take more time to meet her target. Therefore, several factors might determine how good a conversational recommender system is. These factors include the pruning rate (Pruning), the running time (Time), the dialogue length (Steps) and the shortfall (Fall). The running time records, in milliseconds (ms), the time spent in checking the dominance of the products. The shortfall expresses how far is the preference of the product the user ended up with from the best product (in the database) the user could have obtained (in percentage). Table 1 and Table 2 give the results of the experiments with the true preferences of the simulated users represented

Table 1. Averages (over 1,000) of the pruning rates, the computation time, the number of steps per dialogue and the shortfalls for each induction method and each combination of induction methods (users as weights vectors)

Induction methods	Pruning (%)	Time (ms)	Steps	Fall (%)
Basic	3.03	0.017	6.06	0.068
Lex-Basic	27.73	0.03	5.59	0.064
Every-Selected-Value	55.09	0.027	4.94	0.052
Cond-Selected-Value	0.42	0.008	6.1	0.069
Basic + Every-Selected-Value	55.09	0.036	4.94	0.052
Basic + Cond-Selected-Value	5.17	0.018	6	0.069
Lex-Basic + Every-Selected-Value	55.09	0.049	4.94	0.052
Lex-Basic + Cond-Selected-Value	30.42	0.033	5.52	0.065

as weights vectors and cp-trees respectively. The measures shown are averaged over 1, 000 dialogues.

Table 1 shows that, the amount of pruning increases as the preference statements induced become less conservative (from *Basic* to *Every-Selected-Value*). For example, pruning goes from 3.03% *Basic* to 55.09% *Every-Selected-Value*. *Lex-Basic* has also significantly improved its pruning rate with regards to *Basic* (27.73% versus 3.03%) after unconditioning the preference statements that were conditional in *Basic*. The exception to this is *Cond-Selected-Value* case (0.42%) which is probably due to the fact that the system induces much less preference information about the user. In fact, experiments have shown that one feature value that is seen in the chosen product is likely to be in at least one of the rejected products which makes the system refrain from inducing a preference statement that involves that feature value when *Cond-Selected-Value* is adopted. Thus, *Cond-Selected-Value* implies preference statements that would be satisfied by a quite large set of models which makes the inference weaker. *Cond-Selected-Value* has the smallest pruning rate with no positive effect on the shortfall. *Every-Selected-Value* distinguishes itself by having the best pruning capability (55.09%) and the shortest dialogue (4.94) that did not prevent it from having the best shortfall (0.052%).

When combined with a more conservative method as *Basic* or *Lex-Basic*, *Every-Selected-Value* takes longer period of time (0.036ms and 0.049ms versus 0.027ms) even though the pruning and the dialogue length are still the same. We can also see the running time is increasing with the pruning rate. In fact, this conversational recommender system keeps retrieving products from the database and trying to gather a predefined number of undominated products. A high pruning rate usually indicates that the number of products retrieved is quite large. This involves more pairwise comparisons between products and so takes more time.

Table 2 gives the results of the experiments with the true preferences of the simulated users represented as cp-trees. The measures shown are averaged over 1,000 dialogues. A look at Table 2 shows that we can infer similar conclusions to the deductions made from results in Table 1. We can see that all the pruning rates are higher than the pruning percentages in Table 1 as well as the shortfall percentages. These differences can be explained by the nature of the user's true preferences and the way the user satisfaction is computed for both preference models (i.e., weights vectors and cp-trees). The shortfalls are all very small. It may be that *Basic* has the smallest shortfall (i.e., 0.187) in the second setting because it is the most cautious, i.e., it makes the weakest assumptions on the preferences.

Table 2. Averages (over 1,000) of the pruning rates, the computation time, the number of steps per dialogue and the shortfalls for each induction method and each combination of induction methods (users as cp-trees)

Induction methods	Pruning (%)	Time (ms)	Steps	Fall (%)
Basic	18.68	0.049	9.558	0.187
Lex-Basic	64.83	0.045	5.482	0.642
Every-Selected-Value	74.74	0.039	4.383	0.646
Cond-Selected-Value	1.28	0.015	11.211	0.640
Basic + Every-Selected-Value	74.74	0.048	4.383	0.646
Basic + Cond-Selected-Value	21.18	0.052	9.52	0.642
Lex-Basic + Every-Selected-Value	74.74	0.064	4.383	0.646
Lex-Basic + Cond-Selected-Value	65.19	0.046	5.435	0.641

7   Conclusions and Perspectives

Recommender systems are gaining momentum in the e-commerce applications market to face the "information overload" problem. This progressively reveals an increasing need to enable those recommender systems with suitable preference formalisms and dominance engines that can efficiently handle and reason with the user preferences while conversing with her (see, e.g., [10]). We specify new preference induction methods based on a recently developed preference language (i.e., comparative preference theories). We implemented these methods for a conversational recommender system to handle the user's preferences when recommending multi-valued feature products. We showed that these methods allow the system to capture preference nuances and various forms of preferences without giving up the attractive computational properties of the preference dominance relation.

As a continuation of this work, we will consider similar preference induction methods to be integrated with different critiquing-based recommender systems. Extending the conclusion to a more general scope, in the future, we intend to look for more elaborate and intuitive preference elicitation formalisms that we can prove to be efficient in practice with conversational recommenders. These formalisms will adapt with the different dialogue strategies the conversational recommenders go through. They can be part of an intelligent query selection strategy to drive the elicitation process in the recommenders.

References

1. Zaslow, J.: If tivo thinks you are gay, here's how to set it straight. The Wall Street Journal (2002)
2. Fishburn, P.C.: Lexicographic orders, utilities, and decision rules: A survey. Management Science 20(11), 1442–1471 (1974)
3. Öztürk, M., Tsoukiàs, A., Vincke, P.: Preference modelling. In: Bosi, G., Brafman, R.I., Chomicki, J., Kießling, W. (eds.) Preferences. Dagstuhl Seminar Proceedings, vol. 04271, IBFI, Schloss Dagstuhl (2004)
4. Stefanidis, K., Koutrika, G., Pitoura, E.: A survey on representation, composition and application of preferences in database systems. ACM Transactions on Database Systems 36(3), 19 (2011)
5. Bridge, D.G., Göker, M.H., McGinty, L., Smyth, B.: Case-based recommender systems. The Knowledge Engineering Review 20(3), 315–320 (2005)
6. Chen, L., Pu, P.: Survey of preference elicitation methods. In: Technical Report IC/200467 (2004)
7. McGinty, L., Reilly, J.: On the evolution of critiquing recommenders. In: Ricci, F., Rokach, L., Shapira, B., Kantor, P.B. (eds.) Recommender Systems Handbook, pp. 419–453. Springer (2011)
8. McGinty, L., Smyth, B.: Comparison-based recommendation. In: Craw, S., Preece, A.D. (eds.) ECCBR 2002. LNCS (LNAI), vol. 2416, pp. 575–589. Springer, Heidelberg (2002)
9. Bridge, D.G., Ricci, F.: Supporting product selection with query editing recommendations. In: Konstan, J.A., Riedl, J., Smyth, B. (eds.) RecSys, pp. 65–72. ACM (2007)
10. Trabelsi, W., Wilson, N., Bridge, D.G., Ricci, F.: Preference dominance reasoning for conversational recommender systems: a comparison between a comparative preferences and a sum of weights approach. International Journal on Artificial Intelligence Tools 20(4), 591–616 (2011)
11. Wilson, N.: Efficient inference for expressive comparative preference languages. In: Boutilier, C. (ed.) IJCAI, pp. 961–966 (2009)
12. Yaman, F., Walsh, T.J., Littman, M.L., desJardins, M.: Democratic approximation of lexicographic preference models. Artificial Intelligence 175(7-8), 1290–1307 (2011)
13. Shani, G., Gunawardana, A.: Evaluating recommendation systems. In: Recommender Systems Handbook, pp. 257–297 (2011)

Budgeted Personalized Incentive Approaches for Smoothing Congestion in Resource Networks

Pradeep Varakantham[1], Na Fu[1], William Yeoh[2],
Shih-Fen Cheng[1], and Hoong Chuin Lau[1]

[1] School of Information Systems, Singapore Management University, Singapore
{pradeepv,nafu,sfcheng,hclau}@smu.edu.sg
[2] New Mexico State University, United States
wyeoh@cs.nmsu.edu

Abstract. Congestion occurs when there is competition for resources by selfish agents. In this paper, we are concerned with smoothing out congestion in a network of resources by using personalized well-timed incentives that are subject to budget constraints. To that end, we provide: *(i)* a mathematical formulation that computes equilibrium for the resource sharing congestion game with incentives and budget constraints; *(ii)* an integrated approach that scales to larger problems by exploiting the factored network structure and approximating the attained equilibrium; *(iii)* an iterative best response algorithm for solving the unconstrained version (no budget) of the resource sharing congestion game; and *(iv)* theoretical and empirical results (on an illustrative theme park problem) that demonstrate the usefulness of our approach.

1 Introduction

Competition for resources by autonomous agents typically leads to congestion if the agents access these resources in an uncoordinated fashion [1]. It is hence common for a network to experience congestion even when the average demand for a resource is much less than its capacity. Researchers have generally taken three approaches to address this issue. The first approach is to use the theory of *mechanism design*, where a central authority designs rules of agent interactions [2–4] by taking agent incentives into account . By designing appropriate rules, the central authority can obtain desirable goals such as maximizing social welfare. This assumes that the central authority defines and controls the rules of interaction. However, in this paper, we consider scenarios where the basic settings (rules) of the environment cannot be modified (like preferences of people going to a theme park or theme park configuration or communication protocols in a computer network).

Secondly, researchers have investigated the use of *penalties or incentives* on certain resources to discourage or encourage interactions that will lead to desirable goals. A central authority can alter the demand for certain resources by tweaking the amount of penalty or incentive for those resources. Much of the initial work in this area, especially in transportation applications [5, 6], assumes

P. Perny, M. Pirlot, and A. Tsoukiàs (Eds.): ADT 2013, LNAI 8176, pp. 375–386, 2013.
© Springer-Verlag Berlin Heidelberg 2013

that every agent using the same resource will get the same penalty or incentive. A good example is the use of toll gates on roads. [7] and [8] provide further examples of settings where using external penalties or incentives affect the utilities involved. More recently, researchers have relaxed this assumption and implemented penalties or incentives that are probabilistic in nature [9]. For example, a public radio listener will be entered in a draw for a free iPad if he/she donates to the radio station.

Finally, Monderer and Tennenholtz have studied the problem of minimizing incentive needed to sufficiently incentivize agents to take desirable strategies (that are inputs to the problem) [10]. While there are similarities, we differ from this work in multiple ways: (1) Our focus is on finding an equilibrium strategy that is closest to the set of desired strategies given a budget; (2) we assume that the total amount of incentives that can be used must be within a given budget; and (3) our desirable strategies are specified at an aggregate level with respect to a set of agents. For instance, "no more than 10 agents can consume resource 3" as opposed to "agent 2 should take strategy 3". These differences preclude the applicability of their approach on problems with budget constraints and large number of agents.

These differences are motivated by a crowd congestion control problem in an actual theme park. Through interviews with park operators, we learnt that they can provide well-timed incentives to specific patrons through mobile devices to change their behavior and thereby ease congestion (long queues at certain attractions). Naturally, the (monetary) incentives must be within a given budget. Lastly, the park operators are interested in specifying aggregated desirable levels of congestion instead of individualized desirable strategies.

More precisely, we are interested in the problem on how best to distribute incentives among different agents at different time points so that certain resource congestion thresholds are satisfied at equilibrium and that the incentives distributed are within a given budget. We make the following contributions:

(1) We introduce a non-linear mathematical programming formulation and show how it can be linearized into a mixed-integer linear program (MILP) to compute the equilibrium for a networked congestion game with incentives and budget constraints.
(2) We exploit the factored network structure to drastically reduce the complexity of enumerating the space of agent strategies and provide an enhancement to compute approximation equilibria to scale up the MILP.
(3) We provide a scalable iterative best response algorithm to solve a version of the game without budgets while minimizing the overall incentive required.
(4) Lastly, we provide theoretical and empirical results showing that congestion is reduced at equilibrium on an illustrative theme park problem.

2   Model: NRSG

We provide the *Network Resource-Sharing Game* (NRSG), which builds on the Resource Sharing (RS) model [11] and network cost-sharing games [12]. A NRSG is similar to a network cost-sharing game except for positive rewards in NRSGs compared to positive penalties in cost-sharing games. An NRSG is the tuple:

$$\langle N, \mathcal{V}, \mathcal{E}, \{U_v^i\}_{i \in N, v \in \mathcal{V}}, \{s^i\}_{i \in N}, H \rangle$$

$N = \{1, 2, \ldots, n\}$ represents the set of agents.

$\mathcal{V}$ represents the resources and also the vertices in a graph that are connected by the edges in $\mathcal{E}$. This graph constrains certain orders of consuming resources or connections between resources.

U_v^i represents the utility obtained by agent i when it consumes one unit of resource v. For a joint strategy $\mathbf{a} = \langle a^1, a^2, \cdots, a^i, \cdots a^n \rangle$, where a^i is the action of agent i, the utility obtained by agent i is given by

$$u^i(a^1, \cdots, a^i, \cdots, a^n) = \frac{U_{a^i}^i}{\sigma_{\mathbf{a}}(a^i)} \tag{1}$$

where $\sigma_{(a)}(a^i) = \sum_{k \leq n} I(a^k = a^i)$, with $I(a^k = a^i) = 1$ if $a^k = a^i$ and 0 otherwise. While we focus on this definition of utility, our approaches can be trivially modified to work with any non-increasing function over number of agents consuming a resource.

s^i represents the starting vertex for agent i.

H represents the time horizon of the problem.

The goal in an NRSG is to find Nash equilibrium strategies for all individual agents, that is, no agent has an incentive to deviate from its strategy. It should be noted that this repeated game cannot be represented by a single-shot decision-making problem [12] because a resource selection path (of length H) cannot be considered as an independent resource. Also, it should be noted that this is not a single stage game repeated multiple times due to the following reason: (a) Utility can change over time (e.g., preferences for rollercoasters before and after lunch are different). (b) There exists a network structure on how resources can be utilised. (c) There can be domain-specific constraints (e.g., each resource/attraction can only be visited once or should visit at least 3 of my 5 preferred attractions). Note that these constraints are all linear.

A *pure strategy* for an agent i is the sequence of resources selected at each time step, and the set of all pure strategies is given by $\Pi^i = \{\pi^i \mid \pi^i = \langle a_1^i, a_2^i, \cdots, a_H^i \rangle, \forall t : a_t^i \in \mathcal{V}\}$. We do not have edges as part of the strategy, because, given a source and destination vertex, the edge is uniquely determined. A *mixed strategy* can be defined as a probability distribution over all possible pure strategies $\Delta(\Pi^i)$. To provide better understanding of the concepts, we will use the following toy example throughout the paper.

Example 1. *We consider a theme park with four attractions (resources) $\mathcal{A} = \{A1, A2, A3, A4\}$ that is being visited by eight patrons (agents) $\mathcal{P} = \{P1, \cdots, P8\}$. For ease of explanation, we assume that the service rate of each attraction d_i is 1 for all attractions. Let the utility for all patrons in getting serviced at an attraction is the same, which is as follows: $\mathcal{U} = \{2, 3, 5, 7\}$. The horizon H for decision making is 1 and the ideal minimum queue length γ_i^* desired by the theme park operator is 2 for all attractions i.*

3   Incentivized Budget Constrained Equilibrium

In this section, we represent the problem of finding a Nash equilibrium in an NRSG with incentives and budget constraints as an optimization problem. Traditionally, iterative best response mechanisms such as fictitious play [13] have been used to compute equilibrium solutions in congestion game models. The presence of budget constraints and desired congestion levels preclude the application of such methods.

Our approach provides personalized incentives constrained by a budget so as to achieve certain properties of resource congestion like ensuring that all queue lengths at attractions are no less than a minimum queue length or no greater than a maximum queue length. Examples of personalized incentives are freebies at an attraction if it is visited at a certain time. The key assumption in our approach is that such incentives increase the utility for individual agents.

We use the following notation to describe the optimization problem, where lower case letters such as x represent variables, bold letters such as $\mathbf{x}$ represent vectors, bold and upper case letters such as $\mathbf{X}$ represent sets of vectors:

U_j^i is the utility at resource j for agent i. U_j is the utility at resource j (if it is the same for all agents i).

$x_{j,t}^i$ is a binary variable indicating whether agent i has selected $(= 1)$ resource j at time t.

$\mathbf{x}^i$ is the strategy of agent i:

$$\begin{pmatrix} x_{1,1}^i & x_{1,2}^i & \cdots & x_{1,H}^i \\ x_{2,1}^i & x_{2,2}^i & \cdots & x_{2,H}^i \\ \cdots & \cdots & \cdots & \cdots \\ x_{|\mathcal{V}|,1}^i & x_{|\mathcal{V}|,2}^i & \cdots & x_{|\mathcal{V}|,H}^i \end{pmatrix}$$

where $|\mathcal{V}|$ and H are number of resources and horizon, respectively.

$\mathbf{x}$ is the strategy profile of all players over all resources and the entire horizon:
$\mathbf{x} = (\mathbf{x}^1, \mathbf{x}^2, \dots, \mathbf{x}^n)$

$\mathbf{X}^i$ is the set of all possible strategies for agent i:
$\mathbf{X}^i = \{\mathbf{x}^i \mid \sum_j x_{j,t}^i \leq 1, x_{j,t}^i \in \{0,1\}, \forall t \leq H\}$

$\mathbf{\Delta}$ is the matrix of incentives of all agents, $\mathbf{\Delta} = (\mathbf{\Delta}^1, \dots \mathbf{\Delta}^n)$,

$$\mathbf{\Delta}^i = \begin{pmatrix} \delta_{1,1}^i & \delta_{1,2}^i & \cdots & \delta_{1,H}^i \\ \delta_{2,1}^i & \delta_{2,2}^i & \cdots & \delta_{2,H}^i \\ \cdots & \cdots & \cdots & \cdots \\ \delta_{|\mathcal{V}|,1}^i & \delta_{|\mathcal{V}|,2}^i & \cdots & \delta_{|\mathcal{V}|,H}^i \end{pmatrix}$$

where $\delta_{j,t}^i$ is a decision variable representing the incentive agent i obtained at resource j time t.

B is a constant representing the total amount of budget available for incentives.

m is index of a policy of an agent in the set $\mathbf{X}^i$.

$$\min_{\mathbf{X}, \mathbf{\Delta}} \Gamma \qquad \textbf{such that}$$

$$u_{j,t}^i = \frac{x_{j,t}^i \cdot U_j}{\max\left\{\sum_k x_{j,t}^k, 1\right\}} + x_{j,t}^i \cdot \delta_{j,t}^i \qquad \forall i, j, t \qquad (2)$$

$$u_{j,t}^{m,i} = \frac{x_{j,t}^{m,i} \cdot U_j}{\max\left\{\sum_{k \neq i} x_{j,t}^k + x_{j,t}^{m,i}, 1\right\}} + x_{j,t}^{m,i} \cdot \delta_{j,t}^i \qquad \forall m, i, j, t \qquad (3)$$

$$\sum_{j,t} u_{j,t}^i \geq \sum_{j,t} u_{j,t}^{m,i}, \qquad \forall m, i \qquad (4)$$

$$\sum_{i,j,t} f_j(\delta_{j,t}^i) \leq B \qquad (5)$$

$$\Gamma \geq \gamma_j^* - \sum_i x_{j,t}^i \qquad \forall j, t \qquad (6)$$

$$\sum_j x_{j,t}^i \leq 1 \qquad \forall i, t \qquad (7)$$

$$x_{j,t}^i \leq x_{k,t-1}^i \qquad \forall (k,j) \in \mathcal{E} \qquad (8)$$

$$x_{j,t}^i \in \{0, 1\} \qquad \forall i, j, t \qquad (9)$$

Fig. 1. Non-Linear Optimization Problem

$x_{j,t}^{m,i}$ is a value representing if agent i chooses resource j and at time t under agent i's m^{th} policy.

γ_j^* is a constant representing the preferred number of agents selecting resource j at any time step.

Figure 1 shows the optimization problem formulated as a non-linear mixed-integer program. For ease of explanation, we assume that all agents consuming a resource get the same utility U_j. However, the optimization problem and the proceeding linearization can be trivially adapted to have a different utility for each agent U_j^i. The key aspects of the optimization problem are:

- **No Incentive to Deviate:** Constraint 4 ensures that when all agents follow their equilibrium strategies, the overall utility (including the allocated incentive) $u_{j,t}^i$ of agent i obtained by following its equilibrium strategy is no less than the utility $u_{j,t}^{m,i}$ obtained by any other strategy m for all resources j and time steps t.

- **Budgeted Incentives:** Constraint 5 ensures that the total amount of all incentives is bounded by the budget B. One key assumption here is that the function f_j is a linear function and $\delta^{max} = \sum_j \delta_j^{max}$ is a constant computed from the following expression: $\sum_j f_j(\delta_j^{max}) = B$.

- **Desired Resource Congestion Properties:** These properties are inputs to the problem and can be constraints on the minimum or maximum number of agents consuming a resources. Constraint 6 represents the constraint for

$$u_{j,t}^i = w_{j,t}^i + \delta_{j,t}^i \qquad\qquad \forall i,j,t \qquad (10)$$

$$0 \le w_{j,t}^i \le x_{j,t}^i \cdot U_j \qquad\qquad \forall i,j,t \qquad (11)$$

$$0 \le \delta_{j,t}^i \le x_{j,t}^i \cdot \delta^{max} \qquad\qquad \forall i,j,t \qquad (12)$$

$$w_{j,t}^i - w_{j,t}^k \le (2 - x_{j,t}^i - x_{j,t}^k) \cdot U_j \qquad\qquad \forall i,j,t,k \qquad (13)$$

$$w_{j,t}^k - w_{j,t}^i \le (2 - x_{j,t}^i - x_{j,t}^k) \cdot U_j \qquad\qquad \forall i,j,t,k \qquad (14)$$

$$\sum_k w_{j,t}^k = U_j \cdot \alpha_{j,t} \qquad\qquad \forall j,t \qquad (15)$$

$$\frac{\sum_k x_{j,t}^k}{N} \le \alpha_{j,t} \le \sum_k x_{j,t}^k, \qquad\qquad \forall j,t \qquad (16)$$

$$\alpha_{j,t} \in \{0,1\} \qquad\qquad \forall j,t \qquad (17)$$

Fig. 2. Linearization Constraints for Constraint 2

the minimum number of agents γ_j^* at any resource j, where Γ represents the maximum deviation from the desired consumption.

- **Deviation Minimization:** The maximum deviation from the desired congestion properties Γ is minimized in the objective.
- **Network Structure:** Consraint 8 enforces the network structure.

While this optimization problem can model incentives accurately, there are two key issues: (1) Non-linear constraints in constraints 2 and 3 prevent scalability to larger problems, and (2) enforcing the equilibrium for each agent requires enumerating over all possible pure strategies possible for each agent, which can be exponential in the horizon and the number of resources. To address these issues, we propose three methods that increase the scalability considerably.

3.1 Linearizing the Non-linear Constraints

As indicated earlier, the utility function can be any non-increasing piecewise constant or piecewise linear function over number of agents for us to employ similar linearization tricks on the utility function that will be explained in this section. Figure 2 shows the equivalent linear constraints to the non-linear constraints in constraint 2. The same techniques can be applied to linearize constraint 3. Using these linearized constraints, the optimization problem in Figure 1 can be represented as a mixed-integer linear program (MILP). Furthermore, for each agent, we introduce new variables $w_{j,t}^i$ and $\delta_{j,t}^i$ to represent the unincentivized utility and incentive, respectively. Thus they sum up to the overall utility $u_{j,t}^i$ (constraint 10). The intuitions for the linearization constraints are as follows:

- Constraints 11, 12: If an agent i is not consuming resource j at time t ($x_{j,t}^i = 0$), then the unincentivized utility $w_{j,t}^i$ and incentive $\delta_{j,t}^i$ are zero.
- Constraints 13, 14: If an agent i is consuming resource j at time step t ($x_{j,t}^i = 1$), then its unincentivized utility $w_{j,t}^i$ is equal to the unincentivized utility

$w^k_{j,t}$ of any other agent k that consumes the same resource at the same time $(x^k_{j,t} = 1)$.

- Constraints 15-17 account for the "max" in the denominator of constraint 2.

Example 2. *At equilibrium, the number of agents at attractions A1,A2,A3 and A4 is 1, 1, 2 and 4, respectively for Example 1. That is to say, attraction A4 is more crowded than any other attractions. We can use the optimization problem above to help reduce the congestion at A4. Suppose the theme park operator provided the minimum queue length γ_a, which is 2 for all attractions a, and the budget B, which is 5. Then, the resulting equilibrium (along with the incentives in terms of utility that is same for all agents selecting the same attraction) is*

$$A1 = \{P2, P3\}, \delta^{P2}_{A1,1} = 1.33; A3 = \{P6, P7\}, \delta^{P6}_{A3,1} = 0$$
$$A2 = \{P4, P5\}, \delta^{P4}_{A2,1} = 0.83; A4 = \{P1, P8\}, \delta^{P1}_{A4,1} = 0$$

The number of agents at each attraction now is 2, which satisfies the minimum queue length, and so is the criterion for equilibrium.

3.2   Exploiting Factored Structure

We exploit the factored structure of the NRSG graph to solve the MILP faster. This efficiency comes about due to the reduction in the number of elements in the set $\mathbf{X}^i$ for every agent i and thus the number of equilibrium constraints (Constraint 4). The basic definition for X^i is given by:

$$\mathbf{X}^i = \{\mathbf{x}^i \mid \sum_j x^i_{j,t} \le 1, x^i_{j,t} \in \{0,1\}, \forall t \le H\}$$

We can update the expression to exploit the graph structure:

$$\mathbf{X}^i = \{\mathbf{x}^i \mid \sum_j x^i_{j,t} \le 1, x^i_{j,t} \le \sum_{k \mid (k,j) \in \mathcal{E}} x^i_{k,t}, x^i_{j,t} \in \{0,1\}, \forall t \le H\}$$

Furthermore, if the graph is fully connected, that is, agents can consume any resource at any time step (a reasonable assumption for theme parks, where patrons can go to any attraction at any time step), then the equilibrium constraints on constraint 4 can be replaced with $\sum_j u^i_{j,t} \ge \sum_j u^{m,i}_{j,t}$ for all m, i and t. The key difference is that the new constraints sums over resources j only as opposed to over resources j and time steps t. This difference yields a reduction in number of equilibrium constraints from $|\mathcal{V}|^H$ to $|\mathcal{V}| \cdot H$.

3.3   Finding ε-Nash Equilibrium Solutions

The MILP representation provides the flexibility to compute an approximate Nash Equilibrium. If we modify the equilibrium constraints on Line 4 to $\epsilon + \sum_{j,t} u^i_{j,t} \ge \sum_{j,t} u^{m,i}_{j,t}$ for all m and i. Then, the resulting Nash equilibrium is an ε-Nash equilibrium, where each agent has an incentive of at most ϵ to deviate from the equilibrium strategy.

Example 3. *An ϵ-equilibrium strategy with $\epsilon = 0.1$ to the problem in Example 2 is given by*

$$A1 = \{P2\}, u^{P2}_{A1,1} = 2; A3 = \{P6, P7, P8\}, u^{P6}_{A3,1} = 1.66$$

$$A2 = \{P3\}, u^{P3}_{A2,1} = 3; A4 = \{P1, P4, P5\}, u^{P1}_{A4,1} = 2.33$$

This solution is not a true equilibrium because patron P8 can switch to attraction A4 to gain an additional 0.09 units of utility but it is an ϵ-Nash Equilibrium because the gain by changing strategy for each agent is less than $\epsilon = 0.1$.

4   Incentivized Unconstrained Equilibrium

In this section, we provide a technique for solving the problem where: (1) there is no constraint on the budget; (2) there are hard constraints on the desired consumption of resources; and (3) the goal is to minimize the total amount of incentive required to achieve the equilibrium. This problem is similar to the problem solved by the k-implementation approach [10]. However, the main differences are that we assume that agents can be individually incentivized and desired strategies are specified at an aggregate level in our work (e.g., "no more than 200 agents can consume resource 3") as opposed to specific strategies in [10] (e.g., "agent 2 should take strategy 5").

The optimization problem mentioned in Figure 1 with the linearized constraints can be easily modified to solve the unconstrained problem. Here, we provide another approach that is more scalable and based on the more typical iterative best response mechanism that also allows for mixed strategies in the equilibrium. Figure 3 shows the best response linear program for each agent i. We use the following additional variables:

$p^i_{j,t}$ is the probability of agent i choosing resource j at time t.

$\mathbf{p}^i$ is the mixed strategy of agent i similar to how $\mathbf{x}^i$ is the pure strategy of agent i in the previous MILP.

u^{*i} is the utility of best response strategy of agent i given strategies of other agents.

u^i is the utility of a strategy of agent i given the strategies of other agents.

In this approach, at each iteration and for each agent i, we fix the policies of all other agents and compute the best response strategy $\mathbf{p}^i$ that satisfies the constraint on desired resource consumption (constraint 21) and the required incentive δ^i to incentivize agent i to take that strategy. The incentives are computed in constraint 18 as the difference in utility for the best strategy u^{*i} (without the constraint on desired resource consumption) and the current utility (with the constraint on desired resource consumption) u^i. We continue this process until convergence. We do not yet have a proof for guaranteed convergence. However, if the iterative best response process converges, then we obtain an equilibrium strategy.

$$\min_{\mathbf{P^i}} \quad \delta^i \qquad \textbf{such that}$$

$$\delta^i = u^{*i} - u^i \tag{18}$$

$$u^{*i} = \max_{\mathbf{x}^i} \sum_{j,t} \frac{x^i_{j,t} \cdot U_j}{\sum_{k \neq i} p^k_{j,t} + x^i_{j,t}} \tag{19}$$

$$u^i = \sum_{j,t} \frac{p^i_{j,t} \cdot U_j}{\sum_{k \neq i} p^k_{j,t} + 1} \tag{20}$$

$$\sum_k p^k_{j,t} \leq \gamma^*_j \qquad \forall j, t \tag{21}$$

$$\sum_j p^i_{j,t} \leq 1 \qquad \forall t \tag{22}$$

$$0 \leq p^i_{j,t} \leq 1 \qquad \forall j, t \tag{23}$$

Fig. 3. Best Response Linear Program for Agent i

5 Theoretical Results

We now show that the welfare of any equilibrium solution is at least one half of the optimal social welfare in an NRSG in two steps: (1) We show that the social utility function in NRSGs is sub-modular, and (2) we show that the NRSG game is a utility system [14] and, hence, the Price of Anarchy, PoA (Ratio of social welfare for the worst equilibrium solution to the optimal social welfare) is at least $\frac{1}{2}$. Note that an NRSG with incentives is an NRSG and hence the bounds hold even when we are providing incentives.

Proposition 1. *Social utility for a joint strategy $\mathbf{x}$ in NRSG is sub-modular.*

Proof Sketch: Let $\mathcal{Q}$ be the set of all agents and $\mathcal{F} : 2^{\mathcal{Q}} \to \mathbb{R}$ be the social utility function. The social utility for a joint policy $\mathbf{x}$ given a set $\mathcal{Q}$ of agents is:

$$\mathcal{F}(\mathcal{Q}) = \sum_{i \in \mathcal{Q}} \frac{\sum_{j,t} x^t_{i,j} \cdot U_j}{\sum_{k \in \mathcal{Q}, j,t} x^t_{k,j}}$$

For $A \subseteq B$, we show that $\mathcal{F}(A \cup \{p\}) - \mathcal{F}(A) \geq \mathcal{F}(B \cup \{p\}) - \mathcal{F}(B)$. ∎

Definition 1. *A utility system represents a game where:*

- *Social and private utilities are in the same standard unit;*
- *Social utility function is sub-modular; and*
- *Private utility of an agent $\geq$ change in social utility if the agent declined to participate in the game.*

Proposition 2. *PoA for NRSGs is at least $\frac{1}{2}$.*

Proof Sketch. We show that NRSG is a utility system and hence from Vetta et al. [14], PoA for utility systems is at least $\frac{1}{2}$. ∎

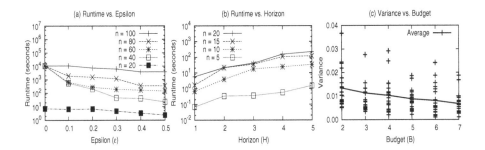

Fig. 4. Results for Incentivized Budget Constrained Problems

6   Experimental Results

We now describe our experimental results for the problem with incentives on the theme park problem described in Example 1. We performed two sets of experiments: one on problems with incentive budgets and one on problems without. For problems with incentive budgets, we only have the linearized optimization problem of Figure 1 (referred to as BC-MILP). For problems without incentive budgets and desired congestion levels, we have the iterative best response (referred to as IBR) and we compare it to a modified BC-MILP-Mod .

There are a number of different parameters that we experimented with, namely the number of agents n, the horizon H, the number of resources $|\mathcal{V}|$, budget B, desired maximum or minimum consumption of any attraction γ^* and finally the approximation parameter ϵ. If not explicitly stated, the default values for some of the parameters are as follows: $H = 1$, $|\mathcal{V}| = 4$, $B = 2$, $\gamma^* = \frac{n}{|\mathcal{V}|} \pm p \cdot n$ (depending on whether we have constraints on maximum or minimum consumption) with a default value of 10% for p and $\epsilon = 0$. Due to space constraints, we only show representative results. We conducted our experiments on a machine with a 2.40GHz CPU and 6GB of RAM.

6.1   Incentivized Budget Constrained Problems

In this set of problems, we demonstrate some of the key results with the BC-MILP. Figure 4(a) shows the runtimes, where we vary n and ϵ. We only show the runtimes for one combination of budget and γ^* parameter as the trends here are similar for other parameters. We make two observations:

(1) As the number of agents increases, the runtime increases as expected. With the increase in the number of agents, the number of variables and constraints in the MILP increases and hence the increase in runtime. However, we are able to solve problems with up to 100 agents with the BC-MILP approach. By exploiting homogeneity in agents (future work), we hope to increase this significantly.

(2) As ϵ increases, even by a small value, the runtime decreases significantly.[1] With the increase in ϵ, the problem becomes simpler as the MILP can

---

[1] Note that ϵ is an absolute error on utility and not a percentage error.

Table 1. BC-MILP-Mod vs. IBR

No. of agents (γ^*)	BC-MILP-Mod			IBR		
	runtime (sec)	incentive	social welfare	runtime (sec)	incentive	social welfare
10 (3)	8.42	2.25	19.25	0.41	6.80	19.14
12 (3)	44.01	5.75	22.75	0.42	9.79	22.54
14 (4)	123.64	4.00	21.00	0.70	7.55	20.90
16 (4)	1105.45	6.80	23.80	0.48	10.07	23.67
18 (5)	4001.52	5.16	22.16	0.67	8.10	22.10
20 (5)	8312.05	7.50	24.50	0.54	10.26	24.42

return solutions with larger deviations from the Nash equilibrium. Thus, these results show the tradeoff between computation time and solution quality in terms of distance from the optimal solution.

Figure 4(b) shows the runtimes, where we vary n and H with $\epsilon = 0.3$. As expected, the runtime increases with increasing horizon. We are able to solve problems with 20 agents and horizon 5 in less than 4 minutes.

Figure 4(c) shows the variance in resource consumption of each agent (as a percentage of n), where we vary n from 10 to 30, and B from 2 to 7. The nice observation from this result is that the average variance decreases as the budget increases. In other words, *we have a better load balance, even at equilibrium strategies, when the budget B increases.*

6.2   Incentivized Unconstrained Problems

We first show the performance comparison of the IBR algorithm with the BC-MILP algorithm modified to suit the unconstrained budget and incentive minimization setting (referred as BC-MILP-Mod). The BC-MILP-Mod thus finds a solution with the least required incentive. Table 1 shows the results. IBR converged and that implies that an equilibrium solution is found in both cases. The results show that IBR is at least one order of magnitude faster than BC-MILP-Mod but finds solutions that requires much higher incentive and slightly lower social welfare, thus highlighting the tradeoff between the two approaches.

To demonstrate the scalability of the IBR algorithm, we increased the number of agents up to 500 and we were still able to solve the problem within 20 seconds. We also computed runtimes for the IBR approach while varying resources, however, there was no significant change in runtime when the number of resources was less than or equal to 10. Finally, we computed the the overall incentive required as a mapping of the ideal resource consumption (γ^*) parameter p. We varied p from 5%-15% and computed the overall incentive required with n varying between 100 to 300. As expected, the incentive decreased as the constraint on resource consumption was relaxed.

7   Conclusion

Congestion is common in resource networks that exist in domains as varied as transportation, computer networks and theme parks. In this paper, we aim to smooth out the congestion by using well-timed incentives that are constrained by a budget and are personalized to resource consumers. To that end, we provided an efficient mixed-integer linear formulation that can exploit network structure and is amenable to bounded approximation schemes. We also provide a scalable alternative to solve the incentives problem when there is no constraint on the budget and the goal is to find an equilibrium strategy with the least incentive. Our experimental results demonstrate the scalability of our approaches and on an illustrative problem, we also show that there is less congestion when the budget increases and the incentive required increases as the constraints on resource congestion become tighter.

Acknowledgement. This research/project is supported by the Singapore National Research Foundation under its International Research Centre @ Singapore Funding Initiative and administered by the IDM Programme Office.

References

1. Axelrod, R.: The Complexity of Cooperation: Agent-Based Models of Competition and Collaboration. Princeton University Press (1997)
2. Shoham, Y., Tennenholtz, M.: Social laws for artificial agent societies: Off-line design. Artificial Intelligence 73 (1995)
3. Conitzer, V., Sandholm, T.: Complexity of mechanism design. In: Proceedings of the Conference on Uncertainty in Artificial Intelligence (UAI), pp. 103–110. Morgan Kaufmann Publishers Inc. (2002)
4. Nisan, N., Ronen, A.: Algorithmic mechanism design. In: Proceedings of the ACM Symposium on Theory of Computing (STOC), pp. 129–140 (1999)
5. Bergendorff, P., Hearn, D., Ramana, M.: Congestion toll pricing of traffic networks. Lecture Notes in Economics and Mathematical Systems, pp. 51–71 (1997)
6. Hearn, D., Ramana, M.: Solving congestion toll pricing models. In: Marcotte, P., Nguyen, S. (eds.) Equilibrium and Advanced Transportation Modeling, pp. 109–124. Kluwer Academic Publishers (1997)
7. Dybvig, P., Spatt, C.: Adoption externalities as public goods. Journal of Public Economics 20, 231–247 (1983)
8. Segal, I.: Contracting with externalities. The Quarterly Journal of Economics 2, 337–388 (1999)
9. Leyton-Brown, K., Porter, R., Prabhakar, B., Shoham, Y., Venkataraman, S.: Incentive mechanisms for smoothing out a focused demand for network resources. Computer Communications 26(3), 237–250 (2003)
10. Monderer, D., Tennenholtz, M.: K-implementation. Journal of Artificial Intelligence Research 21, 37–62 (2004)
11. Law, L.M., Huang, J., Liu, M.: Price of anarchy for congestion games in cognitive radio networks. IEEE Transactions on Wireless Communications PP(99), 1–10 (2012)
12. Chen, H.L., Roughgarden, T., Valiant, G.: Designing network protocols for good equilibria. SIAM Journal on Computing 39(5), 1799–1832 (2010)
13. Monderer, D., Shapley, L.S.: Fictitious play property for games with identical interests. Journal of Economic Theory 68(1), 258–265 (1996)
14. Vetta, A.: Nash equilibria in competitive societies, with applications to facility location, traffic routing and auctions. In: Proceedings of the IEEE Symposium on Foundations of Computer Science (FOCS), pp. 416–425 (2002)

Optimization Approaches for Solving Chance Constrained Stochastic Orienteering Problems

Pradeep Varakantham[1] and Akshat Kumar[2]

[1] School of Information Systems, Singapore Management University
[2] IBM Research India

Abstract. Orienteering problems (Ops) are typically used to model routing and trip planning problems. OP is a variant of the well known traveling salesman problem where the goal is to compute the highest reward path that includes a subset of nodes and has an overall travel time less than the specified deadline. Stochastic orienteering problems (SOPs) extend OPs to account for uncertain travel times and are significantly harder to solve than deterministic OPs. In this paper, we contribute a scalable mixed integer LP formulation for solving *risk aware* SOPs, which is a principled approximation of the underlying stochastic optimization problem. Empirically, our approach provides significantly better solution quality than the previous best approach over a range of synthetic benchmarks and on a real-world theme park trip planning problem.

1 Introduction

Motivated by competitive orienteering sports, Orienteering Problems (OPs) [15] represent the problem of path selection, where the reward accumulated by visiting a subset of nodes in the path is the maximum for the condition that overall travel time to traverse the path does not violate the deadline. While OPs have been used to represent problems like vehicle routing [7] and production scheduling [2], in this work, we are motivated by the problem of tourist trip design problems similar to the one described in Vansteenwegen *et al.* [16,1]. Specifically, we address the problem of providing risk sensitive route guidance to visitors at theme parks [10], where the presence of queues at nodes lead to stochastic travel time between nodes.

As OP assumes deterministic edge lengths, they are insufficient to represent the route guidance problem at theme parks. Thus, researchers have extended OPs to stochastic OPs (SOPs), where edge lengths are now random variables that follow a given distribution. The goal is to find a sequence that maximizes the sum of utilities from vertices in the sequence [4]. In this paper, we consider the risk aware SOP [10], where the goal is to compute a path that maximizes the overall reward while enforcing a risk aware deadline constraint. That is, we compute paths where the probability of violating the deadline is less than a given risk parameter, α.

Lau *et al.* [10] introduced a local search approach for solving such risk aware SOPs. While such an approach is scalable, it is adhoc and does not provide any

P. Perny, M. Pirlot, and A. Tsoukiàs (Eds.): ADT 2013, LNAI 8176, pp. 387–398, 2013.

a priori or posteriori guarantees with respect to optimal solution. To address these limitations, we provide a principled optimization based approach that employs ideas from the sample average approximation technique to solve stochastic optimization problems [12].

In order to illustrate the utility of our approach, we provide comparisons with the local search approach on a synthetic benchmark set introduced in the literature [4] and also on a real theme park navigation problem, where the travel times are computed from a year-long data set of travel times at a popular theme park in Singapore. The results are quite encouraging—our approach provides significant and consistent increase in solution quality (more than 50% for some synthetic benchmarks and more than 100% for real-world problems) when compared against the local search approach of [10].

2   Background: OPs and SOPs with Chance Constraints

The *orienteering problem* (OP) [15] is defined by a tuple $\langle V, E, T, R, v_1, v_n, H \rangle$, where V and E denote the vertices and edges respectively of the underlying graph. $T : v_i \times v_j \to \mathbb{R}^+ \cup \{0, \infty\}$ specifies a finite non-negative travel time between vertices v_i and v_j if $e_{ij} \in E$ and ∞ otherwise; and $R : v_i \to \mathbb{R}^+ \cup \{0\}$ specifies a finite non-negative reward for each vertex $v_i \in V$. A solution to an OP is a Hamiltonian path over a subset of vertices including the start vertex v_1 and the end vertex v_n such that the total travel time is no larger than H. Solving OPs optimally means finding a solution that maximizes the sum of rewards of vertices in its path. Researchers have shown that solving OPs optimally is NP-hard [7]. In this paper, we assume that the end vertex can be any arbitrary vertex. The start and end vertices in OPs are typically distinct vertices.

Researchers have proposed several exact branch-and-bound methods to solve OPs [9] including optimizations with cutting plane methods [11,6]. However, since OPs are NP-hard, exact algorithms often suffer from scalability issues. Thus, constant-factor approximation algorithms [3] are necessary for scalability. Researchers also proposed a wide variety of heuristics to address this issue including sampling-based algorithms [15], local search algorithms [7,5], neural network-based algorithms [17] and genetic algorithms [14]. More recently, Schilde et al. developed an ant colony optimization algorithm to solve a bi-objective variant of OPs [13].

The assumption of deterministic travel times is not a valid one in many real-world settings and thus researchers have extended OPs to *Stochastic OPs* (SOPs) [4], where travel times become random variables that follow a given distribution. The goal is to find a path that maximizes the sum of expected utilities from vertices in the path. The random variables are assumed to be independent of each other.

Existing research has focussed on two different objectives in obtaining solutions for a SOP. The first objective by Campbell *et al.* [4] is to maximize sum of expected utilities of visited nodes. The expected utility of a vertex is the difference between the expected reward and expected penalty of the vertex. The expected reward (or penalty) of a node is the reward (or penalty) of the vertex times the probability that the travel time along the path thus far is no larger (or larger) than H. More formally, the expected utility $U(v_i)$ of a vertex v_i is

$$U(v_i) = P(a_i \leq H)\,R(v_i) - P(a_i > H)\,C(v_i)$$

where the random variable a_i is the arrival time at vertex v_i (that is, the travel time from v_1 to v_i), $R(v_i)$ is the reward of arriving at vertex v_i before or at H and $C(v_i)$ is the penalty of arriving at vertex v_i after H. Campbell et al. have extended OP algorithms to solve SOPs including an exact branch-and-bound method and a local search method based on variable neighborhood search [4]. Gupta et al. introduced a constant-factor approximation algorithm for a special case of SOPs, where there is no penalty for arriving at a vertex after H [8].

The approach by [4] suffers from many limitations. Firstly, it is a point estimate solution which does not consider the "risk" attitude with respect to violating the deadline. By "risk", we refer to probability of completing the path within the deadline. In other words, a risk-seeking user will be prepared to choose a sequence of nodes that have a large utility, but with a higher probability of not completing the path within the deadline, compared to a risk-averse user who might choose a more "relaxed" path with lower utility. Secondly, the underlying measurement of expected utility is not intuitive in the sense that a utility value accrued at each node does not usually depend on the probability that the user arrives at the node by a certain time; but rather, the utility is accrued when the node is visited.

Given the above consideration, Lau *et al.* [10] proposed a second objective where we maximize accumulated reward while satisfying a chance constraint to account for the risk of exceeding the horizon. This allows the user to tradeoff risk against total utility. More precisely, given a value $0 \leq \alpha \leq 1$, we are interested in obtaining a path, where the probability of failing to complete the entire path within a deadline H is less than α. Formally,

$$prob(a_n > H) \leq \alpha \tag{1}$$

where a_n is the arrival time at the last vertex of the path.

3 Deterministic Approximation for Chance Constrained Optimization

In this section, we provide a brief overview of the *sample average approximation* (SAA) technique for solving stochastic optimization problems [12]. The stochastic orienteering problem is an instance of the *stochastic optimization* problem, where the risk sensitive behavior is often encoded in the form of chance constraints. An example of such an optimization problem is given below:

$$\min_{x \in X} \{g(x) := \mathbb{E}_P\left[G(x, W)\right]\} \tag{2}$$

$$\text{s.t. } prob\{F(x, W) \leq 0\} \geq 1 - \alpha \tag{3}$$

where X is the feasible parameter space, W is a random vector with probability distribution P and $\alpha \in (0, 1)$. The above stochastic optimization problem is called a chance constrained problem [12]. Notice that the objective function is an expectation due to the unobserved random variable W. Similarly, the constraint function $F(\cdot)$ is also a random variable due to its dependence on W. The parameter α can be interpreted as the parameter to tune the risk seeking or risk averse behavior.

It may seem that such an optimization problem is too unwieldy to solve. Fortunately, a number of techniques do exist that transform such stochastic optimization problem into a deterministic problem in a principled manner. One such technique is called *sample average approximation* [12] or *SAA*. We describe a brief outline below; further details can be found in [12]. Interestingly, the SAA technique can also provide stochastic bounds on the solution quality and thus, provides a principled approximation.

The main idea behind the SAA is to generate a number of samples for the random vector W. Let us denote these samples as W^i. First, we define the following indicator-like function that returns 1 if the argument is positive and 0 otherwise.

$$\mathbb{I}(t) = \begin{cases} 1 & if \ t > 0 \\ 0 & if \ t \leq 0 \end{cases} \tag{4}$$

We generate N samples for the random variable W. Based on these samples, we define the approximate probability of constraint violation for a particular point x as follows:

$$\hat{p}_N(x) = \frac{1}{N} \sum_{i=1}^{N} \mathbb{I}\big(F(x, W^i)\big) \tag{5}$$

Now the stochastic optimization is reformulated (approximately) as the following deterministic optimization problem:

$$\min_{x \in X} \frac{1}{N} \sum_{i=1}^{N} G(x, W^i) \tag{6}$$

$$\text{s.t. } \hat{p}_N(x) \leq \alpha' \tag{7}$$

The parameter α' plays the role of α in the above optimization problem. Typically, we set $\alpha' < \alpha$ to get a feasible solution. Often, the above optimization problem can be formulated as a mixed-integer program and thus, can be solved using CPLEX. Based on the number of samples and the parameter α', several bounds for the solution quality and feasibility can be derived [12].

4   Solving SOP with Chance Constraints

In this section, we first formulate the SOP problem with chance constraints (see Section 2) as an optimization problem. We then employ the SAA scheme to get a

Table 1. Formulation of chance constrained SOP as a mathematical program

$$\max_{\pi} \sum_{i,j} \pi_{ij} R_i \tag{8}$$

s.t.

$$\pi_{i,j} \in \{0, 1\}, \qquad\qquad \forall v_i, v_j \in V \tag{9}$$

$$\sum_j \pi_{ji} \leq 1, \qquad\qquad \forall v_i \in V \tag{10}$$

$$\sum_j \pi_{ij} \leq 1, \qquad\qquad \forall v_i \in V \tag{11}$$

$$\sum_j \pi_{1j} = 1; \sum_j \pi_{jn} = 1 \tag{12}$$

$$\sum_j \pi_{ij} - \sum_j \pi_{ji} = \begin{cases} 1 & \text{if } i = 1; \\ -1 & \text{if } i = n; \\ 0 & \text{otherwise}; \end{cases} \qquad , \qquad \forall v_i \in V \tag{13}$$

$$r_i \leq r_j - 1 + (1 - \pi_{ij}) * M \qquad \forall v_i, v_j \in V \tag{14}$$

$$r_1 = 1, \ r_n = n, r_i \in [1, n] \qquad \forall v_i \in V \tag{15}$$

$$\Pr\left(\sum_{i,j} \pi_{ij} T_{ij} \geq H\right) \leq \alpha \tag{16}$$

deterministic approximation. For each directed edge (v_i, v_j), the binary variable π_{ij} denotes whether the edge (v_i, v_j) is in the final path. The random variable T_{ij} denotes the travel time for traversing the directed edge (v_i, v_j). We assume that the underlying distribution for each variable T_{ij} is provided as input. The parameter R_i represents the reward obtained on visiting the node v_i.

Table 1 shows the mathematical program for chance constrained SOPs. We next describe its structure. We designate the start node with id 1 and the destination node with n. The objective function seeks to maximize the overall reward obtained based on nodes visited. Constraints (10)-(11) specify that there is a single incoming and outgoing active edge for each node. Constraint (13) denotes the flow conservation.

To ensure that there are no cycles in the path, we introduce a new set of variables r_i for each node v_i to denote its rank in the final path. For instance, if the rank of the source node is 1, then any node connected immediately from source will be ranked greater than 1 and so on. Such monotonically increasing ranking of nodes will enforce that no cycles are generated. The constraint (14) models this ranking scheme. The parameter M is a large constant used to maintain the consistency of the constraint.

Constraint (16) denotes the chance constraint. The total duration of the SOP is denoted as $\sum_{i,j} \pi_{ij} T_{ij}$, which is a random variable as each T_{ij} is a random variable. The parameter H denotes the input deadline. The chance constraint states that the probability of violating the deadline should be no greater than $\alpha \in (0, 1)$, another input parameter. This constraint is not linear and in general, a closed form expression is not readily available. We next show how to determinize this constraint using SAA in the MIP framework.

For each edge of the graph, we generate Q samples for the duration random variable T_{ij}, denoted by t_{ij}^q. We represent the function $\mathbb{I}(\cdot)$ of Eq. (4) using the following linear constraints:

$$z^q \geq \frac{\sum_{ij} \pi_{ij} t_{ij}^q - H}{H} \qquad \forall q \in Q \qquad (17)$$

$$z^q \in \{0, 1\} \qquad \forall q \in Q \qquad (18)$$

where we have introduced auxiliary integer variables z^q for each sample q. Using these auxiliary variables, the constraint (7) is represented as:

$$\frac{\sum_q z^q}{Q} \leq \alpha' \qquad (19)$$

where α' is a parameter that is set by the user and is generally smaller than the parameter α as used in constraint (16). The setting of α' is critical and we will provide a detailed discussion about the same in our experimental results section. To summarize, we get a deterministic mixed-integer program corresponding to the stochastic program of Table 1 by replacing the stochastic constraint (16) using Q samples for each random variable corresponding an edge, introducing auxiliary integer variables z^q for each SAA sample and using linear constraints (17), (18) and (19). The following theoretical results establish the convergence guarantees for the SAA technique:

Theorem 1 ([12]). *Let $v^\star$ be the optimal solution quality, $\hat{v}_N$ be the quality of the SAA problem, $x^\star$ be the optimal solution, $\hat{x}_N$ be the SAA solution and the parameter $\alpha' = \alpha$, then $\hat{v}_N \to v^\star$ and $\hat{x}_N \to x^\star$ as $N \to \infty$.*

The next theorem provides convergence results regarding the feasibility of the solution w.r.t. the chance constraint.

Theorem 2 ([12]). *If $\hat{x}_N$ be the feasible solution of the SAA problem and $\alpha' < \alpha$, then the probability that $\hat{x}_N$ is a feasible solution of the true problem approaches one exponentially fast with the increasing number of samples N.*

5   Experimental Results

To illustrate the effectiveness of our approach, referred to as MILP-SAA, for solving the SOP, we provide experimental results on a synthetic benchmark set employed in the literature [4] and a real world theme park decision support problem introduced by Lau *et al.* [10]. We measure the performance of our approach with respect to the solution quality and the probability of constraint violation by varying problem parameters. Our results are quite encouraging—the MILP-SAA approach provides significant increase in solution quality (more than 50% for some synthetic benchmarks and more than 100% for real-world problems) when compared against the local search approach of [10], all the while keeping the probability of constraint violation within the specified limit ($=\alpha$).

5.1   Synthetic Benchmark Set

Firstly, we provide the comparison on the benchmark set introduced by Lau *et al.* [10]. In this set, we vary the following key problem parameters:

- The graph structures are taken from existing work [4] and the number of nodes ($|V|$) in these graphs vary in the following range: $\langle 20, 32, 63 \rangle$. The reward obtained by visiting a node is chosen randomly between 1 and 10.

- The probability of constraint violation or the α parameter of Eq. 16 is varied as: $\langle 0.3, 0.25, 0.2, 0.15, 0.11 \rangle$. Corresponding to each setting of α, we use the parameter α' ($\leq \alpha$, see Eq. (19)) from the values $\langle 0.2, 0.15, 0.1, 0.05, 0.01 \rangle$.

- As in the previous work [10], we employ a gamma distribution, $f(x; k, \theta)$, for modeling the travel time of an edge or the random variable T_{ij}.

$$f(x; k, \theta) = \frac{1}{\theta^k} \frac{1}{\Gamma(k)} x^{k-1} e^{\frac{x}{\theta}}, \ \ x > 0, \ k, \theta > 0$$

The k parameter is randomly selected for different edges, the theta (θ) parameter is varied as: $\langle 1, 2, 3 \rangle$.

- Finally, we also test for a range of the deadlines H. For each instance, we calculate approximately the total time required to visit all the nodes and then set the deadline H to be the following fraction of the total time: $\langle 20\%, 25\%, 30\%, 35\% \rangle$.

We do not modify the parameters of the local search algorithm provided in Lau *et al.* [10], as it was shown to work across a wide variety of problems. With the MILP-SAA, we compute a 90% optimal solution to ensure easy scalability to larger problems. Also, to understand the performance of our approach better, we employ the following settings for the algorithm:

- The number of samples (Q) used by MILP-SAA is varied as: $\langle 25, 30, 35, 40 \rangle$

- The number of sample sets generated for each problem is 15. This corresponds to the initial random seeds used to sample the travel time from the gamma distribution.

While we obtained results for all the combination of parameters, we only show a representative set of results due to space constraints. We show results where one parameter is modified while keeping other parameters set to their default value. The default values for different parameters are as:

$$\theta = 1; \alpha = 0.3; \alpha' = 0.1; H = 25\%; Q = 40; \tag{20}$$

The local search approach always provides a solution with the specified limit α. For the MILP-SAA, we empirically determine the actual probability of constraint violation for a particular solution π, say β, by generating 1000 complete samples for edge duration and computing the fraction of samples for which the solution violated the deadline H. Ideally, the probability β should be less than α for the solution to be valid, which is indeed the case for most problem instances for MILP-SAA.

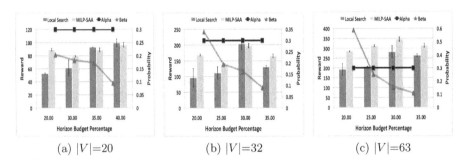

(a) $|V|$=20   (b) $|V|$=32   (c) $|V|$=63

Fig. 1. Effect on reward as the horizon budget is varied

Runtime: In this paper, we do not provide detailed results on run-time[1] because both approaches were able to solve all the problems within less than 10 minutes. Local search was able to obtain solutions on the most difficult of problems within a few seconds. On the other hand, we were able to compute solutions by using MILP-SAA approach within 10 minutes on the most difficult problem (63 nodes, $H = 20\%, Q = 70, \alpha' = 0.01, \theta = 3$). When number of samples is less than or equal to 40, we obtain solutions within 2 minutes. There exist expected patterns, such as run-time increasing with decreasing horizon budget and increasing number of samples, however, due to space constraints, we would not be going through those in this paper.

Horizon Budget: Figure 1 shows the effect of varying horizon on the overall reward for the three graph configurations. The X-axis shows the horizon as the percentage of total time required to visit all the nodes. A 20% horizon budget indicates that on an average, only about 20% of all the nodes can be traversed. The primary Y-axis (left side) indicates the reward obtained and the secondary Y-axis (right side) indicates the probability of constraint violation. The bars indicate the reward obtained by local search and MILP-SAA. In addition, the two lines represent the probability of constraint violation. The legend 'Alpha' denotes the α parameter and 'Beta' denotes the empirically computed probability of constraint violation for the MILP-SAA solution using 1000 samples. We make the following observations:

- MILP-SAA outperforms the local search in terms of reward consistently and significantly for several cases. In addition, this difference in performance is significant and consistent in the 63 node case. For instance, for the 25% horizon budget case for 63 nodes in Figure 1(c), the reward difference is close to 100, indicating about 50% improvement over the local search. This also implies the traversal of an additional 10 nodes[2] in the worst case and 20 nodes in the average case.

---

[1] We conduct our experiments on an Intel Core i5 machine with 1.8 GHz processor speed and 8 GB RAM.

[2] Reward for nodes is drawn from a uniform distribution with minimum value of 1 and maximum value of 10.

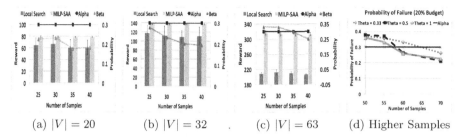

(a) $|V| = 20$ (b) $|V| = 32$ (c) $|V| = 63$ (d) Higher Samples

Fig. 2. Effect on reward as the number of samples is varied

- In most of the cases, the variance in performance of local search is much higher than the variance of MILP-SAA. This is an important observation, especially for the few cases where local search dominates MILP-SAA. Thus, MILP-SAA was highly consistent in providing good solution quality.

- As the horizon budget is increased, the problem becomes less constrained and the difference in the reward values between the two approaches reduces. This is as expected.

- As the horizon budget is decreased, the problem is more constrained and hence the actual probability of constraint violation (β) increases. Specifically, 20% horizon budget is a difficult problem to solve for the 32 and 63 node problems when MILP-SAA employs 40 samples only. This is reflected in the β values, which are greater than the $\alpha = 0.3$ threshold. As we show later in this section, this can be addressed by increasing the number of samples (> 40) or reducing the α' value employed (< 0.1).

Number of Samples: We now show the effect of increasing the number of SAA samples on reward in Figure 2, while setting all the other parameters to their default value as in Eq. (20). We make the following key observations:

- As the number of SAA samples is increased, the β value reduces. This is as expected as with the increasing number of samples, the SAA approximation becomes tighter.

- There is only a minor reduction in reward values obtained by MILP-SAA with the increasing number of samples. This shows that MILP-SAA can find a good solution that minimizes the probability of constraint violation even with increased problem complexity with the higher number of samples.

- Figure 2(d) shows the effect of increasing the number of samples for 20% horizon budget setting. We see that with 60 SAA samples, the probability of constraint violation β is smaller than α. Thus, increasing the number of samples can provide a feasible solution.

Alpha$'$ and Theta: Figures 3 and 4 indicate the impact of changing α' and the θ parameter of the gamma distribution on the performance of MILP-SAA in comparison to local search. The remaining parameters are fixed to their default setting (20).

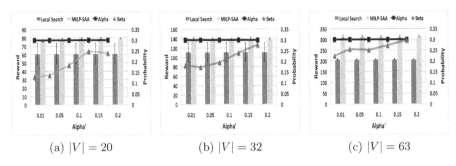

(a) $|V| = 20$ (b) $|V| = 32$ (c) $|V| = 63$

Fig. 3. Effect on reward as α' is varied

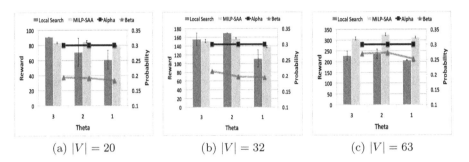

(a) $|V| = 20$ (b) $|V| = 32$ (c) $|V| = 63$

Fig. 4. Effect on reward as θ is varied

- As expected, with the increase in α', the empirical probability of constraint violation, β, increases. However, the increase in accumulated reward is minimal for increased α' values. This shows that a smaller value of α' is preferable to limit the probability of constraint violation.

- For a fixed budget percentage, as we increase the θ parameter of the gamma distribution, local search on average performs slightly better (albeit with higher standard deviation) than MILP-SAA in smaller problems (20 and 32 nodes). However, on the 63 node problems, we see that MILP-SAA is significantly better over all the values of θ.

5.2   Real World Theme Park Problem

Lau *et al.* [10] introduced the route guidance problem for experience management at theme parks. Based on a year long data set of wait times at attractions in the theme park, they constructed best fit gamma distributions for travel times between attractions. In their work, the problem was formulated as a dynamic SOP and hence, there was a different travel time distribution for different time interval of the day. In contrast, we model the problem as a SOP and based on the same data set, we compute best fit gamma distributions for travel times between nodes over the entire time horizon. Extending our approach for dynamic SOPs remains an important area for future work.

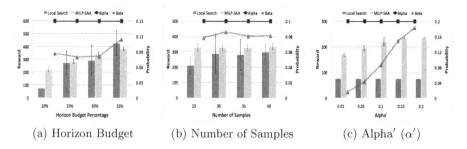

(a) Horizon Budget   (b) Number of Samples   (c) Alpha$'$ (α')

Fig. 5. Solution quality comparisons on real-world theme park SOP

Figure 5 provides the results on the real world data set when horizon budget, number of samples and α' parameters are varied. Due to space constraints, we are unable to show the results where only one parameter is varied in each set of graphs. In fact, we show results with different values of α (0.1,0.15, 0.2) to indicate that, unlike in the synthetic data set, we do not get cases where β exceeds α.

- MILP-SAA consistently obtains higher average reward solutions in comparison to local search. In some cases, the reward improvement in using MILP-SAA was more than 100%. For instance, if we consider the case with 20% horizon budget in Figure 5 (a), the actual reward improvement is more than 125 and the simulated probability of constraint violation, β, is well below the α. Similarly, in Figure 5(c), we obtain more than 100% improvement in solution quality over the local search approach, all the while keeping the β within the limit.

- In most cases, the standard deviation in the solutions obtained with local search is significantly higher in comparison with MILP-SAA.

- Even with 25 samples, we obtain sufficiently stable solutions where the empirical probability of failure, β, is less than the α.

- As the parameter α' employed by MILP-SAA approach is increased in Figure 5(c), as expected, the overall reward accumulated and probability of constraint violation increases.

To summarize, using extensive experiments, we analyzed a number of important properties of the MILP-SAA approach, such as the number of samples required and the effect of parameter α' on the feasibility and quality of the solution. Our approach provided significantly better results than the local search technique for both synthetic and real-world benchmarks.

6   Summary

In this paper, we have presented a new optimization based approach for solving risk aware Stochastic Orienteering Problems, where chance constraints represent risk attitude towards violating the given deadline. By approximating chance constraints with deterministic linear constraints, we provide a scalable approach that provides confidence based guarantees on solution quality. In addition, we show that our approach provides significantly superior strategies in comparison to an existing local search approach over a wide range of real world and synthetic problems.

References

1. Archetti, C., Feillet, D., Hertz, A., Speranza, M.: The capacitated team orienteering and profitable tour problems. Journal of the Operational Research Society 13(1), 49–76 (2008)
2. Balas, E.: The prize collecting traveling salesman problem. Networks 19, 621–636 (1989)
3. Blum, A., Chawla, S., Karger, D., Lane, T., Meyerson, A., Minkoff, M.: Approximation algorithms for orienteering and discounted-reward TSP. SIAM Journal on Computing 37(2), 653–670 (2007)
4. Campbell, A., Gendreau, M., Thomas, B.: The orienteering problem with stochastic travel and service times. Annals of Operations Research 186(1), 61–81 (2011)
5. Chao, I.-M., Golden, B., Wasil, E.: Theory and methodology – the team orienteering problem. European Journal of Operational Research 88, 464–474 (1996)
6. Fischetti, M., González, J.J.S., Toth, P.: Solving the orienteering problem through branch-and-cut. INFORMS Journal on Computing 10, 133–148 (1998)
7. Golden, B., Levy, L., Vohra, R.: The orienteering problem. Naval Research Logistics 34(3), 307–318 (1987)
8. Gupta, A., Krishnaswamy, R., Nagarajan, V., Ravi, R.: Approximation algorithms for stochastic orienteering. In: Proceedings of the ACM-SIAM Symposium on Discrete Algorithms, SODA (2012)
9. Laporte, G., Martello, S.: The selective traveling salesman problem. Discrete Applied Mathematics 26, 193–207 (1990)
10. Lau, H.C., Yeoh, W., Varakantham, P., Nguyen, D.T., Chen, H.: Dynamic stochastic orienteering problems for risk-aware applications. In: Proceedings of the International Conference on Uncertainty in Artificial Intelligence, UAI (2012)
11. Leifer, A., Rosenwein, M.: Strong linear programming relaxations for the orienteering problem. European Journal of Operational Research 73, 517–523 (1994)
12. Pagnoncelli, B., Ahmed, S., Shapiro, A.: Sample average approximation method for chance constrained programming: theory and applications. Journal of Optimization Theory and Applications 142, 399–416 (2009)
13. Schilde, M., Doerner, K., Hartl, R., Kiechle, G.: Metaheuristics for the bi-objective orienteering problem. Swarm Intelligence 3(3), 179–201 (2009)
14. Tasgetiren, M.F.: A genetic algorithm with an adaptive penalty function for the orienteering problem. Journal of Economic and Social Research 4(2), 1–26 (2001)
15. Tsiligrides, T.: Heuristic methods applied to orienteering. Journal of Operation Research Society 35(9), 797–809 (1984)
16. Vansteenwegen, P., Oudheusden, D.V.: The mobile tourist guide: An OR opportunity. OR Insights 20(3), 21–27 (2007)
17. Wang, Q., Sun, X., Golden, B.L., Jia, J.: Using artificial neural networks to solve the orienteering problem. Annals of Operations Research 61, 111–120 (1995)

Thompson Sampling for Bayesian Bandits with Resets

Paolo Viappiani

CNRS-LIP6 and Univ. Pierre et Marie Curie, France
paolo.viappiani@lip6.fr

Abstract. Multi-armed bandit problems are challenging sequential decision problems that have been widely studied as they constitute a mathematical framework that abstracts many different decision problems in fields such as machine learning, logistics, industrial optimization, management of clinical trials, etc. In this paper we address a non stationary environment with expected rewards that are dynamically evolving, considering a particular type of drift, that we call *resets*, in which the arm qualities are re-initialized from time to time. We compare different arm selection strategies with simulations, focusing on a Bayesian method based on Thompson sampling (a simple, yet effective, technique for trading off between exploration and exploitation).

1 Introduction

The multi-armed bandit problem is a general framework that can represent several different sequential decision problems. Essentially, a multi-armed bandit is a a slot machine with n arms (representing possible decisions), each associated with a different and unknown expected payoff (reward). The problem is to select the optimal (or near-optimal) sequence of arms to pull in order to maximize reward (in expectation). Previous rewards obtained from earlier steps are taken into account in order to identify arms that are associated with high payoff; but since reward is uncertain, several pulls of the same arms are usually necessary to assess the quality of an arm with some confidence. A key concept in bandit problems is the trade-off between *exploitation* (pulling the arm with the highest estimated expected payoff) and *exploration* (focusing on getting more information about the expected payoffs of the other arms).

A large number of works have addressed bandit problems [12,11,2,9,14,10]. In particular, large attention has been given to methods (including heuristics) that are computationally fast and produce a decision about the next arm to pull in very short time. This include action-value strategies, but also the family of UCB methods [1].

Thompson sampling is a simple and effective strategy for multi armed bandits. It can be implemented very efficiently and it is based on principled Bayesian reasoning. Essentially, this strategy maintain a probabilistic estimate on the value of the arms, and select arms according to their probability of being the optimal arm (it is a randomized selection strategy). This idea was first described eighty years ago [13]. However, it has been surprisingly neglected until it has been recently rediscovered [7,8,5,9], showing its effectiveness in a number of different settings.

Most of the works on bandits assume a stationary distribution of rewards. In many situations, however, one cannot expect the quality of the different arms to be constant.

P. Perny, M. Pirlot, and A. Tsoukiàs (Eds.): ADT 2013, LNAI 8176, pp. 399–410, 2013.

If the values of the different possibilities change with time (non stationarity) the situation is that of dynamic bandits (also called *restless* in the literature [3]). In this work, in particular, we consider the situation in which drastic changes in the quality of an arm occur (we say that the arm is "reset"). This can model situations of drastic drift, but also situations in which an arm is substituted with a new one (for example in the problem of choosing a seller in electronic commerce, where vendors can suddenly disappear and new ones arrive).

The goal of this work is to show how probability matching with Thompson sampling can be efficiently implemented in non stationary domains (in particular in the case of resets) and to evaluate its performance compared to a number of classic bandit methods.

2   Bayesian Bandits

In bandit problems, one is accumulating rewards from an unknown distribution. There are different possible views on this problem, mainly distribution-free methods (such as the UCB family of strategies [1]) aiming at providing bounds on the worst-case performance, and Bayesian methods (aiming at optimizing average performance). Bayesian bandit strategies maintain an estimation of the goodness of the different arms in term of probability distributions (known as "beliefs") on the value of the arm. A prior (usually uninformative) is given, that is updated using Bayes rules every time an arm is pulled and a reward is observed.

In order to specify a Bayesian approach for bandit problems, we need to address two issues. First, we need to represent distribution information in a practical way. Second, we need to define a strategy that based on the current belief picks the next arm to pull.

More formally, at each round t we have to choose an action (arm) $a \in \mathcal{A}$, where $|\mathcal{A}| = n$, obtaining reward r_t; each arm is associated with a probability density $P(r|a)$ that dictates an average reward $\mu_a = \mathbb{E}_{P(r|a)}[r|a]$; the "best" arm is the one associated with $\mu^* = \max_{a \in \mathcal{A}} \mu_a$. Since the true distribution $P(r|a)$ is not known with certainty, one will often pull suboptimal arms. Let $a[t]$ be the arm pulled at time t. One can state that the goal of a strategy for bandit problems is to maximize long-term (either discounted or undiscounted) cumulative reward $\sum_{t=1,..,T} r_t$, or alternatively, minimize cumulative expected regret, defined as $\sum_{t=1,..,T} \mu^* - \mu_{a[t]}$. Expected regret is the difference between the expected reward associated with the best arm and that of choice made; cumulative expected regret is the sum of this quantity over time.

The Bayesian approach maintains a belief on the possible reward distributions. Assuming a particular type of distribution, we write $P(r|a; \theta)$ to explicitly express the dependency over a set of parameters θ. The belief is then a distribution $P(\theta)$ over the possible instantiations of the parameters θ; that is updated whenever a new pair action-reward (a, r) is observed. The belief $Bel_t(\theta)$ at time t is the probability $P(\theta|(a[1], r_1), .., (a[t-1], r_{t-1}))$ conditioned to the whole history of pulls and rewards. The posterior $Bel(\theta|a[t], r_t)$ becomes the new belief $Bel_{t+1}(\theta)$ (taking the role of "prior") for the timestep $t + 1$.

When considering a Bayesian approach, as we do here, a fundamental issue is that to update the beliefs whenever a new reward is obtained (this essentially mean to apply Bayes theorem). A practical way to do that is to choose distributions of particular forms, so that they are easy to update. In this paper we focus on Bernoulli bandits, where the reward associated to an arm follows a Bernoulli distribution. Bernoulli bandits can for instance model click behavior and purchase activity in electronic commerce settings [5]. The sequence of rewards/penalties obtained from each arm forms a Bernoulli process with (unknown) probability q_a of "success" (and probability $1 - q_a$ of "failure"). Thus, in the Bernoulli case, possible rewards are in $\{0, 1\}$ and the parameters θ of the model are the elements of the vector $q = \{q_1, ..., q_n\}$ of the success probabilities for each of the n arms; q_a is also the expected reward of the arm ($\mu_a = \mathbb{E}[r|a] = q_a$) and $q^* = \max_a q_a$ is the (true) optimal arm. The q values are not known with certainty, and a belief is maintained. Let $t_a(k)$ be the timestep in which arm a was used for the k-th time; the belief $Bel_t(q_a)$ is the probability $P(q_a \,|\, r_{t_a(1)}, r_{t_a(2)}, ...)$ conditioned to the rewards obtained when using arm a.

In order to facilitate the operation of Bayesian update, we use the Beta distribution for representing beliefs. The Beta distribution is a conjugate prior for the Binomial distribution; we can therefore implement Bayesian reasoning very efficiently. We maintain two sets of hyper-parameters $(\alpha_1, ..., \alpha_n)$ and $(\beta_1, ..., \beta_n)$. The distribution Beta(α_i, β_i) is the prior belief for arm i. Whenever a success is observed (reward is 1) after pulling arm i, we increment the corresponding α_i; if, on the contrary, we observe a failure (reward is 0) we increment the corresponding β_i. If one assumes an uniform prior, then the initial α_0 and β_0 are set to 1, but a different choice is possible. [1]

Based on the current information about the value of the arms, a strategy needs to select the next arm to pull. Traditional methods, such as action-value strategies and the UCB-1 method, based their selection on the empirical mean alone [2]. Bayesian strategies use the current belief distribution $P(q) = P(q_1, ..., q_n)$ to select the next arm. We now discuss, in the next Section, probability matching with Thompson sampling as a method for arm selection. Then, in Section 4, we adapt this method for environments with resets.

3   Probability Matching with Thompson Sampling

The idea of Thompson sampling is to choose the arm that maximizes the expected reward with respect to a randomly drawn belief. It is a Bayesian method because the current belief (about the q values of the arms) is used directly in order to decide which arm to pull. This technique is also known as "probability matching" and it is based on the intuition that if the number of pulls for a given arm matches its (estimated) probability of being the optimal arm, one can have a good compromise between exploitation and exploration. It is consistent with intuition: if one arm has very low chances of being a good arm (probability of optimality close to zero), it will be (almost) never pulled; similarly if an arm is very likely to be the best, it will be pulled very often. Thompson sampling has been showed to be effective in the context of stationary Bernoulli bandits [7,5]. Moreover, it has been showed [8] to be effective in the presence of Brownian motion.

---

[1] α and β are often called "pseudo-counts" for this reason.
[2] UCB-tuned considers the sample variance as well; however in Bernoulli events the variance is a simple function of the mean.

Algorithm 1. Thompson sampling: general case

$Bel_0(\theta) \leftarrow P(\theta)$ (Initialize belief to some initial distribution);
Set $t \leftarrow 0$;
while *true* **do**

> Sample $\hat{\theta} \sim Bel_t(\theta) \ \forall i \ 1 \le i \le n$;
> Select arm $a \leftarrow \arg\max_i \mathbb{E}[r|\hat{\theta}_i]$;
> Observe reward r_t;
> Bayesian update: $Bel_{t+1}(\theta) \leftarrow Bel_t(\theta|r_t)$;
> $t \leftarrow t + 1$;

end

Algorithm 2. Bayesian Bernoulli bandits with Thompson sampling

---

Initialize pseudo-counts: $(\alpha_i, \beta_i) \leftarrow (\alpha_0, \beta_0)$;
while *true* **do**

> Sample $\hat{q}_i \sim Beta(\alpha_i, \beta_i) \ \forall i \in \{1, ..., n\}$;
> Select arm $a \leftarrow \arg\max \hat{q}_i$;
> Observe reward r_t;
> $\alpha_a \leftarrow \alpha_a + r_t$;
> $\beta_a \leftarrow \beta_a + (1 - r_t)$;

end

Formally, for Bernoulli bandits, assume indicator variable $I_i^{opt}(\boldsymbol{q})$ to yield 1 iff $q_i = q^*$ and 0 otherwise. Probability matching with Thompson sampling is a randomized strategy that consists, at any time t, in pulling arm i with probability equal to $P^{opt}(i)$, being the probability that the arm i is optimal according to the current beliefs. For each arm i, this is

$$P^{opt}(i) = \int I_i^{opt}(\boldsymbol{q})P(\boldsymbol{q})d\boldsymbol{q} = \int_0^1 \ldots \int_0^1 I_i^{opt}(q_1, ..., q_n) \prod_i P(q_i) \, dq_1...dq_n. \quad (1)$$

(where we use the fact the q values are probabilistically independent) Since the value P^{opt} might not be easy to compute, in practice, the rule is implemented in the following way (that is what is more strictly referred as *Thompson sampling*). In each round, a set of parameter $\hat{q}$ is sampled from the posterior $P(q|r_1, ..., r_t)$, and the arm with highest value $\hat{q}^*$ is chosen. Conceptually, this means that the agent instantiates the value of the arms randomly according to his beliefs in each round, and then he acts optimally according to this instantiation (the general algorithm is shown in Algorithm 1 and the algorithm specific to Bernoulli bandits in Algorithm 2). An advantage of Thompson sampling is the absence of tuning parameters, in contrast with most (if not all) heuristic methods, where setting the right value for the parameters is crucial.

Thompson sampling is particularly simple to implement in the case of Bernoulli bandits assuming Beta priors. The set of hyper-parameters α_i and β_i are initially initialized according to prior information (setting them all to 1 coincides to an uniform prior). Bayesian update consists in updating the hyper-parameter in the following way: when

---

Algorithm 3. Particle Filter reset-aware Thompson for Bayesian Bernoulli bandits

---

Data: Prior hyper-parameters (pseudo-counts): α_0, β_0
Initialize set of particles $\mathcal{P}_i = (q_i^1, ..., q_i^L)$ for each i uniformly in $Beta(\alpha_0, \beta_0)$;
Set time $t \leftarrow 0$;
while *true* **do**
    for *each arm i in 1,...,n* **do**
        for *each particle j in 1,...,L* **do**
            while *rand*() $< p_{reset}$ **do**
                | resample $q_i^j \sim Beta(\alpha_0, \beta_0)$;
            end
        end
        Sample a particle from each particle set: $\hat{q}_i \sim \mathcal{P}_i \ \forall i$;
    end
    Select arm $a \leftarrow \arg\max_i \hat{q}_i$;
    Observe reward r_t;
    $\mathcal{P}_a \leftarrow$ ImportanceSampling$(\mathcal{P}_a, r_t)$;
    $t \leftarrow t + 1$;
end

---

arm i is pulled, α_i is incremented if a positive reward ($r = 1$) is observed, otherwise in the case of no reward ($r = 0$) β_i is incremented. The decision of the arm to pull consists in sampling the values for the q_i according from Beta with the current value hyper-parameters, and selecting the highest one.

4 Reset-Aware Thompson Sampling

In this paper, we consider dynamic (restless) bandits, focusing on the particular of rewards that can drastically change from time to time (resets). This can model situations like a new user of a electronic commerce website, or when a supplier changes ownership. In our analysis, we assume that there exists a (fixed) probability p_{reset} under which the payoffs are changed, and reset rewards are re-sampled according to the prior $Beta(\alpha_0, \beta_0)$, with hyper-parameters α_0 and β_0 (prior pseudo-counts).

The principle of Thompson sampling can be extended to the case of presence of resets. The key issue in using Thompson is being able to sample from the posterior. The problem is now that we cannot anymore represent the posterior by simply maintaining a vector of pseudo-counts. We will show however that we can still use Thompson sampling from the posterior distribution, considering two different methods.

4.1 Particle Filter Thompson

This method (Algorithm 3) computes an unbiased estimation of the belief distribution of the quality of each arm using particles. $Bel_t(\boldsymbol{q})$ is approximated by a set of particles: a set of L particles is associated to each arm (each particle is a scalar between 0 and 1, representing a particular hypothesis about the value q_i). The approximation is unbiased,

---

Algorithm 4. Geometric-Beta reset-aware Thompson sampling for Bayesian Bernoulli bandits

---

Data: Prior pseudocount: α_0, β_0
Initialize history log;
Set time $t \leftarrow 0$;
while *true* **do**
    for *each arm i in 1,...,n* **do**
        $\bar{t}_i \leftarrow lastUse(i,t)$;
        while *rand()* $< 1 - p_{reset}$ **do**
           | $\bar{t}_i \leftarrow lastUse(i, \bar{t}_i)$;
        end
        $\alpha_i \leftarrow \alpha_0 +$ number of sucessful pulls of arm i between time $\bar{t}_i$ and t;
        $\beta_i \leftarrow \beta_0 +$ number of unsucessful pulls of arm i between time $\bar{t}_i$ and t;
        Sample $\hat{q}_i \sim Beta(\alpha_i, \beta_i)$;
    end
    Select arm $a \leftarrow \arg\max \hat{q}_i$;
    Observe reward r_t;
    Log results;
    $t \leftarrow t + 1$;
end

---

meaning that for a large number of particles, the mean of the distribution converges to the true mean. The belief is propagated during time using a particle filer, composed of two parts: a *transition model* (that simulates the dynamics; i.e. the possibility that an arm can be reset) and an *observation filtering*, where using importance sampling, a new set of particles is selected from the old ones, favoring the particles that better explain the observed reward.

The transition model (the innermost loop in Algorithm 3) iterates over the particles of each arm and substitutes a particle with a new one (sampled from the prior) with probability p_{reset}. In the observation filtering model, particles are weighted by the likelihood of the observed reward given their hypothesis: q_i if $r_t = 1$ and $1 - q_i$ otherwise (if $r_t = 0$). The weights are used to resample a new set of L particles; the new particles represent an approximation of the posterior distribution. At any given time-step, Thompson sampling is realized by sampling exactly one particle for each arm uniformly at random from the associated particles, and pulling the arm whose sampled particles has the highest value.

4.2 Geometric-Beta Thompson

The key insight of this method is what it matters is the last time that the arm was reset. The belief distribution of the value of the arm *assuming the arm was reset at time* $\hat{t}$ is a Beta distribution with hyper-parameters dictated by the number of successes and failures since $\hat{t}$. Let x_t^i be 1 if the arm i is pulled at time t (0 otherwise), and r_t be the reward received at time t. Following our intuition, and assuming Beta priors and a probability of drift p_{reset}, we approximate the probability distribution (belief) $Bel_t(q)$ of the value of an arm i at time t with the following:

$$\sum_{k=1}^{t} p_{reset}(1-p_{reset})^k \cdot f_{Beta}\left(q \; ; \alpha_0 + \sum_{m=k}^{t} r_m \, x_m^i, \; \beta_0 + \sum_{m=k}^{t} (1-r_m) \, x_m^i\right)$$

where $f_{Beta}(x; \alpha, \beta)$ is the density of the Beta distribution with parameters α and β.

We now adapt Thompson sampling for the case of resets by sampling, independently, for each arm, the time of the last reset. At each time step, for each arm, we independently sample a time-step $\bar{t}_i$ from a geometric distribution, parameterized by p_{reset} the reset probability, in order to decide "how far in the past we should go" to set the temporal "window" that it will be used to compute the hyper-parameters α_i and β_i. Let $lastUse(i,t)$ be a function that returns the last time-step t' before t in which arm i was pulled. We repeat the following procedure: we sample a uniform random number between 0 and 1 and, while it is lower than p_{reset}, we set $\hat{t}_i \leftarrow lastUse(i, \hat{t}_i)$ and repeat the loop. The sampled time-step will be used to derive the pseudo-counts that will be used for Thompson sampling.

Considering the history of previous pulls, we consider the interval from $\hat{t}_i$ to the current t, and count the number of successful and unsuccessful pulls (for arm i). The belief distribution of the value of the arm i *assuming the arm was reset at time* $\hat{t}_i$ can be modeled as a Beta distribution with hyper-parameters dictated by the number of success and failures since $\hat{t}$. Using our notation, the meta-parameters are $\alpha_i = \alpha_0 + \sum_{t'=\bar{t}_i}^{t} r_{t'} x_{t'}^i$ and $\beta_i = \beta_0 + \sum_{t'=\bar{t}_i}^{t} (1-r_{t'}) x_{t'}^i$. The complete algorithm is shown in Algorithm 4; the computation of pseudo-counts can be made more efficient by maintaining a cumulative sum at each time step. We note that this is a (biased) approximated method: in general the belief (posterior probability) of an arm being reset at a time-step is not independent from the belief for the arm value (for example, if an arm has given reward 1 for 10 times and then we observe reward 0 for several times since then, our estimation for a reset at timestep 11 becomes much higher).

5   Experiments

In order to evaluate the effectiveness of different methods, we need to define some evaluation metrics. In principle, one would like to be able to accumulate as much reward as possible. Cumulative reward is therefore a natural criterion. It is also interesting to compare the reward obtained following a policy with that of always pulling the "best" arm (that of course is not known with certainty by the bandit strategy). Expected regret for a single pull of a Bernoulli bandit is $R = q^* - q_a$, and cumulative expected regret is $\sum_{t=1,..,T} q^* - q_{a[t]}$ for a particular run of the algorithm. We remark that all algorithms make choices (on which arm to pull) based on the history of rewards obtained, often including some explicit randomization (that is the case of epsilon-greedy methods, and also Thompson sampling). In theory, each strategy could be measured according to its *expected expected regret* (in expectation over possible history of pulls and rewards obtained) and *expected expected reward*, but these measures are extremely hard to calculate analytically. Therefore, with simulations, we compare the different strategies according to their average performance, in particular the values obtained for

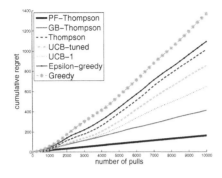

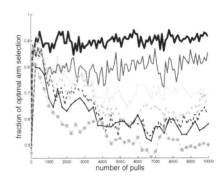

Fig. 1. Regret; $p_{reset} = 0.001$, $n = 2$

Fig. 2. Fraction of optimal arm selection; $p_{reset} = 0.001$, $n = 2$

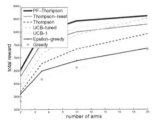

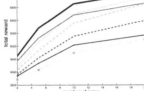

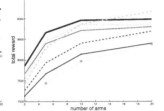

Fig. 3. Reward vs number of arms; $p_{reset} = 0.001$; $(\alpha, \beta) = (1, 1)$

Fig. 4. Reward vs number of arms; $p_{reset} = 0.001$; $(\alpha, \beta) = (2, 1)$

Fig. 5. Reward vs number of arms; $p_{reset} = 0.001$; $(\alpha, \beta) = (1, 2)$

cumulative expected reward averaged over a large number of runs (this choice allows to directly compare the degradation of total reward when p_{reset} increases).

In the following experiments we evaluate Thompson sampling and some classic bandit strategies in presence of resets, in a variety of settings. We are interested to verify whether probability matching, implemented with Thompson sampling as presented in this paper, is an effective strategy for balancing exploration and exploration. At any time-step, let $\bar{\mu}_i$ be the empirical mean and $\bar{\sigma}_i$ the sample variance of the rewards observed when pulling arm i. We compare the following strategies for choosing the next arm to pull:

Thompson sampling with Particle-filter: (indicated as *PF-Thompson* in the plots below) our strategy, described in Algorithm 3, using a particle filter to estimate the belief distribution (we use 10000 particles in our simulations).

Geometric-Beta Thompson sampling: (indicated as *Thompson-reset* in the plots) our probability matching reset-aware strategy using Thompson sampling in two steps, sampling first from a Geometric distribution and then from Beta (see Algorithm 4).

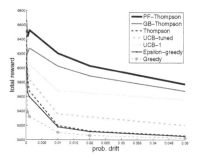
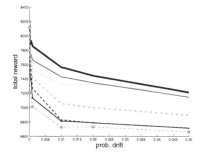

Fig. 6. Reward vs p_{reset}; $n = 2$, $(\alpha, \beta) = (1, 1)$ **Fig. 7.** Reward vs p_{reset}; $n = 2$, $(\alpha, \beta) = (2, 1)$

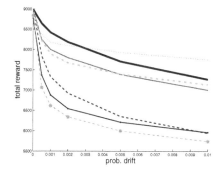
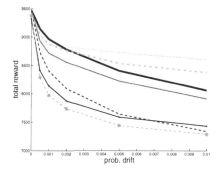

Fig. 8. Reward vs p_{reset}; $n = 10$, $(\alpha, \beta) =$ **Fig. 9.** Reward vs p_{reset}; $n = 10$,$(\alpha, \beta) =$
$(1, 1)$ $(2, 1)$

(Standard) Thompson sampling: probability matching with Thompson sampling. A set of hyper parameters α and β (a pair for each arm) are maintained as explained above. A value $q_i \sim Beta(\alpha_i, \beta_i)$ is sampled for each arm and the arm $a = \arg\max_i q_i$ with maximal value is pulled (see Algorithm 2).

UCB-tuned: The strategy is a modification of UCB-1 that is claimed to be more effective in practice; the index associated to each arm is the following:

$$\tau_i = \bar{\mu}_i + \sqrt{\frac{ln(t)}{m_i} \min\left(0.25, \ \bar{\sigma}^2 + \sqrt{\frac{2ln(t)}{m_i}}\right)}$$

where m_i is the number of times arm i has been pulled.

UCB-1: Each arm i is associated with an index $\tau_i = \bar{\mu}_i + \sqrt{\frac{\log 2t}{m_i}}$; the arm $i^* = \arg\max \tau_i$ with highest index is picked.

Greedy strategy: This basic strategy always selects the currently best performing arm according to the empirical mean: $i^* = \arg\max \bar{\mu}$. Greedy always exploits, therefore it will often lead to suboptimal choices. In order to force some exploitation, especially at the beginning, the empirical mean of each arm is biased by adding 1 to both to the numerator and the denominator.

Epsilon-Greedy: This random strategy selects the currently best performing arm i^* (wrt empirical mean) with probability $1-\epsilon$ and any other arm $j \neq i^*$ is selected with probability $\frac{\epsilon}{n}$ (with $\epsilon = 0.05$ in the simulations below).

The initial rewards are sampled from a Beta prior; we consider the following possible values for the prior hyper-parameters (α_0, β_0) shared by all arms: $(1, 1)$ equivalent to a uniform distribution, $(2, 1)$, and $(1, 2)$. We simulate the behavior of each strategy; at each step of the simulation our Bayesian strategies choose which arm to pull based on the current belief, a reward is sampled and the belief is updated; the other non-Bayesian methods only keep aggregate information (sample mean and variance of each arm). We let simulations proceed for a relatively long time (10000 steps) in order to be able to witness the effect of (possibly multiple) resets. In order to compare the strategies in the fairest possible way (and reducing noise), at each run a complete history of the reward dynamics is generated beforehand, and experienced by all strategies in the same manner. Experimental results are averaged over 300 runs.

First, let's consider a specific setting with 2 arms and $p_{reset} = 0.001$ (this means we can expect approximately around 10 resets per arm during each simulation). We show the cumulative regret obtained by each of the strategy as function of the number of pulls in Figure 1. PF-Thompson clearly dominates all other methods; the approximated strategy Thompson-reset beats the other methods, but it is significantly worse than PF-Thompson. Action-value methods and UCB methods fail to effectively adapt to the reward dynamics. We also note that regret is practically stable for PF-Thompson, but it actually increases for the other strategies (this is due to the drift). We also show the fraction of optimal arm selection (Figure 2): the Thompson strategies are constantly selecting the true highest performing arm with high probability. The fraction of optimal arm selection decreases over time for action value methods, only UCB-1 remains somewhat competitive.

We evaluated the impact of the number of arms on total reward. The results are shown in Figures 3, 4, and 5 for prior hyper-parameters (α_0, β_0) set to $(1, 1)$, $(2, 1)$ and $(1, 2)$, respectively. The reset-aware Thompson strategy with particle filter (PF-Thompson) is dominating in most of the settings. The Geometric-Beta version is also very efficient in many settings (in particular when considering 2 arms); however, when considering 10 arms, it seems that either UCB-1 (in some settings) or UCB-tuned (in other settings) are better.

We also considered the impact of changing p_{reset} on the total cumulative reward (obtained with 10000 pulls) when fixing the number of arms (Figures 6, 7, 8, and 9). Surprisingly, UCB-1 is the best performing strategy when considering 10 arms and a high reset probability. When prior hyper-parameters are optimistic ($\alpha_0 = 2$, $\beta_0 = 1$), also UCB-tuned becomes very effective for $p_{reset} \geq 0.003$ and 10 arms.

Overall, our reset-aware Thompson sampling is an effective technique for bandit problems with presence of resets. In particular, the version based on particle-filtering is particularly effective, and it is dominating the other strategies in most of the settings we tested. However, it is not the best strategy in all the cases; UCB-1 and UCB-tuned are surprisingly competitive in a small number of settings (but perform very poorly in others).

6   Discussion and Conclusion

Multi-armed bandits are the quintessential problem facing the exploration/exploitation tradeoff. In this paper we addressed the problem of non-stationary multi armed bandits with resets, a form of drift that will typically occur (relatively) rarely but it is associated with drastic changes in the value of the choices. In our specific setting, a reset occur with a fixed i.i.d. probability at each step and, in case of reset, the value of the arm is reassigned to a random value sampled from a prior distribution.

We compared different strategies for multi armed bandits, aimed at achieving a good compromise between exploitation and exploration in presence of resets, evaluating them with respect to cumulative reward and regret. We simulated a number of bandit problems, with different values for the reset probability p_{reset}, the number of arms and the initial priors. In particular we showed how Thompson sampling can be effective in case of resets. Differently from drift-aware techniques based on computing pseudo-counts in fixed temporal windows [8], our is a principled solution that use Thompson sampling form the right posterior.

We stress that, while much interest in the Bandit community is about theoretical bounds, we are instead particularly interested in practical efficacy. We show, with simulations, how our strategy based on Thompson sampling is effective in practical circumstances. Thompson sampling is particularly appealing because of its simplicity and (when using conjugate-prior distributions) its efficient implementation. A practical problem is that generally one cannot assume that the reset probability is known a priori. Moreover, realistic domains will possibly include different type of drifts at the same type; for instance, one could consider dynamic values generated by random walks. We plan to investigate techniques for simultaneously learning the "drift" (the reset probability p_{reset} in our case) while estimating the value of the arms.

Our work is related to the Mortal bandit problem of Chakrabarti et al. [4], where arms may "die" and become unavailable, while at the same time new arms may appear. In our model, an arm that is reset could be viewed as dead arm substituted with a new one, but our problem is more challenging as we do not observe which arm dies. Gittins indices [6] are a solution to the bandit problems, and in principle can be used for dynamic settings as well. However they are computationally intensive to compute.

Other directions for future works include bandits with continuous rewards, further experimental evaluation, theoretical analysis of the worst-case.

References

1. Auer, P., Cesa-Bianchi, N., Fischer, P.: Finite-time analysis of the multiarmed bandit problem. Machine Learning 47(2-3), 235–256 (2002)
2. Bhulai, S., Koole, G.: On the value of learning for Bernoulli bandits with unknown parameters. IEEE Transactions on Automatic Control 45(11), 2135–2140 (2000)
3. Bubeck, S., Cesa-Bianchi, N.: Regret Analysis of Stochastic and Nonstochastic Multi-armed Bandit Problems. Foundations and Trends in Machine Learning 5(1), 1–122 (2012)
4. Chakrabarti, D., Kumar, R., Radlinski, F., Upfal, E.: Mortal Multi-Armed Bandits. In: Advances in Neural Information Processing Systems 21. Proceedings of the Twenty-Second Annual Conference on Neural Information Processing Systems (NIPS 2008), pp. 273–280 (2008)

5. Chapelle, O., Li, L.: An Empirical Evaluation of Thompson Sampling. In: Advances in Neural Information Processing Systems 24: 25th Annual Conference on Neural Information Processing Systems (NIPS 2011), pp. 2249–2257 (2011)
6. Gittins, J., Glazebrook, K., Weber, R.: Multi-armed Bandit Allocation Indices, 2nd edn. Wiley (March 2011)
7. Granmo, O.-C.: Solving Two-Armed Bernoulli Bandit Problems Using a Bayesian Learning Automaton. International Journal of Intelligent Computing and Cybernetics (IJICC) 3, 207–234 (2010)
8. Gupta, N., Granmo, O.-C., Agrawala, A.: Thompson Sampling for Dynamic Multi-armed Bandits. In: Fourth International Conference on Machine Learning and Applications, vol. 1, pp. 484–489 (2011)
9. Kaufmann, E., Korda, N., Munos, R.: Thompson Sampling: An Asymptotically Optimal Finite-Time Analysis. In: Bshouty, N.H., Stoltz, G., Vayatis, N., Zeugmann, T. (eds.) ALT 2012. LNCS, vol. 7568, pp. 199–213. Springer, Heidelberg (2012)
10. Lin, C.-T., Shiau, C.J.: Some optimal strategies for bandit problems with beta prior distributions. Ann. Inst. Stat. Math. 52(2), 397–405 (2000)
11. Macready, W.G., Wolpert, D.: Bandit problems and the exploration/exploitation tradeoff. IEEE Trans. Evolutionary Computation 2(1), 2–22 (1998)
12. Ryzhov, I.O., Powell, W.B.: The value of information in multi-armed bandits with exponentially distributed rewards. Procedia CS 4, 1363–1372 (2011)
13. Thompson, W.R.: On the likelihood that one unknown probability exceeds another in view of the evidence of two samples. Biometrika 25(3/4), 285–294 (1933)
14. Valizadegan, H., Jin, R., Wang, S.: Learning to trade off between exploration and exploitation in multiclass bandit prediction. In: Proceedings of the 17th ACM SIGKDD International Conference on Knowledge Discovery and Data Mining (KDD 2011), pp. 204–212 (2011)

Robust Optimization of Recommendation Sets with the Maximin Utility Criterion

Paolo Viappiani[1] and Christian Kroer[2]

[1] CNRS-LIP6 and Univ. Pierre et Marie Curie, France
paolo.viappiani@lip6.fr
[2] Carnegie Mellon University, USA
ckroer@cs.cmu.edu

Abstract. We investigate robust decision-making under utility uncertainty, using the *maximin* criterion, which optimizes utility for the worst case setting. We show how it is possible to efficiently compute the maximin optimal recommendation in face of utility uncertainty, even in large configuration spaces. We then introduce a new decision criterion, *setwise maximin utility (SMMU)*, for constructing optimal recommendation sets: we develop algorithms for computing SMMU and present experimental results showing their performance. Finally, we discuss the problem of elicitation and prove (analogously to previous results related to regret-based and Bayesian elicitation) that SMMU leads to myopically optimal query sets.

1 Introduction

Reasoning about preferences [9] is an important component of many systems, including decision support and recommender systems, personal agents and cognitive assistants. Because acquiring user preferences is expensive (with respect to time and cognitive cost), it is essential to provide techniques that can reason with partial preference (utility) information, and that can effectively elicit the most relevant preference information. Adaptive utility elicitation [3] tackles the challenges posed by preference elicitation by representing the system knowledge about the user in the form of *beliefs*, that are updated following user responses. Elicitation queries can be chosen adaptively given the current belief. In this way, one can often make good (or even optimal) recommendations with sparse knowledge of the user's utility function.

Since utility is uncertain, there is often value in recommending a *set* of options from which the user can choose her preferred option. Retrieving a "diverse" set of recommended options increases the odds that at least one recommended item has high utility. Intuitively, such a set of "shortlisted" recommendations should include options of high utility relative to a wide range of "likely" user utility functions (relative to the current belief) [10]. This stands in contrast to some recommender systems that define diversity relative to product attributes. "Top k" options (those with highest expected utility) do not generally result in good recommendation sets.

Recommendation systems can be classified according to the way they represent the uncertainty about the user preferences (encoded by an utility function) and how such uncertainty is aggregated in order to produce recommendations that are believed to have

P. Perny, M. Pirlot, and A. Tsoukiàs (Eds.): ADT 2013, LNAI 8176, pp. 411–424, 2013.

high utility. A common approach [4,10,3,14] is to consider a distribution over possible user preferences, and make recommendations based on expected utility. Another line of work [1,2,3,13] assumes no probabilistic prior is available (the *strict* uncertainty setting) and provides recommendations using the minimax regret criterion. The latter approach makes robust recommendations; regret measures the worst-case loss in utility the user might incur by accepting a given recommendation instead of the true optimal option.

In this paper, we take an approach similar to the second setting; we cast decision-making as a problem of optimization under strict uncertainty and we produce robust recommendations based on the *maximin* utility criterion. Maximin [15,11] is the most pessimistic decision criterion; the recommended decision or option is the one that leads to the highest utility in the worst case. It is a well-known concept and we believe it is worth studying it from an utility elicitation perspective. While we recognize that maximin might not be the right decision criterion in many circumstances due to its intrinsic pessimism, we argue it is apt for high-stakes decisions requiring the strongest guarantees. As in works on regret-based utility elicitation, in our setting the uncertainty over possible utility functions is encoded by a set of constraints (usually obtained through some form of user feedback, such as responses to elicitation queries of the type: *"Which of these products do you prefer ?"*). Differently from regret, the recommendation ensures that a certain level of utility is attained.

In order to provide recommendation sets that efficiently cover the uncertainty over possible user preferences, we define a new *setwise maximin* utility criterion, formalizing the idea of providing a set of recommendations that optimize the utility of the user-selected option in the context of our framework. We show how linear and mixed integer programming techniques can be used to efficiently optimize both singleton recommendations and sets in large configuration spaces, and more computationally efficient heuristic techniques motivated by our theoretical framework. Finally, we discuss the problem of interactive elicitation (which can be viewed as active preference learning) and how to identify queries that are myopically optimal with respect to a non-probabilistic analogue of value of information.

2   Decision-Making with Maximin Utility

Much work in AI, decision analysis and OR has been devoted to effective elicitation of preferences [1,4,12]. Adaptive preference elicitation recognizes that good, even optimal, decisions can often be recommended with limited knowledge of a user's utility function [1]; and that the value of information associated with elicitation queries is often not worth the cost of obtaining it [3]. This means we must often take decisions in the face of an incompletely specified utility function.

These approaches all represent the uncertainty about the user's utility function explicitly as "beliefs". In the case of strict uncertainty (no probabilistic prior is available), the belief takes the form of a set of possible utility functions, usually implicitly encoded by constraints [1,13]. In this work, we adopt the notion of *maximin utility* as our decision criterion for robust decision making under utility function uncertainty. Maximin utility (like minimax regret [1,2,13]) relies on relatively simple prior information in the form of bounds or constraints on user preferences.

2.1 Basic Setting

The setting is that of [13]: we consider a multi-attribute space as, for instance, the space of possible product configurations from some domain (e.g., computers, cars, apartment rentals, etc.). Products are characterized by a finite set of attributes $\mathcal{X} = \{X_1, ... X_n\}$, each with finite domains $Dom(X_i)$. Let $\mathbf{X} \subseteq Dom(\mathcal{X})$ denote the set of *feasible configurations*. Attributes may correspond to the features of various apartments, such as size, neighborhood, distance from public transportation, etc., with $\mathbf{X}$ defined either by constraints on attribute combinations (e.g., constraints on computer components that can be put together), or by an explicit database of feasible configurations (e.g., a rental database). Let $\mathbf{x} \in \mathbf{X}$ be a feasible configuration, and x_i the value of the i-th attribute.

The user has a *utility function* $u : Dom(\mathcal{X}) \rightarrow \mathbf{R}$. In what follows we will assume either a *linear* or *additive* utility function depending on the nature of the attributes [8]. In both additive and linear models, u can be decomposed as follows[1]:

$$u(\mathbf{x}) = \sum_i f_i(x_i) = \sum_i \lambda_i v_i(x_i)$$

where each *local* utility function f_i assigns a value to each element of $Dom(X_i)$. In classical utility elicitation, these values can be determined by assessing local value functions v_i over $Dom(X_i)$ that are normalized on the interval $[0, 1]$, and importance weights λ_i ($\sum_i \lambda_i = 1$) for each attribute [6,8]. This sets $f_i(x_i) = \lambda_i v_i(x_i)$ and ensures that global utility is normalized on the interval $[0, 1]$. A simple additive model in the rental domain might be: $u(Apt) = f_1(Size) + f_2(Distance) + f_3(Nbrhd)$. [2]

Since a user's utility function is not generally known, we write $u(\mathbf{x}; w)$ to emphasize the dependence of u on parameters that are specific to a particular user. In the additive case, the values $f_i(x_i)$ over $\cup_i\{Dom(X_i)\}$ serve as a sufficient parameterization of u (for linear attributes, a more succinct representation is possible). The optimal product for the user with utility parameters w is $argmax_{\mathbf{x} \in \mathbf{X}} u(\mathbf{x}; w)$. The goal of a decision aid system is to recommend, or help the user find, an optimal, or near optimal, product.

2.2 Singleton Recommendations

Assume that using some prior knowledge, we determine that the user's utility function w lies in some bounded set W.[3] Such prior knowledge might be obtained through some interaction with a user (the exact form of W will be defined in section 3.1). We define:

Definition 1. *Given a set of feasible utility functions W, the minimum utility $MU(\mathbf{x}; W)$ of $\mathbf{x} \in \mathbf{X}$ is defined as:*

$$MU(\mathbf{x}; W) = \min_{w \in W} u(\mathbf{x}; w)$$

---

[1] In our notation, we use bold lowercase for vectors.
[2] Our presentation relies heavily on the additive assumption, though our approach is easily generalized to more general models such as GAI [6,2].
[3] We assume that W is topologically closed. Otherwise one should substitue min and max with inf and sup in the definitions below.

Definition 2. *The* maximin utility $MMU(W)$ *of W and a corresponding* maximin optimal configuration $\mathbf{x}_W^*$ *are defined as follows:*

$$MMU(W) = \max_{\mathbf{x} \in \mathbf{X}} MU(\mathbf{x}; W) = \max_{\mathbf{x} \in \mathbf{X}} \min_{w \in W} u(\mathbf{x}; w)$$
$$\mathbf{x}_W^* = \arg\max_{\mathbf{x} \in \mathbf{X}} MU(\mathbf{x}; W) = \arg\max_{\mathbf{x} \in \mathbf{X}} \min_{w \in W} u(\mathbf{x}; w)$$

Intuitively, $MU(\mathbf{x}; W)$ is the worst-case utility associated with recommending configuration $\mathbf{x}$; i.e., by assuming an adversary will choose the user's utility function $\mathbf{w}$ from W to minimize the utility. The maximin optimal configuration $\mathbf{x}_W^*$ is the configuration that maximizes this minimum utility. Any choice that is not maximin optimal has strictly lower utility than $\mathbf{x}_W^*$ for some $\mathbf{w} \in W$.

In problems where the items or choices are explicitly listed in a database, we can in principle iterate over all candidate items, compute their minimum utility (this requires solving a linear program defined in Section 3.1), and pick the item with the highest value for recommendation. In configuration problems, the product space $\mathbf{X}$ is formulated as a constraint satisfaction problem (CSP) or mixed integer program (MIP). In Section 3.1 we show how computing maximin utility in configuration domains can be formulated as a mathematical programming problem and solved using techniques such as Bender's decomposition and constraint generation, adapting techniques developed for minimax regret optimization [1,2].

2.3   Recommendation Sets: The Setwise Maximin Utility Criterion

Suppose we wish to pick a subset $\mathbf{Z} \subseteq \mathbf{X}$ of size k to present to the user and want to quantify the minimum utility obtained by restricting the user's decision to options in that set. In the maximin utility criterion, we choose the set of k options first, and then the adversary picks the utility function $\mathbf{w}$ such that it minimizes the utility of the best of the k options. We assume $\mathbf{Z}$ is restricted to subsets of $\mathbf{X}$ of cardinality k without making this explicit. In practical circumstances, constraints on the user interface design might lead to the choice of k.

Definition 3. *Let W be a feasible utility set, $\mathbf{Z} \subseteq \mathbf{X}$. Define:*

$$SMU(\mathbf{Z}; W) = \min_{w \in W} \max_{\mathbf{x} \in \mathbf{Z}} u(\mathbf{x}; w)$$
$$SMMU(W) = \max_{\mathbf{Z} \subseteq \mathbf{X}} \min_{w \in W} \max_{\mathbf{x} \in \mathbf{Z}} u(\mathbf{x}; w)$$
$$\mathbf{Z}_W^* = \arg\max_{\mathbf{Z} \subseteq \mathbf{X}} \min_{w \in W} \max_{\mathbf{x} \in \mathbf{Z}} u(\mathbf{x}; w)$$

The *setwise minimum utility(SMU)* of a set $\mathbf{Z}$ of k options reflects the intuitions above. *Setwise maximin utility (SMMU)* is SMU of the minimax optimal set $\mathbf{Z}_W^*$, i.e., the set that maximizes $SMU(\mathbf{Z}, W)$.

Setwise maximin utility has some intuitive properties. SMU is monotone with respect to set inclusion: adding new items to a recommendation set cannot decrease SMU (Observation 1). Incorporating options that are known to be dominated given W does not change setwise maximin utility (Observation 2).

Observation 1. $SMU(\mathbf{A} \cup \mathbf{B}; W) \geq SMU(\mathbf{A}; W)$.

Observation 2. *If* $u(\mathbf{a}, w) > u(\mathbf{b}, w)$ *for some* $\mathbf{a}, \mathbf{b} \in \mathbf{Z}$ *and all* $w \in W$, *then* $SMU(\mathbf{Z} \cup \{\mathbf{b}\}; W) = SMU(\mathbf{Z}; W)$.

Observation 3. *MU and SMU can be explicitly expressed as the minimization over different utility spaces*

$$MU(\mathbf{A}; W_1 \cup W_2) = \min\{MU(\mathbf{A}; W_1), MU(\mathbf{A}; W_2)\}$$
$$SMU(\mathbf{A}; W_1 \cup W_2) = \min\{SMU(\mathbf{A}; W_1), SMU(\mathbf{A}; W_2)\}$$

The computation of SMU is made with respect to the item $\mathbf{x} \in \mathbf{Z}$ with highest utility, when the utility value is computed according to $w \in W$. Due to this, the different choices of $\mathbf{x} \in \mathbf{Z}$ define a partition of the utility space, where a partition with respect to a given $\mathbf{x}$ is the region of W where the utility of $\mathbf{x}$ is highest among the options in $\mathbf{Z}$. More formally,

$$W[\mathbf{Z} \to \mathbf{x}_i] = \{\mathbf{w} \in W : u(\mathbf{x}_i; w) \geq u(\mathbf{x}_j; w) \; \forall j \neq i, 1 \leq j \leq k\}$$

That is, $W[\mathbf{Z} \to \mathbf{x}_i]$ is the region of $\mathbf{w}$ where the utility of $\mathbf{x}_i$ is at least as high as any other option in $\mathbf{Z}$. (the regions $W[\mathbf{Z} \to \mathbf{x}_i]$, $\mathbf{x}_i \in \mathbf{Z}$, partition W if one ignores ties). We call this the $\mathbf{Z}$-*pseudo-partition*[4] *of* W. Using the $\mathbf{Z}$-pseudo-partition, we can rewrite SMU (this will be useful for optimization):

Observation 4. *Let* $\mathbf{Z} = \{\mathbf{x}_1, \ldots, \mathbf{x}_k\}$. *Then*

$$SMU(\mathbf{Z}, W) = \min_{\mathbf{x} \in \mathbf{Z}} \min_{w \in W[\mathbf{Z} \to \mathbf{x}]} u(\mathbf{x}, w) = \min_{i=1 \leq \ldots \leq k} MU(\mathbf{x}_i; W[\mathbf{Z} \to \mathbf{x}_i])$$

We use a similar notation to express the combination of two partitions: $W[\mathbf{Z}_1 \to \mathbf{x}_i, \mathbf{Z}_2 \to \mathbf{x}_j] = W[\mathbf{Z}_1 \to \mathbf{x}_i] \cap W[\mathbf{Z}_2 \to \mathbf{x}_j]$.

We introduce a transformation that modifies a given recommendation set $\mathbf{Z}$ in such a way that SMU cannot decrease and usually increases. This will be used as a heuristic for efficiently generating recommendation sets. It will also be useful when discussing elicitation. Define the transformation T to be a mapping that updates a given recommendation set $\mathbf{Z}$ in the following way: (a) First we construct the $\mathbf{Z}$-pseudo-partition of W; (b) we then compute the *single recommendation* that has maximin utility in each region of the pseudo-partition of W; (c) finally, we let $T(\mathbf{Z})$ be the new recommendation set consisting of these new recommendations. Note that $T(\mathbf{Z})$ may have cardinality less than $|Z| = k$.

Definition 4. *Let* $\mathbf{Z} = \{\mathbf{x}_1, \ldots, \mathbf{x}_k\}$. *We define* $T(\mathbf{Z}) = \{\mathbf{x}^*_{W[\mathbf{Z} \to \mathbf{x}_1]}, \cdots \mathbf{x}^*_{W[\mathbf{Z} \to \mathbf{x}_k]}\}$

We will discuss optimization of a recommendation set below in Section 3.2.

---

[4] The definition of the $\mathbf{Z}$-partition first appeared in [13], in the context of recommendations based on the minimax regret crierion.

3   Maximin Utility Optimization

In this section we formalize the problem of generating recommendations (both single recommendations and setwise recommendations) using mathematical programming techniques (linear programming models and mixed integer programming models). These optimization techniques are adaptation of techniques previously proposed [1,13] for minimax regret. We note that maximin is faster to compute than minimax regret.

3.1   Optimization of Singleton Recommendations

In the following we assume the utility to be linear in w: $u(\mathbf{x}; w) = w \cdot \mathbf{x}$. In this case W is a convex polytope effectively represented by a set of constraints (whenever the user answers a query, new constraints are added) that we denote with *Constraints(W)*.

MU(x, W).   Given configuration $\mathbf{x}$ and the space of possible utility functions W (encoded by linear constraints), the minimum utility of x can be found by minimizing the function $\mathbf{w} \cdot \mathbf{x} = \sum_{1 \leq i \leq n} x_i \cdot w_i$, subject to *Constraints(W)*, and $w_i^{\perp} \leq \mathbf{w}_i \leq w_i^{\top}$ for all $i \in \{1 \ldots n\}$, which is solvable by linear programming.

MMU(W).   Given the possible utility functions W (encoded by linear constraints), the problem is to find the configuration $\mathbf{x}_W^*$ that is associated with maximin utility. In order to "break" the maximin optimization, we make use of Benders decomposition:

$$\max_{\mathbf{x}, \delta} \quad \delta$$
$$\text{s.t. } \delta \leq \mathbf{w} \cdot \mathbf{x} \qquad\qquad \forall \mathbf{w} \in GEN \qquad (1)$$

In this model, δ corresponds to the *maximin utility* of the optimal recommendation $\mathbf{x}_W^*$. Constraint 1 ensures that δ is less than the utility of choice $\mathbf{x}$ for each $\mathbf{w}$. The optimization is exact when $GEN = W$ in constraint 1. However, all the constraints over W need not be expressed for each of the (continuously many) $\mathbf{w} \in W$. Since maximin utility is optimal at some vertex of W, we only need to add constraints for all vertices $Vert(W)$ of W, but they can still be exponential. We apply constraint generation in order to solve the MIP much more efficiently, as very few of the vertices are usually needed. This procedure works by solving a relaxed version of the problem above—the *master problem*— using only the constraints corresponding to a small subset $GEN \subset Vert(W)$. We then test whether any constraints are violated in the current solution. This is accomplished by computing the *minimum utility* of the returned solution (the *slave problem*). If MU is lower than what was found in the master problem, a constraint was violated. The constraint (corresponding to the choice w^a of the adversary) is added to the master problem, tightening the MIP relaxation. The master problem is recomputed, and this process is repeated until no violated constraint exists.

3.2   Optimization of Recommendation Sets

SMU(Z, W).   Given a set Z and a space of possible utility functions W, by observation 4 the setwise minimum utility of Z can be found by solving k (k being the cardinality of Z) optimization problems. Using the Z-partition of W, we compute $MU(\mathbf{x}, W[\mathbf{Z} \to \mathbf{x}])$ for each $\mathbf{x} \in \mathbf{Z}$, using the LP model shown above. We then take the (arithmetic) minimum of the results: $\min_{x \in \mathbf{Z}} MU(\mathbf{x}, W[\mathbf{Z} \to \mathbf{x}])$.

SMMU(W). Given utility space W, we can compute the maximin optimal set (of cardinality k) using the following MIP.

$$\max \quad \delta$$

$$\text{s.t. } \delta \leq \sum_{1 \leq j \leq k} v_w^j \qquad \forall \mathbf{w} \in GEN \qquad (2)$$

$$v_w^j \leq \mathbf{w} \cdot \mathbf{x}^j \qquad \forall j \leq k, w \in GEN \qquad (3)$$

$$v_w^j \leq w^\top I_w^j \qquad \forall j \leq k, w \in GEN \qquad (4)$$

$$\sum_{1 \leq j \leq k} I_{\mathbf{w}}^j = 1 \qquad \forall \mathbf{w} \in GEN \qquad (5)$$

$$I_{\mathbf{w}}^j \in \{0, 1\} \qquad \forall j \leq k, \mathbf{w} \in GEN$$

Decision variables: $\mathbf{x}^j$, δ, $\mathbf{I_w}$, $\mathbf{v}_w$

In this model, δ corresponds to the *setwise maximin utility* of the optimal set $\mathbf{Z}_W^*$. $w^\top$ is some upper bound on the values taken by the weight parameters. Constraints 2, 3 and 4 ensures that δ is less than the utility of the best option in $\{\mathbf{x}^1, ..., \mathbf{x}^k\}$ for each $\mathbf{w}$, by introducing a variable v (for each w and each element of the set) to represent the value of minimum utility for the item selected, and indicators $\mathbf{I}_w$ to represent the selection. Only one $\mathbf{v}_w$ will be different from zero for each w, and since the objective function is maximized, the optimization will set $v_w^j = \mathbf{w} \cdot \mathbf{x}^j$ for the j such that $I_w^j = 1$; constraint 4 enforces 0 in the other cases. Constraint 5 ensures that only one of the k items is selected for each utility function w.

We employ constraint generation in a way analogous to the single item case. At each step of the optimization, we compute the *setwise minimum utility*, solved using a series of LPs (as discussed above).

Alternative Heuristics. Setwise optimization requires solving a large number of MIPs using constraint generation strategies. We present a number of heuristic strategies that are computationally less demanding.

- The *current solution strategy* (CSS) proceeds as follows. Consider $\mathbf{w}^a$, the adversary's utility parameters minimizing the utility of $\mathbf{x}_W^*$, the current maximin optimal recommendation; $u(\mathbf{x}_W^*; \mathbf{w}^a) = MU(\mathbf{x}_W^*; W)$. Let's further consider $\mathbf{x}^a = \arg\max_{\mathbf{x} \in \mathbf{X}} u(\mathbf{x}; w^a)$. CSS will return the set $\mathbf{Z}_{CSS} = \{\mathbf{x}_W^*, \mathbf{x}^a\}$. We extend this to sets with cardinality greater than two. Considering a set $\mathbf{Z}$, define $\mathbf{w}^a(\mathbf{Z}) = \arg\min_{\mathbf{w} \in W} \max_{\mathbf{x} \in \mathbf{Z}} u(\mathbf{x}; w)$ and $\mathbf{x}^a(\mathbf{Z}) = \arg\max_{\mathbf{x} \in \mathbf{X}} u(\mathbf{x}; \mathbf{w}^a(\mathbf{Z}))$. We start by initializing $\mathbf{Z}$ to be $\mathbf{Z}_{CSS}$, the set of size two returned by the current solution strategy, and then iteratively add one element ($k - 2$ times) by setting $\mathbf{Z} := \mathbf{Z} \cup \mathbf{x}^a(\mathbf{Z})$.
- The *query iteration strategy* (QIS) directly applies the T operator until a fixed point is reached. A fixed point is such that $SMU(T(\mathbf{Z}); W) = SMU(\mathbf{Z}; W)$. We start QIS with the solution found by CSS.

3.3   Evaluation of Optimization Strategies

Using randomly generated elicitation data we ran a number of experiments using the algorithms described above. For all experiments, we generated constraints on the

Table 1. Computation times and utility values for our strategies (averaged over 30 instances). On the rightmost columns, setwise minimum utility values of the sets retrieved by each strategy. A dash means that one or more instances for this set of parameters timed out. Note that for set size 1, all three approaches reduce to MMU in this case.

n. of features	set size	Computation time			Setwise utility		
		exact SMMU	CSS	QIS	exact SMMU	CSS	QIS
5	1	0.1s	n.a.	n.a.	0.188	n.a.	n.a.
5	2	0.1s	0.1s	0.2s	0.349	0.321	0.347
5	3	0.1s	0.1s	0.3s	0.366	0.356	0.366
5	4	0.2s	0.2s	0.4s	0.366	0.363	0.366
5	5	0.2s	0.2s	0.5s	0.366	0.366	0.366
10	1	0.1s	n.a.	n.a.	0.218	n.a.	n.a.
10	2	0.2s	0.1s	0.3s	0.375	0.305	0.332
10	3	0.3s	0.1s	0.3s	0.389	0.369	0.376
10	4	0.5s	0.2s	0.5s	0.391	0.385	0.385
10	5	-	0.3s	0.6s	-	0.392	0.392
15	1	0.1s	n.a.	n.a.	0.213	n.a.	n.a.
15	2	0.5s	0.1s	0.4s	-	0.268	0.290
15	3	-	0.2s	0.5s	-	0.314	0.322
15	4	-	0.2s	0.7s	-	0.359	0.368
15	5	-	0.3s	0.8s	-	0.371	0.375

possible options using random binary constraints of the form $\neg f_1 \vee \neg f_2$ where f_1 and f_2 are features. We also assume some prior knowledge of user preferences, represented by random utility constraints of the form $\mathbf{w} \cdot \mathbf{x_k} \geq \mathbf{w} \cdot \mathbf{x_l}$, where $\mathbf{x_k}$ and $\mathbf{x_l}$ are random assignments $\in [0, 1]^m$ (not necessarily feasible options) sampled with uniform probability over all possible assignments. The user's preference values $w_1 \ldots w_n$ are random and normalized such that $\sum_i w_i = 1$. All experiments were run on a laptop with a 2.5GHz Core 2 Duo processor and 2GB ram, with all mathematical programs solved with CPLEX, version 12.2.

First, we ran experiments to determine how the runtime of the algorithms are affected by increasing the problem size (number of features) and the size (cardinality) of the recommendation sets. This was done by running the algorithms on instances ranging from 5 to 15 features, with 30 experiments performed on each size. [5] As seen in Table 1, runtime of exact SMMU computation becomes rapidly higher, and we were unable to perform experiments with more than 15 features, as several of the 30 experiments per size would time out. In contrast to this we see that the runtime of the CSS and QIS algorithms increase much more gradually. In another experiment, we compared the average runtimes of our strategies when fixing the set size, and varying the number of features in the problem domain (Figure 1).

We then investigated whether the computational effort is worth it in terms of utility increase. In the rightmost column of Table 1, we show how our strategies perform on different problem settings. Showing a set of items, instead of a single top item, is very beneficial: the minimum utility roughly doubles with five items instead of a single one. Moreover, the set-wise utility values of our approximate strategies (CSS and QIS) are very close to optimal SMMU.

---

[5] For comparison, we include the "degenerate" case of set size equal 1, corresponding to retrieving the single best recommendation according to maximin.

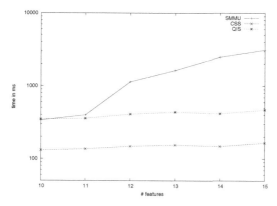

Fig. 1. Average runtime of maximin utility optimization for an increasing number of features. Averaged over 30 instances per size, with set size k = 3.

4   Utility Elicitation

Usually, utility information is not readily available, but must be acquired through an elicitation process. Since elicitation can be costly, it is important to ask queries eliciting the most valuable information. Our setwise criterion can be used directly for this purpose, implementing a form of preference-based diversity. This stands in contrast to "product diversity" typically considered in many recommender systems. And unlike recent work in polyhedral conjoint analysis [12], which emphasizes volume reduction of the utility polytope W, our maximin utility-based criterion is sensitive to the range of feasible products and does not reduce utility uncertainty for its own sake.

4.1   Optimal Myopic Elicitation

In general, there is a tension between recommending the best options to the user, and acquiring informative feedback from the user. While in the recommendation task the goal is to retrieve the best possible options to show to an user, in the elicitation task the objective is to identify candidate queries with high information value, so that better recommendations can be made when the user's response is incorporated in the model. The two tasks, recommendation and elicitation, considered separately in classic decision theory, are interleaved in decision aid tools such as conversational recommender systems where the user is in control.

Here, we consider *choice queries* requiring a user to indicate which choice/product is preferred from a set of k options. Hence, we can view any set of products as either a recommendation set or query (or choice) set. Given a set, one can evaluate the value of the set as a recommendation set and as a query set. Recently, Viappiani and Boutilier [14,13] showed how these two problems are connected to each other, under both a Bayesian framework and when using *minimax regret*. In the following we show the same connection with maximin utility used as criterion.

Any set $\mathbf{Z}$ can be interpreted as a choice query: we simply allow the user to state which of the k elements $\mathbf{x}_i \in \mathbf{Z}$ she prefers. We refer to $\mathbf{Z}$ interchangeably as a *query* or a *choice set*. The choice of some $\mathbf{x}_i \in \mathbf{Z}$ refines the set of feasible utility functions W by imposing the $k - 1$ linear constraints $u(\mathbf{x}_i; \mathbf{w}) > u(\mathbf{x}_j; \mathbf{w}), j \neq i$.

When treating $\mathbf{Z}$ as a choice set (as opposed to a recommendation set), we are not interested in its maximin utility, but rather in *how much a query response will increase maximin utility*. In our distribution-free setting, the most appropriate measure is *posterior maximin utility*, a measure of the value of information of a query. We define:

Definition 5. *The worst case posterior maximin utility (WP) of* $\mathbf{Z} = \{\mathbf{x}_1, \ldots, \mathbf{x}_k\}$ *is*

$$WP(\mathbf{Z}, W) = \min[MMU(W[\mathbf{Z} \rightarrow \mathbf{x}_1]), \ldots, MMU(W[\mathbf{Z} \rightarrow \mathbf{x}_k])]$$

which can be rewritten as: $WP(\mathbf{Z}, W) = \min_{\mathbf{x} \in \mathbf{Z}} \max_{\mathbf{x}' \in \mathbf{X}} \min_{\mathbf{w} \in W[\mathbf{Z} \rightarrow \mathbf{x}]} u(\mathbf{x}', \mathbf{w})$.
An optimal query set *is any* $\mathbf{Z}$ *that maximizes worst case posterior maximin utility*
$MaxWP(W) = \max_{\mathbf{Z} \subseteq \mathbf{X}} WP(\mathbf{Z}, W)$

Intuitively, each possible response $\mathbf{x}_i$ to the query $\mathbf{Z}$ gives rise to updated beliefs about the user's utility function. We use the worst-case response to measure the quality of the query (the updated W with lowest maximin utility). The optimal query is the query that maximizes this value. We observe:

Observation 5. $WP(\mathbf{Z}; W) \geq SMU(\mathbf{Z}; W)$.

We now consider the transformation T introduced earlier (see Definition 4). Using Observation 3 and Observation 4, we prove the following.

Observation 6. *Let* $\mathbf{Z} = \{\mathbf{x}_1, \ldots, \mathbf{x}_k\}$. *Let* $W^1, \ldots, W^l$ *be any partition of* W.

$$WP(\mathbf{Z}, W) = \min_i MMU(W[\mathbf{Z} \rightarrow \mathbf{x}_i]) = \min_i MU(\mathbf{x}^*_{W[\mathbf{Z} \rightarrow \mathbf{x}_i]}, W[\mathbf{Z} \rightarrow \mathbf{x}_i])$$

$$= \min_{i,j} \{MU(\mathbf{x}^*_{W[\mathbf{Z} \rightarrow \mathbf{x}_i]}, W[\mathbf{Z} \rightarrow \mathbf{x}_i] \cap W^j)\}$$

In particular, if we consider $T(\mathbf{Z}) = \{\mathbf{x}'_1, \ldots, \mathbf{x}'_k\}$ *where* $\mathbf{x}'_i = \mathbf{x}^*_{W[\mathbf{Z} \rightarrow \mathbf{x}_i]}$ *and its induced partition on* W, *the expression above becomes the following.*

$$WP(\mathbf{Z}, W) = \min_{i,j} \{MU(\mathbf{x}^*_{W[\mathbf{Z} \rightarrow \mathbf{x}_i]}, W[\mathbf{Z} \rightarrow \mathbf{x}_i; T(\mathbf{Z}) \rightarrow \mathbf{x}'_j])\}$$

Using this, we can now prove the following lemma:

Lemma 1. $SMU(T(\mathbf{Z}), W) \geq WP(\mathbf{Z}, W)$

From observation 5 and lemma 1 it follows that $SMU(T(\mathbf{Z}), W) \geq SMU(\mathbf{Z}, W)$, supporting our use of T as a local search optimization strategy.

Theorem 7. *Let* $\mathbf{Z}^*_W$ *be a maximin optimal recommendation set. Then* $\mathbf{Z}^*_W$ *is an optimal choice set:* $WP(\mathbf{Z}^*_W, W) = MaxWP(W)$.

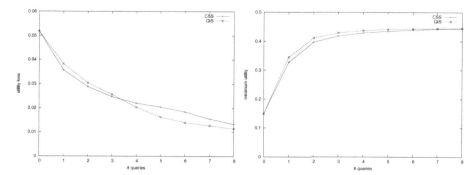

Fig. 2. Average utility loss with respect to the optimal recommendation

Fig. 3. Comparison of CSS and QIS. Results averaged over 30 instances, with set size $k = 3$.

4.2 Evaluation of the SMMU Criterion for Preference Elicitation

We simulated the interaction between a recommender system and a user; at each run a new utility function is randomly sampled. At each cycle the system asks a choice query presenting a set of items to the simulated user and the item with highest utility is selected; this is used by the system to produce a new set of items in the next cycle. Due to the high computation time of the exact method, we focused on CSS and QIS. [6]

In figure 2 we present the "true" utility loss by comparing, after each query answer, the utility of the optimal singleton recommendation according to MMU with utility of the true optimal recommendation (according to the utility function of the simulated user), as a function of the number of queries. The CSS and QIS algorithms have comparable performance, both improving utility loss when more queries are asked, but stalling approximately after 8 queries. In figure 3 we plot the minimum utility guarantee as a function of the number of queries. It quickly increases with the first 4-5 queries, but after that there is little improvement. While our theoretical results show that there is a connection between the problem of generating recommendations and queries, our experiments seem to indicate that maximin is generally unable to effectively elicit useful utility information beyond the first cycles. This is due to the extremely pessimistic nature of maximin and the "absorbing" nature of worst-case posterior maximin utility: in many situations there is no query leading to an improvement of maximin in the worst case (one needs to rely on ad-hoc tie-breaking strategies in these cases). It might be useful to adopt a non-myopic approach, or an alternative decision criterion. Further investigation is required to make long-term preference elicitation with maximin effective and avoiding stalling. [7]

---

[6] These were performed using larger instances, with 30 features per instance, 40 binary feature constraints and 40 utility constraints.

[7] We note that it is of course possible to use maximin as a decision criterion, while resorting to other strategies (perhaps based on regret or on probabilistic methods) for elicitation.

5   Discussion

In this paper we have developed a novel formalization for decision-making under utility uncertainty using the maximin utility criterion. This approach provides the highest degree of robustness, as the recommendation is guaranteed to yield the highest utility in the worst case. We addressed the problem of generating recommendation sets introducing a new decision criterion, setwise maximin utility, that formalizes the intuition that the best recommendation set is the one that is maximally "diverse" in a decision-theoretic way. Adapting ideas from [1,13], we developed computational methods for optimizing sets according to this new criterion, as well as heuristics useful for large problem domains.

Following analogous models available for the minimax regret and Bayesian frameworks [13,14], we showed the connection between the problem of generating optimal recommendation sets and myopically optimal elicitation queries. Our setwise maximin criterion (a natural extension of maximin to sets), in addition to providing robust recommendation sets, also serves as a means of generating myopically optimal choice queries (asking the user to pick his most preferred option in a set). In our experiments we evaluated the performance of our optimization methods on randomly generated data, showing that there is often value in recommending a set of options (instead of a single recommendation) to the user, and that recommendation sets can be efficiently optimized in practice. We also experimented with interactive utility elicitation, with elicitation driven by the (set-wise) maximin optimization; however in this setting the myopic elicitation with choice queries is not very effective (differently from [13,14]).

In this work we consider the value of a set with respect to its capacity of "covering" the uncertainty associated with the partially known user utility function. The underlying assumption is that the user is looking for a single item to pick or purchase, and a set is shown to increase the chance that at least one item has high utility. We underline that there are works[5,7] that consider recommendation sets with a different semantics (the problem of recommending a set of options accounting for positive or negative synergies between options). It would be interesting to consider their setting with a principled decision-theoretic view.

We conclude with a remark about the choice of the decision criterion. Maximin is very pessimistic; indeed expected utility may yield better recommendations in many cases. However, when a decision maker requires guarantees on the worst-case performance (e.g. in critical-decisions with high stakes), she must be willing to sacrifice "average" utility. This is the price to pay for the (strong) worstcase guarantees of maximin.

Acknowledgments. The fist author would like to acknowledge Craig Boutilier for discussion about recommendation sets, minimax regret and utility elicitation. While he was not involved in this paper, his works on minimax regret (including joint works with the first author) laid down fundamental ideas that have been adapted to the maximin criterion here. We also thank the anonymous reviewers for valuable comments and suggestions.

References

1. Boutilier, C., Patrascu, R., Poupart, P., Schuurmans, D.: Constraint-based optimization and utility elicitation using the minimax decision criterion. Artifical Intelligence 170(8-9), 686–713 (2006)
2. Braziunas, D., Boutilier, C.: Minimax regret-based elicitation of generalized additive utilities. In: Proc. of UAI 2007, pp. 25–32. Vancouver (2007)
3. Braziunas, D., Boutilier, C.: Elicitation of factored utilities. AI Magazine 29(4), 79–92 (2008)
4. Chajewska, U., Koller, D., Parr, R.: Making rational decisions using adaptive utility elicitation. In: Proc. of AAAI 2000, Austin, TX, pp. 363–369 (2000)
5. des Jardins, M., Eaton, E., Wagstaff, K.: Learning user preferences for sets of objects. In: International Conference on Machine Learning (ICML), pp. 273–280 (2006)
6. Fishburn, P.C.: Interdependence and additivity in multivariate, unidimensional expected utility theory. International Economic Review 8, 335–342 (1967)
7. Guo, Y., Gomes, C.P.: Learning optimal subsets with implicit user preferences. In: Boutilier, C. (ed.) IJCAI 2009, pp. 1052–1057 (2009)
8. Keeney, R.L., Raiffa, H.: Decisions with Multiple Objectives: Preferences and Value Trade-offs. Wiley, New York (1976)
9. Peintner, B., Viappiani, P., Yorke-Smith, N.: Preferences in interactive systems: Technical challenges and case studies. AI Magazine 29(4), 13–24 (2008)
10. Price, R., Messinger, P.R.: Optimal recommendation sets: Covering uncertainty over user preferences. In: Proc. of AAAI 2005, pp. 541–548 (2005)
11. Salo, A., Hämäläinen, R.P.: Preference programming multicriteria weighting models under incomplete information. In: Handbook of Multicriteria Analysis. Applied Optimization, vol. 103, pp. 167–187 (2010)
12. Toubia, O., Hauser, J., Simester, D.: Polyhedral methods for adaptive choice-based conjoint analysis, 4285–03 (2003)
13. Viappiani, P., Boutilier, C.: Regret-based optimal recommendation sets in conversational recommender systems. In: Proceedings of the 3rd ACM Conference on Recommender Systems (RecSys 2009), New York, pp. 101–108 (2009)
14. Viappiani, P., Boutilier, C.: Optimal bayesian recommendation sets and myopically optimal choice query sets. In: Advances in Neural Information Processing Systems 23 (NIPS), pp. 2352–2360 (2010)
15. Wald, A.: Statistical Decision Functions. Wiley, New York (1950)

Appendix

Proof of Observation 5. Considering the definition of $WP(\mathbf{Z}, W)$ and the equation for $SMU(\mathbf{Z}, W)$ in observation 4, we see that they are the same except that $WP(\mathbf{Z}, W)$ picks a maximizing $\mathbf{x}' \in \mathbf{X}$ after $\mathbf{x} \in \mathbf{Z}$ has been picked. Since $\mathbf{X}$ includes all options, $\mathbf{x}'$ can at worst be $\mathbf{x}$. ∎

Proof of Lemma 1. Let $T(\mathbf{Z}) = \{\mathbf{x}'_1, \ldots, \mathbf{x}'_k\}$ where $\mathbf{x}'_i = \mathbf{x}^*_{W[\mathbf{Z} \to \mathbf{x}_i]}$. The previous observations allow to write WP and SMU compactly

$$WP(\mathbf{Z}, W) = \min_{i,j}[MU(\mathbf{x}'_i, W[\mathbf{Z} \to \mathbf{x}_i, T(\mathbf{Z}) \to \mathbf{x}'_j])] \qquad (6)$$

$$SMU(T(\mathbf{Z}), W) = \min_{i,j}[MU(\mathbf{x}'_j, W[\mathbf{Z} \to \mathbf{x}_i, T(\mathbf{Z}) \to \mathbf{x}'_j])] \qquad (7)$$

We now compare the two expressions componentwise. Consider the utility space $W[\mathbf{Z} \to \mathbf{x}_i, T(\mathbf{Z}) \to \mathbf{x}'_j]$: if $i = j$ then the two MU components are the same. If $i \neq j$, consider any $w \in W[\mathbf{Z} \to \mathbf{x}_i, T(\mathbf{Z}) \to \mathbf{x}'_j]$. Since $w \in W[T(\mathbf{Z}) \to \mathbf{x}'_j]$, we must have $u(\mathbf{x}'_j; w) > u(\mathbf{x}'_i; w)$. Therefore $MU(\mathbf{x}'_j, W[\mathbf{Z} \to \mathbf{x}_i, T(\mathbf{Z}) \to \mathbf{x}'_j]) \geq MU(\mathbf{x}'_i, W[\mathbf{Z} \to \mathbf{x}_i, T(\mathbf{Z}) \to \mathbf{x}'_j])$. In the expression of $SMU(T(\mathbf{Z}))$ (Eq. 7), each element is no less than its correspondent in the $WP(\mathbf{Z})$ expression (Eq. 6). Thus $SMU(T(\mathbf{Z}), W) \geq WP(\mathbf{Z}, W)$. ∎

Proof of Theorem 7. Suppose $\mathbf{Z}^*_W$ is not an optimal query set, i.e., there is some $\mathbf{Z}'$ such that $WP(\mathbf{Z}', W) > WP(\mathbf{Z}^*_W, W)$. If we apply transformation T to $\mathbf{Z}'$ we obtain a set $T(\mathbf{Z}')$, and by the results above we have: $SMU(T(\mathbf{Z}'), W)) \geq WP(\mathbf{Z}', W) > WP(\mathbf{Z}^*, W) \geq SMU(\mathbf{Z}^*_W, W)$. This contradicts the (setwise) maximin optimality of $\mathbf{Z}^*_W$. If $T(\mathbf{Z}')$ has lower cardinality than the initial set, then a set of the original cardinality can be constructed in arbitrary way, since MMU is montone. ∎

Possible Winner Problems on Partial Tournaments: A Parameterized Study

Yongjie Yang* and Jiong Guo**

Universität des Saarlandes,
Campus E 1.7, D-66123 Saarbrücken, Germany
{yyongjie,jguo}@mmci.uni-saarland.de

Abstract. We study possible winner problems related to uncovered set and Banks set on partial tournaments from the viewpoint of parameterized complexity. We first study the following problem, where given a partial tournament D and a subset X of vertices, we are asked to add some arcs to D such that all vertices in X are included in the uncovered set. Here we focus on two parameterizations of the problem: parameterized by $|X|$ and parameterized by the number of arcs to be added to make all vertices of X be included in the uncovered set. In addition, we study a parameterized variant of the problem to decide whether we can make all vertices of X be included in the uncovered set by reversing at most k arcs. Finally, we study some parameterizations of a possible winner problem on partial tournaments, where we are given a partial tournament D and a distinguished vertex p, and asked whether D has a maximal transitive subtournament with p being the 0-indegree vertex. These parameterized problems are related to Banks set. For all these parameterized problems studied in this paper, we achieve $\mathcal{XP}$ results, $\mathcal{W}$-hardness results as well as $\mathcal{FPT}$ results along with a kernelization lower bound.

1 Introduction

A tournament can be expressed as a directed graph where between every pair of vertices there is exactly one arc. Tournaments play a significant role in voting systems due to their nice expression ability in many winner determination problems. For example, tournaments can perfectly illustrate the Condorcet winner determination problem (when the number of voters is odd): create a vertex for each candidate and add an arc (v, u) between two vertices v and u if more than half of the voters prefer v to u. Then, the Condorcet winner is the candidate who has an arc to every other candidate. Several other winner determination methods are also based on tournaments, such as Banks, Slater, and Schwartz winners [14,4]. However, in practical settings, we might not be able to access the full information of an election to build the tournament. For example, the number of candidates is too huge to give a full preference at once, or, consider an online voting where in each time only part of the votes is submitted. In these cases, a partial tournament may be a useful tool, and thus, the problems of deciding which candidates have positive possibility to win the election should be of particular importance

---

* Supported by the DFG Excellence Cluster (MMCI) and the China Scholarship Council (CSC).
** Supported by the DFG Excellence Cluster (MMCI).

P. Perny, M. Pirlot, and A. Tsoukiàs (Eds.): ADT 2013, LNAI 8176, pp. 425–439, 2013.

(A partial tournament is a tournament with some arcs missing). Partial tournaments also appear in settings, where a given election cannot specify a relationship between two candidates. For example, we have an election to select the Condorcet winner. If the number of voters is even, then, it is possible that for two candidates v and u, exactly half of the voters prefer v to u and the others prefer u to v.

Tournament solutions have wide applications in decision-making problems and in social choice area. Informally, a *tournament solution* maps a tournament to a non-empty set of vertices in the tournament. Banks set and uncovered set are two of the most important tournament solutions which have been extensively studied from the viewpoints of game theory, economics, computational complexity, etc. Banks set is named by its introducer Banks [2]. Given a tournament, a candidate (a vertex in the tournament) v is a Banks winner, if there is a maximal transitive subtournament with v being the 0-indegree vertex. Here, "transitive" means that for every three vertices v, u, w in a tournament D, the existence of arcs (v, u) and (u, w) in D implies that (v, w) is in D. The Banks set then contains all Banks winners. Clearly, if there is a Condorcet winner, then the Banks winner coincides with the Condorcet winner. The uncovered set of a tournament is a maximal subset C of candidates such that no candidate outside C dominates a candidate in C. Here, a candidate v dominates a candidate u if all out-neighbors of u are also out-neighbors of v. Thus, an uncovered set includes exactly all vertices each of which can reach any other vertex in no more than two steps (a precise definition is in the next section). The vertices in an uncovered set are called kings from the viewpoint of graph theory. It is well-known that every tournament contains at least one king. Moreover, if the Condorcet winner exists, then the uncovered set contains only the Condorcet winner. Uncovered set has some advantages compared with Banks set. For example, determining whether a candidate is a Banks winner is $\mathcal{NP}$-hard [19], while computing the uncovered set is solvable in polynomial time [14]. Selecting the elements from the uncovered set as the winners of the given tournament has been independently suggested by Fishburn [12] and Miller [15].

1.1   Parameterized Complexity

Parameterized complexity was introduced by Downey and Fellows [9] as a tool to deal with hard problems. A *parameterized problem* is a language $\Sigma^* \times \Sigma^*$, where Σ is a finite alphabet. The first component is called the *main part* of the problem while the second component is called the *parameter*. Throughout this paper, the parameter is a positive integer. Parameterized problems have the following main hierarchy:

$$\mathcal{FPT} \subseteq \mathcal{W}[1] \subseteq \mathcal{W}[2], ..., \subseteq \mathcal{XP}$$

where $\mathcal{FPT}$ includes all parameterized problems which admit $O(f(k) \cdot |I|^{O(1)})$-time algorithms, while $\mathcal{XP}$ includes all parameterized problems which admit $O(f(k) \cdot |I|^{g(k)})$-time algorithms. Here, I is the main part of the instance, k is the parameter, and $f(k)$ and $g(k)$ are computable functions depending only on k. There are also parameterized problems beyond $\mathcal{XP}$. For example, the k-colorable problem which is to determine whether an undirected graph admits a proper k-coloring of the vertices has no algorithm of the form $O(f(k) \cdot |I|^{g(k)})$, unless $\mathcal{P} = \mathcal{NP}$ [10]. Finally, classes between $\mathcal{FPT}$ and $\mathcal{XP}$ are defined based on $\mathcal{FPT}$-reductions.

Given two parameterized problems Q and Q', an $\mathcal{FPT}$-*reduction* from Q to Q' is an algorithm that takes as input an instance (I, k) of Q and outputs an instance (I', k') of Q' such that

(1) the algorithm runs in $f(k) \cdot |I|^{O(1)}$ time, where f is a computable function in k;

(2) (I, k) is a true-instance of Q if and only if (I', k') is a true-instance of Q'; and

(3) $k' \leq g(k)$, where g is a computable function in k.

A problem is $\mathcal{W}[i]$-hard if all problems in $\mathcal{W}[i]$ can be $\mathcal{FPT}$-reducible to the problem. From the practical point of view, $\mathcal{W}[1]$ is the basic class of parameterized problems which unlikely admit $\mathcal{FPT}$-algorithms.

Kernelization is a main technique to derive $\mathcal{FPT}$ algorithms. Formally, a *kernelization* for a parameterized problem Q is a polynomial-time algorithm that reduces a given instance (I, k) of Q to a new instance (I', k') of Q such that

(1) (I, k) is a true-instance if and only if (I', k') is a true-instance;

(2) $k' \leq k$; and

(3) $|I'| \leq f(k)$, where f is a computable function in k.

The new instance (I', k') is called the *problem kernel*, while the function $f(k)$ is the *kernel size*. Moreover, if f is a polynomial function, we call (I', k') a *polynomial kernel*. Intuitively, a kernelization shrinks the original instance to a new equivalent and sized-bounded instance without changing the solvability. It is folklore that a parameterized problem is in $\mathcal{FPT}$ if and only it has a kernelization. For more background on kernelization, we refer to [13,3].

1.2 Motivation and Our Contribution

In this paper we study some parameterized problems related to uncovered set and Banks set on partial tournaments. We first study the possible winners of uncovered set problem [1]: given a partial tournament and a subset X of vertices, we are seeking for a completion of D such that all vertices in X become kings, or equivalently, all vertices in X are in the uncovered set. For convenience, in the following we use the terminology "kings" instead of "uncovered set". We study the problem with the size of X as the parameter. The motivation is based on the observation that in practical settings, one is mostly interested to make few vertices, which correspond to candidates, to become winners. We prove that this problem is in $\mathcal{XP}$; thus, when the size of X is bounded by a constant, it can be solved in polynomial time. In addition, we study two variations of possible winners of uncovered set problem where we are asked to make all vertices of X kings by modifying few number of arcs. We study two kinds of modifications: adding arcs and reversing arcs. In the "adding arcs" case we are allowed to add at most k arcs to the partial tournament, while in the "reversing arcs" case we are allowed to reverse at most k arcs in the partial tournament. For both problems, k is the parameter. These two parameterized variations could illustrate a bribery strategic behavior. For example, consider a politician in a political election who wants to make one of his accomplices win the election. Then, the arc reversal and arc addition problems illustrate the case where the politician has limited money and to bribe voters to change the pairwise compared relationship between every two candidates needs a cost. We prove that, somewhat surprising, both the variations are $\mathcal{W}[2]$-hard, even when X

contains only a single vertex. Furthermore, our $\mathcal{W}[2]$-hardness proof for the "reversing arcs" case applies to the special case where the input is a tournament and X contains only a single vertex. These results imply that the problems of finding the minimum number of arcs which are needed to add (resp. to reverse) to make all vertices of X kings are beyond $\mathcal{XP}$, when consider the size of X as the parameter. Finally, we study a possible winner problem related to Banks set on partial tournaments, where we are given a partial tournament D and a distinguished vertex p, and asked whether D has a maximal transitive subtournament with p being the 0-indegree vertex. This problem is a natural generalization of Banks winner to partial tournaments. Here we study three parameterizations. The first parameter we study is the size of the subtournament we are looking for. We prove that this parameter leads to a $\mathcal{W}[2]$-hardness result. Then, we study the parameter defined as the number of candidates who defeat p. We show that the problem is $\mathcal{W}[1]$-hard with this parameter. Finally, we consider the Copeland score of p (the number of candidates defeated by p) as the parameter. Different from the previous results, we show that the problem with the Copeland score of p as the parameter is in $\mathcal{FPT}$. Furthermore, we prove that the problem does not have a polynomial kernel unless the polynomial hierarchy collapses to the third level.

1.3 Preliminaries

A *directed graph* D is a pair (V, A) where V is the set of vertices and A is the set of arcs. An arc from a vertex v to a vertex u is denoted by (v, u). We say v is the *tail* of (v, u) and u is the *head* of (v, u). For simplicity, we also use $A(D)$ and $V(D)$ to denote the set of arcs and the set of vertices of D, respectively. For a vertex v, we use $N^-(v)$ and $N^+(v)$ to denote its *in-neighbors* and *out-neighbors*, respectively, that is, $N^-(v) = \{u \mid (u, v) \in A(D)\}$ and $N^+(v) = \{u \mid (v, u) \in A(D)\}$. The *in-degree* and *out-degree* of v, denoted by $d^-(v)$ and $d^+(v)$, are the sizes of $N^-(v)$ and $N^+(v)$, respectively. Meanwhile, we say that v is a $d^-(v)$-*indegree vertex* or a $d^+(v)$-*outdegree vertex*. The subgraph induced by a subset $S \subseteq V(D)$, denoted by $D[S]$, is $D[S] = (S, \{(u, v) \mid u \in S, v \in S, (u, v) \in A(D)\})$.

A *partial tournament* is a directed graph such that $|\{(v, u), (u, v)\} \cap A(D)| \leq 1$ for all $v, u \in V$ and $(v, v) \notin A(D)$ for all $v \in V$. If there is no arc between two vertices v and u in D, then we call (v, u) and (u, v) *missing arcs*. A *tournament* is a partial tournament without missing arcs. A tournament D is a *completion* of a partial tournament D' if $V(D) = V(D')$ and $A(D') \subseteq A(D)$.

A tournament D is *transitive* if there is an ordering $(v_1, v_2, ..., v_n)$ of $V(D)$ such that there is no arc (v_j, v_i) with $j > i$ (or, equivalently, for every three vertices v, u, w, $(v, u) \in A(D)$ and $(u, w) \in A(D)$ implies $(v, w) \in A(D)$). Clearly, there is a unique 0-indegree vertex in every transitive tournament (the first one in the ordering). For a partial tournament and a subset $S \subseteq V(D)$, we say $D[S]$ is a *maximal transitive subtournament* of D if $D[S]$ induces a transitive tournament and no other vertices outside S can be added to S to form a bigger induced transitive tournament.

For two vertices v and u, we say v *can reach* u if $(v, u) \in A(D)$ or there is a $w \in V(D) \setminus \{v, u\}$ with $(v, w) \in A(D)$ and $(w, u) \in A(D)$. In the former case we say v *reaches* u *directly*, while in the latter case we say that v *reaches* u *by (or through)* w. A *king* in a directed graph is a vertex which can reach all other vertices. For a subset

$X \subseteq V(D)$ and a vertex $v \in V(D)$, v is a *serf with respect to* X if v can be reached by all vertices in $X \setminus \{v\}$.

In the following, when we say "adding an arc", we mean to add an arc between two vertices which have no arc between them. Thus, adding an arc to a partial tournament still results in a partial tournament. *Reversing an arc* $(v, u) \in A(D)$ is the operation that firstly deletes (v, u) from D, and then adds a new arc (u, v) to D. The parameterized problems studied here are defined as follows.

Possible Winners of Uncovered Set (PWU)
Input: A partial tournament $D = (V, A)$ and a subset $X \subseteq V$.
Parameter: $|X|$.
Question: Is there a completion of D such that all vertices in X are kings?

PWU-ADD (resp. PWU-REVERSE)
Input: A partial tournament $D = (V, A)$ and a subset $X \subseteq V$.
Parameter: A positive integer k.
Question: Can we add (resp. reverse) at most k arcs such that all vertices in X are kings?

Transitive Winner on Partial Tournaments (TW)
Input: A partial tournament $D = (V, A)$ and a vertex $p \in V$.
Parameter: A positive integer k.
Question: Is there a subset $S \subseteq V$ of size k such that $D[S]$ is a maximal transitive tournament with p being the 0-indegree vertex?

TW-INDEGREE (resp. TW-OUTDEGREE)
Input: A partial tournament $D = (V, A)$ and a vertex $p \in V$.
Parameter: $|N^-(p)|$ (resp. $|N^+(p)|$).
Question: Is there a subset $S \subseteq V$ such that $D[S]$ is a maximal transitive tournament with p being the 0-indegree vertex?

1.4 Related Work

In [1], the authors studied possible and necessary winner problems in partial tournaments for diverse tournament solution concepts. They mainly considered three topics: deciding whether a given candidate is a possible (resp. a necessary) winner, and deciding whether a given subset of candidates equals the set of winners in some completion. For the possible winners of uncovered set (PWU) [1] defined as above, they proved that this problem is $\mathcal{NP}$-hard by a reduction from SAT. However, the problems of deciding whether a given candidate is a possible winner or a necessary winner for uncovered set are both polynomial-time solvable [1]. Moreover, computing the uncovered set is polynomial-time solvable [14].

As for the problems related to Banks set, in spite of the polynomial-time solvability of computing a Banks winner, deciding whether a distinguished candidate is a Banks

---

[1] In their paper, they use PSW_{UC} to denote the problem.

winner is $\mathcal{NP}$-hard [19]. The latter problem is also related to the DUAL DIRECTED FEEDBACK VERTEX SET (DUAL-DFVS) problem. In DIRECTED FEEDBACK VERTEX SET (DFVS), we are given a directed graph D and a positive integer parameter k, and asked to decide whether there is a subset of vertices of size k whose removal results in a directed graph without a cycle. In DUAL-DFVS, we are given a directed graph and a positive integer parameter k, and asked whether there is a subgraph of size k containing no cycle. DFVS has been proved $\mathcal{FPT}$ [5] over a long time of studying. In particular, when restricted to tournament, DFVS has an $O(k^3)$ kernel [6]. By a dichotomy theorem from [18], DUAL-DFVS is $\mathcal{W}[1]$-hard. However, when restricted to tournaments this problem is $\mathcal{FPT}$ [17]. It is well-known that a tournament contains no cycle if and only if it is transitive. These problems are also related to Slater set problems, where the main task is to reverse minimum number of arcs so that a given tournament become transitive. We refer to [14] for detailed complexity results about problems on Slater set.

2   Problems Related to Uncovered Set

It is easy to see that all problems except PWU defined above are in $\mathcal{XP}$: try all possibilities of selecting a subset of size k in V, A or $\{(v, u) \mid (v, u) \notin A(D)\}$, where k is the parameter of the corresponding problem. All these algorithms run in $O(|I|^{2k})$ time, where I is the size of the given partial tournament and k is the related parameter; and thus, these problems are in $\mathcal{XP}$. However, showing a problem is in $\mathcal{XP}$ is not always an easy work, as stated by Downey, Fellows and Stege in their seminal paper [10].

"Knowing that a problem is in $\mathcal{XP}$ has some practical value and can be difficult to show."

In the following, we show that PWU is also in $\mathcal{XP}$.

Theorem 1. PWU *is in* $\mathcal{XP}$.

Proof. We prove the theorem by giving an $\mathcal{XP}$-algorithm. The following lemma is useful for illustrating our algorithm. Let $\mathcal{E} = (D = (V, A), X)$ be an instance of PWU.

Lemma 2. *Let* $v \in X$ *be a serf with respect to* X *in* D *and* $\mathcal{E}' = (D' = (V, A'), X)$ *be a new instance with* $A' = A \cup \{(v, u) \mid \{(v, u), (u, v)\} \cap A = \emptyset, u \in V \setminus X\}$, *then* $\mathcal{E}$ *is a true-instance if and only if* $\mathcal{E}'$ *is a true-instance.*

Proof. It is clear that if $\mathcal{E}'$ is a true-instance, then $\mathcal{E}$ must be a true-instance. To prove the other direction, note that adding an arc from some vertex $u \in V \setminus X$ to v is to make v reachable by some vertex $w \in X \setminus \{v\}$ through u. However, since v is already a serf with respect to X, such an arc addition is then unnecessary. However, adding the arc (v, u) for $u \in V \setminus X$ to the partial tournament would make v reach further vertices. □

Our algorithm first tries all possibilities of completions of $D[X]$. Clearly, there can be at most $2^{|X| \cdot (|X|-1)/2}$ such possibilities. In each of the completions, there may have some pairs (u, w) with $(u, w) \in A(D[X])$ such that w does not reach u. For all these pairs, we further try all possibilities of making w reach u by some vertex

$v \in V \setminus X$ (thus, there are at most $|V \setminus X|$ possibilities for each pair and totally at most $|V \setminus X|^{|X| \cdot (|X|-1)/2}$ possibilities for all pairs), by adding one or two new arcs between $\{w, u\}$ and v. Meanwhile, if there is no chance to make w reach u, then we give up the possibility. Clearly, if the given instance is a true-instance, then at least one of the possibilities leads to a "yes" answer. We have totally at most $2^{|X| \cdot (|X|-1)/2} \cdot |V \setminus X|^{|X| \cdot (|X|-1)/2}$ possibilities. Now, in each case, $D[X]$ induces a tournament and every vertex $v \in X$ is a serf with respect to X. Then, due to Lemma 2, we can safely add all missing arcs between X and $V \setminus X$ with tails in X and heads in $V \setminus X$. It remains to add arcs between vertices in $V \setminus X$ to make the vertices in X kings. For convenience, let's give a formal definition of the remaining part first.

$\overline{\text{PWU}}$

Input: A partial tournament $D = (V, A)$ and a subset $X \subseteq V$ such that $D[X]$ induces a tournament, every vertex $v \in X$ is a serf with respect to X in D and there is no missing arcs between X and $V \setminus X$, that is, $\{(v, u), (u, v)\} \cap A \neq \emptyset$ for all $v \in X$ and all $u \in V \setminus X$.

Question: Is there a completion of D such that all vertices in X are kings?

In the following, we prove that $\overline{\text{PWU}}$ is solvable in polynomial time. We begin with a useful observation.

Observation. *Let v and u be two vertices in $V \setminus X$ with missing arcs between them. If there is a vertex $x \in X$ such that x can reach v directly but x cannot reach u, then every true-instance has a solution containing the arc (v, u).*

The observation is correct; since adding an arc (v', u') between $v', u' \in V \setminus X$ to the partial tournament is to make some vertex $w \in X$ reach u' by v'. Since x cannot reach u, all vertices in X which can directly reach u must also directly reach x. Therefore, no vertex in X needs an arc from u to v to reach v; since all such vertices have already reached v by x. Thus, adding (v, u) is the optimal choice.

Based on the above observation, we can solve $\overline{\text{PWU}}$ in polynomial time as showed in Algorithm 1.

Algorithm 1. A polynomial-time algorithm for $\overline{\text{PWU}}$

---

1   **forall the** *vertices $x \in X$* **do**
2   Let $V_x = \{v \in V \setminus X \mid (x, v) \in A(D)\}$ be the set of vertices that x can reach directly;
3   Let $V_{\bar{x}} = \{v \in V \setminus X \mid (v, x) \in A(D), \nexists y \in V \text{ with } (x, y) \in A(D) \text{ and } (y, v) \in A(D)\}$ be the set of vertices that x cannot reach;
4   **if** $V_x = \emptyset$ *and* $V_{\bar{x}} \neq \emptyset$ **then**
5 | Return "No"
6   **else**
7   **forall the** $v \in V_x$ *and* $u \in V_{\bar{x}}$ *with* $\{(v, u), (u, v)\} \cap A(D) = \emptyset$ **do**
8 | Add (v, u) to D
9   **end**
10   **end**
11 **end**
12 Return "Yes" if all vertices in X are kings and return "No" otherwise;

In summery, PWU is in $\mathcal{XP}$; since there are at most $2^{|X|\cdot(|X|-1)/2}\cdot|V\backslash X|^{|X|\cdot(|X|-1)/2}$ instances of $\overline{\text{PWU}}$ and $\overline{\text{PWU}}$ can be solved in polynomial time.   □

With the above theorem, we can trivially get the following result.

Corollary 3. PWU *is polynomial-time solvable if the size of the given subset* X *is bounded by a constant.*

Now we study the problems of deciding whether we can make all vertices of X kings by adding (resp. reversing) at most k arcs. The following results are somewhat interesting compared with Corollary 3. In particular, we prove that both PWU-ADD and PWU-REVERSE are $\mathcal{W}[2]$-hard even when X contains only one single vertex. Furthermore, our $\mathcal{W}[2]$-hardness proof for PWU-REVERSE applies to the case that the input is a tournament and X contains only one single vertex. These results imply that the problem of finding the minimum number of arcs which are needed to add to the given partial tournament (resp. to reverse in the given (partial) tournament) to make all vertices of X kings is beyond $\mathcal{XP}$, in the case that $|X|$ is the parameter.

Theorem 4. PWU-ADD *is* $\mathcal{W}[2]$-*hard even when* $|X| = 1$.

Proof. We prove the theorem by an $\mathcal{FPT}$-reduction from SET COVER which has been proved $\mathcal{W}[2]$-hard (Theorem 13.29 of [16]).

SET COVER
Input: A base set $S = \{s_1, s_2, ..., s_n\}$ and a collection C of subsets of S, $C = \{c_1, c_2, ..., c_m\}$, $c_i \subseteq S$ for $1 \leq i \leq m$, and $\bigcup_{1 \leq i \leq m} c_i = S$.
Parameter: A positive integer t
Question: Is there a subset $C' \subseteq C$ of size at most t which covers all elements in S, that is, $\bigcup_{c \in C'} c = S$.

Given an instance $\mathcal{E} = (S, C, t)$ of SET COVER, we construct an instance $\mathcal{E}' = (D = (V, A), X, k)$ of PWU-ADD as follows.
The partial tournament D contains $n+m$ vertices one to one labeled by the elements in $S \cup C$ together with further two vertices $\{x, y\}$. We further use S and C to denote the sets of vertices labeled by the elements in S and C, respectively. For each $c \in C$, there is an arc $(y, c) \in A(D)$. For each $s \in S$, there is an arc $(s, x) \in A(D)$ and an arc $(s, y) \in A(D)$. For each pair $\{s, c\}$ where $s \in S$ and $c \in C$, there is an arc $(c, s) \in A(D)$ if $s \in c$, and an arc $(s, c) \in A(D)$ otherwise. In addition, there is an arc $(x, y) \in A(D)$. Finally, we add arbitrary arcs in $D[S]$ and $D[C]$ to make both $D[S]$ and $D[C]$ complete (subtournament of D). See Fig. 1. We set $X = \{x\}$ and $k = t$.
Due to the construction, a vertex $c \in C$ can reach a vertex $s \in S$ only if c covers s, that is, $s \in c$. Meanwhile, x can reach every vertices in C by y but cannot reach any vertex in S. In order to make x a king, we must add some arcs from x to C to make x reach all vertices in S. We prove that $\mathcal{E}$ is a true-instance if and only if the new instance $\mathcal{E}'$ is true.
$\Rightarrow$: Suppose that $\mathcal{E}$ is a true-instance and C' be a solution of $\mathcal{E}$. Then, it is easy to verify that we can make x a king by adding arcs (x, c) in D for all $c \in C'$; thus, $\mathcal{E}'$ is a true-instance.

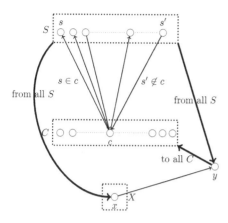

Fig. 1. The graph illustrates the construction for PWU-ADD. Here, $D[S]$ and $D[C]$ are made complete arbitrarily. The thick arcs labeled with "from all S" mean that there is an arc (s, x) and an arc (s, y) for all $s \in S$. The thick arc labeled with "to all C" means that there is an arc (y, c) for all $c \in C$. Finally, there is an arc (c, s) if $s \in c$ and an arc (s, c) otherwise, for every $c \in C$ and $s \in S$.

$\Leftarrow$: Suppose that $\mathcal{E}'$ is a true-instance and B is a solution for $\mathcal{E}'$. Let $C' = \{v \mid (x, v) \in B\}$. Clearly, C' is a subset of C. We claim that C' is a solution for $\mathcal{E}$: the only way to make x reach a vertex $s \in S$ is to add an arc from x to some vertex $c \in C$ which can cover s. Since x is a king after adding all arcs in B to the given instance, x can reach every $s \in S$ by at least one vertex $c \in C'$ which covers s, implying C' is a set cover for $\mathcal{E}$. $\qquad\square$

The following theorem shows the parameterized complexity of the problem to decide whether we can make a certain set of vertices kings by reversing at most k arcs.

Theorem 5. PWU-REVERSE *is* $\mathcal{W}[2]$-*hard even when the input is a tournament and* X *contains only a single vertex.*

Proof. The reduction is from DOMINATING SET ON TOURNAMENTS which has been proved $\mathcal{W}[2]$-hard [8].

Dominating Set on Tournaments (DST)
Input: A tournament T.
Parameter: A positive integer t.
Question: Does T have a dominating set of size at most t? Here, a dominating set C for a tournament T is a subset of the vertices of T such that every vertex outside C has at least one of its in-neighbors in C.

Given an instance $\mathcal{E} = (T, t)$ of DST, we construct an instance $\mathcal{E}' = (T', X = \{x\}, k = t)$ for PWU-REVERSE as follows. T' contains a copy of T, which is denoted by $\bar{T}$, together with a further vertex x having an arc from every vertex in $\bar{T}$, that is, $(\bar{v}, x) \in A(T')$ for all $\bar{v} \in V(\bar{T})$. We will use $\bar{v}$ to refer to the copy of the vertex $v \in V(T)$. It is easy to verify that if T has a dominating set C of size at most t, then reversing the arcs $\{(\bar{v}, x) \mid v \in C\}$ makes x a king. To show the other direction, we first observe that if $\mathcal{E}'$ is a true-instance, then there is a solution such that all reversed arcs are between x and $V(\bar{T})$. The observation is correct since each reversal of an arc $(\bar{v}, \bar{u})$ with $\bar{v}, \bar{u} \in V(\bar{T})$ can be replaced by a reversal of the arc $(\bar{v}, x)$ to form a new solution.

Now suppose that $\mathcal{E}'$ is a true-instance and B is a solution (represented by a set containing all reversed arcs) containing only arcs between x and $V(\bar{T})$. Let T'' be the tournament obtained from T' by reversing all arcs in B. We claim that $C = \{v \mid (\bar{v}, x) \in B\}$ is a dominating set of T (the size of C is clearly at most t). To this end, we need to show that, in the tournament T, every vertex which is not in C has at least one of its in-neighbors in C. Let u be any arbitrary vertex in $V(T) \setminus C$. Due to the construction, there is an arc $(\bar{u}, x)$ in T''. Since x is a king in T'', we know that x reaches $\bar{u}$ by some vertex $\bar{v}$ with $(x, \bar{v}) \in T''$. Due to the construction, $(x, \bar{v})$ is in T'' only if $(\bar{v}, x)$ is in B, or equivalently, $v \in C$. Since $(\bar{v}, \bar{u}) \in A(\bar{T})$ and $\bar{T}$ is a copy of T, $(v, u) \in A(T)$. Therefore, we can conclude that every vertex u outside C has at least one vertex $v \in C$ with $(v, u) \in A(T)$, which completes our proof. $\square$

3   Problems Related to Banks Set

In this section, we study problems of deciding whether a distinguished vertex p is contained in a maximal transitive subtournament with p being the 0-indegree vertex. We first prove that Tw is $\mathcal{W}[2]$-hard by a reduction from a variant of SET COVER, which is defined as follows.

t-MULTICOLORED SET COVER, (t-MSC)
Input: A base set $S = \{s_1, s_2, ..., s_n\}$ and a collection $C = \{c_1, c_2, ..., c_m\}$ of subsets of S, each of which having a color from $\{1, 2, ..., t\}$, and $\bigcup_{1 \leq i \leq m} c_i = S$.
Parameter: t
Question: Is there a subset $C' \subseteq C$ such that C' includes exactly one from the same colored subsets and C' covers all elements of S, that is, $\bigcup_{c \in C'} c = S$. We call such a C' a t-multicolored set cover.

Lemma 6. t-MSC *is* $\mathcal{W}[2]$-hard.

Proof. The proof is by an $\mathcal{FPT}$-reduction from SET COVER. Given an instance $\mathcal{E} = (S, C, t)$ of SET COVER we construct a collection $\bar{C}$ by taking t copies $\bar{c}_1, \bar{c}_2, ..., \bar{c}_t$ of each $c \in C$, and then color each $\bar{c}_i$ with color $i \in \{1, 2, ..., t\}$. The constructed instance for t-MSC is $\mathcal{E}' = (S, \bar{C}, t)$. It is straightforward to verify that $\mathcal{E}$ has a set cover of size t if and only if $\mathcal{E}'$ has a t-multicolored set cover. $\square$

With the $\mathcal{W}[2]$-hardness of t-MSC we now prove the hardness of Tw.

Theorem 7. Tw *is* $\mathcal{W}[2]$-hard.

Proof. We prove the theorem by an $\mathcal{FPT}$-reduction from t-MSC. Given an instance $\mathcal{E} = (C, S, t)$ of t-MSC where C is the colorful collection, S is the base set and t is the parameter, we construct an instance $\mathcal{E}' = (D = (V, A), p, k)$ of Tw as follows. Let C_i be the collection of subsets in C colored by $i \in \{1, 2, ..., t\}$.

D contains $n + m$ vertices one to one labeled by the elements in $S \cup C$ together with the distinguished vertex $\{p\}$. We further use S and C to denote the sets of vertices labeled by the elements in S and C, respectively. For every $s \in S$ and $c \in C$, there is an arc from c to s if $s \in c$ and an arc from s to c otherwise. In addition, there is an arc from s to p for all $s \in S$ and an arc from p to c for all $c \in C$. Finally, there is an arc (c, c') for all $c \in C_i$ and $c' \in C_j$ with $i < j$. See Fig 2. The parameter is set to $k = t + 1$. We now prove that $\mathcal{E}$ is a true-instance if and only if $\mathcal{E}'$ is a true-instance.

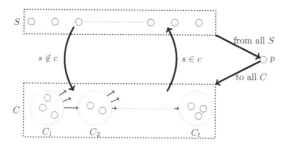

Fig. 2. Illustration of construction for Tw

$\Rightarrow$: Suppose that $\mathcal{E}$ is a true-instance and C' is a solution. Clearly, $C' \cup \{p\}$ induces a transitive tournament with p being the 0-indegree vertex. Due to the construction, for each vertex $s \in S$, C' contains at least one of its in-neighbors; thus, no vertex in S can be added to $C' \cup \{p\}$ to make a bigger transitive tournament (since otherwise, there would be a triangle), implying that $C' \cup \{p\}$ is maximal in D.

$\Leftarrow$: Suppose that $\mathcal{E}'$ is a true-instance and $B \cup \{p\}$ is a solution which induces a maximal transitive tournament with p being the 0-indegree vertex. Clearly, $B \subseteq C$. Due to the maximality of $D[B \cup \{p\}]$, $N^-(s) \cap B \neq \emptyset$ for all $s \in S$, implying that at least one subset in B covers s; thus, B must be a set cover of $\mathcal{E}$. By the construction, there is no arc in $D[C_i]$ for all $i \in \{1, 2, ..., t\}$, thus, exactly one from each C_i can be in B. Therefore, B must be a t-multicolored set cover for D.   □

In the following, we study two further parameterizations of finding a Banks winner in a partial tournament. First, we study the parameter $|N^-(p)|$, that is, the number of candidates who defeat p in a pairwise comparison. We show that this problem is $\mathcal{W}[1]$-hard.

Theorem 8. TW-INDEGREE *is* $\mathcal{W}[1]$-*hard.*

Proof. We prove the theorem by an $\mathcal{FPT}$-reduction from t-MULTICOLORED CLIQUE which has been proved $\mathcal{W}[1]$-hard [11]. An *undirected graph* is a tuple $G = (V, E)$ where V is the vertex set and E is the edge set. An edge between two vertices u and v is denoted by $\{u, v\}$. A *clique* Q (resp. An *independent set* I) in G is a subset of V such that there is an (resp. no) edge between every pair vertices of Q (resp. I).

t-MULTICOLORED CLIQUE
Input: An undirected graph $G = (V, E)$ with each vertex having a color from $\{1, 2, ..., t\}$, such that the vertices with the same color induce an independent set.
Parameter: t.
Question: Does G have a clique including vertices of all t colors?

Let $\mathcal{E} = (G, t)$ be an instance of t-MULTICOLORED CLIQUE. Let V_i be the set of all vertices in G with color i. We construct an instance $\mathcal{E}' = (D = (N^-(p) \cup N^+(p) \cup \{p\}, A), p, |N^-(p)|)$ for TW-INDEGREE from $\mathcal{E}$ as follows.

Firstly, we construct the set of vertices as follows: $N^+(p) = V(G)$, $N^-(p) = \{c_1, c_2, ..., c_t\}$ (corresponds to colors); thus, D has totally $|V(G)| + t + 1$ vertices. For each $v \in N^+(p)$ we add an arc (p, v), and for each $v \in N^-(p)$ we add an arc (v, p). For each c_i, there is an arc (v, c_i) for all $v \in V_i$ and an arc (c_i, v) for all $v \in V_j$ with $j \neq i$. In addition, there are some arcs between V_i and V_j for $i \neq j$. Precisely, for two vertices $v \in V_i$ and $u \in V_j$ with $1 \leq i < j \leq t$, there is an arc (v, u) in D if there is an edge between v and u in G. See Fig. 3. In the following, we prove that $\mathcal{E}$ is a true-instance if and only if $\mathcal{E}'$ is a true-instance.

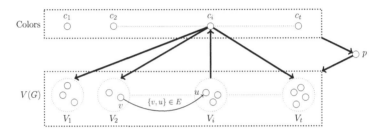

Fig. 3. Illustration of construction for TW-INDEGREE

($\Rightarrow$:) Suppose that $\mathcal{E}$ is a true-instance and Q is a clique including all t colors, that is $\{u, v\} \in E$ for all $u, v \in Q$ and $|Q \cap V_i| = 1$ for all $1 \leq i \leq t$. Due to the construction, Q induces a transitive tournament in D. Moreover, the induced transitive tournament is maximal in $D[N^+(p)]$ since there is no arc in $D[V_i]$ for all $1 \leq i \leq t$. Since $Q \cap V_i \neq \emptyset$ and $V_i = N^-(c_i)$ for all $1 \leq i \leq t$, every c_i has an in-neighbor in Q; thus, $D[Q \cup \{p\}]$ is a maximal transitive tournament in D with p being the 0-indegree vertex.

($\Leftarrow$:) Suppose that $\mathcal{E}'$ is a true-instance and $Q \cup \{p\}$ induces a maximal transitive tournament in D with p being the 0-indegree vertex. Due to the construction, Q induces a clique in G. Since there is no arc in each $D[V_i]$ for $1 \leq i \leq t$, there can be at most one vertex of V_i in Q. Due to the maximality of $D[Q \cup \{p\}]$, at least one vertex of V_i must be in Q for all $1 \leq i \leq t$ (since otherwise, c_i can be added to $D[Q \cup \{p\}]$ to form a bigger transitive subtournament). In summery, we conclude that Q is a clique of G including all colors. □

The last parameter we study is $|N^+(p)|$, that is, the Copeland score of p.

Theorem 9. TW-OUTDEGREE *is in* $\mathcal{FPT}$.

The proof for Theorem 9 is trivial: if there is a solution, it must be totally included in $N^+(p) \cup \{p\}$. Thus, the problem can be solved by trying all $2^{|N^+(p)|}$ subsets of $N^+(p)$ and checking whether at least one of them together with p forms a maximal transitive tournament with p being the 0-indegree vertex. The algorithm implies a $2^{|N^+(p)|}$-size kernel: if the input partial tournament D contains at most $2^{|N^+(p)|}$ vertices then we are done; otherwise, solve the problem in polynomial time (note that $2^{|N^+(p)|} \leq |V(D)|$) and return a trivial true- or false-instance according to the output of the algorithm. One would ask whether the kernel can be improved greatly. The following theorem shows that, however, the kernel size cannot be improved to polynomial unless the polynomial hierarchy collapses to the third level.

Theorem 10. TW-OUTDEGREE *does not admit a polynomial kernel unless the polynomial hierarchy collapses to the third level* ($\mathcal{PH} = \sum_{\mathcal{P}}^{3}$).

To prove the theorem, we need some new definitions. We say that a parameterized problem Q is *polynomial parameter reducible* to a parameterized problem Q', if there exists a polynomial-time algorithm with an instance (I, k) of Q as input, where k is the parameter, and this algorithm outputs an instance (I', k') of Q' such that (1) (I, k) is a true-instance of Q if and only if (I', k') is a true-instance of Q'; and (2) $k' \leq \text{Poly}(k)$, where $\text{Poly}(k)$ is a polynomial function in k.

The following lemma, which has been successfully used for proving non-existence of polynomial kernels for many problems, is the main tool to prove Theorem 10.

Lemma 11. *([7]) Let Q and Q' be two parameterized problems and $\tilde{Q}$ and $\tilde{Q}'$ be the unparameterized versions of Q and Q', respectively. Suppose that $\tilde{Q}$ is $\mathcal{NP}$-hard and $\tilde{Q}'$ is in $\mathcal{NP}$. Moreover, Q is polynomial parameter reducible to Q'. Then, if Q' has a polynomial kernel, then Q has a polynomial kernel.*

In order to show the non-existence of a polynomial kernel for a specific problem Q, it suffices to derive a polynomial parameter reduction from a parameterized problem which does not have a polynomial kernel (under some assumption which is unlikely to happen) to Q.

In the following, we prove Theorem 10 using the above lemma. In fact, the reduction from t-MSC to TW in the proof of Theorem 7 has already implied that TW-OUTDEGREE does not admit a polynomial kernel. This holds because the t-MULTICOLORED SET COVER problem with parameter $|C|$, the size of the collection of subsets, is $\mathcal{FPT}$ but does not admit a polynomial kernel unless the polynomial hierarchy collapses to the third level. Formally, the following problem is $\mathcal{FPT}$ and does not admit a polynomial kernel unless the polynomial hierarchy collapses to the third level.

$|C|$-MULTICOLORED SET COVER, ($|C|$-MSC)
Input: A base set $S = \{s_1, s_2, ..., s_n\}$ and a collection $C = \{c_1, c_2, ..., c_m\}$ of subsets of S, each of which having a color from $\{1, 2, ..., t\}$, and $\bigcup_{1 \leq i \leq m} c_i = S$.
Parameter: $|C|$
Question: Is there a subset $C' \subseteq C$ such that C' includes exactly one from the same colored subsets and C' covers all elements of S.

The following lemma can be derived from the non-existence of polynomial kernels for the colored version of the small universe hitting set problem shown in [7].

Lemma 12. $|C|$-MSC *has no polynomial kernel unless the polynomial hierarchy collapses to the third level.*

The following lemma directly follows from the proof of Theorem 7.

Lemma 13. $|C|$-MSC *is polynomial parameter reducible to* TW-OUTDEGREE.

Lemmas 11, 12 and 13 together then proves Theorem 10.

4 Concluding Remarks

In this paper, we study some possible winner(s) problems related to uncovered set and Banks set on partial tournaments from the viewpoint of parameterized complexity. We show some $\mathcal{XP}$ results, $\mathcal{W}$-hardness results as well as $\mathcal{FPT}$ results along with a kernelization lower bound. See Table. 1 for a summery of our results.

Table 1. A summery of the results. The precise definitions of the problems are in Subsection 1.3.

PWU	$\mathcal{XP}$
PWU-ADD	$\mathcal{W}[2]$-hard even when $|X| = 1$
PWU-REVERSE	$\mathcal{W}[2]$-hard even on tournaments and with $|X| = 1$
TW	$\mathcal{W}[2]$-hard
TW-INDEGREE	$\mathcal{W}[1]$-hard
TW-OUTDEGREE	$\mathcal{FPT}$ but no polynomial kernel unless $\mathcal{PH} = \sum_{\mathcal{P}}^3$

There remain several open problems for future research. For instance, we do not know whether PWU is $\mathcal{FPT}$ or $\mathcal{W}$-hard. In addition, it would be interesting to study further standard parameterizations for problems related to tournament solution.

Acknowledgement. We sincerely thank the anonymous referee(s) of ADT 2013 for their constructive comments.

References

1. Aziz, H., Harrenstein, P., Brill, M., Lang, J., Fischer, F.A., Seedig, H.G.: Possible and necessary winners of partial tournaments. In: AAMAS, pp. 585–592 (2012)
2. Banks, J.S.: Sophisticated voting outcomes and agenda control. Social Choice and Welfare 1(4), 295–306 (1985)
3. Bodlaender, H.L.: Kernelization: New upper and lower bound techniques. In: Chen, J., Fomin, F.V. (eds.) IWPEC 2009. LNCS, vol. 5917, pp. 17–37. Springer, Heidelberg (2009)
4. Brandt, F., Fischer, F.A., Harrenstein, P.: The computational complexity of choice sets. Math. Log. Q. 55(4), 444–459 (2009)
5. Chen, J., Liu, Y., Lu, S., O'Sullivan, B., Razgon, I.: A fixed-parameter algorithm for the directed feedback vertex set problem. In: STOC, pp. 177–186 (2008)
6. Dom, M., Guo, J., Hüffner, F., Niedermeier, R., Truß, A.: Fixed-parameter tractability results for feedback set problems in tournaments. J. Discrete Algorithms 8(1), 76–86 (2010)
7. Dom, M., Lokshtanov, D., Saurabh, S.: Incompressibility through colors and ids. In: Albers, S., Marchetti-Spaccamela, A., Matias, Y., Nikoletseas, S., Thomas, W. (eds.) ICALP 2009, Part I. LNCS, vol. 5555, pp. 378–389. Springer, Heidelberg (2009)

8. Downey, R.G., Fellows, M.R.: Parameterized computational feasibility. In: Feasible Mathematics II, pp. 219–244 (1995)
9. Downey, R.G., Fellows, M.R.: Parameterized Complexity. Springer (1999)
10. Downey, R.G., Fellows, M.R., Stege, U.: Parameterized complexity: A framework for systematically confronting computational intractability. In: Contemporary Trends in Discrete Mathematics: From DIMACS and DIMATIA to the Future, pp. 49–99. Springer (1999)
11. Fellows, M.R., Hermelin, D., Rosamond, F.A., Vialette, S.: On the parameterized complexity of multiple-interval graph problems. Theor. Comput. Sci. 410(1), 53–61 (2009)
12. Fishburn, P.C.: Condorcet social choice functions. SIAM Journal on Applied Mathematics 33(3), 469–489 (1977)
13. Guo, J., Niedermeier, R.: Invitation to data reduction and problem kernelization. ACM SIGACT News 38(1), 31–45 (2007)
14. Hudry, O.: A survey on the complexity of tournament solutions. Mathematical Social Sciences 57(3), 292–303 (2009)
15. Miller, N.R.: A new solution set for tournaments and majority voting: Further graph-theoretical approaches to the theory of voting. American Journal of Political Science 24(1), 68–96 (1980)
16. Niedermeier, R.: Invitation to Fixed-parameter Algorithms. Oxford University Press Inc., Oxford University Press Inc (2006)
17. Raman, V., Saurabh, S.: Parameterized algorithms for feedback set problems and their duals in tournaments. Theor. Comput. Sci. 351(3), 446–458 (2006)
18. Raman, V., Sikdar, S.: Parameterized complexity of the induced subgraph problem in directed graphs. Inf. Process. Lett. 104(3), 79–85 (2007)
19. Woeginger, G.J.: Banks winners in tournaments are difficult to recognize. Social Choice and Welfare 20(3), 523–528 (2003)

Author Index

Printed in the United States
By Bookmasters